Microbiology-2.0 Update for a Sustainable Future

Juhi Gupta · Akarsh Verma
Editors

Microbiology-2.0 Update for a Sustainable Future

Springer

Editors
Juhi Gupta
Amity Institute of Environmental Sciences
Amity University
Noida, India

Akarsh Verma
Department of Mechanical Engineering
University of Petroleum and Energy Studies
Dehradun, India

ISBN 978-981-99-9616-2 ISBN 978-981-99-9617-9 (eBook)
https://doi.org/10.1007/978-981-99-9617-9

This Springer imprint is published by the registered company Springer Nature Singapore Pte Ltd.
The registered company address is: 152 Beach Road, #21-01/04 Gateway East, Singapore 189721, Singapore

Paper in this product is recyclable.

Contents

MICROBIOLOGY: The Science to the World of Microbes

Juhi Gupta and Akarsh Verma

1 Introduction

The fascinating world of microorganisms is capable of pushing the boundaries of human resilience (Mun et al. 2019). This narrative is linked to the Giant gypsum crystals cave located in the Chihuahuan Desert, Mexico, notable for its intense heat of 50°C and extreme 99% humidity, situated over a volcanic magma chamber. This hostile environment is such that humans can't survive for longer than half an hour, even when equipped with specialized apparatus and protective clothing. Nonetheless, scientists have isolated more than 80 distinct bacterial strains from the cave's crystal fluid, each one being a novel addition to Earth's microbiome (Black and Black 2018). The crystals provide an inviting home for these extraordinary microorganisms by creating cavities in their lattice structure. Remarkably, these microbes flourish in the dark, living solely on the inherent minerals found in the cave (Sun et al. 2020).

This naturally raises the question: Do similar habitable conditions exist in other unexpected corners of the Universe? Research suggests that Europa and Titan, the moons of Jupiter and Saturn respectively, might harbour environments suitable for microbial life. Wherever life has been discovered in the Universe's most surprising locations, it has largely been microbial in nature (Gupta et al. 2020). It's clear that the field of microbiology is likely to uncover more astonishing revelations in the future.

From ancient times, microorganisms have fascinated humans as these invisible creatures occupy a significant position in Brahma's Creation (Patil 2022). Microbes

J. Gupta (✉)
Amity Institute of Environmental Sciences, Amity University, Noida 201303, India
e-mail: jgupta2@amity.edu

A. Verma
Department of Mechanical Engineering, University of Petroleum and Energy Studies, Dehradun 248007, India

Department of Mechanical Science and Bioengineering, Osaka University, Osaka 560-8531, Japan

J. Gupta and A. Verma (eds.), *Microbiology-2.0 Update for a Sustainable Future*,
https://doi.org/10.1007/978-981-99-9617-9_1

are known for their ability to produce disease as well as their beneficial role in nature. They are causative agents of many diseases (Mun et al. 2019). The field of microbiology began with the establishment of the germ theory of disease by Robert Koch and Louis Pasteur, marking the beginning of a golden period of microbiology in the Western hemisphere. However, in Vedic Bharat (ancient India), the Germ theory of diseases was first established by Vedic Rishis and documented in the Vedas. The Vedas are notably the first texts globally to highlight the relationship between microbes and disease (Patil 2022). The role of microorganisms in causing various infectious diseases has been detailed in Ayurveda since the Vedic Period.

From the Vedic period, ancient Indians recognized the presence of microorganisms and their role in causing diseases. The Vedas mention microbes and infectious diseases, referring to them as "Krumi" and "Krumi Rogas" (Patil 2022). Additionally, the morphology and functional anatomy of these microbes has been astonishingly detailed in the Vedic texts. It's plausible that contemporary pathologists might have derived insights into the germ theory from Ayurveda. When European interactions with India began, numerous significant scientific and cultural manuscripts were shared with the West for interpretation (Patil 2022). This gives even more reasons to study the ancient texts comprehensively and deduce the unknown in this contemporary world (Sun et al. 2020). This fascinating world has unveiled marvellous contributions from one of the brightest minds of our globe (see Table 1).

Contemporary science describes Microbiology as the examination of organisms too tiny to be seen without magnification. The word 'Microorganism' was introduced by Sedillot in 1878. Organisms that are a millimetre or smaller in size necessitate observation under a microscope (Gupta et al. 2019a, b). Microorganisms, often invisible to the naked eye, play a pivotal role in shaping our environment, health, and industries (Gupta et al. 2021a, b). Their vast diversity and adaptability make them suitable for a range of applications, from bioremediation, food production, wastewater treatments, cosmetics to dye corrections (Rathour et al. 2018). Various avenues through which the potential of microbes is being harnessed for the betterment of society would be discussed. The microbial world, though minuscule in size, holds immense potential that can be tapped into for various applications (refer to Fig. 1). As research progresses, it's evident that microbes will continue to play an integral role in shaping the future of industries and the environment (Gupta et al. 2021b). Harnessing their potential requires a multidisciplinary approach, combining microbiology, genetics, and engineering, to unlock solutions to some of the world's most pressing challenges. From the first observations of tiny "animalcules" to the modern-day exploration of complex microbial communities, the field has witnessed revolutionary discoveries that have reshaped medicine, agriculture, and biotechnology (Rathour et al. 2020).

Microbiology predominantly comprises of two major techniques: (a) Culture-Dependent Microbiology: It relies on culturing microorganisms in the laboratory by using specific nutrients, temperatures, pH levels, and other conditions. It is advantageous for detailed study of individual species, including their morphology, physiology, and genetic characteristics. This technique upholds a limitation as >90% microbes are unculturable (Gupta et al. 2022; Gupta et al. 2019a, b). (b) Culture-Independent Microbiology: This technique doesn't rely on culturing organisms in the

Table 1 Noble Laureates and their contributions in the field of Microbiology

Year	Laureate(s)	Contribution
1905	Robert Koch	Work on tuberculosis
1928	Charles Nicolle	Work on typhus
1945	Sir Alexander Fleming, Ernst Boris Chain, Sir Howard Florey	Discovery of penicillin and its curative effect in various infectious diseases
1952	Selman Waksman	Discovery of streptomycin, the first antibiotic effective against tuberculosis
1969	Max Delbrück, Alfred D. Hershey, Salvador E. Luria	Discoveries concerning the replication mechanism and the genetic structure of viruses
2005	Barry J. Marshall, J. Robin Warren	Discovery of the bacterium Helicobacter pylori and its role in gastritis and peptic ulcer disease
2008	Harald zur Hausen; Françoise Barré-Sinoussi and Luc Montagnier	Discovery of human papilloma viruses causing cervical cancer Discovery of human immunodeficiency virus
2015	William C. Campbell and Satoshi Ōmura; Tu Youyou	Discoveries concerning a novel therapy against infections caused by roundworm parasites Her discoveries concerning a novel therapy against Malaria
2020	Harvey J. Alter, Michael; Houghton and Charles M	Discovery of Hepatitis C virus
2022	Svante Pääbo	Discoveries concerning the genomes of extinct hominins and human evolution

Data collected from https://www.nobelprize.org/prizes/lists/all-nobel-laureates

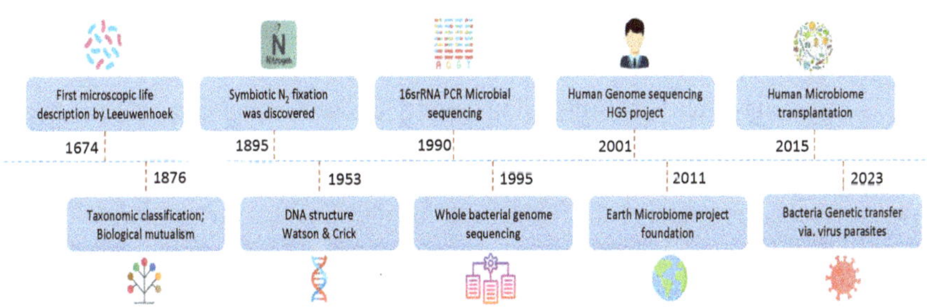

Fig. 1 Milestones in the field of microbiology

lab. Instead, it uses molecular techniques like metagenomics. This provides a broader view about the entire community's structure, function, and dynamics. It inherits the complexity of analytical techniques and equipment (Gupta et al. 2022).

Culture-independent techniques continue to reshape our understanding of microbiology, illuminating a richer microbial diversity than that captured by traditional culturing methods (Table 2). Many microorganisms in environmental samples remain elusive to conventional cultivation due to their specific or uncharted growth conditions (Lloyd et al. 2018). Advanced methods such as metagenomic sequencing and 16S rRNA gene sequencing enable direct genetic analysis from environmental samples without the necessity for cultivation. These approaches have unveiled a wide array of hitherto unknown microbial taxa, deepening our grasp of their ecological roles and metabolic capacities (Rinke et al. 2013). Furthermore, culture-independent strategies present a more comprehensive snapshot of microbial community structures, dynamics, and functional potential across diverse habitats, from terrestrial biomes to intricate human-associated microbiomes (Human Microbiome Project Consortium 2012). A recent exploration also highlighted the potential of these techniques in uncovering microorganisms' roles in biogeochemical cycles, emphasizing their broader implications in environmental and climate sciences (Thompson et al. 2020). The present chapter is going to highlight the importance of various microbes in multiple applications via. culture dependent and independent techniques (Gupta et al. 2022).

2 Diversity of Microbial Life

2.1 Bacteria:

Bacteria are single-celled microorganisms that lack a nucleus and other membrane-bound organelles. They are among the earliest forms of life on Earth e.g., *Cyanobacteria, Chloroflexi, Chloroflexi, Thermotogae* and *Aquificae.* Bacteria exhibit diverse morphologies, including cocci, bacilli, spirilla, and more, each reflecting functional adaptations. The advent of electron microscopy has further refined our understanding of bacterial cell structures like flagella and pili, essential for motility and adherence (Dobro et al. 2017).

In terms of classification, bacterial taxonomy has seen significant advancements through molecular methods. Genetic sequencing has facilitated a more precise phylogenetic categorization, allowing researchers to identify distinct bacterial phyla, genera, and species (Parks et al. 2018). Traditional techniques such as Gram staining remain useful but are now complemented by genomic data. Bacterial reproduction primarily occurs through binary fission. The process of division and the subsequent growth phases (lag, exponential, stationary, and death) are complex, regulated by various environmental factors like temperature and nutrient availability (Dobro et al. 2017). Monitoring these growth curves is pivotal in many applications, including medical diagnostics and biotechnology.

2.2 Viruses:

Viruses are complex organisms at the nexus between living and non-living, exhibit a vast array of structural designs e.g., Orthomyxovirus, Human Immunodeficiency

Table 2 Major potential applications fields for microorganisms

Avenue	Microbial application	Examples
Bioremediation	Using microorganisms to degrade or transform pollutants into less harmful substances	1. *Deinococcus radiodurans* for radioactive waste. 2. *Pseudomonas putida* for oil spills. 3. Mycoremediation using fungi for heavy metals
Biofuels	Production of sustainable fuels through microbial processes	1. Algae for biodiesel. 2. *Clostridium* for bioethanol. 3. *Methanobrevibacter smithii* for biogas
Agriculture	Enhancing crop yields and health through microbial interactions	1. *Rhizobium* for nitrogen fixation. 2. *Trichoderma* fungi for plant growth promotion. 3. *Bacillus thuringiensis* as a biopesticide
Food and Beverage Industry	Utilizing microbes for fermentation and production of food-related compounds	1. Yeasts for bread. 2. *Lactobacillus* for yogurt. 3. *Saccharomyces pastorianus* for lager beer
Medicine	Discovery and production of medical compounds and treatments	1. *Penicillium* for penicillin. 2. *Streptomyces* for tetracycline. 3. Probiotics for gut health
Bioplastics	Production of biodegradable plastics as alternatives to conventional plastics	1. *PHA* producing bacteria. 2. *Bacillus subtilis* for polyglutamic acid. 3. Yeasts for polysaccharide-based plastics
Biotechnology	Customizing microbes for specific industrial tasks using genetic engineering	1. E. coli for insulin production. 2. Yeast for bioethanol production. 3. Modified bacteria for enzyme production

Adapted and modified from Gupta et al. (2022)

Virus (HIV), Hepatitis C Virus (HCV), Varicella-Zoster Virus (VZV) etc. Primarily, they contain genetic material, DNA or RNA, cloaked in a protein coat known as the capsid. Some, such as the SARS-CoV-2, also feature a lipid bilayer called an envelope, furthering their complexity (Denison et al. 2019; Walls et al. 2020). A typical viral life cycle initiates with the attachment to host cells, followed by entry, replication, assembly, and eventual release. Throughout this cycle, the virus exploits the host's cellular mechanisms for its propagation, often leading to cell damage (Denison et al. 2019).

In terms of classification, the International Committee on Taxonomy of Viruses (ICTV) has recognized a vast diversity of viruses. This categorization, based on nucleic acid type, replication strategies, morphology, and other factors, underscores the significance of molecular and genetic data in understanding viral taxonomy (Walls et al. 2020). Grasping the intricacies of viral structure, life cycle, and classification is integral to forging effective therapeutic and preventive strategies against viral diseases.

2.3 Fungi:

Fungi, an incredibly diverse kingdom, consist of multicellular mushrooms, molds, and the unicellular yeasts e.g., *Acremonium cucurbitacearum, Agaricus campestris, Albugo candida* etc. Characteristically, fungi possess a complex cell wall primarily made of chitin, distinguishing them from plants that use cellulose (Nagahashi et al. 2016). Furthermore, they are heterotrophic, deriving nutrients through absorption after external digestion. Taxonomically, fungi are divided into several major phyla, such as Ascomycota, Basidiomycota, and Zygomycota, based on reproductive structures and other morphological features (Fisher et al. 2018). Molecular techniques have further refined fungal taxonomy, providing insights into their evolutionary relationships.

The significance of fungi cannot be understated. Ecologically, they are crucial as decomposers, returning vital nutrients to the ecosystem. Medically, fungi have furnished us with antibiotics like penicillin, while in industries, yeast is vital for fermentative processes in bread and alcohol production. However, they can also pose threats, causing diseases in plants, animals, and humans (Fisher et al. 2018).

2.4 Protozoa:

Protozoa are a diverse group of unicellular eukaryotic organisms. Morphologically, they range from amorphous amoebae, like *Amoeba proteus*, to more structured forms like the flagellated *Trypanosoma* species (Talman et al. 2016). Most protozoa have specialized cellular structures for movement, such as cilia, flagella, or pseudopodia.

Their life cycles can be complex, often involving both asexual and sexual reproduction. For instance, *Plasmodium* spp., the causative agents of malaria, have a life cycle involving both human and mosquito hosts. Within these hosts, the protozoa undergo various developmental and reproductive stages (Talman et al. 2016). Protozoa play significant roles in ecosystems as primary consumers, feeding on bacteria and other microscopic organisms. They contribute to nutrient cycling and can be essential food sources for higher trophic levels. However, their significance isn't limited to positive contributions. Many protozoan species, such as *Giardia* and *Entamoeba histolytica*, are pathogenic to humans, causing diseases that can be severe (López-Escamilla et al. 2018).

2.5 *Algae:*

Algae, predominantly photosynthetic, encapsulate a diverse array of eukaryotic organisms, ranging from microscopic phytoplankton to macroscopic seaweeds. They're primarily distinguished by their ability to perform photosynthesis and produce oxygen, but unlike higher plants, they lack roots, stems, and leaves (Mishra et al. 2021). Their cellular structures can be simple, as in unicellular *Chlorella*, or complex, as in multicellular kelps.

Algal classification, historically based on color (green, brown, red, etc.), has been redefined using molecular techniques, revealing a more nuanced phylogenetic understanding. Broadly, they can be categorized (Table 3) into:

- Chlorophyta (green algae)
- Phaeophyta (brown algae)
- Rhodophyta (red algae)
- Bacillariophyta (diatoms)
- Dinophyta (dinoflagellates), among others (Leliaert et al. 2018).

The advent of molecular tools has dramatically altered our understanding of algal evolution and relationships, with some previously distinct groups being merged, and others split into multiple entities.

3 Microbial Genetics and Molecular Biology

Microbial genetics highlights the principles governing cellular functions, evolution, and interactions among microorganisms. At its essence, it probes the intricacies of microbial genomes, with the operations of DNA, RNA, and protein synthesis forming its backbone (Turner et al. 2019).

3.1 *DNA: The Molecular Foundation*

DNA (Deoxyribonucleic acid) remains the quintessential genetic material in a majority of microbes. It's characterized by its double helix structure, wherein two strands are interwoven. Every strand is a sequence of nucleotides: adenine (A), thymine (T), guanine (G), and cytosine (C). Through complementary base pairing: A with T, G with C, DNA ensures accurate replication, pivotal for cell division (Alberts et al. 2019).

Bacteria, representative microbes, typically harbour a singular circular chromosome localized in the nucleoid. Moreover, many house plasmids, minuscule DNA fragments that can host genes crucial for survival, such as those conferring antibiotic resistance (Lopatkin et al. 2020).

Table 3 Different types of algae and their characteristics

Type of Algae	Characteristics	Distinguishing Feature	Significance	Examples	Locations in India
Chlorophyta (Green Algae)	Green pigments; unicellular to multicellular	Chlorophylls a and b	Oxygen production, biofuel potential	*Chlorella, Spirogyra, Cladophora*	Freshwater bodies, ponds, and lakes like Dal Lake, Loktak Lake
Phaeophyta (Brown Algae)	Brown pigments; mainly multicellular	Fucoxanthin	Agar production; coastal ecosystem stabilizers	*Sargassum, Fucus*	Coastal regions, notably the coasts of Tamil Nadu and Gujarat
Rhodophyta (Red Algae)	Red pigments; mostly multicellular	Phycoerythrin	Agar and carrageenan production	*Gracilaria, Porphyra*	Coastal areas, especially around the Gulf of Mannar and Lakshadweep Islands
Bacillariophyta (Diatoms)	Silica cell walls; unicellular	Frustules (silica cell walls)	Oxygen production; oil production for biofuel	*Cyclotella, Navicula*	Freshwater and marine habitats, including the Bay of Bengal and freshwater lakes in Kashmir
Dinophyta (Dinoflagellates)	Unicellular; two flagella	Cell covering of cellulose plates	Some are bioluminescent; red tide causatives	*DinophysisNoctiluca*	Coastal waters, especially during red tide events in the Arabian Sea

Adapted from Leliaert et al. (2018)

3.2 RNA: The Conduit and the Facilitator

RNA (Ribonucleic acid) diverges from DNA in function and form. Three principal RNA types are indispensable for protein synthesis (Lodish et al. 2019):

- mRNA (Messenger RNA): Functions as the protein synthesis template. RNA polymerase-driven transcription crafts mRNA, referencing DNA as its guide.
- rRNA (Ribosomal RNA): Constitutes ribosomes, cellular machineries executing protein synthesis. Notably, microbial ribosomes differ slightly from their eukaryotic kin, a disparity exploited by specific antibiotics targeting bacterial pathogens.
- tRNA (Transfer RNA): Acts as an intermediary, delivering specific amino acids to ribosomes during translation, thus translating the mRNA's message into a protein sequence.

3.3 Protein Synthesis: Translating the Genetic Code

The molecular biology central dogma encapsulates the processes wherein DNA undergoes transcription to produce RNA, which subsequently directs protein synthesis via translation. This cascading mechanism ensures the gene-to-protein information relay remains intact (Lodish et al. 2019). During transcription, selected genes convert into mRNA sequences. In translation, ribosomes interpret the mRNA's codons and, in collaboration with tRNAs, facilitate the protein's amino acid chain assembly. For microbes, this mechanism can be remarkably swift, underpinning their rapid adaptability to fluctuating environments (Turner et al. 2019).

3.4 Gene Regulation and Expression

Gene regulation and expression are essential mechanisms that govern cellular functions by determining which genes are activated or suppressed, and to what extent. These processes are especially crucial in microbes, where rapid response to environmental changes is vital for survival (Gupta et al. 2021b). At its core, gene regulation ensures that genes are expressed at the right time and in the appropriate context. For instance, in the presence of lactose, *E. coli* bacteria will induce the expression of genes required for lactose metabolism. This is a classic example of an inducible system, where the presence of a specific molecule activates a set of genes (Liu and Zong 2019). Regulation is achieved primarily at the transcriptional level, although post-transcriptional, translational, and post-translational regulations also play significant roles. Regulatory proteins, often responding to environmental cues, can bind to specific DNA sequences, promoting or inhibiting the binding of RNA polymerase, and thus influencing transcription (Gupta et al. 2021b; Adams and Storz 2020). MicroRNAs (miRNAs) represent another layer of gene regulation. These are short

RNA sequences that can bind to mRNA molecules, preventing their translation or leading to their degradation. In bacteria, similar regulatory roles are played by small RNAs (sRNAs) that typically act by base pairing with target mRNAs and affecting their stability or translation (Storz et al. 2021).

Understanding gene regulation is imperative, not only for basic biological insights but also for potential applications. Manipulating gene expression levels has direct implications for biotechnology, allowing, for instance, the enhanced production of biofuels or therapeutic compounds by genetically engineered microbes (Gupta et al. 2021b). Conclusively, gene regulation and expression are dynamic processes that ensure the correct genes are expressed in response to internal and external stimuli, allowing cells, especially microbes, to adapt and thrive in changing environments (Storz et al. 2021).

4 Microbial Metabolism and Physiology

Microbial metabolism provides the necessary energy and building blocks to sustain life. It encompasses a vast array of chemical reactions and pathways, each tailored to harness energy from diverse sources. Two primary mechanisms, respiration and fermentation, are central to microbial energy production (Schönheit et al. 2019).

4.1 Respiration: Oxidative Phosphorylation and Electron Transport

Respiration is an aerobic process, wherein oxygen acts as the final electron acceptor, although in some microorganisms, alternative molecules like nitrate or sulfate might serve this role. Central to respiration is the electron transport chain (ETC) found in microbial membranes. As electrons traverse this chain, from one protein complex to another, protons are pumped across the membrane, creating an electrochemical gradient. This gradient, known as proton motive force (PMF), drives ATP synthesis through ATP synthase, an elegant molecular mechanism (Berry et al. 2020).

Facultative anaerobes, like certain strains of *E. coli*, can perform both aerobic and anaerobic respiration, adapting to the availability of oxygen. This metabolic flexibility allows them to thrive in diverse environments, from oxygen-rich to oxygen-depleted niches (Milo and Phillips 2019).

4.2 Fermentation: Energy Without Oxygen

In the absence of external electron acceptors like oxygen, some microbes resort to fermentation, an anaerobic energy production pathway. Instead of the ETC, fermentation pathways regenerate NAD+ by transferring electrons from NADH to internal organic molecules, producing various organic products as end results, such as lactic acid, ethanol, or butyric acid. Different microbes have evolved distinct fermentation pathways, each characterized by its unique end products. For instance, *lactobacilli* ferment sugars to produce lactic acid, a process harnessed in dairy fermentation, while *Saccharomyces cerevisiae* generates ethanol and CO_2, used in beer and wine production (Eisenstein 2019). It's worth noting that fermentation is less efficient than respiration, yielding lesser ATP per glucose molecule. However, it allows microbial survival in anaerobic habitats and can occur rapidly, supporting rapid growth in certain conditions (Schuchmann and Müller 2020). The intricate microbial metabolism underpins the ability of these tiny organisms to colonize nearly every conceivable niche on Earth. From oxygen-rich atmospheres to the anoxic depths of marshlands, microbes, through respiration and fermentation, have developed strategies to extract energy and flourish.

5 Biogeochemical Pathways

Biogeochemical pathways refer to the natural processes through which elements, such as carbon, nitrogen, phosphorus, and sulphur, cycle between living organisms, the atmosphere, soil, and water. Microorganisms play crucial roles in these pathways, facilitating the movement and transformation of these elements. In the present chapter, Nitrogen cycle would be demonstrated and discussed. The nitrogen cycle involves a series of processes through which nitrogen is transformed and cycled through various forms in the environment. Microorganisms play crucial roles in each step of the nitrogen cycle (Fig. 2).

Nitrogen Fixation:

Nitrogen fixation is the process by which atmospheric nitrogen (N_2) is converted into ammonia (NH_3) or ammonium (NH_{4+}), which can be used by plants. Certain microorganisms, known as nitrogen-fixing bacteria, are responsible for this process. These bacteria include free-living species like *Azotobacter* and *Clostridium*, as well as symbiotic bacteria like *Rhizobium* that form nodules on the roots of leguminous plants. These bacteria possess an enzyme called nitrogenase that enables them to convert N_2 into ammonia. Nitrogen fixation provides a critical source of nitrogen for ecosystems.

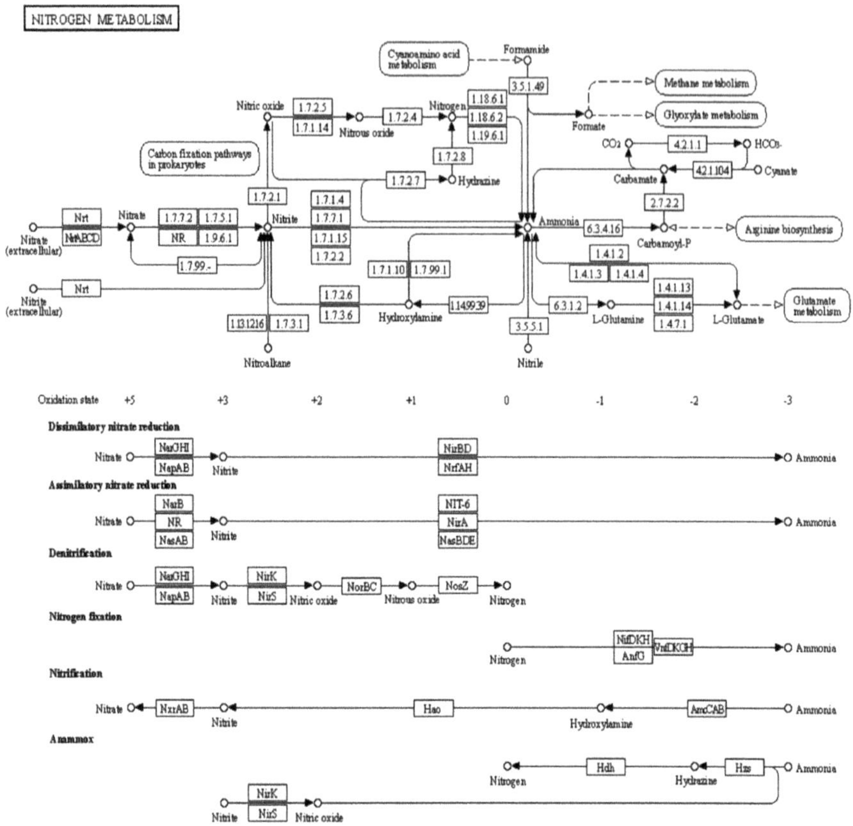

Fig. 2 Typical Nitrogen metabolism pathway and associated genes (sourced from KEGG: Kyoto Encyclopaedia of Genes and Genomes)

Ammonification:

Ammonification is the process of converting organic nitrogen compounds (such as dead plants and animals) into ammonium ions (NH_{4+}). Microorganisms like bacteria and fungi are responsible for breaking down complex organic nitrogen compounds into simpler forms. During this process, enzymes released by these microorganisms break down proteins and other nitrogen-containing molecules, releasing ammonia as a byproduct.

Nitrification:

Nitrification is a two-step process that converts ammonium (NH_{4+}) into nitrite (NO_{2-}) and then into nitrate (NO_{3-}). Nitrifying bacteria carry out these steps. The first group of bacteria, called ammonium-oxidizing bacteria, convert ammonium into nitrite. The second group, nitrite-oxidizing bacteria, further convert nitrite into nitrate. This process makes nitrogen available in the form of nitrate, which plants can readily absorb.

Assimilation:

Plants and microorganisms take up nitrate and ammonium from the soil to build their own organic molecules, such as amino acids and proteins. Microorganisms play a critical role in facilitating this process by incorporating nitrogen compounds into their cells as they grow and reproduce.

Denitrification:

Denitrification is the process in which nitrate (NO_{3-}) and nitrite (NO_{2-}) are converted back into nitrogen gas (N_2) or nitrous oxide (N_2O) under anaerobic (low oxygen) conditions. Denitrifying bacteria are responsible for this process. These bacteria use nitrates and nitrites as electron acceptors in the absence of oxygen, releasing nitrogen gas back into the atmosphere.

Anammox (Anaerobic Ammonium Oxidation):

In some environments, like sediments and wastewater treatment systems, a unique process called anammox occurs. This process involves the anaerobic oxidation of ammonium (NH_{4+}) with nitrite (NO_{2-}), resulting in the production of nitrogen gas (N_2). Anammox bacteria carry out this process, and it is a crucial mechanism for removing excess nitrogen from certain ecosystems. Microorganisms are central to the nitrogen cycle, mediating the conversion of nitrogen between various forms and facilitating its movement through ecosystems. This cycle is essential for maintaining the availability of nitrogen, a vital nutrient for both plants and animals.

6 Conclusions

Microbiology, since its inception, has unveiled a myriad of secrets held within the microscopic world, altering our understanding of life and ecosystems. From the earliest observations of "animalcules" under rudimentary lenses to contemporary advancements in genomics and biotechnology, this domain has consistently pushed the frontiers of scientific knowledge. Each discovery, whether it pertains to pathogenesis, mutualistic symbioses, or metabolic versatility, not only amplifies our comprehension of microbial processes but also reshapes interdisciplinary fields, from medicine to environmental science.

The dynamism inherent to microbiology is underscored by the adaptability and resilience of microbes themselves. As we unearth extremophiles thriving in conditions once deemed inhospitable or decode intricate communication networks between microbial communities, we are reminded of the depth of complexity yet to be explored. Furthermore, the rise of antimicrobial resistance and the continuous emergence of novel pathogens spotlight the ever-evolving challenges within the field, demanding perpetual vigilance and innovation.

As the field of microbiology progresses, its intersection with technology and other scientific disciplines promises unprecedented advancements with profound societal implications. The potential of metagenomics, for instance, in analysing environmental DNA, allows us to catalogue the vast microbial diversity without needing to culture them in labs. Recent studies suggest that less than 10% of microbial species have been cultured, leaving a vast reservoir of undiscovered potential. Furthermore, the synthesis of artificial life forms, such as the minimal bacterial cell created by the J. Craig Venter Institute in 2016, could revolutionize industries from biofuel production to waste management. The expanding field of gut microbiome research underscores another pivotal avenue. Preliminary findings highlight connections between gut bacteria and myriad health conditions, from mental health disorders like depression to systemic diseases like diabetes. As personalized medicine evolves, custom probiotics might become a mainstay in therapeutic regimens, optimizing individual health based on one's microbiome.

However, these advancements are not without challenges. The surge in antimicrobial resistance, with WHO 2019 data indicating that by 2050, drug-resistant diseases could cause 10 million deaths annually worldwide necessitates a reimagined approach towards antibiotics. Microbiology's future trajectory holds promises and challenges, each shaping societal paradigms, health perspectives, and global policies. Looking forward, as tools become more refined and methodologies more sophisticated, microbiology promises deeper insights into cellular mechanisms, community dynamics, and global impacts. It requires collaborative efforts, combining traditional techniques with cutting-edge technologies, to harness microbial potential for societal benefits while safeguarding against potential threats. In essence, microbiology remains a testament to the inexhaustible wonders of the microscopic realm and its overarching influence on the macroscopic world.

Acknowledgements Author "Akarsh Verma" is grateful to the monetary support provided by the University of Petroleum and Energy Studies (UPES)-SEED Grant program.

Declarations

Conflict of Interest The authors declare that there is no conflict of interest. Their primary aim is the dissemination of information.

Ethical Approval The authors hereby state that the present work is in compliance with the ethical standards.

Competing Interests The authors declare no competing interests.

Availability of Data and Materials Not applicable.

References

Adams, P. P., & Storz, G. (2020). Prevalence of small base-pairing RNAs derived from diverse genomic loci. Biochimica et Biophysica Acta (BBA) - Gene Regulatory Mechanisms, 1863(5), 194514.

Alberts, B., Johnson, A., Lewis, J., Morgan, D., Raff, M., Roberts, K., & Walter, P. (2019). Molecular biology of the cell. Garland science.

Berry, E. A., Huang, L. S., Saechao, L. K., & Pon, N. G. (2020). Structure and function of cytochrome bc complexes. Annual Review of Biochemistry, 89, 605–630.

Black, J. G. and Black, L. J., 2018. Microbiology: principles and explorations. John Wiley & Sons.

Denison, M. R., et al. (2019). Coronaviruses: An Overview of Their Replication and Pathogenesis. Methods in Molecular Biology, 1282, 1–23.

Dobro, M. J., et al. (2017). Electron cryotomography of bacterial cells. Journal of Visualized Experiments, 123, e55428.

Eisenstein, M. (2019). Exploring the yeast biodiversity of the wine-making microbiome. Nature Reviews Microbiology, 17(9), 519–527.

Fisher, M. C., et al. (2018). Emerging fungal threats to animal, plant and ecosystem health. Nature, 484(7393), 186–194.

Gupta, J., Rathour, R., Dupont, C., Mishra, A. and Thakur, I. S., 2022. Biogeochemical profiling and taxonomic characterization of municipal landfill site by metagenomic sequencing. Bioresource Technology, 351, p. 126936.

Gupta, J., Rathour, R., Dupont, C. L., Kaul, D. and Thakur, I. S., 2021b. Genomic insights into waste valorized extracellular polymeric substances (EPS) produced by Bacillus sp. ISTL8. Environmental Research, 192, p. 110277.

Gupta, J., Rathour, R., Singh, R. and Thakur, I. S., 2019b. Production and characterization of extracellular polymeric substances (EPS) generated by a carbofuran degrading strain Cupriavidus sp. ISTL7. Bioresource Technology, 282, pp. 417–424.

Gupta, J., Rathour, R., Maheshwari, N. and Thakur, I. S., 2021a. Integrated analysis of Whole genome sequencing and life cycle assessment for polyhydroxyalkanoates production by Cupriavidus sp. ISTL7. Bioresource Technology, 337, p. 125418.

Gupta, J., Rathour, R., Medhi, K., Tyagi, B. and Thakur, I. S., 2020. Microbial-derived natural bioproducts for a sustainable environment: a bioprospective for waste to wealth. In Refining biomass residues for sustainable energy and bioproducts (pp. 51–85). Academic Press.

Gupta, J., Tyagi, B., Rathour, R. and Thakur, I. S., 2019a. Microbial treatment of waste by culture-dependent and culture-independent approaches: Opportunities and challenges. Microbial Diversity in Ecosystem Sustainability and Biotechnological Applications: Volume 1. Microbial Diversity in Normal & Extreme Environments, pp. 415–446.

Human Microbiome Project Consortium. (2012). Structure, function and diversity of the healthy human microbiome. Nature, 486(7402), 207–214.

Leliaert, F., et al. (2018). Chloroplast phylogenomic analyses reveal the deepest-branching lineage of the Chlorophyta, Palmophyllophyceae class. nov. Scientific Reports, 8(1), 1–11.

Liu, L., & Zong, W. (2019). Elucidating the role of transcription in shaping the 3D structure of the bacterial genome. Nucleic Acids Research, 47(10), 5099–5110.

Lloyd, K. G., Steen, A. D., Ladau, J., Yin, J., & Crosby, L. (2018). Phylogenetically novel uncultured microbial cells dominate Earth microbiomes. Nature Communications, 9(1), 1–10.

Lodish, H., Berk, A., Zipursky, S. L., Matsudaira, P., Baltimore, D., & Darnell, J. (2019). Molecular cell biology. WH Freeman.

Lopatkin, A. J., Meredith, H. R., Srimani, J. K., Pfeiffer, C., Durrett, R., & You, L. (2020). Persistence and reversal of plasmid-mediated antibiotic resistance. Nature Communications, 8(1), 1689.

López-Escamilla, E., et al. (2018). Prevalence and molecular identification of Entamoeba histolytica in a rural area of Oaxaca, Mexico. Parasitology International, 67(1), 166–170.

Milo, R., & Phillips, R. (2019). Cell Biology by the Numbers. Garland Science.

Mishra, A., Gupta, J., Kumari, T., Pal, R. and Thakur, I. S., 2021. Unravelling the attributes of novel cyanobacteria Jacksonvillea sp. ISTCYN1 by draft genome sequencing. Bioresource Technology, 337, p. 125473.

Mun, Y., Kim, M. K. and Oh, J. Y., 2019. Ten-year analysis of microbiological profile and antibiotic sensitivity for bacterial keratitis in Korea. PLoS One, 14(3), p. e0213103.

Nagahashi, G., et al. (2016). Characteristics of the Outer Layer of the Fungal Wall. Fungal Biology and Biotechnology, 3(1), 7.

Parks, D. H., et al. (2018). A standardized bacterial taxonomy based on genome phylogeny substantially revises the tree of life. Nature Biotechnology, 36(10), 996–1004.

Patil, G., 2022 Vedas and Knowledge of our Ancestors: True origin of microbiology. International Journal of Current Science 12,4 ISSN: 2250–1770

Rathour, R., Gupta, J., Tyagi, B., Kumari, T. and Thakur, I. S., 2018. Biodegradation of pyrene in soil microcosm by Shewanella sp. ISTPL2, a psychrophilic, alkalophilic and halophilic bacterium. Bioresource Technology Reports, 4, pp. 129–136.

Rathour, R., Gupta, J., Mishra, A., Rajeev, A. C., Dupont, C. L. and Thakur, I. S., 2020. A comparative metagenomic study reveals microbial diversity and their role in the biogeochemical cycling of Pangong lake. Science of the Total Environment, 731, p. 139074.

Rinke, C., Schwientek, P., Sczyrba, A., Ivanova, N. N., Anderson, I. J., Cheng, J. F., ... & Lucas, S. (2013). Insights into the phylogeny and coding potential of microbial dark matter. Nature, 499(7459), 431–437.

Schönheit, P., Buckel, W., & Martin, W. F. (2019). On the origin of heterotrophy. Trends in Microbiology, 27(2), 88–98.

Schuchmann, K., & Müller, V. (2020). Energetics and application of heterotrophy in acetogenic bacteria. Applied and Environmental Microbiology, 82(14), 4056–4069.

Storz, G., Wolf, Y. I., & Ramamurthi, K. S. (2021). Small proteins can no longer be ignored. Annual Review of Biochemistry, 89, 409–435.

Sun, Y., Kumar, M., Wang, L., Gupta, J. and Tsang, D. C., 2020. Biotechnology for soil decontamination: opportunity, challenges, and prospects for pesticide biodegradation. Bio-based materials and biotechnologies for eco-efficient construction, pp. 261–283.

Talman, A. M., et al. (2016). Malaria parasite and vector dynamics: insights from population biology and landscape ecology. Acta Tropica, 157, 103–115.

Thompson, L. R., Sanders, J. G., McDonald, D., Amir, A., Ladau, J., Locey, K. J., ... & Earth Microbiome Project Consortium. (2020). A communal catalogue reveals Earth's multiscale microbial diversity. Nature, 551(7681), 457–463.

Turner, P. E., Cooper, V. S., & Lenski, R. E. (2019). Evolutionary genomics of host–pathogen interactions. Trends in Genetics, 35(6), 436–450.

Walls, A. C., et al. (2020). Structure, Function, and Antigenicity of the SARS-CoV-2 Spike Glycoprotein. Cell, 181(2), 281–292.e6.

Revolutionizing Water Quality Through Microbial Interactions in Aquatic Ecosystems for Pollution Abatement

Harshini Sivakumar, Juhi Gupta, and Akarsh Verma

1 Introduction

In the recent years, aquatic pollution has garnered considerable attention from everyone around the world since it has reached the depths of the ocean even unexplored by humans. And with each day, new findings suggest the exponential increase in its intensity which comes as a repercussion of human actions (Bar-On and Milo 2019). There are millions of species at the micro as well as macro level which form the aquatic ecosystem. Pollution in several ways such as plastic, chemical discharge, oil spill, etc., induce a chain of chemical and biological reactions which affects different life forms, thus disrupting the balance in nature (Ando et al. 2021). With increasing concern on aquatic pollution coupled with the need and commitment to fulfill the sustainable goal 14: 'Life Below Water', scientists and environmentalists have been constantly involved in the research to come up with effective ways to mitigate it so that we can also ensure a quality life for the aquatic organisms and sustain the aquatic resources. First approach is to prevent pollution which is being implemented by actions such as banning single-use plastics, more strict norms on industrial discharge limits (zero liquid discharge), etc. But it is equally or rather more important to reverse the existing pollution as it has already crossed the threshold limits.

H. Sivakumar
Department of Environmental Studies, The Maharaja Sayajirao University of Baroda, Gujarat 390002, India

J. Gupta (✉)
Amity Institute of Environmental Sciences, Amity University, Noida 201303, India
e-mail: jgupta2@amity.edu

A. Verma
Department of Mechanical Engineering, University of Petroleum and Energy Studies, Dehradun 248007, India

Department of Mechanical Science and Bioengineering, Osaka University, Osaka 560-8531, Japan

J. Gupta and A. Verma (eds.), *Microbiology-2.0 Update for a Sustainable Future*,
https://doi.org/10.1007/978-981-99-9617-9_2

Though several ways of mitigating aquatic pollution have been researched upon, the positive impact of microbes on pollution reversal has made it a key area of focus for research (Ando et al. 2021). Microbes always have been playing a crucial role in maintaining balance in the nature. It is an interesting fact that though not visible to the naked eye, microbes contribute to about half of the biomass weight of all living organisms in the ocean (Bar-On and Milo 2019). And many among these microorganisms serve as entities which are responsible for the nutrition cycle in aquatic ecosystem (Ando et al. 2021), cleaning of waste matter naturally, preventing the propagation of other deadly microbes. Also, they are known to withstand some of the extreme environments which shows their adaptability for survival. Use of microbes is an effective way to abate pollution due to its primary advantage of being a natural source which ensures minimal or no side-effects of using this approach (Gupta et al. 2019a; b). Another advantage of using microbes is its tendency to proliferate and form colonies at a very high rate. While other ways of abating water pollution are in general energy intensive, use of microbes is highly efficient way to be implemented without use of much energy.

This chapter focuses on the review of recent literatures on the different mechanisms to abate aquatic pollution using microbes. Firstly, it is important to understand the roles and interactions of microbes in any aquatic ecosystem. This provides the fundamental functioning of microbes based on which the different mechanisms have been derived and can be further innovated. Source of pollution is another important factor to be considered while devising the mechanism for mitigation as it involves physical, chemical and biological alterations to achieve the desired effectiveness of pollution removal. Finally, the mitigations mechanisms which have been broadly discussed in the chapter includes biofilms, phytoremediation, allelopathy, genetically engineered microorganisms (GEM) and biofloc technology.

2 Understanding Roles and Interactions of Microbes in Aquatic Ecosystem

As already mentioned briefly that there is a ubiquitous presence of microorganisms in this earth and hence in all the aquatic ecosystems (Gupta et al. 2021b), the spectrum of their roles and interactions are also very vast and vital for a sustained aquatic ecosystem. Microorganisms' way of deriving and consuming energy are main determinants of biogeochemical cycling and the trophic balance (Paerl and Pinckney 1996; Porter et al. 1988). Ranging from fixing nitrogen which is an essential element required by plants for its metabolism it plays a major role in carrying out biogeochemical cycles in the system. For example, the organic matter when decomposed in oxygen deficient sediments in a water body produces ammonia/ammonium (Paerl and Pinckney 1996), oxidization of which is called as nitrification that takes place as follows:

$$NH_3/NH_4^+ + O_2 \rightarrow NO_2 + 3H^+ + 2e^-(\text{Nitrosomonas : Chemoautotrophs})$$

When it comes to carbon cycle (Paerl and Pinckney 1996), huge number of processes like methanogenesis, anaerobic and aerobic mineralization, methane oxidation, fermentation and respiration are carried out by the microbes. Even, the contribution to the oxygen flux can be seen in (Cotner and Biddanda 2002) where total community oxygen demand of up to 90% is generated by prokaryotic heterotrophs in an oligotrophic system. If we take phosphorus cycle, then some part of organic-c bound phosphorus is assimilated and some part of it is liberated as dissolved inorganic phosphorus during the mineralization (decomposition) process of organic matter (Donald et al. 2005). Likewise, microbes play a major role in cycling of sulphur too in anoxic conditions (Gupta et al. 2021a, 2022).

Now, when the microbes decompose the organic matter, ultimately, they derive nutrients out of it which may be a part of biogeochemical cycle as already stated but also a major source of food for other organisms. It also serves as food for other organisms like protozoa (Joshi 2016) hence it is responsible for material cycling through a system called microbial loop, which is a system functioning for the production and cycling of carbon in the trophic chain through biomass production while also cycling and the other essential and limiting nutrients. Here, the microbes are basically the picoplankton which can be either an autotroph like cyanobacteria or some heterotroph like bacteria where autotrophs can initiate the cycle with significant amount of primary production also known as phytoplankton biomass (Meyers 2000). This biomass is consumed by the zooplanktons as part of secondary production. On the other hand, the organic matter released by these phytoplankton, zooplankton and also other trophic levels in the form of particulate organic carbon (faecal pellets like that of sea urchins, turtles, etc.) (Fenchel and Jørgensen 1977) and dissolved organic carbon together constituting what is called as detritus is fed upon by the heterotrophic bacteria also known as detritivores and then the organic matter is decomposed. These detritivores are then consumed by protozoa which are also consumed by zooplankton (Fig. 1) and they in turn are consumed by some larval fish (Porter et al. 1988) or crustaceans, and the level goes up through the trophic chain. This part of food chain where chemical energy derived from the detrital organic carbon is mobilized across the biota is known as detritus food chain (Fenchel and Jørgensen 1977). In this way the energy flux is maintained by the microbial interactions in aquatic ecosystems. Also, that chemical energy can be mobilized by inorganic compounds like H_2S or NO_2^-, etc., in the so-called detrital electron flux (Fenchel and Jørgensen 1977).

The microbes are found in various communities as detrital mat, biofilm and planktonic forms of assemblages (Srivastava et al. 2017). There are also some interactions associated with plants where microbes form an association in the form of biofilm with the aquatic plants that mainly consists of a mutualistic relationship in terms of transfer of minerals and defensive immunity (Srivastava et al. 2017) is provided to the plants by microbes and in return receives oxygen and organic carbon. For example, Srivastava et al. (2017), such a microbial biofilm association with the plant *Lemna minor* prevented the Cadmium metal ions from being consumed by the plant and hence prevented it from being intoxicated. In the aquatic ecosystem, there are

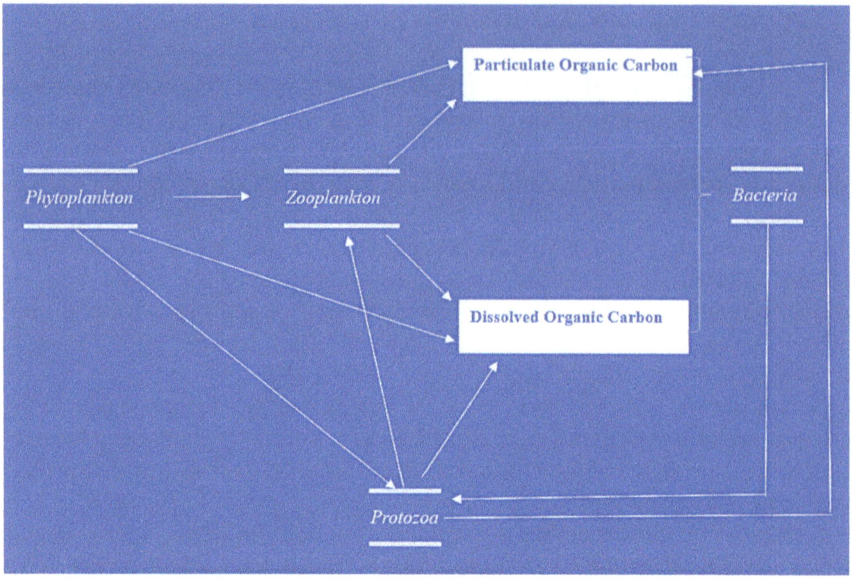

Fig. 1 Representing a part of detritus food chain

many substances suspended or submerged which are known as surfaces like gravel, rock or sediments which are inorganic and animal, algae and leaf litter surfaces which are organic (Bonnineau et al. 2021). These surfaces are quickly colonized by the microbes and form what is called as biofilm. Biofilms consist of some matrix of extracellular polymeric substances (EPS) that are produced by themselves, and it is composed of exopolysaccharides, different variety of proteins, glycoproteins, glycolipids along with high quantities of extracellular DNA, SPMs, and detritus (Bonnineau et al. 2021). Many charged or ionized (Dang and Lovell 2016), and hydrophobic substances get accumulated on these surfaces and detritus particles like that of phytoplankton, zooplankton, etc., which are rich in organic matter having more availability of inorganic nutrients, carbon, and some energy sources too, making them very important sites for biogeochemical cycle related processes. The decomposition of those particles which are sinking or settling down like biogenic aggregates facilitates the carbon flux through biological pump system that transports the photosynthetically fixed carbon from surface (or euphotic) waters to the deep waters. In this way this also helps modulating the climate variability globally as a significant part of those particulate organic carbon present in the marine aggregates are sequestered from the atmosphere (Dang and Lovell 2016). The microbial biofilms often tend to interact with the contaminants in the aquatic ecosystem and accumulate them and as they are primary source of food, there occurs the bioaccumulation across the trophic levels through food webs and it may have significant impact on organisms like insects, fish, shrimps, and gastropods as such organisms consume microbes of specific taxa according to their preference (Bonnineau et al. 2021). Apart from the above-mentioned purposes the surface association of microbes are also involved in providing sites for settling larvae of some invertebrates and in decomposing xenobiotic compounds (Dang and Lovell 2016). The microbes are also often seen and used and indicators of changes in an aquatic ecosystem. With their abundance they can readily identify any positive or negative changes in the environment and bring it to our notice, and hence its always considered a good indicator that can show early signals of pollution or changes that indicate any faulty land use or climate change and a range of biotic and abiotic changes (Paerl et al. 2003; Zaghloul et al. 2020). The consortium of microbes has great capacity to modify the operational levels, group composition and biomass to manage the ecosystem pollutants. The microbial toxins are capable of being efficiently monitored and used as an indicator of pollutant. Bioluminescent bacteria are also often used to identify some ecosystem toxins as microbial markers and then it is further studied or used in many ways (Zaghloul et al. 2020). An example of the microbial indicator is vast algal blooms that are mainly cyanobacteria in the upward flow of oligohaline water and dinoflagellates cryptomonads in the downward flow of mesohaline water which indicate the eutrophication of the water body which later leads to hypoxia (deficiency of oxygen) and then anoxia (absence of oxygen) in the water body that together influence microbe community's structure and function (Paerl et al. 2003). Some other examples of microbial indicators are coliform bacteria like *Escherichia coli* and non-coliform bacteria like *Streptococcus bovis* and some *Clostridia sp.,* whose presence in water indicates fecal pollution. Also, when some bacteria are exposed to contaminants like cadmium or benzene, they secrete some stress proteins that can indicate any warning for pollution of that ecosystem (Zaghloul et al. 2020). Some fungal species are also used as indicators like *Hylocomium splendens* was used to indicate PTEs in the tundra ecosystems in northwestern Alaska (Zaghloul et al. 2020). Microbes like *Rhodobacter* and *Rhodoferax* are also found to provide the resistance towards turbidity to the primary producers of the aquatic ecosystems (Zhang et al. 2022), improve the heat resistance capacity of corals, etc. With carrying all such crucial roles, microbes also play an important role in mitigating the pollution by pesticides, herbicides, oil spills, plastics, contamination by heavy metals, pathogens, etc., which will be detailed in a different section.

3 Sources of Pollutants in Aquatic Ecosystems

As two thirds of our planet earth is covered up by the aquatic ecosystems, it plays a key role in providing habitats for a wide range of organisms as well as serves as an important system for anthropological activities and in the recent times the activity has increased manifold pertaining to the population explosion due to which there is also a severe increase in the pollution of aquatic ecosystems. There are several types of pollutants, the source of few of which have been discussed here. The extraction of crude oil from oceans, its transportation or storage in ships, are main causes of oil spills which are mainly composed of petroleum hydrocarbons in marine ecosystems. The corals and its inhabitants are highly affected and the organisms like mammals with fur suffer as they lose their buoyancy and warmness needed for survival (Hakeem et al. 2020). For example, the case like Exxon Valdez in Alaska in 1989, indicates that spills caused by storms, earthquakes, human negligence, mechanical errors elevate the problem and almost 47% of oil spilt into oceans is by natural seepage and around 600,000 metric tonnes are spilt yearly (Häder et al. 2020). Some of the atmospheric pollutants also end up in the water bodies like POPs, brominated flame retardants, synthetic fragrances, and the atmospheric travel of substances like mercury or any other nutrients from agriculture consisting of phosphorus and nitrogen from causes problems like eutrophication and extreme toxication of aquatic species like fish and other wildlife and ultimately affects humans too who consume them (Issue 2004). The synthetic dyes coming from textile industries also pose great challenges. They are also used in paper, staining of wood, agriculture, leather, pharmaceuticals, etc. (Chakravarty et al. 2015). The industry uses around 1.6 million litres of water each day and a huge portion of it is wasted that contains emulsifiers, heavy metals, levelling agents, dispersing agents, azo dyes, xanthenes, diphenylamine dyes, etc. (Sharma et al. 2021a). Some of the azo dyes and their biotransformation products have been found to be very toxic and it may be highly carcinogenic and mutagenic also (Influences of New Azo Dyes to the Aquatic Ecosystem n.d.). For example, chromium released into water which is a mordant is used to bind colour to the fibre (Chakravarty et al. 2015). Another serious problem is caused by the emerging pollutants called as microplastics (<5 mm) which are reactive and recalcitrant pollutants and serve as substrates for many organic matter, heavy metals, pesticides, xenobiotics, antibiotics etc., are directly released by many industries and by the physical, chemical, and biological degradation of plastic wastes also (Arias-Andres et al. 2018). Even the storm events which occur frequently in this climate changing world, collects, and brings a lot of plastic and microplastic wastes into the aquatic ecosystems (Hitchcock 2020). The persistent organic pollutants which are also recalcitrant like PAHs, PCBs, etc., released by various activities like consuming obsolescent oil, destruction of buildings, manufacture of cement, incineration, chlor-alkali plants, furnace activity, fly ash stowage, etc., are brought to the aquatic ecosystems in the form of liquids, fuels, oil, sediment, ash with the runoffs from roads, agricultural lands and diffusion from atmosphere also (Ashraf 2017). These are also carcinogenic and may often cause endocrine disruption (Harmon 2015). The Concentrated Animal Feeding Operations (CAFOs) for the livestock acts as wastes emerging from the livestock fields or waste lagoons which contains nitrogen and phosphorus are released into the rivers, estuaries or some streams causing eutrophication that may result in toxic blooms (Mallin and Cahoon 2003). The sewage from municipal is often directly released into the water bodies in developing and underdeveloped countries. It contains lot of pathogens like fecal coliforms that makes the water highly unfit for drinking, bathing like purposes and it also leads to many water-borne diseases like diarrhoea, typhoid, etc. (Rochelle-Newall et al. 2015) (see Table 1).

Table 1 Summarization of pollutants, its sources and activities and their effects on the aquatic ecosystems and human beings

Pollutants	Sources	Activities	Effects	References
Petroleum hydrocarbons: BTEX (Benzene, Toluene, Ethylene, Xylene), PAHs, Pyrene, *n*-alkanes, etc.	Petroleum extraction and refinery plants, Cargo ships	Leakages from oil tanks, pipes, ship accidents, Ballast discharge, etc.	Humans: Nausea, cancer, CNS issue Marine organisms: Loss of buoyancy, warmth, intoxication	Hakeem et al. (2020)
Persistent Organic Pollutants: PAHs, PCBs, PCDFs, PCDDs, etc.	Industries, petroleum refineries, Waste treatment facilities	Demolishment of buildings, cement production, incineration, Clor-alkali processes, fly ash storage, etc.	Carcinogenic, endocrine disrupters	Edwards and Kjellerup (2013)
Agrochemicals: Nitrates and Phosphates	Agricultural fields	Agricultural runoffs	Causes eutrophication, subsequent loss of DO, blockage of sunlight, etc.	Häder et al. (2020)
Dyes and other synthetic compounds: azo dyes, xanthenes, diphenylamine, etc.	Textile and paper industries, leather industries, etc.	Colouring, dispersing, levelling processes, etc.	Causes cancer, mutation, affects liver, CNS, Kidney, etc.	Chakravarty et al. (2015), Influences of New Azo Dyes to the Aquatic Ecosystem (n.d.)
Heavy metals: Cr, Hg, As, Cd, Pb, Ni, Fe, etc.	Batteries and electrical industries, fertilizers, pesticides, refiners, alloys, and solders, etc.	Burning of fossil fuels, petroleum, weathering or erosion of rocks, smelting, mining, etc.	Neurotoxicological problems, DNA damage, carcinogenesis, production of ROS, oxidative stress, etc.	Ghorab (2018)

(continued)

Table 1 (continued)

Pollutants	Sources	Activities	Effects	References
Microplastics	Municipal wastes, industries, Wastewater sewage treatment facilities, etc.	Degradation of fragmentation of macroplastics into microplastics	Biomagnifies in the food chain, chokes the gut of tiny organisms, adsorbs harmful contaminants which affects the organisms that ingest it, etc.	Afianti et al. (2022), Delacuvellerie et al. (2021)
Fecal Organisms: *E. coli*, enterococci, fecal coliforms, etc.	Sewage wastes	Sewage treatments, open defecation and associated runoffs from roads, high rainfall, and floods	Water borne diseases like cholera, typhoid, diarrhoea, etc.	Rochelle-Newall et al. (2015)

4 Mitigation Mechanisms

With the ongoing industrialization at a pace never seen before, the problem of pollution and contamination of aquatic ecosystems has worsened now, and it needs to be abated to keep the balance of ecosystems intact as well as healthy. When it comes to the microbes, they exhibit a range of methods and mechanisms to interact with those pollutants, making reactions with them and ultimately deciding their fate in the system that is often observed as the degradation or transformation of the pollutants. There comes the phenomenon called bioremediation which basically remediates a polluted system using biological organisms (here, microbes). It is basically the process to degrade or transform pollutants into less toxic or non- toxic substances and restore the original condition of the ecosystem (Chaturvedi et al. 2015; Ojuederie and Babalola 2017). The same is also often called as biotransformation. Some specialized processes like (Chaturvedi et al. 2015) bio-stimulation where nutrients, oxygen and some electron acceptors or donors are supplied to make the population of microbe flourish in the system and bioaugmentation where microbial cultures are added to the system to carry out bioremediation. The principle of bioremediation is to modify the chemistry and microbiology of water using the organisms to reduce the solubility of pollutants by altering the pH, carrying out some redox reactions and adsorption of pollutants (Ojuederie and Babalola 2017) which can be both in- situ or ex-situ. Microbes do this through various channels and mechanisms out of which the major types are detailed below.

4.1 Biofilms

Biofilms are those structures or layers where microorganisms are adhered to some submerged or floating surfaces and form microcolonies where they live in sessile mode (Flemming 1993; Jain et al. 2023). Following the adhesion, certain cells initiate the production of exopolysaccharide or glycocalyx after which they form those microcolonies that are basic units of biofilm where cells that have similar morphotypes are present in the thick matrix of polysaccharide (Costed 1995; Costerton and Lappin-Scott 2009) also called as microbial slime regulating an efficient contact with fluid phase. Both prokaryotes and eukaryotes are involved in the biofilm formation. Initially, those sessile bacteria modify their phenotype where they are present as single and multispecies microcolonies and along with the matrix production different bacteria of biofilm get a specific microbial niche (Fig. 2) that respond to environmental conditions differently and in that way, they mature (William Costerton et al. 1995). After a point of time the matured ones get detached from old surface to adhere to new ones. The general conditions or factors responsible for the colonization by the microbes vary widely across physical, chemical, and biological aspects like temperature, pH, salinity, composition of nutrients, availability of light, tides, depth of water, etc. Also initially, the wettability of substrates, its hydrophobicity, and many more properties of the substrate also matters to the adhesion of bacteria to the substrates (Caruso 2020). The substances that are generally adhered to by the bacteria are both living surfaces as well as artificial surfaces preferably hydrophobic surfaces like Teflon, silicone, plastics, etc. (Caruso 2020). The advantage of bacterial biofilms over the planktonic forms is mainly that the EPS (Extracellular Polymeric Substance) provides a considerable shelter to the bacteria along with the ability to express homeostasis, prevents antimicrobial entities (for example, antibiotics) from entering the biofilm, and also safeguards the bacteria from a range of stresses like exposure to UV radiation, desiccation, shift in pH, etc. (Kokare et al. 2009). This ultimately makes it suitable for bioremediating the pollutants present in aquatic ecosystems even if the surrounding is of harsh conditions. The EPS with reduced concentration of O_2 around the centre driving some aerobic and anaerobic heterotrophic organisms with nitrifying organisms and sulphate- reducing organisms with those that are sulphate-oxidizers towards each other which aids in better and quicker degradation of various pollutants (Sharma et al. 2019). A good number of heavy metals are removed or degraded by the biofilms or reduced to non-toxic products like CO_2 and H_2O and some microbe species degrade some contaminants through fermentation also (Yadav and Chandra 2019). Cyanobacteria produced EPS also serves for biosorption and removes aqueous heavy metals whereas those EPS consisting of surfactants can solubilize hydrophobic substances also (Sharma et al. 2019). Basically, the metal ions are immobilized because of the binding capacity of EPS and other components of the cell membrane where the non-aqueous fluid part of biofilm helps in maintaining static metabolism and mobility of ions in the matrix that aids in enhancing the amount of toxic metals that are bioavailable and hence the bioremediation process too (Mohapatra et al. 2019). For example, a study in which a biofilm having sulphate-reducing species based on the oxidation and reduction potentials to make sulphides of copper, zinc, nickel and iron precipitate along with concomitant precipitation of arsenic gave results as 98% of zinc, copper, nickel, 82% of iron and 78% of arsenic was removed efficiently (Edwards and Kjellerup 2013). As they have high biomass density, they can easily bioaccumulate and also reduce some of the dyes to less toxic levels using their enzymes (Mohapatra et al. 2019). Biofilm also deals well with the persistent organic pollutants which are way too toxic like polychlorinated biphenyls (PCBs), PAH, toluene, pentachlorophenol, trinitrotoluene (TNT), polychlorinated ethenes (PCEs) which are hydrophobic can accumulate in food chain and biomagnified (Edwards and Kjellerup 2013; Yadav and Chandra 2019). It can be seen in the examples like Toluene being degraded by hollow-fiber membrane biofilm reactor and in an observation where increase in solubility of PAHs by biofilms following their mass transfer from recalcitrant crystals into cells for their biotransformation was found. With increasing petroleum wastes some strains like *Pseudomonas, Arthrobacter, Bacillus, Rhodococcus,* etc., have been seen as potential ones for forming biofilms to degrade those wastes (Yadav and Chandra 2019). As a part of ex-situ bioremediation, bioreactors are employed where biofilm is grown on a solid inert substrate that carries out some catabolic activities for biodegrading odorous contaminants and VOCs (Sharma et al. 2019). Some of their types are fixed bed reactor, packed bed reactor and fluidized bed reactor (Mohapatra et al. 2019). One specific type of biofilm is the phototrophic biofilm where phototrophs like cyanobacteria, diatoms and green algae form the biofilm with continuous exposure to light where the photosynthesis process fuels the other processes and metabolism in the biofilm and provides organic substrate, oxygen and produces energy (Roeselers et al. 2008). They are helpful in wastewater treatments as they assimilate compounds like ammonium, nitrate, urea and some other nitrogen compounds and also in degrading oil where the free radicals produced in oxygenic photosynthesis can improve the photo-chemical oil degradation (Roeselers et al. 2008). Microbial mat which is a multilayer with multiple microbe communities with vertical stratification consisting of both phototroph and heterotroph bacteria also play similar role in mitigating aquatic pollution like those which are employed in remediating aquaculture systems (Paniagua-Michel 2017). When it comes to microplastics in water bodies, the trap-and-release strategy of Liu et al. (2021) where microplastics are trapped and biologically aggregated in the sticky EPS following which treatment with inducible stimulus can dissemble the biofilm and the microplastics are released out safely. The microplastics offer a new microbial niche to the microbes where they involve in biodegradation of those microplastics as for example, the *Alteromonadaceae* and *Burkholderiales* in biofilms of PHBH are representative of bacteria that majorly degrade PHBH and *Alcanivorax borkumensis* involve in the degradation of low-density polyethylene (Yang et al. 2020). There are several advantages of biofilm form of pollution mitigation where in the process they also involve in generating fuel as they convert organic waste into electricity leading to a sustainable form of energy and also they are employed to mitigate the pollution by radioactive and explosive substances (Yadav and Chandra 2019). The immediate structural and physiological change of biofilm when exposed to toxicants makes them a good and early sign or indicator of pollution of the aquatic ecosystems and

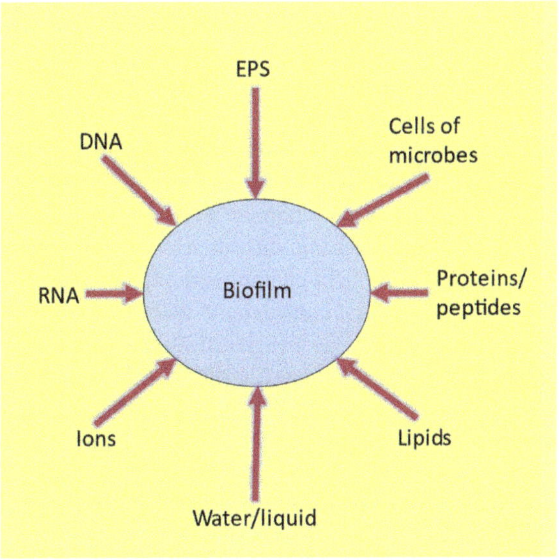

Fig. 2 Different components of biofilm making a unique microbial niche

hence can be used for monitoring (Sharma et al. 2019). They also don't produce any extra wastes when they are decontaminated, rather they are reused in detritus food chain as energy "sources". When the phototrophic biofilms are used the separation of algal biomass that is suspended in water is evited as the nitrogen compounds in it can be used as fertilizers in agriculture (Roeselers et al. 2008). Some of the points in the limitations (Dang and Lovell 2016; Edwards and Kjellerup 2013) side of for biofilm are that there is very less microbe's potential of degrading pollutants, lack of electron donors or acceptors, lack of access to pollutants because of adsorption, biofouling, biocorrosion and possible transmission of pathogens.

4.2 Phytoremediation

It is that process which uses plants in association with microbes where the pollutants and contaminants are removed or cleaned up or transformed into other lesser or non-toxic substances. Here both natural or genetically engineered microbe species can be employed where the plants with the help of microbes carry out some of their routine physiological processes in which the pollutants in the surrounding are either removed or transformed into less toxic forms. Generally, the plants use carry out precipitation, adsorption, and absorption like processes to remove the pollutants. Transformation is generally applicable to inorganic pollutants like Cu, Mn, Hg, Cd, etc., as they cannot be completely removed since they become highly toxic to the plants and microbes while the organic pollutants like TNT, oil, gasoline, PAHs, etc., are mineralized into substances like CO_2, chlorine, nitrate, etc. (Pilon-Smits 2005). The basic idea is to collect and sequester the pollutants from the surrounding in the root system, thus preventing it from leaching or spreading to other parts of the ecosystem (Kuiper et al. 2004). In the root system of plants, the presence of microbes makes the rhizosphere, where they have a symbiotic relationship with the plants as they get nutrients and shelter from them in turn for which they provide support in growth and development of plants, mobilizing nutrients for them and securing them from abiotic stress. In that way they happen to assist in the remediation of the pollutants as a part of phytoremediation with their own metabolism and often prevent the plants from uptake of harmful contaminants (Kumar et al. 2019b). In freshwater systems, plant–microbe interactions depend on factors like pH, salt concentrations, dissolved organic matter, DO, etc.) (Srivastava et al. 2017). There are certain major mechanisms of phytoremediation like Phytostabilization, Phytoextraction, Phytodegradation and Phytovolatilization. In phytostabilization, the contaminants are immobilized by plant induced changes that make contaminants adsorbed or precipitated on the root surfaces whereas in phytoextraction the contaminants are absorbed by the roots following which translocation takes place and later we can harvest down those plants (Kumar et al. 2019b). The major role of microbes is played in phytodegradation. The vast root system gives large surface area where the roots of plants release out some exudates that contains nutrient material like sugar, amino acids, organic acids, etc., enabling them to flourish more and aid in degradation of contaminants using some unique enzymes (Praveen et al. 2019; Wenzel and Adriano 1999) calling it as rhizoremediation or rhizosphere remediation and it also prevents the plants from being intoxicated. It is a passive process as there is also passive adsorption of both and the plant and microbe induced enzymes or biosurfactants improves the solubility of organic pollutants in the water (Pilon-Smits 2005). For example, *Zostera marina* also known as eelgrass has shown a significant remediation of marine sediments that are submerged as they removed PCBs and PAHs (Kumar et al. 2019b) and *Vetiver* is also seen to remediate Pb and Zn efficiently (Praveen et al. 2019). Generally, the plant species with suitable properties for rhizoremediation or phytostimulation is legumes and grasses (Pilon-Smits 2005; Praveen et al. 2019) or any species that has fibrous roots. Some of the microbes perform in phytoremediation by being present inside the plant itself without having any negative effect on them which are known as endophytic bacteria which in some cases provide high tolerance to plants towards pathogens, heavy metals, etc., by engaging in nitrogen fixation, enhancing mineral and water uptake, and inducing system of resistance (Behera 2014). The degradation of pollutants mainly the organic ones, by microbes in association with plants into volatile substances denotes the photovolatilization (Wenzel and Adriano 1999). A similar mechanism called rhizofiltration that employs plant roots which filters out contaminants from the water in a hydroponic or some man-made wetland systems where aquatic plant species are utilized and here the contaminated water flows through the cycle multiple times till it gets considerably remediated (Behera 2014). An arrangement called as hydraulic barriers which are generally tall trees pump up the water in large volumes through their roots and then remediate it (Haldar and Ghosh 2020). In some cases, the fungal structures i.e., the mycorrhizal associations are also found to sequester toxicants and degrade them, an example of which is *Glomus intraradices* which can tolerant cadmium stress (Kumar et al. 2019b). With the rising instances of oil spills in aquatic ecosystems, this phytoremediation by aquatic plants proves very promising. An example is the water hyacinth whose fibrous roots have large surface area and consist of thick rhizosphere which remediate the petroleum composed of hydrocarbon to a significant amount (Yavari et al. 2015). In the estuaries and coastal areas where high chances of frequent flooding and high salinity is there, the halophytes stand as efficient ones in remediating the contamination mainly the heavy metals (Oliveira et al. 2015). While noting about the pesticide pollution, the phytoextraction process is seen as effective in remediation the endosulfan (Bhat et al. 2020). With specific reference to heavy metals contamination, the phytoremediation is carried out either in the form of stabilization by plants called as excluders or in the form of extraction performed by plants known as hyperaccumulators (Haldar and Ghosh 2020). There are a lot of beneficial side of phytoremediation like it is a low-cost method and can be employed in-situ that prevents the contamination of surrounding environment or the human bodies (Pilon-Smits 2005). Those plants that are harvested post phytoextraction can be treated to recover and recycle the metals or can be burnt to produce energy (Kumar et al. 2019b). And some of the limitations are like the limited root's depth of plants, plant's tolerance towards toxicity and the bioavailability of pollutants (Ansari et al. 2015).

4.3 Allelopathy

It is that phenomena which includes processes in which the metabolic substances released out from organisms, here plants and microbes, that affects the growth and development of other biological systems i.e., one species has an inhibiting or stimulating influence in another species (Allelochemicals: Sustainable Ecosystems Development 2003). Though the huge diffusion of substances in water and its continuous flow in water bodies results in reduced accumulation of toxins (Macías et al. 2008), the microbes in aquatic environments are also seen to have great potential of allelopathy in form of interactions such as competition or the ecological succession as they secrete various substances like alkaloids, cyclic peptides, organic acids, etc., as part of their metabolism (Gomes et al. 2017). It saves them and helps to sustain in adverse environmental conditions, maintaining the composition and balance of the ecosystem. The metabolites conferring defence from stress conditions like lack of nutrients are produced, for example the Prymnesins which is an allelopathic toxin produced by *Prymnesium parvum* in nitrogen or phosphorus deficient environment induces more toxicity (Macías et al. 2008). The general mechanisms through which the allelopathy is achieved by microbes are cell lysis, inhibition of growth by limiting protein activities, altering physiological processes, or blistering, and factors responsible for production of allelochemicals are observed as the lack of nutrients, pH, and temperature (Legrand et al. 2003). The allelopathy in aquatic ecosystem has its effect on bacteria, fungi, and algae (Legrand et al. 2003) which can be native or foreign bodies also. Most important problem in aquatic environment pertains to the toxic algal blooms especially in freshwater known as harmful algal blooms which produce their own secondary metabolites that affects the zooplankton, shellfish and many more organisms (Legrand et al. 2003), like in the case of cyanobacteria, the allelochemicals produced are identified as chlorinated aromatic compounds, fatty acids, alkaloids, polyketides, etc. (Leão et al. 2009). Many phytoplankton species release allelochemicals that causes inhibition of electron transport during photosynthesis in their competitors or inhibition of carbon fixation (Hulot and Huisman 2004). It is noted that allelopathy in phytoplankton have an effect on the species succession too the study of which renders the potential to manage and curb the toxic blooms (Gomes et al. 2017) phytoplankton bloom in prairie ponds was supressed using *Chara* (Dakshini 1994). The green tides and red tides during which high amounts of organic compounds are present in water leads to the toxicity in zygotes, larvae of oysters, large animals and even humans (Gomes et al. 2017; Macías et al. 2008). As allelopathy accounts for the toxic blooms it also enables for its control (Dakshini 1994). In a study, the allelopathic effect of *Gomphosphaeria aponina* and *Nannoehloris oeulata* through cell lysis, causes inhibition of *Karenia brevis* which is a dinoflagellate that causes neurotoxicity in shellfish and huge number of deaths in coastal zone of Florida, but still, both those species are freshwater or brackish water inhabitants while *Karenia brevis* is a marine water inhabitant which questions their reliability for practical applications (Legrand et al. 2003). Some research on corals and sponges as involved in allelopathy has also been reported (Macías et al. 2008). One of the advantages of allelopathy is that it enables an organism to overcome any adverse change in its surrounding by making them as benefits and thus are responsible for the co-evolution there (Macías et al. 2008). And the main limitation is that allelopathy leads to a very decreased growth of population (Gomes et al. 2017). However more solid research on allelopathy in aquatic ecosystems and the extent of its reliability in mitigation the contaminants or any other foreign bodies has still not been up to the mark.

4.4 Genetically Engineered Microorganisms

The microbes which are altered, i.e., whose genetic material is modified by genetic engineering are known as genetically engineered microorganisms (GEM). They can also be called as genetically modified microorganisms (GMM). This method is also called as recombinant DNA technology. The 4 main ideas for developing the GEMs for bioremediation purpose are that (1) To modify specificity and affinity of enzymes, (2) To construct pathway and regulate it, (3) To develop bioprocess, monitor and control and (4) Some systems for sensing chemical substances, reducing toxicity and for analysing the end point (Tahri et al. 2013). In some instances, the natural attenuation by the microbes is either not suitable or are not at efficient rates and in such cases the genetically engineered microorganisms through recombinant DNA technology that has been practiced for 25 years are considered suitable to be used to remediate the polluted or contaminated sites (Garbisu and Alkorta 1999) in an in-situ way. Once the GEMs are introduced into the environment the evolution that follows increases their likelihood of sustenance in that environment and techniques to monitor the microbes for any risks by their introduction is looked up to (Garbisu and Alkorta 1999). The basic steps taken for the engineering purpose is given (Fig. 3) (Liu et al. 2019). The dyes that are used in paper industries, fabrics, drugs etc., affects the water quality which causes serios carcinogenicity and toxicity. and the GEMs prove to be efficient at remediating them. Bacteria in anaerobic conditions are observed to convert the dyes into colourless substance using azoreductase enzyme and some electron donors (Kumar et al. 2019a). For example, *Escherichia coli* was seen to increase the activity of azoreductase, and the Acid Red was efficiently removed from wastewater and the same bacteria also degrades azo dye Direct Blue with high rates of decolorization (Kumar et al. 2019a). The coral reefs which form an important part of marine ecosystems, where several prokaryotic species are the inhabitants which aid in mobilizing nutrients, fixing carbon and nitrogen, providing defence opportunities, etc., are getting deteriorated by the increasing global warming and subsequent coral bleaching, which seeks for the need to be flexible to adverse environmental conditions (Damjanovic et al. 2017). Here, the GEMs can be used to provide a microbial inocula to the coral that can withstand the effects of climate change or any other pollution. For example, *Mussismilia hartii* is a coral and the bacteria assembled from it were cultured on a specific medium from which the strain having the capacity to degrade water soluble oil fraction was isolated and when it was exposed to a similar condition to that of an oil spill, the inocula synthesized were found to have less negative impact than the sample that was not inoculated (Damjanovic et al. 2017). The engineered microorganisms which are endophytes are also capable to improve or enhance the degradation of volatile and organic compounds that are water soluble as a mechanism of phytoremediation (Barac et al. 2004). Due to the metal ions being non- biodegradable they often cause severe toxicity in the microbes, plants and get biomagnified to affect the larger animals and humans. The genetically engineered microbes are efficiently able to bind the proteins and peptides that can bind well the metals (or any target metal ion) to the surface of bacteria and remove them from the environment and they can also engineer the catabolic pathways for speedy transformation of metal ions (Jan et al. 2014). For example, *Rhodopseudomonas palustris* is a photosynthetic bacterium produced through recombinant DNA technology which expresses the mercury transport system and metallothionine for removing Hg^{2+} from water polluted with heavy metals (Sharma et al. 2021b). Another example is the fusion of human MT to *araB* in *Escherichia coli* has shown a 3–5 times increase in the bioaccumulation of cadmium and copper (Valls and Victor De Lorenzo 2002). However, there are considerable limitations for this technique such as (1) the engineered growing cultures in small amounts there is a question for its bulk performance (2) the GEMs often experience a hostile environment in the field to be bioremediated (3) horizontal gene transfer of GEMs to other bacteria and can affect the microflora adversely, etc. (Verma and Kuila 2019). And the shortcomings can be overcome using some monitoring methods like using detectable markers, using defective transposons which once integrate into a chromosome can't be done again in absence of it, etc. (Davison 2005).

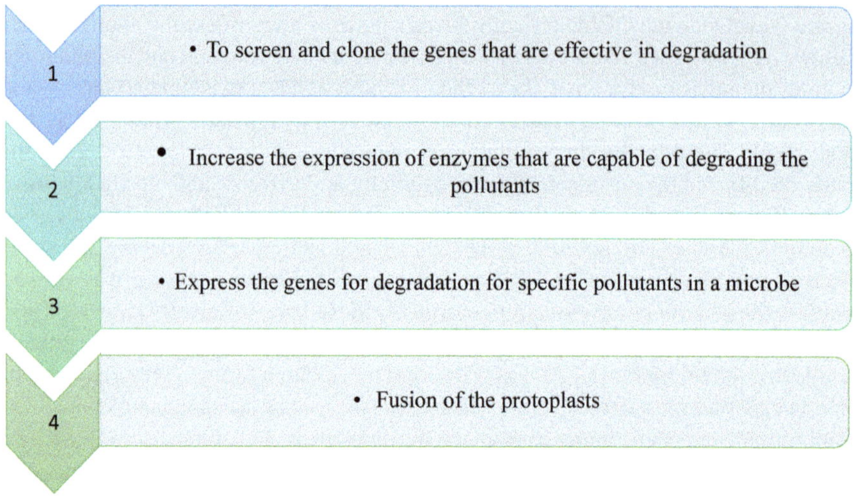

Fig. 3 Basic steps for the engineering of microbes for remediating polluted sites

4.5 Biofloc Technology

When it comes to aquatic ecosystems, the biofloc technology is known to be proven to remediate as well maintain the water quality, especially in the aquaculture. It is that low cost and sustainable technology (Khanjani and Sharifinia 2020) which involves the in-situ production of microorganisms in the form of macroaggregates i.e., the biofloc and the major work is done in (1) maintaining the water quality by consuming the nitrogenous compounds producing in-situ proteins (2) Competing with the pathogens and (3) to reduce the feed conversion ratio and increase culture feasibility and decrease feed costs (Emerenciano et al. 2017). It has been noted that this system has contributed to the removal of total ammonia nitrogen (TAN) along with nitrite that reduces the waste generated in water and it also reduces the amount of water utilized (Khanjani and Sharifinia 2020). The controlled supply of carbohydrates is the cause of the growth of the bacteria, and the subsequent production of proteins consumes up the nitrogen (Padeniya et al. 2022). Several organisms are seen repelling the polymeric substances of humic, polysaccharides and proteins that cover the outer surface of those organisms that acts as sticky slimy layer and causes other particles and cells to adhere to it and the gravitational and electrostatic forces also influence this biofloc formation (Khanjani and Sharifinia 2020). Generally, the biofloc is irregularly shaped, fine, highly porous, and the particle size is broadly distributed (Crab et al. 2012). It minimises the water used for exchange (Deswati et al. 2022). Also, the sludge that is collected down from aquaculture contains the floc- forming microflora that has the potential to mitigate or remediate an oil spilt environment in the water (Syakti et al. 2017). The microbes in the biofloc consists of compounds like organic acids that is potentially resistant to the growth of pathogenic microorganisms and thereby increases immunity, defence and survival rate of fish and shrimp species (Padeniya et al. 2022). However, one major thing to be taken care of in this technology is the amount of total suspended solids which needs to be in suspension for proper aeration and mixing should be done periodically either by exchanging the water or draining the sludge (Crab et al. 2012; Padeniya et al. 2022).

5 Conclusion

When we take the case of aquatic ecosystems, the microbes constitute the major role in balancing the processes, functions, and services of ecosystem where it forms the base for the whole material and energy flux. A range of roles have been noted like providing defensive immunity, providing shelter for some larvae like creatures, serving as indicators of change, etc. Among all the roles, their role in mitigating the aquatic pollution have been emphasized pertaining to the fact that there are a vast number and a wide range of sources of pollution like petroleum hydrocarbons, POPs, agrochemicals, dyes, heavy metals, microplastics and fecal organisms. Five mechanisms such as using biofilms, phytoremediation, allelopathy, genetic engineered microbes, and biofloc technology which are all the forms of bioremediation have been studied. These mechanisms can be applied both in-situ and ex-situ though the success rates differ, and the efficiency is still being questioned and studied. In all the mechanisms the microbes either alone or in association with some other organisms like plants or other microbes remediate the polluted water. Overall, the microbes by remediating the polluted water, support the survival of other organisms and along with all the other roles played by them, they can be considered as promising organisms in enhancing and maintaining the quality of aquatic ecosystems.

Acknowledgements Author "Akarsh Verma" is grateful to the monetary support provided by the University of Petroleum and Energy Studies (UPES)-SEED Grant program.

Declarations

Conflicts of Interest The authors declare that there is no conflict of interest. Their primary aim is the dissemination of information.

Ethical Approval The authors hereby state that the present work is in compliance with the ethical standards.

Competing Interests The authors declare no competing interests.

Availability of Data and Materials Not applicable.

References

Afianti, N. F., Rachman, A., Hatmanti, A., Yogaswara, D., Anggiani, M., Fitriya, N., & Darmayati, Y. (2022). Microbial Biofilm of Plastic in Tropical Marine Environment and their Potential for Bioremediation of Plastic Waste. Journal of Ecological Engineering, 23(4), 261–275. https://doi.org/10.12911/22998993/145463

Allelochemicals: sustainable ecosystems development (2003).

Ando, N., Barquera, B., Bartlett, D. H., Boyd, E., Burnim, A. A., Byer, A. S., Colman, D., Gillilan, R. E., Gruebele, M., Makhatadze, G., Royer, C. A., Shock, E., Wand, A. J., & Watkins, M. B. (2021). The Molecular Basis for Life in Extreme Environments. Annual Review of Biophysics, 50(1), 343–372.https://doi.org/10.1146/annurev-biophys-100120-072804

Ansari, A. A., Gill, S. S., Gill, R., Lanza, G. R., & Newman, L. (Eds.). (2015). Phytoremediation. Springer International Publishing. https://doi.org/10.1007/978-3-319-10969-5

Arias-Andres, M., Klümper, U., Rojas-Jimenez, K., & Grossart, H. P. (2018). Microplastic pollution increases gene exchange in aquatic ecosystems. Environmental Pollution, 237, 253–261. https://doi.org/10.1016/j.envpol.2018.02.058

Ashraf, M. A. (2017). Persistent organic pollutants (POPs): a global issue, a global challenge. In Environmental Science and Pollution Research (Vol. 24, Issue 5, pp. 4223–4227). Springer Verlag. https://doi.org/10.1007/s11356-015-5225-9

Barac, T., Taghavi, S., Borremans, B., Provoost, A., Oeyen, L., Colpaert, J. V., Vangronsveld, J., & Van Der Lelie, D. (2004). Engineered endophytic bacteria improve phytoremediation of water-soluble, volatile, organic pollutants. Nature Biotechnology, 22(5), 583–588. https://doi.org/10.1038/nbt960

Bar-On, Y. M., & Milo, R. (2019). The Biomass Composition of the Oceans: A Blueprint of Our Blue Planet. Cell, 179(7), 1451–1454. https://doi.org/10.1016/j.cell.2019.11.018

Behera, K. K. (2014). Phytoremediation, Transgenic Plants and Microbes (pp. 65–85). https://doi.org/10.1007/978-3-319-00915-5_4

Bhat, S. A., Qadri, H., Cui, G., & Li, F. (2020). Remediation of Pesticides Through Microbial and Phytoremediation Techniques. In Fresh Water Pollution Dynamics and Remediation (pp. 235–245). Springer Singapore. https://doi.org/10.1007/978-981-13-8277-2_13

Bonnineau, C., Artigas, J., Chaumet, B., Dabrin, A., Faburé, J., Ferrari, B. J. D., Lebrun, J. D., Margoum, C., Mazzella, N., Miège, C., Morin, S., Uher, E., Babut, M., & Pesce, S. (2021). Role of Biofilms in Contaminant Bioaccumulation and Trophic Transfer in Aquatic Ecosystems: Current State of Knowledge and Future Challenges. In Reviews of Environmental Contamination and Toxicology (Vol. 253, pp. 115–153). Springer. https://doi.org/10.1007/398_2019_39

Caruso, G. (2020). Microbial colonization in marine environments: Overview of current knowledge and emerging research topics. Journal of Marine Science and Engineering, 8(2). https://doi.org/10.3390/jmse8020078

Chakravarty, P., Bauddh, K., & Kumar, M. (2015). Remediation of Dyes from Aquatic Ecosystems by Biosorption Method Using Algae. In Algae and Environmental Sustainability (pp. 97–106). Springer India. https://doi.org/10.1007/978-81-322-2641-3_8

Chaturvedi, A. D., Pal, D., Penta, S., & Kumar, A. (2015). Ecotoxic heavy metals transformation by bacteria and fungi in aquatic ecosystem. World Journal of Microbiology and Biotechnology, 31(10), 1595–1603. https://doi.org/10.1007/s11274-015-1911-5

Costed, J. W. (1995). Overview of microbial biofilms. In Journal of Industrial Microbiology (Vol. 15). https://academic.oup.com/jimb/article/15/3/137/5988574

Costerton, J. W., & Lappin-Scott, H. (2009). Introduction to Microbial Biofilms. In Microbial Biofilms (pp. 1–12). Cambridge University Press. https://doi.org/10.1017/cbo9780511525353.002

Cotner, J. B., & Biddanda, B. A. (2002). Small players, large role: Microbial influence on biogeochemical processes in pelagic aquatic ecosystems. In Ecosystems (Vol. 5, Issue 2, pp. 105–121). https://doi.org/10.1007/s10021-001-0059-3

Crab, R., Defoirdt, T., Bossier, P., & Verstraete, W. (2012). Biofloc technology in aquaculture: Beneficial effects and future challenges. In Aquaculture (Vols. 356–357, pp. 351–356). https://doi.org/10.1016/j.aquaculture.2012.04.046
Dakshini, K. (1994). Algal Allelopathy. In THE BOTANICAL REVIEW (Vol. 60, Issue 2).
Damjanovic, K., Blackall, L. L., Webster, N. S., & van Oppen, M. J. H. (2017). The contribution of microbial biotechnology to mitigating coral reef degradation. Microbial Biotechnology, 10(5), 1236–1243. https://doi.org/10.1111/1751-7915.12769
Dang, H., & Lovell, C. R. (2016). Microbial Surface Colonization and Biofilm Development in Marine Environments. Microbiology and Molecular Biology Reviews, 80(1), 91–138. https://doi.org/10.1128/mmbr.00037-15
Davison, J. (2005). Risk mitigation of genetically modified bacteria and plants designed for bioremediation. Journal of Industrial Microbiology and Biotechnology, 32(11–12), 639–650. https://doi.org/10.1007/s10295-005-0242-1
Delacuvellerie, A., Benali, S., Cyriaque, V., Moins, S., Raquez, J. M., Gobert, S., & Wattiez, R. (2021). Microbial biofilm composition and polymer degradation of compostable and non-compostable plastics immersed in the marine environment. Journal of Hazardous Materials, 419. https://doi.org/10.1016/j.jhazmat.2021.126526
Deswati, D., Safni, S., Khairiyah, K., Yani, E., Yusuf, Y., & Pardi, H. (2022). Biofloc technology: water quality (pH, temperature, DO, COD, BOD) in a flood & drain aquaponic system. International Journal of Environmental Analytical Chemistry, 102(18), 6835–6844. https://doi.org/10.1080/03067319.2020.1817428
Donald E., C., Erik, K., & Bo, T. (2005). Aquatic Geomicrobiology. https://doi.org/10.1016/S0065-2881(05)48017-7
Edwards, S. J., & Kjellerup, B. V. (2013). Applications of biofilms in bioremediation and biotransformation of persistent organic pollutants, pharmaceuticals/personal care products, and heavy metals. In Applied Microbiology and Biotechnology (Vol. 97, Issue 23, pp. 9909–9921). https://doi.org/10.1007/s00253-013-5216-z
Emerenciano, M. G. C., Martínez-Córdova, L. R., Martínez-Porchas, M., & Miranda-Baeza, A. (2017). Biofloc Technology (BFT): A Tool for Water Quality Management in Aquaculture. In Water Quality. InTech. https://doi.org/10.5772/66416
Fenchel, T. M., & Jørgensen, B. B. (1977). Detritus Food Chains of Aquatic Ecosystems: The Role of Bacteria (pp. 1–58). https://doi.org/10.1007/978-1-4615-8219-9_1
Flemming, H.-C. (1993). Biofilms and Environmental Protection. In Wat Sci. Tech (Vol. 27, Issue 8). https://iwaponline.com/wst/article-pdf/27/7-8/1/103171/1.pdf
Garbisu, C., & Alkorta, I. (1999). Review Utilization of genetically engineered microorganisms (GEMs) for bioremediation.
Ghorab, M. A. (2018). Environmental Pollution by Heavy Metals in the Aquatic Ecosystems of Egypt. Open Access Journal of Toxicology, 3(1). https://doi.org/10.19080/oajt.2018.03.555603
Gomes, M. P., Souza Garcia, Q., Carvalho Barreto, L., Pinheiro, L., Pimenta, S., Matheus, M. T., Figueredo, C. C. (2017). Allelopathy: An overview from micro-to macroscopic organisms, from cells to environments, and the perspectives in a climate-changing world. https://doi.org/10.1515/biolog-2017-00019
Gupta, J., Rathour, R., Dupont, C.L., Kaul, D. and Thakur, I.S., 2021a. Genomic insights into waste valorized extracellular polymeric substances (EPS) produced by Bacillus sp. ISTL8. Environmental Research, 192, p. 110277.
Gupta, J., Rathour, R., Maheshwari, N. and Thakur, I.S., 2021b. Integrated analysis of Whole genome sequencing and life cycle assessment for polyhydroxyalkanoates production by Cupriavidus sp. ISTL7. Bioresource Technology, 337, p. 125418.
Gupta, J., Rathour, R., Dupont, C., Mishra, A. and Thakur, I.S., 2022. Biogeochemical profiling and taxonomic characterization of municipal landfill site by metagenomic sequencing. Bioresource Technology, 351, p. 126936.

Gupta, J., Rathour, R., Singh, R. and Thakur, I.S., 2019a. Production and characterization of extracellular polymeric substances (EPS) generated by a carbofuran degrading strain Cupriavidus sp. ISTL7. Bioresource Technology, 282, pp. 417–424.

Gupta, J., Tyagi, B., Rathour, R. and Thakur, I.S., 2019b. Microbial treatment of waste by culture-dependent and culture-independent approaches: Opportunities and challenges. Microbial Diversity in Ecosystem Sustainability and Biotechnological Applications: Volume 1. Microbial Diversity in Normal & Extreme Environments, pp. 415–446.

Häder, D. P., Banaszak, A. T., Villafañe, V. E., Narvarte, M. A., González, R. A., & Helbling, E. W. (2020). Anthropogenic pollution of aquatic ecosystems: Emerging problems with global implications. In Science of the Total Environment (Vol. 713). Elsevier B.V. https://doi.org/10.1016/j.scitotenv.2020.136586

Hakeem, K. R., Bhat, R. A., & Qadri, H. (2020). Bioremediation and biotechnology: Sustainable approaches to pollution degradation. In Bioremediation and Biotechnology: Sustainable Approaches to Pollution Degradation. Springer International Publishing. https://doi.org/10.1007/978-3-030-35691-0

Haldar, S., & Ghosh, A. (2020). Microbial and plant-assisted heavy metal remediation in aquatic ecosystems: a comprehensive review. In 3 Biotech (Vol. 10, Issue 5). Springer. https://doi.org/10.1007/s13205-020-02195-4

Harmon, S. M. (2015). The Toxicity of Persistent Organic Pollutants to Aquatic Organisms. In Comprehensive Analytical Chemistry (Vol. 67, pp. 587–613). Elsevier B.V. https://doi.org/10.1016/B978-0-444-63299-9.00018-1

Hitchcock, J. N. (2020). Storm events as key moments of microplastic contamination in aquatic ecosystems. Science of the Total Environment, 734. https://doi.org/10.1016/j.scitotenv.2020.139436

Hulot, F. D., & Huisman, J. (2004). Hulot, Florence D., and Jef Huisman. Allelopathic interactions between phytoplankton species: The roles of heterotrophic bacteria and mixing intensity. Limnol. Oceanogr., 49(4, part 2), 2004, 1424–1434. In Limnol. Oceanogr (Vol. 49, Issue 4).

Influences of New Azo Dyes to the Aquatic Ecosystem. (n.d.).

Issue, E. (2004). Impacts of Atmospheric Pollution on Aquatic Ecosystems Issues in Ecology.

Jain, N., Sharma, P., Verma, A. and Gupta, J., 2023. Enhancement of thermo-mechanical, creep-recovery, and anti-microbial properties in PVA-based biodegradable films through cross-linking with oxalic acid: implications for packaging application. Biomass Conversion and Biorefinery, pp. 1–10. DOI: https://doi.org/10.1007/s13399-023-04748-y

Jan, A. T., Azam, M., Ali, A., & Haq, Q. M. R. (2014). Prospects for exploiting bacteria for bioremediation of metal pollution. Critical Reviews in Environmental Science and Technology, 44(5), 519–560. https://doi.org/10.1080/10643389.2012.728811

Joshi, P. (2016). Microbial Diversity of Aquatic Ecosystem and its Industrial Potential. Journal of Bacteriology & Mycology: Open Access, 3(1). https://doi.org/10.15406/jbmoa.2016.03.00048

Khanjani, M. H., & Sharifinia, M. (2020). Biofloc technology as a promising tool to improve aquaculture production. In Reviews in Aquaculture (Vol. 12, Issue 3, pp. 1836–1850). Wiley-Blackwell. https://doi.org/10.1111/raq.12412

Kokare, C. R., Chakraborty, S., Khopade, A. N., & Mahadik, K. R. (2009). Biofilm: Importance and applications. In Indian Journal of Biotechnology (Vol. 8).

Kuiper, I., Lagendijk, E. L., Bloemberg, G. V, & Lugtenberg, B. J. J. (2004). Rhizoremediation: A Beneficial Plant-Microbe Interaction. In Molecular Plant-Microbe Interactions MPMI (Vol. 17, Issue 1).

Kumar, A., Kumar, A., Singh, R., Singh, R., Pandey, S., Rai, A., Singh, V. K., & Rahul, B. (2019a). Genetically engineered bacteria for the degradation of dye and other organic compounds. In Abatement of Environmental Pollutants: Trends and Strategies (pp. 331–350). Elsevier. https://doi.org/10.1016/B978-0-12-818095-2.00016-3

Kumar, M., Muthusamy, A., Kumar, V., & Bhalla-Sarin, N. (2019b). In vitro plant breeding towards novel agronomic traits: Biotic and abiotic stress tolerance. In In vitro Plant Breeding towards

Novel Agronomic Traits: Biotic and Abiotic Stress Tolerance. Springer Singapore. https://doi.org/10.1007/978-981-32-9824-8

Leão, P. N., Vasconcelos, M. T. S. D., & Vasconcelos, V. M. (2009). Allelopathy in freshwater cyanobacteria Allelopathy in freshwater cyanobacteria Pedro N. Leo et al. In Critical Reviews in Microbiology (Vol. 35, Issue 4, pp. 271–282). https://doi.org/10.3109/10408410902823705

Legrand, C., Rengefors, K., Fistarol, G. O., Graneli, E., Legrand, C., Rengefors, K., Fistarol, G., & Graneli, E. (2003). Allelopathy in phytoplankton-biochemical, ecological and evolutionary aspects. In Phycologia (Vol. 42, Issue 4).

Liu, L., Bilal, M., Duan, X., & Iqbal, H. M. N. (2019). Mitigation of environmental pollution by genetically engineered bacteria — Current challenges and future perspectives. In Science of the Total Environment (Vol. 667, pp. 444–454). Elsevier B.V. https://doi.org/10.1016/j.scitotenv.2019.02.390

Liu, S. Y., Leung, M. M. L., Fang, J. K. H., & Chua, S. L. (2021). Engineering a microbial 'trap and release' mechanism for microplastics removal. Chemical Engineering Journal, 404. https://doi.org/10.1016/j.cej.2020.127079

Macías, F. A., Galindo, J. L. G., García-Díaz, M. D., & Galindo, J. C. G. (2008). Allelopathic agents from aquatic ecosystems: Potential biopesticides models. In Phytochemistry Reviews (Vol. 7, Issue 1, pp. 155–178). https://doi.org/10.1007/s11101-007-9065-1

Mallin, M. A., & Cahoon, L. B. (2003). Industrialized Animal Production-A Major Source of Nutrient and Microbial Pollution to Aquatic Ecosystems.

Meyers, S. P. (2000). The early years Developments in aquatic microbiology.

Mohapatra, R. K., Behera, S. S., Patra, J. K., Thatoi, H., & Parhi, P. K. (2019). Potential application of bacterial biofilm for bioremediation of toxic heavy metals and dye-contaminated environments. In New and Future Developments in Microbial Biotechnology and Bioengineering: Microbial Biofilms Current Research and Future Trends in Microbial Biofilms (pp. 267–281). Elsevier. https://doi.org/10.1016/B978-0-444-64279-0.00017-7

Ojuederie, O. B., & Babalola, O. O. (2017). Microbial and plant-assisted bioremediation of heavy metal polluted environments: A review. In International Journal of Environmental Research and Public Health (Vol. 14, Issue 12). MDPI. https://doi.org/10.3390/ijerph14121504

Oliveira, V., Gomes, N. C. M., Almeida, A., Silva, A. M. S., Silva, H., & Cunha, Â. (2015). Microbe-Assisted Phytoremediation of Hydrocarbons in Estuarine Environments. In Microbial Ecology (Vol. 69, Issue 1). Springer Science and Business Media, LLC. https://doi.org/10.1007/s00248-014-0455-9

Padeniya, U., Davis, D. A., Wells, D. E., & Bruce, T. J. (2022). Microbial Interactions, Growth, and Health of Aquatic Species in Biofloc Systems. In Water (Switzerland) (Vol. 14, Issue 24). MDPI. https://doi.org/10.3390/w14244019

Paerl, H. W., & Pinckney, J. L. (1996). A Mini-Review of Microbial Consortia: Their Roles in Aquatic Production and Biogeochemical Cycling. In JVHORORIAI ECOLOGY (Vol. 31, Issue 3). Springer-Verlag New York Inc.

Paerl, H. W., Dyble, J., Moisander, P. H., Noble, R. T., Piehler, M. F., Pinckney, J. L., Steppe, T. F., Twomey, L., & Valdes, L. M. (2003). Microbial indicators of aquatic ecosystem change: Current applications to eutrophication studies. FEMS Microbiology Ecology, 46(3), 233–246. https://doi.org/10.1016/S0168-6496(03)00200-9

Paniagua-Michel, J. (2017). Wastewater treatment using phototrophic-heterotrophic biofilms and microbial mats. In Prospects and Challenges in Algal Biotechnology (pp. 257–275). Springer Singapore. https://doi.org/10.1007/978-981-10-1950-0_9

Pilon-Smits, E. (2005). Phytoremediation. In Annual Review of Plant Biology (Vol. 56, pp. 15–39). https://doi.org/10.1146/annurev.arplant.56.032604.144214

Porter, K. G., Paerl, H., Hodson, R., Pace, M., Priscll, J., Riemann, B., Scavia, D., & Stockner, J. (1988). Microbial Interactions in Lake Food Webs.

Praveen, A., Pandey, V. C., Marwa, N., & Singh, D. P. (2019). Rhizoremediation of Polluted Sites. In Phytomanagement of Polluted Sites: Market Opportunities in Sustainable Phytoremediation (pp. 389–407). Elsevier. https://doi.org/10.1016/B978-0-12-813912-7.00015-6

Rochelle-Newall, E., Nguyen, T. M. H., Le, T. P. Q., Sengtaheuanghoung, O., & Ribolzi, O. (2015). A short review of fecal indicator bacteria in tropical aquatic ecosystems: Knowledge gaps and future directions. Frontiers in Microbiology, 6(MAR). https://doi.org/10.3389/fmicb.2015.00308

Roeselers, G., Loosdrecht, M. C. M. V., & Muyzer, G. (2008). Phototrophic biofilms and their potential applications. In Journal of Applied Phycology (Vol. 20, Issue 3, pp. 227–235). https://doi.org/10.1007/s10811-007-9223-2

Sharma, A., Jamali, H., Vaishnav, A., Giri, B. S., & Srivastava, A. K. (2019). Microbial biofilm: An advanced eco-friendly approach for bioremediation. In New and Future Developments in Microbial Biotechnology and Bioengineering: Microbial Biofilms Current Research and Future Trends in Microbial Biofilms (pp. 205–219). Elsevier. https://doi.org/10.1016/B978-0-444-64279-0.00015-3

Sharma, J., Sharma, S., & Soni, V. (2021a). Classification and impact of synthetic textile dyes on Aquatic Flora: A review. In Regional Studies in Marine Science (Vol. 45). Elsevier B.V. https://doi.org/10.1016/j.rsma.2021.101802

Sharma, P., Sirohi, R., Tong, Y. W., Kim, S. H., & Pandey, A. (2021b). Metal and metal(loids) removal efficiency using genetically engineered microbes: Applications and challenges. Journal of Hazardous Materials, 416. https://doi.org/10.1016/j.jhazmat.2021.125855

Srivastava, J. K., Chandra, H., Kalra, S. J. S., Mishra, P., Khan, H., & Yadav, P. (2017). Plant–microbe interaction in aquatic system and their role in the management of water quality: a review. In Applied Water Science (Vol. 7, Issue 3, pp. 1079–1090). Springer Verlag. https://doi.org/10.1007/s13201-016-0415-2

Syakti, A. D., Arofah, N., Purnomowati, R., & Salim, A. (2017). Sludge Valorization from Biofloc-Based Aquaculture Systems for Bioremediation of Crude Oil-Contaminated Sediment. Waste and Biomass Valorization, 8(3), 561–572. https://doi.org/10.1007/s12649-016-9649-3

Tahri, N., Bahafid, W., Sayel, H., & El Ghachtouli, N. (2013). Biodegradation: Involved Microorganisms and Genetically Engineered Microorganisms. In Biodegradation - Life of Science. InTech. https://doi.org/10.5772/56194

Valls, M., & Victor De Lorenzo, V. (2002). Exploiting the genetic and biochemical capacities of bacteria for the remediation of heavy metal pollution. www.fems-microbiology.org

Verma, S., & Kuila, A. (2019). Bioremediation of heavy metals by microbial process. In Environmental Technology and Innovation (Vol. 14). Elsevier B.V. https://doi.org/10.1016/j.eti.2019.100369

Wenzel, W. W., & Adriano, D. C. (1999). A Plant-Microbe-Based Remediation System.

William Costerton, J., Lewandowski, Z., Caldwell, D. E., Korber, D. R., & Lappin-Scott, H. M. (1995). Microbial Biofilms. www.annualreviews.org

Yadav, S., & Chandra, R. (2019). Biofilm-mediated bioremediation of pollutants from the environment for sustainable development. In New and Future Developments in Microbial Biotechnology and Bioengineering: Microbial Biofilms Current Research and Future Trends in Microbial Biofilms (pp. 177–203). Elsevier. https://doi.org/10.1016/B978-0-444-64279-0.00014-1

Yang, Y., Liu, W., Zhang, Z., Grossart, H. P., & Gadd, G. M. (2020). Microplastics provide new microbial niches in aquatic environments. In Applied Microbiology and Biotechnology (Vol. 104, Issue 15, pp. 6501–6511). Springer. https://doi.org/10.1007/s00253-020-10704-x

Yavari, S., Malakahmad, A., & Sapari, N. B. (2015). A Review on Phytoremediation of Crude Oil Spills. In Water, Air, and Soil Pollution (Vol. 226, Issue 8). Kluwer Academic Publishers. https://doi.org/10.1007/s11270-015-2550-z

Zaghloul, A., Saber, M., Gadow, S., & Awad, F. (2020). Biological indicators for pollution detection in terrestrial and aquatic ecosystems. Bulletin of the National Research Centre, 44(1). https://doi.org/10.1186/s42269-020-00385-x

Zhang, W., Zhou, P., Pan, S., Li, Y., Lin, L., Niu, L., Wang, L., & Zhang, H. (2022). The role of microbial communities on primary producers in aquatic ecosystems: Implications in turbidity stress resistance. Environmental Research, 215, 114353. https://doi.org/10.1016/j.envres.2022.114353

Microbial Communities in Glacial Environments: Key Players in Cryosphere Carbon Cycling and the Emergence of CECs

Aanchal Eusebius, Claudia Jacob, Anshdha Nandra, and Juhi Gupta

1 Introduction

Glaciers are known for their ability to delay discharge by directly blocking the flow of water from precipitation (Jansson et al. 2003). The melting of glacial snow and ice makes a significant contribution to the freshwaters of surface waters around Antarctica (Dierssen et al. 2022). The supraglacial setting concerns the glacier's surface, incorporating ice, snow, and debris. It plays a pivotal role in influencing rates of melting, albedo effects, and overall glacier dynamics. This region holds significance for the examination of interactions between ice and climate (Ben, Evans 2010). Subglacial environments denote the area beneath glaciers where meltwater engages with the bedrock beneath, impacting the movement of ice and shaping the terrain through mechanisms such as erosion and the transportation of sediment (Bennett, Glassier 2011). The englacial domain encompasses the internal ice structure within a glacier, featuring complex networks of crevices and passages that facilitate the flow of meltwater. Research into this realm yields an understanding of ice deformation and the pathways taken by meltwater (Paterson 1994). The process of subglacial hydrological access involves the accumulation of meltwater formed on the glacial surface in supraglacial streams and lakes and eventually flowing into crevices and moulins (Kohler et al. 2020). Melting glaciers reveal increased regions of unconsolidated and unstable sediments in the proglacial area (i.e., formerly sections of ice-covered glacier foreland) and expose growing debris of supraglacial on their surface (Hohensinner et al. 2021). As meltwater accumulates inside bedrock basins, sustained glacier recession is causing the formation of proglacial lakes at glacier termini (Sutherland et al. 2020). Microbial abundance and movement were observed in spiral holes in the

A.E., C.J. & A.N. authors have equally contributed to the current work.

A. Eusebius · C. Jacob · A. Nandra · J. Gupta (✉)
Amity Institute of Environmental Sciences, Amity University, Noida 201303, India
e-mail: jgupta2@amity.edu

J. Gupta and A. Verma (eds.), *Microbiology-2.0 Update for a Sustainable Future*,
https://doi.org/10.1007/978-981-99-9617-9_3

ice and a total of 73 water samples were collected (Edwards 2021). Environmental disturbances in Cryospheric biomes affect the nutrient supply of bacteria, viruses, protozoa, and algae, all of which have different biogeographical distributions (Anesio and Laybourn-Parry 2012). Glaciers and ice sheets have been identified as unique ecosystems with potentially vital global carbon cycle consequences (Anesio and Laybourn-Parry 2012). This can determine atmospheric amounts of carbon dioxide or methane. The microbial element of the Earth's system must either explicitly sequester and/or emit carbon, or indirectly determine carbon uptake or release in different areas of the Earth's systems (Wadham et al. 2019). Climate change is no longer exempted to any sphere and can affect physical, chemical, and specific ecological processes both directly and indirectly in the environment. Global Climate change even has the potential to alter the uptake and subsequent fate of Persistent organic pollutants (POPs) in our biota through multifarious mechanisms. Increasing temperatures, the retreat of sea ice, slumping of permafrost, glacial loss and altering mass balance, alterations in sea ice regimes, and changing precipitation patterns affect the distribution of CECs to determine their fate in the glacial environments, especially the Arctic ecosystems (Davies 2022; Hung et al. 2022). The Arctic cryosphere both marine and terrestrial is rapidly declining and the most recent assessment by IPCC suggests that warming occurs faster in the Arctic region as compared to the other parts of the world (Portner et al. 2019).

2 Microbial Patterns in Arctic Glacial Meltwater

Interactions between the glacial cryosphere and downstream ecosystems occur via water flow. Glacial environments release melt and matter from subglacial and supraglacial regions. In this way, microbial transport can be an indicator of ecological and hydrological conditions in glacial ecosystems (Kohler et al. 2020). Conducted a comparative analysis of microbial communities from 24 glacial meltwater beds in six (sub) Arctic regions, including Disko Island in western Greenland, Svalbard in Iceland, western Norway, the western Greenland Ice Sheet, and eastern Alaska and elsewhere (as shown in Table 1). Glaciers vary in size, lithology, and weather characteristics. The combination of the meltwater mixture gives an idea of the area and the environment. The phyla such as *Proteobacteria, Bacteroidetes, Actinomycetes,* and *Verrucobacteria* dominated 16S rRNA gene sequences in all samples, and aerobic and anaerobic oxygen heterotrophs with important genera such as *Polaromonas, Methylophilus,* and *Nitrotoga* (Anesio and Laybourn-Parry 2012).

Unique sulfur oxidizers like *Sulfuricurvum* sp. and *Thiobacillus* sp. The sites were very abundant in Svalbard, accounting for 16–32% of site-specific OTUs, indicating a unique metabolic potential. The difference between regions was evident in their rare taxa. The alpha variation between meltwater decreased with latitude, and many analyses showed significant regional clusters (Pattyn 2010). Distance-based replication and correlation analysis further revealed the relationship between the entire assembly and specific OTUs whose changes were associated with the sample. However, some

Table 1 Attributes of meltwater streams in terms of their physical traits

Glacier	Region	Glacier area (km sq.)	Temperature (°C)	OTUs	References
Eagle	Alaska	53a	3.7	3,239	Hood and Berner (2009)
Herbert	Alaska	60 a	0.3	3,239	Hood and Berner (2009)
Lemon	Alaska	15a	5.3	3,239	Hood and Berner (2009)
Mendenhall	Alaska	127a	0.2	3,239	Hood and Berner (2009)
Qinnguata Kuussua	GrIS	1800b	0	697	Cameron et al. (2017),
Russell	GrIS	120c	0	697	Kohler et al. (2017)
Leverett	GrIS	1200b	0	697	Bjornsson et al. (2000), Tweed et al. (2005)
Eyjabakkajokull	Iceland	110 g	NA	1,076	Bjornsson et al. (2000), Tweed et al. (2005)
Kaldalonsjokull	Iceland	37 g	NA	1,076	Bjornsson et al. (2000), Tweed et al. (2005)
Skaftafellsjokull	Iceland	100f	NA	1,076	Bjornsson et al. (2000), Tweed et al. (2005)
Solheimasjokull	Iceland	47e	NA	1,076	Bjornsson et al. (2000), Tweed et al. (2005)
Austerdalsbreen	Norway	19.85 h	0.3	1,116	Mateos-Rivera et al. (2016), Andreassen et al. (2012)
Boverbreen	Norway	6.75 h	0.7	1,116	Mateos-Rivera et al. (2016), Andreassen et al. (2012)
Styggedalsbreen	Norway	2.02 h	0.7	1,116	Mateos-Rivera et al. (2016), Andreassen et al. (2012)
Glacier 6	Qeqertarsauq	1.5i	0.1	994	Zarsky et al. (2018)
Glacier 10	Qeqetarsauq	7i	0.1	994	Zarsky et al. (2018)
Glacier 11	Qeqetarsauq	9.7i	2.2	994	Zarsky et al. (2018)
Glacier 13	Qeqetarsauq	18i	0.1	994	Zarsky et al. (2018)
Ebbabreen	Svalbard	25i	0.6	934	Hagen et al. (1993)
Foxfonna	Svalbard	3.95i	0.1	934	Hagen et al. (1993)
Midtre Loverbreen	Svalbard	6i	0.9	934	Hagen et al. (1993)
Nasenbreen	Svalbard	45.1i	0.9	934	Hagen et al. (1993),
Sefstrombreen	Svalbard	155i	0.8	934	Hagen et al. (1993)
Longyearbreen	Svalbard	4i	NA	934	Mateos-Rivera et al. (2016)

OTUs showed poor metabolic processes associated with common features such as nitrification and inorganic nitrogen concentrations. Thus, the microbial production clusters collected changes in large water flow, presenting a specific area. In contrast, OTUs can display local microbial-mediated processes that were not entirely via the host process. These collective findings lead to a better understanding of the interconnection between the earth's cryosphere and underlying ecosystems (Kohler et al. 2020).

2.1 *Microbial Biomass Export from Western Greenland Ice Sheet*

A significant reservoir of microbial life was found within the Greenland Ice Sheet, yet the precise amount of biomass transported from its surface to downstream environments remained uncertain. Researchers quantified the microbial abundance and cellular biomass flow across the sunlit zone of the near-surface weathering crust in the western section of the ice sheet (Chandler et al. 2013). The study demonstrated that the movement of interstitial water was gradual (around 10^{-2} m/day), and flow cytometry analysis revealed that this process carried approximately 500 million cells per square meter per day into supraglacial streams. This translates to a carbon flux of up to 250 g per square kilometer per day using groundwater estimation methods. The accumulation of cellular carbon within the weathering crust surpassed the export through river systems, promoting the storage of biomass, improved carbon cycling, and a reduction in biological albedo. The western weathering crust environment flushes out as much as 37 kg of cellular carbon per square kilometer. Throughout each summer, the Greenland Ice Sheet contributed a substantial flow that supported heterotrophic organisms, and even methanogenesis is observed at its base (Edwards 2021).

3 Microbial Mobility and Abundance

The examination of 73 water samples obtained from auger holes in exposed ice, which were either fully or partially refilled, indicated an average cell count of 2.28×10^4 cells per milliliter. Despite no noteworthy association with the contemporaneous melt rate, a minor negative connection with the time elapsed since the daily peak melt was evident at a 95% confidence level. These findings provide indicators of daily energy intake and the formation of weathering crust through both melt-related factors. The predominant cell size identified by SYBR Gold staining was approximately 1–2 μm, constituting around 50% of the suspended microbial abundance (Edwards 2021). The category of cells smaller than 1 μm represented an average of 19% of the total microbial abundance, potentially overestimating the true cell count but potentially including large viruses. Notably, moderate to high correlations in microbial abundance were observed across all six size classes. The study delved into 73 water samples taken from auger holes in exposed ice, which were either fully or partially refilled. The results revealed an average microbial abundance of 2.28×10^4 cells mL^{-1}, with a standard deviation of $\pm 1.91 \times 10^4$ cells mL^{-1}. Furthermore, the analysis established a noteworthy relationship between microbial abundance and the duration since the daily peak melt ($\rho = -0.26$, $p = 0.03$), suggesting potential impacts of daily energy input and weathering crust development on microbial populations within the samples (as shown in Table 2).

Conducted experiments involving 26 completed recharge trials were combined with concurrent microbial enumerations, unveiling a reciprocal link between hydraulic conductivity and microbial abundance. This association was characterized by an exponential decay function, aimed at investigating the connection between microbial abundance and ice surface hydrology (Chandler et al. 2013).

A defined threshold of 10 μm was utilized to distinguish between bacteria and larger algae. A similar nonlinear, inverse correlation was observed when analyzing the connection between hydraulic conductivity (K) and abundance separately for these designated bacteria and algae categories (as shown in Fig. 1). However, it's noteworthy that while the hydraulic conductivity increased, the reduction in algal abundance occurred at a slower pace (Edwards 2021).

Table 2 Microbial quantities within weathering crust water samples categorized by cell size

Parameter	Value
Water sample source	Fully and partially recharged auger holes in bare-ice
Number of samples	73
Mean microbial abundance	2.28×10^4 cells mL^{-1}
Standard deviation of abundance	$\pm 1.91 \times 10^4$ cells mL^{-1}
Correlation with melt rate (ρ)	0.07
p-value for correlation with melt rate (p)	0.53
Correlation with time since daily peak melt (p)	−0.26
p-value for correlation with time since daily peak melt (p)	0.03
Dominant size of the stained cells	1–2 μm
Dominant size abundance (±standard deviation)	50% (±7.0%)
<1um size abundance (mean ± standard deviation)	19% (±5.5%)
Correlation between microbial size classes (ρ)	$0.28 < \rho < 0.93$ (moderate to highly correlated)
p-Values for correlation between microbial size classes (p)	$p < 0.05$
Excluded samples with abundance $>10^5$ cells mL^{-1}	10 samples

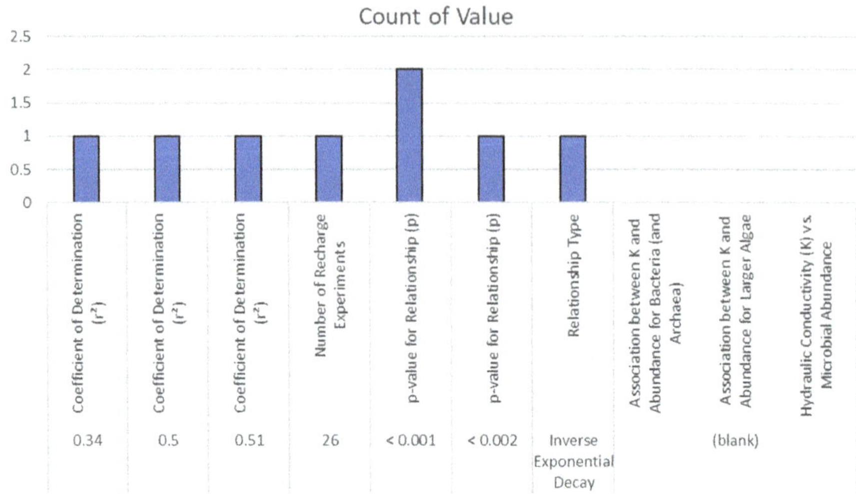

Fig. 1 Correlations between microbial levels and hydraulic conductivity (K)

3.1 *The Weathering Crust's Microbial Accumulation*

Multiple factors such as sunlight exposure, nutrient availability, and water supply impacted the growth of microbes on top of glaciers. Assumed consistently favorable conditions during the melting season and the absence of a clearly defined exponential growth pattern for bacterial and algal communities, a careful estimation based on proportionality suggested that cellular carbon could accumulate at a rate of 0.07 kg per square kilometer per day, with a doubling time of 4 days. This value, as anticipated, is lower but can still be compared to estimates of microbial productivity for algal communities on the ice surface and mineral-rich cryoconite, which typically covers around 8% of the ice surface. These rates of accumulation don't account for previous concentrations, long exposure times, or new populations emerging over time (Chandler et al. 2013).

The research indicated the potential for the buildup of cellular biomass in the uppermost weathering crust. Consequently, predictions for the outflow of biomass, following the proportionality model, indicated a carbon accumulation rate of only 0.01 kg per square kilometer per day. By combining the weathering crust data with previously published information about algal and cryoconite productivity, this hypothesis gained further support, especially when considering the three main supraglacial interfluve environments. In the study area, it was assumed that about 35% of the ice surface without cover was characterized by active supraglacial streams. The rest of the space between these streams was occupied by the weathering crust and algae, with cryoconite covering about 8%.

The projected maximum carbon release of 0.25 kg per square kilometer per day was far surpassed at the site, leading to near-surface biomass accumulation estimated to range from 0.59 to 0.82 kg per square kilometer per day. Consequently, the weathering crust was expected to experience a significant increase in cellular carbon over the course of the melting season (Edwards 2021).

4 Carbon Dynamics of Glaciers

The organic matter that is associated with the vegetation, marine, and lake sediments, and pre-existing vegetation are merged into sediments beneath the formed ice sheets. The organic matter originally came from terrestrial and anthropogenic sources, and it may operate as a sink or be fixed by autotrophic biological processes in the ice sheets. Few ice sheets contain carbon reservoirs regarded as limitless (Arrigo et al. 2017).

For instance, the particulate organic carbon detained in the sediments beneath Antarctica's ice sheet [60 × 108–21 × 109 Gg C] is up to an order of magnitude larger than the organic carbon linked with the permafrost of the northern hemisphere (Wadham et al. 2019).

This promotes reflection due to the occurrence of thick and widespread sedimentary basins in the subglacial region of Antarctica. This ought to consist of fossil organic matter originating from primeval marine sediments. Tiny carbon reserves are assumed to be found beneath the ice sheet of Greenland where the thinner sediments contain the remnants of ancient sequestered paleosols. It is embedded during the ice sheet expansion. Estimates of carbon storage have higher error bounds due to the uncertainty of sediment thickness and organic carbon content detained in the englacial region. This results in dissolved organic carbon values below 10 Pg C for current ice sheets (Hood et al. 2015). The presence of sedimentary organic matter beneath ice sheets can be confirmed through ancient paleosols in basal sediments of Greenland and bulk meltwaters, as well as the presence of marine sediment and their geochemical influences in West Antarctica and McMurdo Dry Valleys (Kohler et al. 2017). Yearly mass turnover and Yearly mass loss are two components of glacier discharge. Englacial OC, surface deposition and generation, and subglacial carbon all contribute to the OC in glacier discharge. The quantity of water discharged by a glacier that is less than or equal to net yearly precipitation is referred to as annual mass turnover, whereas annual mass loss drains the global store of glacial ice and organic carbon, leading to sea level rise (Hood et al. 2015).

The organic matter existing beneath the ice sheets is hardly known due to the unreachability of the subglacial environment. It mostly included the export and erosion of organic matter of sedimentary rocks for instance marine, lake, and soil sediments from the ice sheets, active respiration release of carbon dioxide or methane through microbial activity, storage in subglacial reservoirs and frozen bed regions (Wadham et al. 2019). The Greenhouse gases (GHG) may probably be taken up by the additional microbial activity i.e., transported in the gas/dissolved phases amidst the hydrological subglacial system or the gas hydrate in sediments stored (CH_4). Therefore, processes of carbon cycling on Earth can create a direct association between the carbon cycle of other atmospheres, oceans, and ice sheets (Wadham et al. 2019).

4.1 *Mountain Glaciers: Hotspots of Organic Carbon and Exported Nutrients*

It has been determined that glaciers and ice sheets are distinct ecosystems with the potential to have significant effects on the global carbon cycle (Anesio and Laybourn-Parry 2012). These ecosystems typically pile up organic carbon from the in situ primary production regions—supraglacial, subglacial, and proglacial (accumulation area of englacial deposits) environment and the exterior terrestrial anthropogenic causes. The carbon dioxide is generated and re-enters the atmosphere by the process of microbes breaking down organic carbon (Stubbins et al. 2012). The balance of organic carbon export from glaciers and ice sheets is directed through meltwater flushing, with unconsumed carbon from microbes being directed into aquatic ecosystems downstream.

Additionally, the high bioavailability of glacial DOC may stimulate the metabolism of DOC that seems to be stable in terrestrial ecosystems downstream. It is likely to provoke the heterotrophic metabolism in aquatic ecosystems downstream and eventually lead to the contribution of the outgassing of carbon dioxide to the atmosphere (Hood et al. 2015) (Fig. 2).

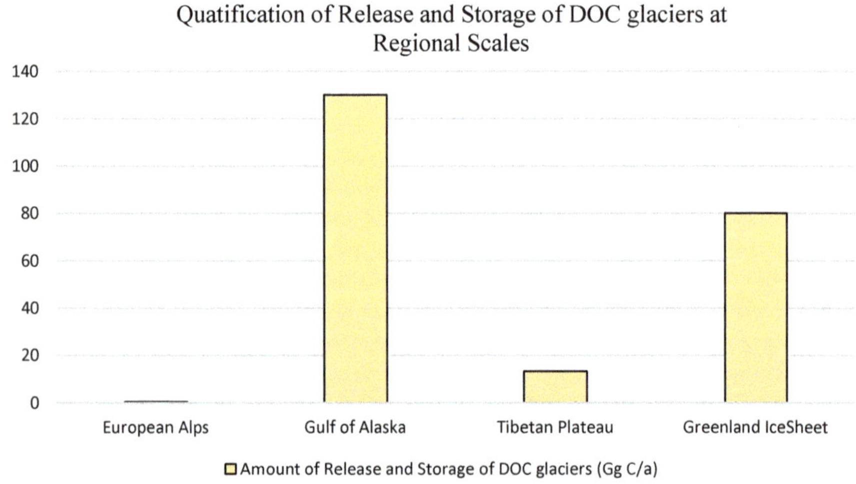

Fig. 2 Graph illustrating the quantification of release and storage of Dissolved Organic Carbon (DOC) at Regional Scales

At the other global scales, the largest supply from the mountain glaciers is from the ice sheets of Greenland and Antarctica through the runoff which liberates around 1,040 Gg C/a of Dissolved Organic Carbon to the region of proglacial ecosystems. The estimation of DOC stored in Chinese glaciers is 3,960 Gg, 70,000 Gg C in the mountain glaciers, and 441×10^4 Gg C in ice sheets, etc. (Li et al. 2018).

Furthermore, recent research reflects that the efficiency of the release and deposition of DOC from the mountainous glaciers in Asia is higher. This is due to the rising glacier melt, precipitation, and stream runoff in Asia considering that the glacial mountain in Asia (ASG) is the prime governor of global loss of Ice volume. Rapid warming had led to the loss of volumes of glaciers, the current day of glacial rate loss has been estimated to be at 369×10^6 Gg/a to 905×10^6 Gg/a (Li et al. 2018).

The quantity of DOC stored in ASG varies from 8,800 to 13,800 Gg C, accounting for 13–20% of the DOC contained in alpine glaciers. The release of DOC from the Asian glaciers (194 Gg C/a) and other glacial areas [Alaska (ALG), Western Canada (CUG) and United States, Svalbard (SJG) and Jan Maye, Arctic Canada North (ACG), Central Europe (CEG)] is 786 Gg C/a, which can be 1.4 times greater (Li et al. 2018).

5 Paradigm Shift

The discovery of microbe populations under a Swiss Alps valley glacier in 1999 revolutionized our understanding of glaciers and ice sheets as abiotic systems. Since then, microbial communities have been found across various glacial systems, from small liquid ice veins to large subglacial water bodies, such as Subglacial Lake Whillans which is located 800 m underneath the Ice Sheet of the West Antarctic (Christner et al. 2014). The survival of these seemingly disparate groups is due to their adaption to a unique combination of physical and geochemical conditions found nowhere else on Earth.

Ice sheets are abundant in liquid water, created by melting surface ice, snow, geothermal heating, and pressure melting at the glacier's bottom. Surface melt on the Greenland Ice Sheet feeds drainage systems near the ice sheet bed, which switch from slow-inefficient to quick efficient channels, with residence times ranging from months to one day (Chandler et al. 2013).

Water running through such extended residence time drainage channels here might take years or decades to reach the ice margin. Second, glacial grinding and crushing produce very fine, highly reactive rock flour when glaciers slide over their bedrock (Wadham et al. 2019).

During glacier and ice sheet development, organic matter from lakes, soil, plant, and marine sediments is buried and deposited onto the ice surface through aeolian processes. This heterogeneous material is augmented by organic carbon fixed in situ by autotrophic bacteria in melting regions, boosting ice melting rates through surface darkening. Ice sheets experience various oxidizing and diminishing environments, ranging from rapidly moving hazardous channels to slow drainage through hypoxic or anoxic sediments and subglacial lakes, due to variations in water flow beneath ice sheets (Christner et al. 2014).

These hydrogeochemical conditions provide niches for microbial colonization and allow for the support of a variety of biogeochemical activities along the redox spectrum, such as denitrification, sulfide sulfate and iron reduction, oxidation, and methanogenesis. The dispersal of icebergs by calving, together with strong hydrological networks, creates the possibility for organic matter and nutrient transfer out of the ice sheet and into aquatic ecosystems downstream (e.g., seas, lakes, rivers, oceans, etc. (Wadham et al. 2019).

Rock-sourced redox pairings and nutrients enable chemolithotrophic and chemoorganotrophic bacteria to metabolize and fix autochthonous carbon in Antarctic Subglacial Lake Whillans. Electron donors, micronutrients, and macronutrients are released by chemical breakdown, with most sediment phylotypes associated with chemolithoautotrophic organisms using sulfur, nitrogen, and iron as energy sources (Christner et al. 2014).

6 Greenhouse Gas Emissions in Glacier Environments

Subglacial microbial activity can penetrate sedimentary organic matter beneath ice sheets, which is neither degraded nor exported from the glacier system. Microbial respiration to greenhouse gases, particularly methane, under anaerobic conditions, is a key fate for trapped organic matter. This may be reinforced by methane from bedrock erosion or deep thermogenic sources in geothermically active regions like West Antarctica (Christiansen and Jørgensen 2018). Recently, microbially supersaturated methane concentrations were found in subglacial discharge flowing from a tiny and far greater catchment in Greenlandic, Antarctic Subglacial Lake Whillans. However, there is a great deal of confusion about the volume of methane sinks and how much methane is oxidized to carbon dioxide before being discharged into the subglacial drainage system (Wadham et al. 2019).

This could lower the positive radiative impact related to ice sheet methane emissions by about a factor of 20. The findings from the ice sheet of Greenland Ice Sheet research show that rapid transportation of subglacial meltwaters can subjugate the oxidative sink, leading to high methane fluxes. This suggests that subglacial hydrology significantly influences methane discharge into the atmosphere. Further investigation into methane generation and export from glacier catchments is needed for a better understanding of the process (Pattyn 2010).

Methane levels in marine sediments across the Antarctic Ice Sheet border are high, with cold seeps found off the Antarctic Peninsula and South Georgia. Recent drilling from the Wilkes Land subglacial basin has also revealed methane levels up to 43,000 ppm. (Escutia et al. 2005). Because of the difficulty in obtaining sediments in subglacial sedimentary basins, there remains substantial uncertainty on the amount of current sub-ice sheet CH_4 hydrate reserves. At the current time, global subglacial methane hydrate reserves are anticipated to be centered around those in Antarctic sedimentary basins (estimated at up to 300 Pg C as methane hydrate and free gas) (Wadham et al. 2012).

Because of their expected placement along the continent's perimeter in sedimentary basins where thinning of ice in a warming environment is likely, Antarctic subglacial methane hydrate deposits are very vulnerable to destabilization. The existence of methane hydrate beneath the Antarctic Ice Sheet and its sensitivity to destabilization in a warming environment is a focus for future ice sheet studies (Wadham et al. 2019).

7 Unravelling the Contaminants of Emerging Concern in the Cryosphere

The past decades have witnessed new contaminants of emerging concern in all the environmental matrices such as air, water, and soil while also inflicting adverse impacts on human health and the system (Yadav et al. 2021). Two significant drivers that put immense pressure on water resources are Population Growth and Climate Change, which will continue to do so in the future. Several regions of our planet suffer water shortages as a challenge especially freshwater, which is crucial for agricultural needs, industrial and recreational requirements, personal care, and foremost- for drinking purposes. The availability of techniques like liquid chromatography-mass spectroscopy specifically, and others have enabled the detection of polar compounds including many metabolites and transformation products, that have not been analyzed to great degrees before. Some may even remain under the cover and our environment is contaminated by many such contaminants. These non-regulated pollutants, emerging pollutants, or even CECs can pose a significant health risk (Llamas-Dios et al. 2021). In certain reviews, the author uses the term 'emerging contaminants' to highlight the detection of some novel chemicals in the water matrix (Gupta et al. 2019a; Kim et al. 2018; Kümmerer 2011).

In the polar ice-capped regions, runoff resulting from terrestrial ecosystems includes meltwater and snow derived from glaciers that could cause seasonal fluctuations in drinking water availability and could be contaminated even (Borgå et al. 2022). Drinking water is an issue of precedence, and a priority environmental policy agenda (Valbonesi et al. 2021). Glacial meltwater and the subsequent runoff contribute to downstream water flow and hinder freshwater availability that drives irrigation, hydropower generation, and ecosystems (Davies 2022).

Arctic amplification is a rapidly spreading scenario of receding glaciers in the polar regions and is attributed to climate change. Even bioaccumulation of persistent organic pollutants and contaminants of emerging concern in the arctic areas affect these systems' food webs and ecosystems and is an extrapolation of climate change's grave effects (Borgå et al. 2022). Large-scale climate patterns that determine atmospheric circulation, wind, oceanic currents, and precipitations affect the transport of POPs along with local and regional changes. Metrics based upon the combination of reflecting broad-scale climate variations like climate oscillation indices and variables are used to accomplish finding the relationship between climate and subsequent ecological changes which also include finding the POPs in the biota. Variations in the hydro-biogeochemical cycles, oceanic circulation, and atmospheric alterations due to climate change have been affecting the fate and distribution of CECs within the Arctic environments (Davies 2022; Rauseo et al. n.d.). Arctic Monitoring and Assessment Programme (AMAP) has identified several chemicals that have received attention from scientists due to their potential toxicity-inducing effects on biodiversity. CECs in the Arctic region of Contaminants of emerging Arctic concern (CEAC) are dynamic entities whose impact and level of toxicity need to be known so that their regulations can help to better know the repercussions they could cause on a global scale and other parts of our planet (Gupta et al. 2019b; Rauseo et al. n.d.). Many of the physical, chemical, and biological processes that are responsible for the distribution and subsequent transformation of these POPs can also be influenced by climate change to various extents. Ambient temperature mostly favors the volatilization of chemicals like POPs and also enhances their mobility in the atmosphere resulting in a prolonged time of stay, turning likely towards their long-range atmospheric transport (Hung et al. 2022).

7.1 *Variety and Risks Associated with CECs*

Internal Arctic redistribution and transformation processes in rapidly fluctuating environments are essential in navigating the fate and legacy of these chemicals like POPs through various processes. Contaminants of emerging concern are defined as the chemicals or certain materials that are found in the environment in trace concentrations that cause a potential, perceived, or "real risk" to health trilogy that includes environment, human, and animal health. Microbial transformation will play a key role in the degradation of persistent organic pollutants (Gupta et al. 2021), however, the alteration due to climate change will modify the structure and pattern of distribution of these microbial communities and the rate at which they can degrade these contaminants (Hung et al. 2022).

Advances in the analytical chemistry domain of environmental sciences have helped in identifying the presence of many chemicals in the Arctic's cryosphere however there has been very limited monitoring. The number of shreds of evidence does not confirm the potential risks associated with such CECs and potential biological effects. Persistent Organic Pollutants (POP) are found to be widely present at levels of exposure that put risk on the health of marine mammals (Sonne et al. 2021). Fluorinated aliphatic compounds such as poly- and perfluoroalkyl substances (PFASs) are global environmental contaminants and are found to be ubiquitous in Arctic wildlife. In the case of the Zeppelin Station at Svalbard, the perfluorooctanoic acid (PFOA) which is a prominent PFA was found to have a declining flux (Muir et al. 2019). Other halogenated hydrocarbons such as Hexachlorobutadiene (HCBD) have similar characteristics to the POPs which led to their listing in Stockholm Convention on POP's global regulation in 2015. Their level of toxicity and long-range transport attributes to the alarming potential of their impact on the cryosphere. Arctic air collected from Canada and Finland suggests that their presence is scarce, but the lack of data for abiotic matrices such as soils, glacier ice, freshwater, seawater, and sediments and less monitoring does not evaluate the associated risks, hence warranting additional research (Balmer et al. 2019a).

Contamination caused by the legacy polychlorinated biphenyls (PCB) is well established but unintentionally produced PCBs in the Arctic call out for concern. Snow samples from Lomonosovfonna collected in the years 2010 and 2014 from a glacial site in Svalbard account for the presence of an unusually high amount of PCB52. Pieces of evidence have been found scattered in biotic and abiotic matrices in the Arctic (Bartlett et al. 2019). Chlorinated Paraffin is yet another class of complex mixtures with varying chains that make their chemical analysis challenging and restrict accurate quantification. Small-chain chlorinates paraffin has been well documented in the Arctic air, Antarctic, and in certain remote regions of the Tibetan plateau serving environmental persistence (Gupta et al. 2022; Vorkamp et al. 2019). Halogenated natural products (HNPs) have been found in the Arctic, Subarctic, and Nordic ecosystems, as confirmed by the Arctic Monitoring and Assessment Program (AMAP 2017). The toxic properties of HNPs are enhanced due to the growing risk of climate change and directly affect the production rate of these organic compounds (Bidleman et al. 2019). A large class of hydrophobic and semi-volatile organic contaminants called Polycyclic aromatic hydrocarbons (PAHs) can enter the environment. Their source of entry can be both natural and anthropogenic. Generation of these followed by a long-rate transport mechanism could deposit them over the surface of glaciers, paving the way for them to enter the supraglacial environments and eventually reach the ice cores and interfere with the microbial assemblage or related dynamics in the region. Efforts through modelling have been employed to extrapolate the understanding of how these contaminants affect the biological, physical, and chemical processes occurring cryosphere (Balmer et al. 2019b).

Many pollutants are found in our environment; however, their potential risks and lack of assessment have moved researchers and scientists to devise ways and strategies that can reduce their ability to cause and 'affect' or 'damage'. The approach focuses on the breakdown of hydrocarbons into much less harmful compounds and many factors such as temperature, oxygen conditions, and nutrient media govern the efficiency of the process. Microbes from various varieties can be selected for the required function such as- Achromobacter. Xanthobacter, Pseudomonas, Alcaligens, Arthrobacter, Mycobacterium, Bacillus, Nitrosomonas, Corynebacterium, Flavobacterium, and many more. Bacteria, fungi, and other microorganisms have a wide potential to degrade organic compounds. Their ability is altered depending on the extracellular enzymatic activity of bacteria, the degradation process for fungi, and the chromosomal genes present (Bala et al. 2022). Sustainable mitigation strategies pose a big question in the minds of researchers as using harsh chemicals on these contaminants is not only an expensive affair, but also raises the disposal problem (Bilal and Iqbal 2020).

8 Conclusion

The research emphasizes the significant importance of microbial communities within glacial ecosystems and their complex interplay with the surrounding environment. Variations in microbial patterns and quantities across diverse regions illustrate unique metabolic capacities and microbe-driven processes. The upper glacier weathering crust emerges as a pivotal site for the accumulation of microbes, potentially impacting carbon flow and ecological changes as glaciers melt. Ice caps and glaciers are enormous stores of debris and biological cells that can stimulate microbial life through ice melting. The transfer, and storage of cellular organic carbon, probable subglacial deposition, and related chemicals through inadequate basal drainage networks increase the periods plus ratios of microbe-water–rock interaction. It offers a sustainable and contributing supply of carbon for the nourishment of heterotrophs and methanogenesis on an ice sheet basis. Changes in surface mass balance and glacier dynamics will have profound effects on the release of glacial organic carbon in the next decades. Emerging CECs pose a great threat to our environment and health. Their identification and subsequent elimination are essential before they become toxic. A wide range of these pollutants are already present in the environment around us. Their potential terrestrial risks have curated interest in extrapolating these assessments on every component and sphere. Microorganisms on the other hand provide a promising future in the degradation and control of these CECs. Pollutants (or contaminants) can be degraded, eliminated, immobilized, and detoxified into less toxic forms which serve as the prime principle of bioremediation, both ex-situ and in situ. Efficient and timely mitigation of CECs is of supreme necessity in the modern world.

Conflict of Interest The authors declare no conflict of interest.

References

Anesio, A.M., Laybourn-Parry, J., 2012. Glaciers and ice sheets as a biome. Trends in Ecology & Evolution 27, 219–225. https://doi.org/10.1016/j.tree.2011.09.012

Andreassen, L.M., Winsvold, S.H., Paul, F., Hausberg, J.E., 2012. Inventory of norwegian glaciers. Rapport 38, 1–236

Arrigo, K.R., Van Dijken, G.L., Castelao, R.M., Luo, H., Rennermalm, Å.K., Tedesco, M., Mote, T.L., Oliver, H., Yager, P.L., 2017. Melting glaciers stimulate large summer phytoplankton blooms in southwest Greenland waters. Geophysical Research Letters 44, 6278–6285. https://doi.org/10.1002/2017GL073583

Arctic Monitoring and Assessment Programme AMAP [WWW Document], n.d. https://www.amap.no/ (accessed 1.24.24).

Bala, S., Garg, D., Thirumalesh, B.V., Sharma, M., Sridhar, K., Inbaraj, B.S., Tripathi, M., 2022. Recent strategies for bioremediation of emerging pollutants: a review for a green and sustainable environment. Toxics 10, 484. https://doi.org/10.3390/toxics10080484

Balmer, J.E., Hung, H., Vorkamp, K., Letcher, R.J., Muir, D.C.G., 2019a. Hexachlorobutadiene (HCBD) contamination in the Arctic environment: A review. Emerging Contaminants 5, 116–122. https://doi.org/10.1016/j.emcon.2019.03.002

Balmer, J.E., Hung, H., Yu, Y., Letcher, R.J., Muir, D.C.G., 2019b. Sources and environmental fate of pyrogenic polycyclic aromatic hydrocarbons (PAHs) in the Arctic. Emerging Contaminants 5, 128–142. https://doi.org/10.1016/j.emcon.2019.04.002

Bartlett, P.W., Isaksson, E., Hermanson, M.H., 2019. 'New' unintentionally produced PCBs in the Arctic. Emerging Contaminants 5, 9–14. https://doi.org/10.1016/j.emcon.2018.12.004

Ben, Evans, D.I., D.J.A., 2010. glaciers and glaciation. BOREAS 40.

BENNETT, GLASSER, M.M., NEIL F., 2011. GLACIAL GEOLOGY: ICE SHEETS AND LANDFORMS, 2nd ed.

Bidleman, T.F., Andersson, A., Jantunen, L.M., Kucklick, J.R., Kylin, H., Letcher, R.J., Tysklind, M., Wong, F., 2019. A review of halogenated natural products in Arctic, Subarctic and Nordic ecosystems. Emerging Contaminants 5, 89–115. https://doi.org/10.1016/j.emcon.2019.02.007

Bilal, M., Iqbal, H.M.N., 2020. Microbial bioremediation as a robust process to mitigate pollutants of environmental concern. Case Studies in Chemical and Environmental Engineering 2, 100011. https://doi.org/10.1016/j.cscee.2020.100011

Bjornsson, H., Palsson, F., Gudmundsson, M.T., 2000. Surface and bedrock topography of the Myrdalsjokull ice cap. Jokull 49, 29–46

Borgå, K., McKinney, M.A., Routti, H., Fernie, K.J., Giebichenstein, J., Hallanger, I., Muir, D.C.G., 2022. The influence of global climate change on accumulation and toxicity of persistent organic pollutants and chemicals of emerging concern in Arctic food webs. Environ. Sci.: Processes Impacts 24, 1544–1576. https://doi.org/10.1039/D1EM00469G

Cameron, K.A., Stibal, M., Hawkings, J.R., Mikkelsen, A.B., Telling, J., Kohler, T.J., et al., 2017. Meltwater export of prokaryotic cells from the Greenland ice sheet. Environ Microbiol 19, 524–534. https://doi.org/10.1111/1462-2920.13483

Chandler, D.M., Wadham, J.L., Lis, G.P., Cowton, T., Sole, A., Bartholomew, I., Telling, J., Nienow, P., Bagshaw, E.B., Mair, D., Vinen, S., Hubbard, A., 2013. Evolution of the subglacial drainage system beneath the Greenland Ice Sheet revealed by tracers. Nature Geosci 6, 195–198. https://doi.org/10.1038/ngeo1737

Christiansen, J.R., Jørgensen, C.J., 2018. First observation of direct methane emission to the atmosphere from the subglacial domain of the Greenland Ice Sheet. Sci Rep 8, 16623. https://doi.org/10.1038/s41598-018-35054-7

Christner, B.C., Priscu, J.C., Achberger, et.al ,A.M., Barbante, C., Carter, S.P., Christianson, K., Michaud, A.B., Mikucki, J.A., Mitchell, A.C., Skidmore, M.L., Vick-Majors, T.J., the WISSARD Science Team, Adkins, W.P., Anandakrishnan, S., Barcheck, G., Beem, L., Behar, A., Beitch, M., Bolsey, R., Branecky, C., Edwards, R., Fisher, A., Fricker, H.A., Foley, N., Guthrie, B., Hodson, T., Horgan, H., Jacobel, R., Kelley, S., Mankoff, K.D., McBryan, E., Powell, R., Purcell, A., Sampson, D., Scherer, R., Sherve, J., Siegfried, M., Tulaczyk, S., 2014. A microbial ecosystem beneath the West Antarctic ice sheet. Nature 512, 310–313. https://doi.org/10.1038/nature13667

Davies, B., 2022. Glaciers as a water resource. AntarcticGlaciers.org. URL https://www.antarcticglaciers.org/glaciers-and-climate/glacier-recession/glaciers-as-a-water-resource/ (accessed 4.14.23).

Dierssen, H.M., Smith, R.C., Vernet, M., 2002. Glacial meltwater dynamics in coastal waters west of the Antarctic peninsula. Proc. Natl. Acad. Sci. U.S.A. 99, 1790–1795. https://doi.org/10.1073/pnas.032206999

Escutia, C., De Santis, L., Donda, F., Dunbar, R.B., Cooper, A.K., Brancolini, G., Eittreim, S.L., 2005. Cenozoic ice sheet history from East Antarctic Wilkes Land continental margin sediments. Global and Planetary Change 45, 51–81. https://doi.org/10.1016/j.gloplacha.2004.09.010

Edwards, I.T., Stevens, A.C., Mitchell, P., Bunting, J.E., Box, K.A., Cameron, J.M., Cook, K., Naegeli, S.M.E., Rassner, J.C., Ryan, M., Stibal, C.J., Williamson, A., Hubbard., 2021. Storage

and export of microbial biomass across the western Greenland Ice Sheet Abstract Nature Communications 12(1). https://doi.org/10.1038/s41467-021-24040-9
Glaciers as a water resource [WWW Document], 2022. URL https://www.antarcticglaciers.org/glaciers-and-climate/glacier-recession/glaciers-as-a-water-resource/
Gupta, J., Rathour, R., Dupont, C., Mishra, A. and Thakur, I.S., 2022. Biogeochemical profiling and taxonomic characterization of municipal landfill site by metagenomic sequencing. Bioresource Technology, 351, p.126936.
Gupta, J., Rathour, R., Maheshwari, N. and Thakur, I.S., 2021. Integrated analysis of Whole genome sequencing and life cycle assessment for polyhydroxyalkanoates production by Cupriavidus sp. ISTL7. Bioresource Technology, 337, p.125418.
Gupta, J., Rathour, R., Singh, R. and Thakur, I.S., 2019b. Production and characterization of extracellular polymeric substances (EPS) generated by a carbofuran degrading strain Cupriavidus sp. ISTL7. Bioresource Technology, 282, pp.417-424.
Gupta, J., Tyagi, B., Rathour, R. and Thakur, I.S., 2019a. Microbial treatment of waste by culture-dependent and culture-independent approaches: Opportunities and challenges. Microbial Diversity in Ecosystem Sustainability and Biotechnological Applications: Volume 1. Microbial Diversity in Normal & Extreme Environments, pp.415–446.
Hagen, J.O., Liestøl, O., Roland, E., Jørgensen, T., 1993. Glacier atlas of Svalbard and Jan Mayen. Nor Polarinst Medd 129, 141
Hood, E., Berner, L., 2009. Effects of changing glacial coverage on the physical and biogeochemical properties of coastal streams in southeastern Alaska. J Geophys Res Biogeosci 114, G03001. https://doi.org/10.1029/2009/2009JG000971
Hohensinner, S., Atzler, U., Fischer, A., Schwaizer, G., Helfricht, K., 2021. Tracing the Long-Term Evolution of Land Cover in an Alpine Valley 1820–2015 in the Light of Climate, Glacier and Land Use Changes. Front. Environ. Sci. 9, 683397. https://doi.org/10.3389/fenvs.2021.683397
Hood, E., Battin, T.J., Fellman, J., O'Neel, S., Spencer, R.G.M., 2015. Storage and release of organic carbon from glaciers and ice sheets. Nature Geosci 8, 91–96. https://doi.org/10.1038/ngeo2331
Hung, H., Halsall, C., Ball, H., Bidleman, T., Dachs, J., De Silva, A., Hermanson, M., Kallenborn, R., Muir, D., Sühring, R., Wang, X., Wilson, S., 2022. Climate change influence on the levels and trends of persistent organic pollutants (POPs) and chemicals of emerging Arctic concern (CEACs) in the Arctic physical environment – a review. Environ. Sci.: Processes Impacts 24, 1577–1615. https://doi.org/10.1039/D1EM00485A
Jansson, P., Hock, R., Schneider, T., 2003. The concept of glacier storage: a review. Journal of Hydrology 282, 116–129. https://doi.org/10.1016/S0022-1694(03)00258-0
Kim, S., Chu, K.H., Al-Hamadani, Y.A.J., Park, C.M., Jang, M., Kim, D.-H., Yu, M., Heo, J., Yoon, Y., 2018. Removal of contaminants of emerging concern by membranes in water and wastewater: A review. Chemical Engineering Journal 335, 896–914. https://doi.org/10.1016/j.cej.2017.11.044
Kohler, T.J., Vinšová, P., Falteisek, L., Žárský, J.D., Yde, J.C., Hatton, J.E., Hawkings, J.R., Lamarche-Gagnon, G., Hood, E., Cameron, K.A., Stibal, M., 2020. Patterns in Microbial Assemblages Exported From the Meltwater of Arctic and Sub-Arctic Glaciers. Front. Microbiol. 11, 669. https://doi.org/10.3389/fmicb.2020.00669
Kohler, T.J., Žárský, J.D., Yde, J.C., Lamarche-Gagnon, G., Hawkings, J.R., Tedstone, A.J., Wadham, J.L., Box, J.E., Beaton, A.D., Stibal, M., 2017. Carbon dating reveals a seasonal progression in the source of particulate organic carbon exported from the Greenland Ice Sheet. Geophys. Res. Lett. 44, 6209–6217. https://doi.org/10.1002/2017GL073219
Kümmerer, K., 2011. Emerging Contaminants, in: Treatise on Water Science. Elsevier, pp. 69–87. https://doi.org/10.1016/B978-0-444-53199-5.00052-X
Li, X., Ding, Y., Xu, J., He, X., Han, T., Kang, S., Wu, Q., Mika, S., Yu, Z., Li, Q., 2018. Importance of Mountain Glaciers as a Source of Dissolved Organic Carbon. J. Geophys. Res. Earth Surf. 123, 2123–2134. https://doi.org/10.1029/2017JF004333
Llamas-Dios, M.I., Vadillo, I., Jiménez-Gavilán, P., Candela, L., Corada-Fernández, C., 2021. Assessment of a wide array of contaminants of emerging concern in a Mediterranean water

basin (Guadalhorce river, Spain): Motivations for an improvement of water management and pollutants surveillance. Science of The Total Environment 788, 147822. https://doi.org/10.1016/j.scitotenv.2021.147822
Mateos-Rivera, A., Yde, J.C., Wilson, B., Finster, K.W., Reigstad, L.J., Ovreas, L., 2016. The effect of temperature change on the microbial diversity and community structure along the chronosequence of the sub-arctic glacier forefield of Styggedalsbreen (Norway). FEMS Microbiol Ecol 92, fnw038. https://doi.org/10.1093/femsec/fiw038
Muir, D., Bossi, R., Carlsson, P., Evans, M., De Silva, A., Halsall, C., Rauert, C., Herzke, D., Hung, H., Letcher, R., Rigét, F., Roos, A., 2019. Levels and trends of poly- and perfluoroalkyl substances in the Arctic environment – An update. Emerging Contaminants 5, 240–271. https://doi.org/10.1016/j.emcon.2019.06.002
PATERSON, W.S.B., 1994. THE PHYSICS OF GLACIERS.
Pattyn, F., 2010. Antarctic subglacial conditions inferred from a hybrid ice sheet/ice stream model. Earth and Planetary Science Letters 295, 451–461. https://doi.org/10.1016/j.epsl.2010.04.025
Rauseo, J., Spataro, F., Baneschi, I., Doveri, M., Giglio, F., Langone, L., Miserocchi, S., Pescatore, T., Patrolecco, L., n.d. Contaminants of emerging concern in the Arctic fjord Kongsfjorden: from glaciers to marine ecosystem.
Sonne, C., Dietz, R., Jenssen, B.M., Lam, S.S., Letcher, R.J., 2021. Emerging contaminants and biological effects in Arctic wildlife. Trends in Ecology & Evolution 36, 421–429. https://doi.org/10.1016/j.tree.2021.01.007
Stubbins, A., Hood, E., Raymond, P.A., Aiken, G.R., Sleighter, R.L., Hernes, P.J., Butman, D., Hatcher, P.G., Striegl, R.G., Schuster, P., Abdulla, H.A.N., Vermilyea, A.W., Scott, D.T., Spencer, R.G.M., 2012. Anthropogenic aerosols as a source of ancient dissolved organic matter in glaciers. Nature Geosci 5, 198–201. https://doi.org/10.1038/ngeo1403
Sutherland, J.L., Carrivick, J.L., Gandy, N., Shulmeister, J., Quincey, D.J., Cornford, S.L., 2020a. Proglacial Lakes Control Glacier Geometry and Behavior During Recession. Geophys. Res. Lett. 47.https://doi.org/10.1029/2020GL088865
Tweed, F.S., Roberts, M.J., Russell, A.J., 2005. Hydrologic monitoring of supercooled meltwater from icelandic glaciers. Quat Sci Rev 24, 2308–2318. https://doi.org/10.1016/j.quacirev.2004.11.020
Valbonesi, P., Profita, M., Vasumini, I., Fabbri, E., 2021. Contaminants of emerging concern in drinking water: Quality assessment by combining chemical and biological analysis. Science of The Total Environment 758, 143624. https://doi.org/10.1016/j.scitotenv.2020.143624
Vorkamp, K., Balmer, J., Hung, H., Letcher, R.J., Rigét, F.F., 2019. A review of chlorinated paraffin contamination in Arctic ecosystems. Emerging Contaminants 5, 219–231. https://doi.org/10.1016/j.emcon.2019.06.001
Wadham, J.L., Arndt, S., Tulaczyk, S., Stibal, M., Tranter, M., Telling, J., Lis, G.P., Lawson, E., Ridgwell, A., Dubnick, A., Sharp, M.J., Anesio, A.M., Butler, C.E.H., 2012. Potential methane reservoirs beneath Antarctica. Nature 488, 633–637. https://doi.org/10.1038/nature11374
Wadham, J.L., Hawkings, J.R., Tarasov, L., Gregoire, L.J., Spencer, R.G.M., Gutjahr, M., Ridgwell, A., Kohfeld, K.E., 2019. Ice sheets matter for the global carbon cycle. Nat Commun 10, 3567. https://doi.org/10.1038/s41467-019-11394-4
Yadav, D., Rangabhashiyam, S., Verma, P., Singh, P., Devi, P., Kumar, P., Hussain, C.M., Gaurav, G.K., Kumar, K.S., 2021. Environmental and health impacts of contaminants of emerging concerns: Recent treatment challenges and approaches. Chemosphere 272, 129492. https://doi.org/10.1016/j.chemosphere.2020.129492
Zarsky, J.D., Kohler, T.J., Yde, J.C., Falteisek, L, Lamarche-Gagnon, G., Hawkings, J.R., et al,. 2018. Prokaryotic assemblages in suspended and subglacial sediments within a glacierized catchment on Qeqertarsauq (Disko Island), west Greenland. FEMS Microbiol Ecol 94, fiy100. https://doi.org/10.1093/femsec/fiy100

Heavy Metal Remediation from Rhizospheric Soil by Using Microbial Consortium

Neha Sharma, Gaurav Yadav, Ajit Varma, Monika Koul, and Arti Mishra

1 Introduction

Soil is an essential resource, but it is limited and non-renewable. Various anthropogenic activities like unsustainable agriculture, overgrazing, industrial pollution, and globalization are the major factors responsible for degrading the quality of the soil. It is observed that more than 24% of the world's land area is deteriorating (Karlen and Rice 2015). The soil quality can also be altered and reduced because of heavy metals (HMs) pollution. Industrialization and inappropriate waste management practices cause the accumulation of HMs in water and soil. It is also because of excessive usage of pesticides and fertilizers. When they enter the soil system in different ways, they accumulate in the soil. Insoluble inorganic metal compounds, exchangeable ions like phosphates and carbonates, soluble metal ions in soil solutions, metal complexes of organic materials, and metals bound to silicate minerals (Marques et al. 2009). HMs are not broken down into chemical harmless products and remain in the environment indefinitely, resulting in environmental damage that is difficult to repair (Anusha et al. 2021). In comparison with other pollutants, HMs are not biologically or chemically degraded. Therefore, the interchangeable fractions of these heavy metals are easily penetrating the plant parts and animals and cause in vivo accumulation. In vivo accumulation of heavy metals causes various disorders and diseases in animals, humans, and plants. Biological restoration of polluted environments is a promising technology that offers the potential to use plants and

N. Sharma · G. Yadav · A. Varma
Amity Institute of Microbial Technology, Amity University, Uttar Pradesh, Noida 201303, India

M. Koul · A. Mishra (✉)
Department of Botany, Hansraj College, University of Delhi, Delhi 110007, India
e-mail: artimishrahrc@gmail.com

A. Mishra
Umeå Plant Science Center, Department of Plant Physiology, Umeå University, Umeå, Sweden

J. Gupta and A. Verma (eds.), *Microbiology-2.0 Update for a Sustainable Future*,
https://doi.org/10.1007/978-981-99-9617-9_4

microorganisms to break down or detoxify various pollutants to repair contaminated areas. These sustainable restoration techniques are a promising way for the natural restoration of soil health. Microorganisms are present both inside and on the plant tissues, mainly on root surface (Hassani et al. 2018). Symbiotic and non-symbiotic microbes, help the host by increasing growth and germination thereby helping the plant to overcome various biotic or abiotic stress. It harbors especially symbiotic rhizomes and/or endogenous microorganisms. Plants are meta-organisms that are closely related to the associated microorganisms (Berg et al. 2016). The synergistic effect that is established in the rhizosphere between the rhizosphere and natural subterranean microbial fauna can eliminate, convert toxic substances in soil, including trace elements (Aken et al. 2010) in less harmful chemicals. HMs are one of the most prevalent soil pollutants in the world, with outbreaks reported in 60% of contaminated soils (Panagos et al. 2013). The presence of higher concentrations of heavy metals compared to that found in nature poses risks due to their toxicity and bioaccumulation and threaten human safety, especially in the most populated and industrialized areas of the world ecosystem's function. If HMs are present in agricultural soils, edible plants are at risk of metal uptake, which could lead to bioaccumulation of these metals in humans and animals, with potentially harmful health repercussions (Deng et al. 2015). This chapter focuses on heavy metal remediation by the various microbes present in the rhizospheric soil. Furthermore, we will also discuss the toxic effects of HMs and how these metals interact with microbes, and how various defense mechanisms are adopted by microbes under stress conditions.

2 Harmful Effects of Heavy Metals on Plants

Metalloids and metals with an atomic number of less than 20 and a density more than 5 g/cm3 are classified as HMs (Ali et al. 2019). Many rocks and minerals contain HMs which are released into the ecosystem by soil erosion and water dissolution. Various microbes, organisms, and plants at higher trophic levels require some heavy metals as essential micronutrients like zinc (Zn), iron (Fe), nickel (Ni), and copper (Cu). Other non-essential HMs include mercury (Hg), lead (Pb), cadmium (Cd) and arsenic (As) (Ali et al. 2019). These HMs are non-biodegradable. Every metal can be harmful if it is released in large quantities than natural levels and or if it is exposed for a longer period (Ali et al. 2019). The most frequent HMs found in polluted soils are Cd, Pb, Cu, Cr, Hg, Zn, Ni, and As (Ali et al. 2019; Shahid et al. 2017). The release of HMs in large amounts is due to anthropogenic activities causing serious contaminant issues (Dixit et al. 2015). Mining, industrial activity, atmospheric deposition, improper waste disposal, fossil fuels, pesticides, fertilizers, and sewage are major sources of anthropogenic HMs release (Ali et al. 2019). Even though the input of the trace metal is minimal, their excessive use has resulted in HM accumulation in soil. Furthermore, irrigating crops with polluted wastewater might be a source of heavy metals in agricultural fields (Dixit et al. 2015; Ali et al. 2019). The fraction of metals that directly interact with live organisms is influenced by heavy metal mobility and bioavailability. Only HMs in a solution are directly accessible to biota and can be absorbed by plant roots, whereas a portion of the total metal concentration in soil is irreversibly bound to soil matrix. (Ali et al. 2019). It has been reported that concentrations of heavy metal in plant tissue were more closely associated with bio-available metals than total metals). As a result, total HMs in soil aren't an appropriate metric for determining the risk of organism accumulation. Many soil variables, including redox potential, pH, texture, clay content, and the presence of soil organic matter, interact with heavy metals and determine their concentration in a soil solution. Heavy metals have a high affinity for negatively charged cellular groups like sulfhydryl's, hydroxyls, and phosphates which cause oxidative stress; they produce reactive oxygen species (ROS) which causes oxidative stress; they compete with essential ions for acquisition, and they disrupt cellular osmotic regulation and ion balance. Biota at all trophic levels including humans, can be hazardous to heavy metals. HMs action mechanisms are comparable among organisms, despite substantial changes in organization and complexity, because they operate predominantly at the cellular level in highly conserved systems. Heavy metals can cause cellular harm by interfering with metabolic functions and normal cellular processes (Fig. 1). Cell membrane modification and lipid peroxidation are caused by the affinity for negatively charged groups at all trophic levels, with ramifications for cell development and division (Amari et al. 2017). In plants, it decreases the rate of photosynthesis by altering the inner membrane of the chloroplast which is essential for photosynthesis (Amari et al. 2017). It also changes the protein structure by binding with the thiol group and leading to the denaturation of protein. Furthermore, both plants and microbes struggle with cellular ionic equilibrium, as changes in cation consumption can cause osmotic difficulties. Zn and Pb contamination in plants, for example, can affect the plant's water balance (Amari et al. 2017). Cellular damage is linked to organism-level damage. The effects of HMs on plants are particularly concerning, as their toxicity and stress can directly damage crops thereby lowering yield.

Various PGPRs including *Pseudomonas*, *Geobacter* and *Bacillus* increase the bioavailability of manganese (Mn) to plants by releasing root exudates and promoting the growth of plants (Dutta and Podile 2010; Miransari 2013). They reduce oxidized Mn^{4+} to Mn^{2+} which is the chemical form and metabolically helpful for plants. PGPRs also increase the availability of Zinc (Zn) and Copper (Cu) in nutrient deficit soil (Badri and Vivanco 2009).

3 Metabolism of Heavy Metals in Plants

Arsenate [As(V)] and Arsenite [As(III)] are two most common forms of arsenic in nature, with less organic arsenic. It has been proposed that As (V) is transported via phosphate transporters, while As (III) is transported via aquaporins. Lsi2 (Low Silicon Rice 2), a silicon efflux transporter in rice, is primarily responsible for root-to-shoot transfer, resulting in HM buildup inside the grain and shoot (Mitani-Ueno et al. 2016). Cd uptake by several divalent transition metal transporters in plants

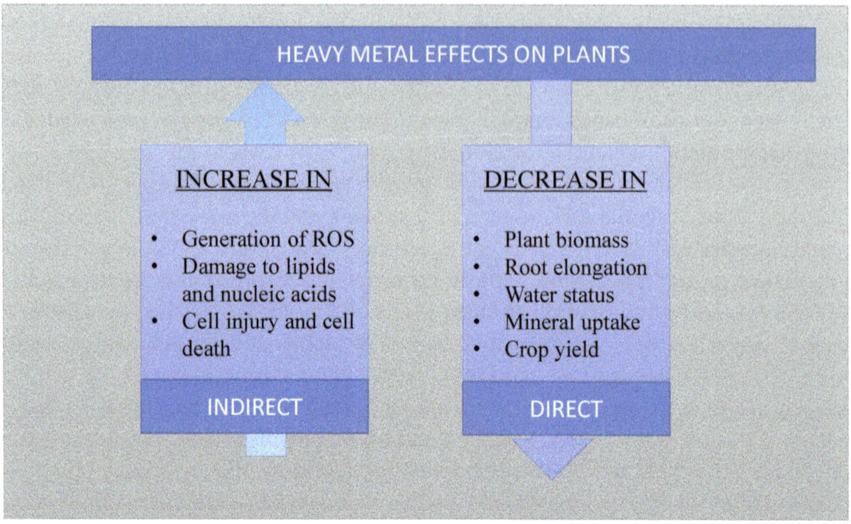

Fig. 1 Impact of heavy metals on plant growth: This leads to increase in production of reactive oxygen species (ROS) and damage of lipid bilayer, cell death and cell injury and leads to decrease in plant biomass, crop yield, water uptake and root elongation

has been described. AtIRT1, an Irt and Zrt-like protein family transporter that transports Fe and Mn, is found in the roots of Arabidopsis thaliana. Fe also aids in the absorption of Cd. The transport of Cd in plants occurs by various transporters of the natural resistance-associated macrophage protein (NRAMP) family (Fig. 1) (Pottier et al. 2015). The Cd influx activity of the *Oryza sativa* natural resistance-associated macrophage protein OsNramp1, an iron transporter localized at the plasma membrane, has been observed in rice (Fig. 1). The xylem-mediated cd transfer into the above-ground section of many plants, including rice, has been proposed as a necessary basis for Cd accumulation in a shoot. (Fontanili et al. 2016). Cd, Ni, Co, and Zn antiporters are members of the cation diffusion facilitator (CDF) protein family found in vacuolar membranes. MTP1's Zn transport activity transports Zn from the cytosol to the vacuoles in a V-ATPase-dependent way. AtMTP1, a Zn^{2+} /H^{+} antiporter that is highly selective for Zn (Lan et al. 2013). In various plants such as *Festuca rubra, Vigna unguiculata* and *Brassica juncea* reports revealed the accumulation of Pb on the surface of roots (Pourrut et al. 2011). The contribution of Nicotiana tabacum calmodulin-binding protein 4 (NtCBP4) in Pb uptake across the plasma membrane has been demonstrated in transgenic tobacco plants overexpressing NtCBP4 (Zeng et al. 2015).

4 Microbial Interaction with Heavy Metals

Elements are harmful if it enters the cells and is bioavailable. There is a structural similarity between divalent cations of HMs. The central metal atom size and shape differ with the ligand and the charges. Thus, chromate structure appears like that of sulfate, and the structure of phosphate is like that of arsenate. Microbes have a unique strategy to overcome HM resistance. The mechanism of heavy metal resistance is present on chromosomes as well as in plasmids like Cd and Hg resistance (Prabhakaran et al. 2016). The efflux of cadmium is mediated through ATPase in *Bacillus* and similarly, arsenic efflux occurs through chromosomal genes of *E. coli*. The resistance genes are present for the halides belonging to VIIA groups of the modern periodic table and these halides are toxic in higher concentrations. It requires transport and active metabolism in the selection system of the genes which make the microbial system resistant to heavy metals. The microbial system brings in a change of oxidation states of HMs and biotransformation (Hennebel et al. 2009).

Sulfate-reducing bacteria efficiently remove metals like Ni, Fe, Pb, Zn by sulfide precipitation technique. The biosurfactant of *Candida sphaerica* UCP0995 could remediate Zn, Pb, and Fe from the soil of battery industries (Luna et al. 2016). In one study by Mishra and Malik 2014a, 2014b, it was reported that *Aspergillus lentutlus* FJ172995 bioaccumulated Cu (56%), Pb (100%) from the waste of electroplating industries (Table 1).

5 Defense Mechanism in Microbes to Cope up with Stress

Soil microbes are subjected to a variety of environmental changes. To survive in such a harsh environment, many soil bacteria evolved adaptive defense mechanisms and structural adaptations. The activities of various proteins decrease or increase under various stress conditions. Microbial response under stressful conditions leads to repression of some proteins which work efficiently under normal conditions and synthesize various proteins which are responsible for their survival in the new environment. Various biochemical changes include morphological changes in the cell and temporary decelerating the cell cycle. Some microbes have resistance mechanisms that enable them to survive under unfavorable environments without the activation of adaptive response while most microbes need adaptive mechanisms resulting in the generation of defense molecules. There are various mechanisms involved in defense processes.

Table 1 Microbial derived remediation of heavy metal

Microorganisms	Class of microorganisms	Heavy metal removal	Sorption efficiency (%)	References
Fungi	Aspergillus versicolor	Cr	99.89	Taştan et al. (2010)
	Aspergillus fumigatus	Pb	96	Ramasamy and Banu (2007)
	Sphaerotilus natans	Cr	60	Xie et al. (2016)
	Gloeophyllum sepiarium	Cr	94	Xie et al. (2016)
	Candida parapsilosis	Hg	80	Goher (2016)
	Piriformospora indica	Cd	–	Su et al. (2010)
	Piriformospora indica	Pb	–	Sartipnia et al. (2013)
	Piriformospora indica	Zn	–	Dianat et al. (2018)
	Aspergillus niger (M1DGR)	Cd, Cr	98, 43	Khan et al (2019)
	Aspergillus fumigatus (M3Ai)	Cd, Cr	79, 69	Khan et al (2019)
Bacteria	*Bacillus firmus*	Zn	61.8	Salehizadeh and Shojaosadati (2008)
	Klebsiella pneumoniae	Hg	28.65	
	Enterobacter cloacae	Co	8	Jafari et al. (2015)
	Cellulosimicrobium sp. (KX710177)	Pb	96. 98	Bharagava and Mishra (2018
	Gemella sp.	Pb	50.99	Marzan et al. (2017)
	B. iodinium	Pb	87	Ramaiah De and Vaedanyan (2008)
	Sporosarcina saromensis (M52)	Cr	82.5	Zhao et al. (2016)
	Sporosarcina pasteurii	Zn, Pb, Cd	–	Jalilvand et al. (2020)
	Stenotrophomonas rhizophila	Zn, Cd, Pb	–	Jalilvand et al. (2020)
	Pseudomonas aeruginosa	Cd	–	Chellaiah et al. (2018)
Algae	*Spirogyra* sp.	Cr	98.23	Mane and Bhosle (2012)
	Chlorella vulgaris	Pb	99.4	Goher (2016)
	Nostoc sp.	Pb	99.6	Kumaran et al (2011)

(continued)

Table 1 (continued)

Microorganisms	Class of microorganisms	Heavy metal removal	Sorption efficiency (%)	References
	Pithophora odeogonia	Pb	97	Singh et al. (2007)
	Spirogyra neglecta	Pb	89	Singh et al. (2007)
	U. lactuca (Green),	Pb	–	Wang, and Chen (2009)
	Undaria pinnatifida (Brown)	Pb	–	Wang, and Chen (2009)
	Lessonia Nigrescens	As^{+}	77	Hansen et al. (2006)
	Sargassum Muticum	Sb^{2+}	50	Ungureanu et al. (2015)
	Spirogyra sp.	Pb^{2+}	30	Gupta.and Rastogi (2008)
	Chlorella Sorokiniana	Cd^{2+}	–	Bayramoğlu et al. (2006)
	Anabaena cylindrica	Cu^{2+}	12.62	Tien et al. (2005)
	Hydrodictyon Reticulatum	Zn^{2+}	3.7	Singh et al. (2007)
Protozoa	*Tetrahymena rostrata*	Hg	40	Muneer et al. (2009)
	Paramecium Multimicronucleatum	Cu	80	Liaqat et al. (2022)
	Vorticella microstoma	Cr^{6+}	48	Rehman et al. (2008)
	Stylonychia Mytilus	Ni^{2+}, Zn^{2+}	28, 48	Shakoori et al. (2011)
Consortium	*Acinetobacter* sp. and *Arthrobacter* sp.	Cr	–	Ramaiah De and Vaedanyan (2008)
	B. subtilis and *P. aeruginosa*	Cr	78	Ramaiah De and Vaedanyan (2008)
	B. subtilis and *S. cerevisiae*	Cr	97.2	Ramaiah De and Vaedanyan (2008)
	C. parapsilosis and *T. rostrata*	Hg	77	Muneer et al. (2009)
	C. parapsilosis and *B. licheniformis*	Hg	85	(Muneer, B et.al 2009)
	P. indica and Rhodococcus Erythropolis	Pb	–	Baghaie and Fereydoni (2019)

(continued)

Table 1 (continued)

Microorganisms	Class of microorganisms	Heavy metal removal	Sorption efficiency (%)	References
	P. indica and Azotobacter chroococcum	Zn	–	Abadi and Sepehri (2016)
	P. indica and Pseudomonas putida	Zn	–	Abadi and Sepehri (2016)
	Bacillus cereus and Bacillus pumilus	Pb^{2+} and Ni^{2+}	95.93, 95.54	Sharma et al. (2021)
	Aspergillus lentulus, Aspergillus terreus and Rhizopus oryzae	Cr^{6+}, Cu^{2+}	–	Mishra et al. (2014)
	Aspergillus flavus (FS4) and Aspergillus fumigatus (FS6),	Cr^{6+}, Cd^{2+}	70, 74	Talukdar et al. (2020)
	B. thuringiensis A1-3, P. aeruginosa A-33, B. cereus A1-5, and B. anthracis A1-7	Cu, Zn and Pb	92.7, 90.3 and 86	Anushaet al (2021)
	Acinetobacter sp. *IrC1, Acinetobacter* sp. *IrC2, and Cupriavidus* sp. *IrC4*	Cu	33	Irawati et al. (2018)
	Aspergillus fumigatus, Aspergillus terreus and Paenibacillus dendritiformis	Cd^{2+}	91.68	Zango et al. (2018)
	Chlorella vulgaris and Enterobacter sp. *MN17*	Cd^{2+}	93	Mubashar et al. (2020)
	Variovorax paradoxus 5C-2, Rhizobium leguminosarum bv. Viciae RCAM1066, and arbuscular mycorrhizal fungus Glomus sp. *1Fo*	Cd^{2+}	–	Belimov et al. (2022)

5.1 *Biofilm Formation*

In environmental stress conditions, biofilm formation is the main mode adopted by microorganisms. It has been reported in previous studies that biofilm formation helps microbes to grow under various stress conditions like extreme temperature, pH, UV radiation, etc. (Sousa et al. 2012). Biofilm formation, either natural or engineered, can be used in heavy metal remediation through the accumulation of toxic metals ions or biochemical modifications. If metals are present in higher concentrations in soils, they can have harmful effects on symbiotic rhizobia by lowering nitrogen-fixing

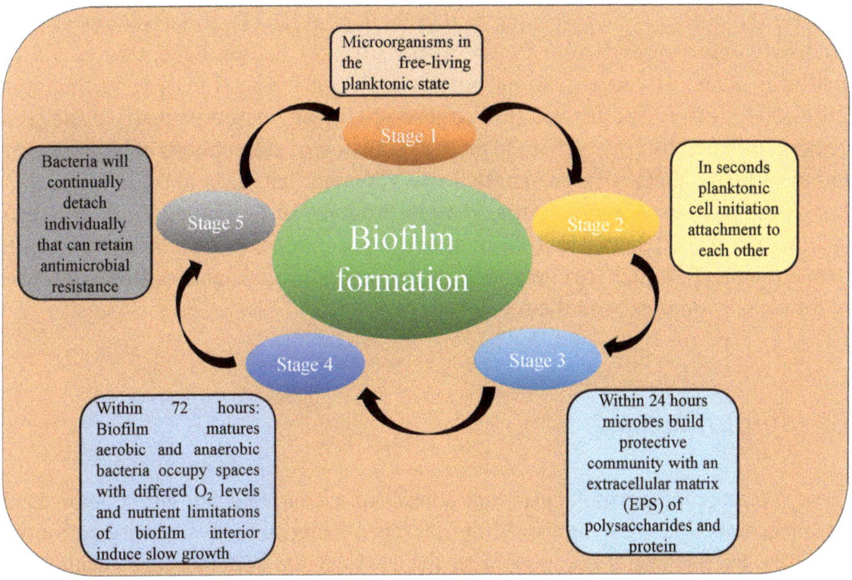

Fig. 2 Stages of biofilm formation: There are five stages of biofilm formation (**a**) stage 1. Involve adhering of microbial cell to solid surface (**b**) stage 2. Planktonic cells start adhering with themselves (**c**) stage 3. Microbes secretes extracellular matrix (EPS) and polysaccharides around them (**d**) stage 4. Maturation of biofilm occurs in which spaces are occupied by aerobic and anerobic bacteria leading to nutrient depletion (**e**) stage 5. Detachment of bacterial cells

efficiency and reducing the survival rate. It has been shown in previous data that EPS from rhizobium cell surface components along with bacterial signals are important for biofilm formation and autoaggregation. The process of biofilm formation is dynamic and complex. In this organized microbial communities are sheathed in an EPS matrix which assembles microbial cells with one other. The components of EPS include proteins, nucleic acids, lipids, and polysaccharides. It forms an extremely hydrated hydrophilic mixture that takes part in the overall scaffolding and 3-D structure of biofilm. There are five stages in biofilm formation including (Fig. 2) (I) Attachment (II) Colonization: (III) Development: (IV) Maturation and (V) Active dispersal (Armbruster and Parsek 2018).

5.2 *Siderophore Production*

Various microorganisms which are associated with plants release iron chelator siderophores in the minute level of iron concentration in soil. Siderophores may form complex with heavy metals like Cd, Al and Cu, etc. which are more stable and helpful in the solubilization of unavailable forms of HMs to available forms and increase the efficacy of phytoextraction (Schalk et al. 2011). Siderophores synthesized by *Streptomyces tendae* F4 are responsible for increasing uptake of Cd by sunflower plants. However, various reports highlight the need to observe the interaction study between plants and siderophore-producing microorganisms in sites that are contaminated by HMs. There is more production of siderophore in high concentration of heavy metals. In one study it was reported that HMs like Cu, Ni, and Cr lead to the synthesis of pyoverdine even in presence of Fe (Braud et al. 2010). In hyperaccumulator plants, the role of siderophore producing microorganisms is to uptake metals. Siderophores help in mobilizing toxic metals and **radionucleotides** by forming a complex with them.

5.3 Biosurfactants

These are amphiphilic molecules that consist of a non-polar tail and a polar head. Biosurfactants that are synthesized by microbes increase bioavailability and metal solubility by forming a complex with the HMs at the soil interface resulting in the desorption of metals from the soil matrix. *Bacillus* sp. J119 which released a biosurfactant was capable of increasing Cd uptake from the soil which was artificially contaminated with varying concentrations of Cd (0 and 50 mg/kg). Biosurfactant uses in inorganic compound recovery require the removal of such ions during the cleaning process, which is mediated through chemical interaction between metal ions and amphipathic molecules (Banat et al. 2010). The applications of biosurfactants in heavy metal remediation have a significant role (Pacwa-Plociniczak et al. 2011). In the bioremediation of heavy metals, these biosurfactants generally form a complex with the metals. Anionic biosurfactants via ionic interactions form complexes with metal ions. These bonds are generally stronger compared to the bonds between metals and soil. These complexes of metal-biosurfactants are then desorbed to the soil solution by the reduction in interfacial tension.

5.4 Biosorption

The microorganisms associated with plants may also contribute via the biosorption process. It is the adsorption of soluble or insoluble pollutants by either an active or passive process (Ma et al. 2011). Biosorption involves the solid and liquid phases (sorbate and solvent, respectively), both of which contain the dissolved species that must be sorbed (Sorbate metal ions). The sorbent's affinity for sorbate is critical and is responsible for the binding of metals on sorbent at various concentrations. It has been found that mycorrhizal fungi act as a filtration barrier against heavy metal transfer from plant roots to shoots. Heavy metal adsorption from soil occurs via large surface area of mycorrhizal fungus. The extracellular as well as intracellular components of fungi like chitin, extracellular slime, metallothioneins, etc. help in the trapping of metals inside the plant roots.

5.5 EPS Production

Various mucopolysaccharides and extracellular polymeric substances (EPS) which are produced by microbes become associated with plants by forming a complex with HMs thereby decreasing their movement in the soil. Rhizospheric bacteria produce more EPS than non-rhizospheric microbes (Seneviratne et al. 2017). It was reported by Joshi and Juwarkar (2009) that EPS synthesized by *Azotobacter* sp. can immobilize Cr and Cd by complex formation (21.9 mg/g of Cr and 15.2 mg/g of Cd) and decreases the uptake of Cr (−0.5) and Cd (−0.4) by *Triticum aestivum*. Other than this EPS play an important role in preserving plants during unfavorable conditions like pathogenesis, dehydration under oligotrophic circumstances. EPS helps in microbial survival by separating nutrients from the environment and it acts as a protective layer by restricting diffusion of some antimicrobial agents into the biofilm through an ion exchanger (Prabhakaran et al. 2016).

5.6 Exopolymer Binding

This is mostly used for reducing the bioavailability of metals. The presence of carbohydrates, polysaccharides, fatty acids and nucleic acids etc. in the niche have important roles in binding of HMs. Microbial exopolymers have an essential role in binding with metals like Cd, Ur, Pb, Cr, etc. In one study, lead was immobilized using exopolymers from *Micrococcus luteus*, *Azotobacter* sp. and *Staphylococcus aureus* (Jarosławiecka and Piotrowska-Seget 2014).

5.7 Transporters in Heavy Metal Resistant Bacteria

Remediation of heavy metals occurs via energy-dependent ion efflux from the cell through membrane proteins. This process is mediated by chemiosmotic proton/cation antiporters. Transporters which are embedded inside, and outer membrane help in the transport of metal ions in the cytosol. Various specific transporters aid in the transport of essential metal ions in the cytosol. It occurs via a chemiosmotic gradient by bacterial cytoplasmic membrane in the case of non-specific transporters. Mainly three classes of efflux transporters are there (Table 2) (a) P-type ATPase (b) Capsule biogenesis/assembly (CBA) transporters and (c) Cation diffusion facilitator (CDF) (Prabhakaran et al. 2016).

Table 2 Various classes of efflux transporters

S. no.	Transporters	Functions	References
1	Capsule biogenesis/ assembly transporters (CBA)	These transporters are present in gram-negative bacteria, and they have three-component trans-envelope pumps which act as chemiosmotic antiporters Resistance nodulation division family transporter (RND) is present in gram-positive bacteria, but capsule biogenesis/assembly (CBA) transporter is not functional in cell walls The presence of RND in the export system depicts differences between CBA transporters and ATP-binding cassette transporters (ABC) transporters RND protein found in the inner membrane is the most important component which combines the bacterial transporters that are required in nodulation and heavy metal resistance	Prabhakaran et al. (2016)
2	Cation diffusion facilitators transporters (CDF)	They function as chemiosmotic ion-proton ex-changers It can be found in both prokaryotes and eukaryotes It is involved in Zn^{2+} transportation and in the transport of other metals like Fe^{2+}, Co^{2+} and Cd^{2+} etc.	Prabhakaran et al. (2016)
3	P-type ATPase	These are incorporated in the inner membranes and uses energy from ATP which helps in transport of metal ions from cytosol to periplasmic space Metal binding domain (MBD) influence specificity of heavy metal translocating ATPase ATPase which takes part in heavy metal translocation is called as CPx-type ATPase (Cysteine-Proline-X) since it contains conserved proline residues (P) followed by cysteine residues	Prabhakaran et al. (2016)

6 Bioremediation of Heavy Metals by Different Microbes

Bioremediation is rising in situ technology for removing environmental contaminant with the help of microorganisms. The process of bioremediation is more economical and environmentally friendly. It is far better than traditional physical and chemical methods. HM remediation can be carried out by immobilizing, concentrating, and partitioning to an environmental compartment, thereby reducing harmful effects (Meliani and Bensoltane 2016). Due to their toxic nature even at minute quantity and non-biodegradative nature, they pose a persistent toxic threat to the environment and life. Heavy metal pollution can cause various problems like skin lesions, learning disabilities mental and physical retardation, cancer, birth defects, kidney and liver damage. In one study, a simultaneous increase in the EPS content of the biofilm was also found, implying that biofilms and EPS play a role in the trapping of metal precipitates. Microbial derived remediation is a process to separate HMs from the polluted sites (Ayangbenro and Babalola 2017). A number of microbes including bacteria, fungus, and algae are known for bio-remediation potential. This process can be expedited by adding nutrients and electron acceptors to polluted sites (Ayangbenro and Babalola 2017). Biotransformation via enzymes, exopolysaccharide (EPS), metallothionein production, and exclusion are major strategies used by diverse microbes to live in heavy metal-polluted environments. Fungi are strong microbes to remove HMs from environment, because of presence of chitin in their cell wall, proteins, lipids, nitrogen-containing polysaccharides, mineral ions and polyphosphates which all can bind to heavy metals. Indeed, the surface of fungal cell walls functions as a ligand for heavy metal binding (Shipra et al. 2011). Lipids and Polysaccharides are two fungal cell wall components that act as heavy metal ligands. Amine (NH_2) groups are the most powerful ligands. The applications of microbes with distinctive features of catabolic potential like biosurfactants and enzymes are a new approach for boosting their remediation efficacy.

It is possible to remove toxic substances including trace elements from the soil by the natural microbiota through their complex and synergistic actions. Heavy metals (HMs) are the major pollutants across the world, and it has been observed that among all the pollutants. HM pollution is more (60%). The higher concentration of heavy metals is lethal because of their toxic, accumulative, and biomagnification characteristics (Ali et al. 2019). These pollutants are a major hazard, particularly in the most industrialized and inhabited areas of the globe, leading to threatening human safety and disrupting ecosystem functions. The presence of HM in soil can cause possible bioaccumulation in animals and humans and deleterious effects on their health.

Till now, various works regarding bacterial communities associated with endophytic and rhizospheric bacteria in hyper-accumulator plants from ultramafic soils, mine tailings have been published. Mostly *Acidobacteria*, *Actinobacteria*, *Proteobacteria* various other taxa have also been detected in soil.

7 Functioning of Metal Transporters in Plants

Metal ions are inevitable in biological processes which are present in cell organelles. The compartmentalized spaces present in cell organelles are specialized for performing functions like respiration, photosynthesis, phytohormone synthesis and metal detoxification. There are various metal transporters present in plants which play important role in sequestering HMs. Among all the transporters, vacuole play important role in detoxification and accumulating metals to protect plant under heavy metal stress (Emamverdian et al. 2015). Mitochondria and chloroplast require transition metals for their functions like photoprotection photosynthesis and so on. Metal homeostasis is essential for the functioning and structural support of mitochondria and chloroplast (Nouet et al. 2011). Figure 3. illustrates the organelles level metal transport for homeostasis.

7.1 *Ar Transporters in Plants*

Arsenic (As) a metalloid causes a difficult issue for the people as well concerning natural wellbeing. It enters the natural system by aggregating in food and feed crops. In soil, they enter by various normal as well as anthropogenic activities like

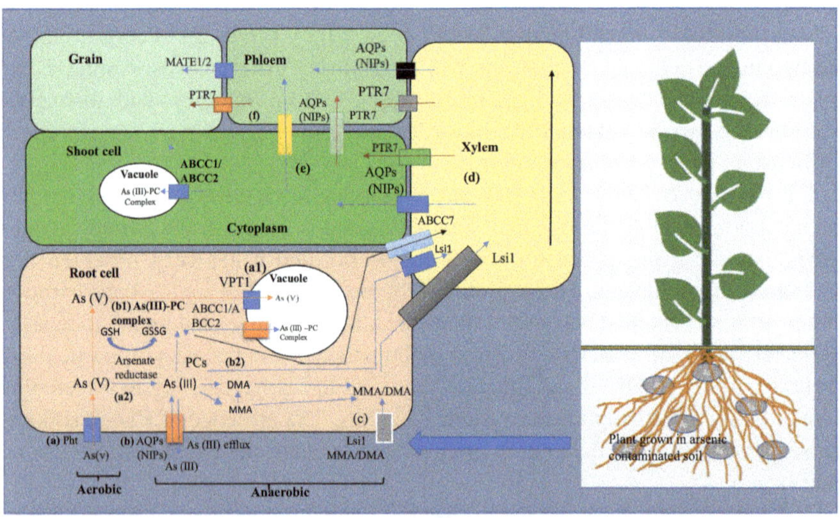

Fig. 3 Mechanism of Arsenic uptake in plant: Plants uptake arsenic (both arsenite (As III) and arsenate (As V) via aquaporins and phosphate transporters respectively and small amount of organic As is taken by some unknown transporters (Dimethyl arsinic acid, DMA and monomethyl arsonic acid, MMA). Conversion of arsenate (As V) to arsenite (As III) occurs through arsenate reductase and then transported to shoot (Tripathi et al. 2007)

the utilization of As based pesticides, and by using water for irrigation purposes previously contaminated with As. Rice very much like wheat is a fundamental crop and a significant dietary source of As. The accumulation rate of arsenic in rice is more than other cereal crops like grain and wheat (Su et al. 2013). Since rice is a hydrophilic plant, it requires a lot of water for its growth and development. Arsenite is a predominant species in paddy conditions however dimethyl arsenic corrosive (DMA), monomethylarsonic corrosive (MMA) and Ar (V) are likewise present in huge amounts. The harmful effects of As results in oxidative stress and preventing ATP formation (Panaullah et al. 2009). There are two components by which arsenic is taken up by roots (Fig. 3) One process incorporates a phosphate (PO_4) transport pathway containing high-affinity phosphate carriers. These phosphate carrier's take-up As (V) from the soil solution and then transport them to aerial parts of plants. In the second route, it is taken up by rice roots via aquaporin channel, which take-up As (III) (silicic acid analogue) and methylated As species (DMA and MMA). Two carriers Lsi 1 and Lsi 2 control Si collection in rice. These carriers Lsi1 and Lsi2 are restricted to the epidermis and endodermis individually. As (III) after entering in roots, most of it is quickly delivered into the rhizospheric soil, this step is carried out by Lsi1 serving as bidirectional channel. Methylated As like MMA and DMA are the minor contributors which add up As in the soil. Though this process isn't extensively studied for inorganic As species. It is observed that harmful effects of inorganic As are much more than pentavalent methylated As species. In one study, it was observed that the movement of DMA in rice occurs to a larger extent than inorganic species.

8 Conclusion

There are various harmful impacts of heavy metals on plant growth and development. They are toxic because of their non-biodegradable nature and persistent nature. There is a need to decontaminate the environment from such persistent pollutants. There is a more toxic effect of heavy metals on agriculture as they reduce crop yield and reduce soil fertility. Various microbes are adapted in such stressful conditions while certain microbes have defense mechanisms to cope with the stress by having efflux pumps and various transporters. Microbes help in the removal of pollutants in an environmentally friendly manner. In the past decade, various research has provided a deeper insight into the various signaling pathways involved in the mechanisms of remediation of heavy metal. However, there is an urgent need for further insights in the mechanisms and understanding of remediation of heavy metals and the various signaling pathways contributing to the process to overcome the stressful conditions.

Conflict of Interest No author has any conflict of interest to declare.

References

Aken, B.V., Correa, P.A., Schnoor, J.L., (2010). Phytoremediation of biphenyls: new trends and promises. Environ. Sci. Technol. 44, 2767–2776.

Ali, H., Khan E., Ilahi I., (2019). Environmental chemistry, and ecotoxicology of hazardous heavy metals: Environmental persistence, toxicity, and bioaccumulation. J. Chem. 6730305.

Amari T., Ghnaya T., Abdelly C. (2017). Nickel, cadmium, and lead phytotoxicity and potential of halophytic plants in heavy metal extraction. S. Afr. J. Bot. 111, 99–110.

Anusha P., Natarajan D. and Narayanan M. (2021). Heavy metal removal competence of individual and bacterial consortium, evolved from metal contaminated soil. Materials Today: Proceedings.

Armbruster, C.R. and Parsek, M.R. (2018). New insight into the early stages of biofilm formation. PNAS, *115*.17, 4317-4319.

Ayangbenro A.S., Babalola O.O. (2017). A new strategy for heavy metal polluted environments: a review of microbial bio sorbents. Int. J. Environ. Res. Public Health 141, 94.

Badri D.V., Vivanco J.M. (2009). Regulation and function of root exudates. Plant Cell Environ. 32, 666–681.

Banat IM, Franzetti A, Gandolfifi I. (2010). Microbial biosurfactants production, applications, and future potential. Appl Microbiol Biotechnol. 87,427–444.

Berg, G., Rybakova, D., Grube, M., Köberl, M. (2016). The plant microbiome explored: Implications for experimental botany. J. Exp. Bot. 67, 995–1002.

Bharagava, R. N., and Mishra, S. (2018). "Hexavalent chromium reduction potential of *Cellulosimicrobium* sp. isolated from common effluent treatment plant of tannery industries," Ecotoxicology and Environmental Safety, 147: 102–109

Braud A, Geoffroy V, Hoegy F, Mislin GLA, Schalk IJ. (2010). The siderophores pyoverdine and pyochelin are involved in *Pseudomonas aeruginosa* resistance against metals: another biological function of these two siderophores. Environ Microbiol Rep 2,419–425.

Chellaiah E.R. (2018). Cadmium (heavy metals) bioremediation by *Pseudomonas aeruginosa*: a minireview. Applied water science, 86,1-10.

De, J., Ramaiah, N., Vardanyan, L. (2008). "Detoxifcation of toxic heavy metals by marine bacteria highly resistant to mercury," Marine Biotechnology, 10. 4, 471–477.

Deng, L., Zeng, G., Fan, C., Lu, L., Chen, X., Chen, M., Wu, H., He, X., He, Y., (2015). Response of rhizosphere microbial community structure and diversity to heavy metal co-pollution in arable soil. Appl. Microbiol. Biotechnol. 99, 8259–8269.

Dianat Maharluei, Z., Yasrebi J., Sepehri M., and Ghasemi R. (2018). Effect of rice husk biochar and *Piriformospora indica* endophytic fungus on corn yield in Zn contaminated soil. Journal of Soil Management and Sustainable Production, 83, 61-78.

Dutta S., Podile A.R. (2010). Plant growth promoting rhizobacteria (PGPR): the bugs to debug the root zone. Crit. Rev. Microbiol. 36, 232–244.

Emamverdian, A., Ding, Y., Mokhberdoran, F. & Xie, Y. (2015) Heavy metal stress and some mechanisms of plant defense response. The Scientific World Journal, 756120, 1–18.

Enayatizamir N., Liu J., Wang L., Lin X., and Fu P. (2020). Coupling Laccase production from *Trametes pubescence* with heavy metal removal for Economic Wastewater Treatment. Journal of Water Process Engineering, 37, 101357.

Fontanili L, Lancilli C, Suzui N, Dendena B, Yin YG, Ferri A, Ishii S, Kawachi N, Lucchini G, Fujimaki S, Sacchi GA, Nocito FF. (2016). Kinetic analysis of zinc/cadmium reciprocal competitions suggest a possible Zn-insensitive pathway for root-to-shoot cadmium translocation in rice. Rice (N Y) 91, 16–2.

Goher, M. E., El-Monem, A. M. A., Abdel-Satar, A. M., Ali, M. H., Hussian, A.-E. M., Napiórkowska-Krzebietke, A. (2016). "Biosorption of some toxic metals from aqueous solution using nonliving algal cells of *Chlorella vulgaris*," Journal of Elementology, vol. 21: 3 703–714.

Gupta V, Rastogi A (2008a). Biosorption of lead from aqueous solutions by green algae *Spirogyra* species: kinetics and equilibrium studies. J Hazard Mater 152:407–414.

Gupta VK, Rastogi A (2008b). Equilibrium and kinetic modeling of cadmium (II) biosorption by non-living algal biomass *Oedogonium* sp. from aqueous phase. J Hazard Mater 153:759–766.

Gupta P., Nicholson F.A., Smith S.R., Alloway B.J., Carlton-Smith C., Chambers B.J. (2003). An inventory of heavy metals inputs to agricultural soils in England and Wales. Sci. Total Environ. 311, 205–219.

Hansen HK, Ribeiro A, Mateus E (2006). Biosorption of arsenic(V) with *Lessonia nigrescens*. Miner Eng 19:486–490.

Hassani, M.A., Durán, P., Hacquard, S., (2018). Microbial interactions within the plant Halobiont. Microbiome. 6, 58.

Hennebel T, Gusseme BD, Boon N, Verstraete W. (2009). Biogenic metals in advanced water treatment. Trends Biotechnol 272, 90–98.

Irawati, W., Ompusunggu, N.P., and Yuwono, T. (2018). August. Influence of bacterial consortium for copper biosorption and accumulation. In AIP conference proceedings (Vol. 2002,1, 020072). AIP Publishing LLC.

Jafari, S. A., Cheraghi, S., Mirbakhsh, M., Mirza, R., Maryamabadi, A. (2015). "Employing response surface methodology for optimization of mercury bioremediation by Vibrio parahaemolyticus PG02 in coastal sediments of Bushehr, Iran," CLEAN - Soil, Air, Water ,43:1, 118–126.

Jahandideh Mahjen Abadi V., Sepehri M., Khoshgoftarmanesh A.H., Eshghizadeh H.R. and Rahmani Iranshahi D. (2015). Inoculation Effects of Endophytic Fungus *Piriformospora Indica* and *Pseudomonas Putida* Bacteria on Growth and Nutrient Uptake of Wheat Plants Under Zinc Deficiency Condition. JWSS-Isfahan University of Technology, 19;71;191-204.

Jalilvand N., Akhgar A., Alikhani H.A., Rahmani H.A. and Rejali F. (2020). Removal of heavy metals zinc, lead, and cadmium by biomineralization of urease-producing bacteria isolated from Iranian mine calcareous soils. Journal of Soil Science and Plant Nutrition, 201;206-219.

Jarosławiecka A, Piotrowska-Seget Z. (2014). Lead resistance in microorganisms. Microbiology 160, 12–25.

Karlen, D.L., Rice, C.W., (2015). Soil degradation: Will humankind ever learn? Sustainability. 7, 12490–12501.

Khan, I., Aftab, M., Shakir, S., Ali, M., Qayyum, S., Rehman, M.U., Haleem, K.S. and Touseef, I., 2019. Mycoremediation of heavy metal (Cd and Cr)–polluted soil through indigenous metallo-tolerant fungal isolates. Environ. Monit. Assess, 191;9, 1-11. https://doi.org/https://doi.org/10.1007/s10661-019-7769-5

Kumaran, N. S., Sundaramanicam, A., Bragadeeswaran, S. (2011). "Adsorption studies on heavy metals by isolated cyanobacterial strain (nostoc sp.) from uppanar estuarine water, southeast coast of India," Journal of Applied Sciences Research 7:11 1609–1615.

Lan HX, Wang ZF, Wang QH, Wang MM, Bao YM, Huang J, Zhang HS (2013). Characterization of a vacuolar zinc transporter OZT1 in rice (*Oryza sativa* L.). Mol Biol Rep 402, 1201–1210.

Liaqat A., Zahra I., Abbas S.Z., Wabaidur S.M., Eldesoky G.E., Islam M.A., Rafatullah M., Shakoori F.R. and Shakoori A.R. (2022). Copper Bioremediation Ability of Ciliate Paramecium *multimicronucleatum* Isolated from Industrial Wastewater. Water, 149;1419.

Luna JM, Rufino RD, Sarubbo LA. (2016). Biosurfactant from *Candida sphaerica* UCP0995 exhibiting heavy metal remediation properties. Process Saf Environ 102,558–566.

Ma Y, Prasad MNV, Rajkumar M, Freitas H. (2011). Plant growth promoting rhizobacteria and endophytes accelerate phytoremediation of metalliferous soils. Biotechnol Adv 29, 248–258.

Mane, P. C., Bhosle, A. B. (2012). "Bioremoval of some metals by living algae *spirogyra* sp. and *spirullina* sp. from aqueous solution," International Journal of Environmental Research, vol. 6:2 571–576.

Marques, A.P., Rangel, A.O., and Castro P.M., (2009). Remediation of heavy metal contaminated soils: phytoremediation as a potentially promising clean-up technology. Crit Rev Environ Sci Technol, 398,622-654.

Marzan, L. W., Hossain, M., Mina, S. A., Akter, Y., and Chowdhury, A. M. M. A. (2017). "Isolation and biochemical characterization of heavy-metal resistant bacteria from tannery effluent in

Chittagong city, Bangladesh: Bioremediation viewpoint," Egyptian Journal of Aquatic Research 43: 65–74.
Meliani, A. and Bensoltane, A., 2016. Biofilm-mediated heavy metals bioremediation in PGPR Pseudomonas. J Bioremediat Biodegrad, 7(370), p.2.
Miransari M. (2013). Soil microbes and the availability of soil nutrients. Acta Physiol. Plant. 35, 3075–3084.
Mishra A. and Malik A. (2014). Metal and dye removal using fungal consortium from mixed waste stream: optimization and validation. Ecological engineering, 69, 226-231.
Mishra, A. and Malik, A. 2014. Metal and dye removal using fungal consortium from mixed waste stream: optimization and validation. Ecol. Eng. 69, 226-231.
Mitani-Ueno N, Yamaji N, Ma JF. (2016). High silicon accumulation in the shoot is required for downregulating the expression of Si transporter genes in rice. Plant Cell Physiol 5712, 2510–2518.
Mohd S., Shukla J., Kushwaha A.S., Mandrah K., Shankar J., Arjaria N., Saxena, P.N., Narayan, R., Roy, S.K. and Kumar, M. (2017). Endophytic fungi *Piriformospora indica* mediated protection of host from arsenic toxicity. Frontiers in Microbiology, 8;754.
Mubashar M., Naveed M., Mustafa A., Ashraf S., Shehzad Baig K., Alamri S., Siddiqui M.H., Zabochnicka-Świątek M., Szota M. and Kalaji H.M. (2020). Experimental investigation of *Chlorella vulgaris* and *Enterobacter* sp. MN17 for decolorization and removal of heavy metals from textile wastewater. Water, 12;11, 3034.
Muneer B., Rehman A., Shakoori F.R. and Shakoori A.R. (2009). Evaluation of consortia of microorganisms for efficient removal of hexavalent chromium from industrial wastewater. Bulletin of environmental contamination and toxicology, 825, 597-600.
Muneer, B., Iqbal, M. J., Shakoori, F. R., Shakoori, A. R. (2013). "Tolerance and biosorption of mercury by microbial consortia: potential use in bioremediation of wastewater," Pakistan Journal of Zoology, vol. 45: 1, 247–254.
Muneer, B., Iqbal, M. J., Shakoori, F. R., and Shakoori, A. R. (2013). "Tolerance and biosorption of mercury by microbial consortia: potential use in bioremediation of wastewater," Pakistan Journal of Zoology 45 247–254.
Kumaran, N. S., Sundaramanicam, A., and Bragadeeswaran, S. (2011). "Adsorption studies on heavy metals by isolated cyanobacterial strain (nostoc sp.) from uppanar estuarine water, southeast coast of India," Journal of Applied Sciences Research, vol. 7: 11 1609–1615, 2011.
Nouet, C., Motte, P. & Hanikenne, M. (2011) Chloroplastic and mitochondrial metal homeostasis. Trends in Plant Science, 16(7), 395–404
Özdemir S., Kılınç E., Poli A. and Nicolaus B. (2013). Biosorption of heavy metals (Cd2+, Cu2+, Co2+, and Mn2+) by thermophilic bacteria, *Geobacillus thermantarcticus* and *Anoxybacillus amylolyticus*: equilibrium and kinetic studies. Bioremediation journal, 172,86-96.
Pacwa-Plociniczak M, Plaza GA, Piotrowska-Seget Z. (2011). Environmental applications of biosurfactants: recent advances. Int J Mol Sci. 12, 633–654.
Panagos, P., Van Liedekerke M., Yigini Y., (2013). *Montanarella* L. Contaminated sites in Europe: Review of the current situation based on data collected through a European network. J. Environ. Public Health, 158-764.
Panaullah GM, Alam T, Hossain MB, Loeppert RH, Lauren JG, Meisner CA et al (2009). Arsenic toxicity to rice (*Oryza sativa* L.) in Bangladesh. Plant Soil 317:31–39.
Pottier M, Oomen R, Picco C, Giraudat J, Scholz-Starke J, Richaud P, Carpaneto A, Thomine S. (2015). Identification of mutations allowing Natural Resistance Associated Macrophage Proteins (NRAMP) to discriminate against cadmium. Plant J 834, 625–637.
Pourrut B, Shahid M, Dumat C, Winterton P, Pinelli E. (2011). Lead uptake, toxicity, and detoxification in plants. Rev Environ Contam Toxicol 213, 113–136.
Prabhakaran P., Ashraf M.A., and Aqma W.S. (2016). Microbial stress response to heavy metals in the environment. Rsc Advances 6111,109862-109877.
Ramasamy K., Banu S.P. (2007). Bioremediation of metals: microbial processes and techniques. Environmental Bioremediation Technologies. Springer, 173 187

Ran, Z.H.A.O., Bi, W.A.N.G., Cai, Q.T., Li, X.X., Min, L.I.U., Dong, H.U., Guo, D.B., Juan, W.A.N.G. and Chun, F.A.N. 2016. Bioremediation of hexavalent chromium pollution by *Sporosarcina saromensis* M52 isolated from offshore sediments in Xiamen, China. Biomed. Environ. Sci. 29(2), 127-136.

Rehman, A., Shakoori, F.R. and Shakoori, A.R. (2008). Heavy metal resistant freshwater ciliate, Euplotes mutabilis, isolated from industrial effluents has potential to decontaminate wastewater of toxic metals. Bioresource technology, 99.9,3890-3895.

Salehizadeh, H., and Shojaosadati, S. A. (2008). "Removal of metal ions from aqueous solution by polysaccharide produced from Bacillus firmus," Water Research, 37:17, 4231–4235.

Sartipnia N., Khavari-Nejad R.A., Babaeizad V., Nejad-Sattari T. and Najafi F. (2013). Effect of *Piriformospora indica* on antioxidant enzymes activity of tomato (Lycopersicon esculentum Mill) under lead stress. Int J Biosci, 3, 55-64.

Schalk IJ, Hannauer M, Braud. (2011). New roles for bacterial siderophores in metal transport and tolerance. Environ Microbiol. 13,2844–2854

Seneviratne, M., Seneviratne, G., Madawala, H.M.S.P. and Vithanage, M., 2017. Role of rhizospheric microbes in heavy metal uptake by plants. In Agro-environmental sustainability (pp. 147–163). Springer, Cham.

Shahid, M., Shamshad, S., Rafiq, M., Khalid, S., Bibi, I., Niazi, N.K., Dumat, C., Rashid, M.I., (2017). Chromium speciation, bioavailability, uptake, toxicity, and detoxification in soil-plant system: A review. Chemosphere. 178, 513–533.

Shakoori F.R., Zafar M.F., Fatehullah A., Rehman, A. and Shakoori, A.R. (2011). Response of glutathione level in a protozoan ciliate, *Stylonychia mytilus*, to increasing uptake of and tolerance to nickel and zinc in the medium. Pakistan Journal of Zoology, 433.

Shakoori, F.R., Zafar, M.F., Fatehullah, A., Rehman, A. and Shakoori, A.R. 2011. Response of glutathione level in a protozoan ciliate, *Stylonychia mytilus*, to increasing uptake of and tolerance to nickel and zinc in the medium. Pak J Zool. 433.

Sharma R., Jasrotia T., Kumar R., Kumar R., Umar A., Alharthi F.A., Alghamdi A.A. and Al-Zaqri N. (2021). An insight into the mechanism of 'symbiotic-bioremoval'of heavy metal ions from synthetic and industrial samples using bacterial consortium. Environmental Technology & Innovation, 21, 101302.

Sharma, R., Jasrotia, T., Kumar, R., Kumar, R., Umar, A., Alharthi, F.A., Alghamdi, A.A. and Al-Zaqri, N. 2021. An insight into the mechanism of 'symbiotic-bioremoval'of heavy metal ions from synthetic and industrial samples using bacterial consortium. Environ. Technol. Innov. 21, 101302.

Shipra J., Dikshit S., Pandey G. (2011). Comparative study of agitation rate and stationary phase for the removal of Cu21 by *A. Lentulus*. Int. J. Pharma Bio Sci. 23, 211- 219.

Singh A., Mehta S.K. and Gaur J.P. (2007). Removal of heavy metals from aqueous solution by common freshwater filamentous algae. World Journal of Microbiology and Biotechnology, 238;1115-1120.

Singh A, Mehta SK, Gaur JP (2007). Removal of heavy metals from aqueous solution by common freshwater filamentous algae. World J Microbiol Biotechnol 23:1115–1120.

Singh, A., Mehta, S.K., Gaur, J.P. 2007. Removal of heavy metals from aqueous solution by common freshwater filamentous algae. World J Microbiol Biotechnol. 23:1115–1120.

Sousa, A.M., Machado, I. and Pereira, M.O., 2012. Phenotypic switching: an opportunity to bacteria thrive.

Su, Y.H., McGrath, S.P., Zhao, F.J. 2010. Rice is more efficient in arsenite uptake and translocation than wheat and barley. Plant Soil. 328(1-2):27–34.

Su Z., Zeng Y., Li X., Perumal A.B., Zhu J., Lu X., Dai M., Liu X., and Lin F. (2021). The endophytic fungus *Piriformospora indica*-assisted alleviation of cadmium in tobacco. Journal of Fungi, 78;675.

Su, Y.H., McGrath, S.P., Zhao, F.J. (2010). Rice is more efficient in arsenite uptake and translocation than wheat and barley. Plant Soil 328(1-2):27–34

Syed T., Batool U., Aslam M., Noreen Z., Farheen I., Gondal A., Usman Shah, S.M., Pucciarelli S. and Bokhari, H. (2019). Bioremediation and decontamination potential of flagellate *Poteriospumella* sp. Bioremediation Journal, 233,142-153.

Talukdar D., Jasrotia T., Sharma, R., Jaglan S., Kumar R., Vats R., Kumar R., Mahnashi M.H. and Umar A. (2020). Evaluation of novel indigenous fungal consortium for enhanced bioremediation of heavy metals from contaminated sites. Environmental Technology & Innovation, 20,101050.

Taştan, B.E., Ertuğrul, S. and Dönmez, G. 2010. Effective bioremoval of reactive dye and heavy metals by *Aspergillus versicolor*. Bioresour technol. 101(3), 870-876.

Tastan, B. E., Ertuğrul, S., and Dönmez, G. (2010). "Effective bioremoval of reactive dye and heavy metals by *Aspergillus versicolor*," Bioresource Technology, vol. 10: 3 870–876.

Tien CJ, Sigee DC, White KN (2005). Copper adsorption kinetics of cultured algal cells and freshwater phytoplankton with emphasis on cell surface characteristics. J Appl Phycol 17:379–389.

Tóth, G., Hermann, T., Da Silva, M.R., (2016). *Montanarella L.* Heavy metals in agricultural soils of the European Union with implications for food safety. Environ. Int. 88, 299–309.

Tripathi R.D., Srivastava S., Mishra S., Singh N., Tuli R., Gupta D.K., Maathuis, F.J.M. (2007). Arsenic hazards: Strategies for tolerance and remediation by plants. Trends Biotechnol. 25, 158–165

Ungureanu G, Santos S, Boaventura R, Botelho C (2015). Biosorption of antimony by brown algae *S. muticum* and *A. nodosum*. Environ Eng Manag J 14:455–463.

Wang J., Chen C. (2009). Biosorbents for heavy metals removal and their future. Biotechnology Advances 27, 195–226.

Xie, Y,. Fan, J., Zhu, W. et al. (2016). "Effect of heavy metals pollution on soil microbial diversity and bermudagrass genetic variation," Frontiers in Plant Science, 7: 775.

Zango U.U., Ahluwalia S.S., and Sharma A.K. (2018). Microbial Consortium of *Aspergillus fumigatus*, *Aspergillus terreus* and *Paenibacillus dendritiformis* in the bioremoval of cadmium. International Journal of Pharmaceutical Research, 103,230-238.

Zeng H, Xu L, Singh A, Wang H, Du L, Poovaiah BW. (2015). Involvement of calmodulin and calmodulin-like proteins in plant responses to abiotic stresses. Front Plant Sci 6, 600.

Zhao, R., Wang, B., Cai, Q. T. et al. (2016). "Bioremediation of hexavalent chromium pollution by *sporosarcina saromensis* m52 isolated from of shore sediments in Xiamen, China," Biomedical and Environmental Sciences, vol. 29:2,127–136.

Microbial Biodiversity and Bioremediation: A Systematic, Biological and Metabolic Engineering Tool

Parveen Kanwar, Nikita Yadav, and Shaili Srivastava

1 Introduction

Microorganisms often known as microbes, are microscopic living organisms that play an important part in forming our environment. Despite their small size, these organisms have a significant impact on a wide range of environmental processes, ecosystems, and even human health. Microbes survive in varied ecosystems and have deep interactions with the environment, from the depths of the ocean to the highest peaks of mountain. This chapter provides an in-depth look at microorganisms, their importance, and their potentiality for bioremediation to clean environment.

Microbes include bacteria, fungi, viruses, and protists. Their microscopic size distinguishes them from a macroscopic form of life which necessitates the use of specialized tools such as microscopes to see. Despite their small size, they have a large and important impact on the environment. Microbes found in practically every imaginable ecosystem on Earth, from corrosive hot springs to deep ocean trenches. This ubiquity shows their feasibility and emphasizes their important role in ecosystem functioning. Microbes are nature's engineers and play critical roles in biogeochemical cycles, which are the pathways of elements such as carbon, nitrogen, and sulphur. Decomposer microorganisms decompose dead organic matter, releasing nutrients such as carbon, nitrogen, and phosphorus back into the environment. This recycling process is critical for keeping nutrients available to plants and other organisms. These cycles are complicated and interwoven, emphasizing the critical role that bacteria play in maintaining environmental balance. While many microorganisms benefit ecosystems and human health, but some can be harmful, producing diseases in a variety of creatures.

P. Kanwar · N. Yadav · S. Srivastava (✉)
Amity School of Earth and Environmental Science, Amity University, Gurugram, Haryana, India
e-mail: ssnigam@ggn.amity.edu

J. Gupta and A. Verma (eds.), *Microbiology-2.0 Update for a Sustainable Future*,
https://doi.org/10.1007/978-981-99-9617-9_5

Bioremediation through microbes is a cost-effective and practical solution for the contaminated environment (Tripathi et al. 2021). Microbes are capable of remediating every possible environment which may range from acidic to basic environment (Krzmarzick et al. 2018; Kour et al. 2021; Chandran et al. 2020; Chatterjee et al. 2011). Recent studies also revealed that the use of different species for the same environment enhances the capability for remediation of the environment. Many technologies are used to enhance the potentiality of microbes (Gaur et al. 2018) different factors are required to maintain the efficiency of microbes toward the decontamination of pollutants (Bhavya et al. 2021; Gaur et al. 2022a, b). This chapter addresses the different approaches and mechanisms used by microbes to detoxify the contaminants of the environment.

2 Strategies Involved in Bioremediation

(a) Pathways and mechanisms involved in microbial bioremediation

There are different strategies and regulatory mechanisms involved in bioremediation through microbes. These are biomineralization, bioventing, bioreactor, bioaugmentation, landfarming, soil vapor evaporation, and landfilling. The main process involves is catalytic action due to the secretion of enzymes that degrade and break down the pollutants in less or nontoxic compounds (Godlewska-Zyłkiewicz 2006; Stentiford 1996). Different type of enzymes is released in the environment by microbes during bioremediation, these are categorised into six different categories. These include Oxidoreductases, Transferases, Hydrolases, Lyases, Isomerases, and Ligases (synthetases) (Liu et al. 2015). Different pathways which are adopted by microbes are divided into two main categories. Ex situ and In situ.

- **Ex situ bioremediation**

When soil or contaminated site is not treated on the respective place due to some unfavourable conditions or insufficient resources then ex-situ treatment is given. This is further divide in different categories which are bio-piling, bioreactor, landfarming, and composting. Bio-piling sometimes also refer to bio cell or bio mounds. This strategy involves excavation, shifting, and heaping of soil into piles. This activity enhances the air supply which improves the respiration rate of microbes. This system involves a proper supply of nutrients, oxygen, pH, and temperature. To avoid the percolation of contamination into the soil, plastic sheets were used as base layer (Delille et al. 2008; Emami et al. 2012; Neill et al. 1999). Land farming is a simple and less technical treatment, in this contaminated soil is excavated and spread over a designed and prepared bed until the pollutants degrade properly. This technique is used to promote the indigenous and local species of microbes. It is a naturally controlled technique. Another approach which is a natural process is a composting technique. Under normal conditions, soil insects like bugs, mites, ants, and beetles, start the breakdown of an organic compound into small particles. During the process of degradation microbes release heat and energy which in turn causes the solubility of pollutants. In this, air supply and C: N ratio is the main component key for regulating the composting.

- **In Situ bioremediation**

In this technique, the remediation process takes place on-site. This process required optimization and control of microbial degradation through the involvement of scientific techniques and engineering disciplines. Different in-situ strategies are bio slurping, bioventing, biosparing, bioaugmentation and microbial-assisted phytoremediation.

Biospraging is the technique in which enough concentration of air is supply to underground water to enhance the degradation by microbes. Basically, it is an aerobic degradation of pollutants, it is very similar to the bioventing process (Leahy and Colwell 1990). Key factors that decide the efficiency of this technique is the permeability of the soil and the nature of the pollutant (Pongratz and Heumann 1999). Bioventing is also an aerobic degradation process, here oxygen is supplied with less pressure, it is a process of pumping air to the contaminated soil above the water table. This process is mainly used for the removal of gasoline, oil etc. Although rate of decontamination depends upon the nature of the pollutant and soil profile (Lee et al. 2002). Bioaugmentation is used the specific microbes, especially those that are genetically engineered. Generally, this technique is used for the treatment of wastewater. Here biomass of microbes is enhanced continuously to improve the genetic pool. In this process, microbes from polluted sites are collected and genetically modified to improve their degradation rate and return back to the contaminated site again. Most commonly it is used for the treatment of sludge in a bioreactor (Lopez et al. 2005). Biostimulation is the process in which specific conditions are provided to the contaminated site for the growth of microbes. Even the presence of a very less number of contaminants can initiate the operon for the enzyme. The presence of specific pH, temperature, and oxygen.

3 Bioremediation of Heavy Metals Using Microbial Process

- **Role of microorganisms in heavy metal bioremediation**

Heavy metals are the most common pollutants which contaminate the environment in a large scale. Cadmium (Cd), Copper (Cu), Argon (Ar), Silver (Ag), Chromium (Cr), Zinc (Zn), Lead (Pb), Uranium (Ur), Ra, Nickel (Ni) etc. are most common and potentially toxic metals. The nonbiodegradable nature of metals makes them persistent in the environment which causes carcinogenic and mutagenic effects on living beings (Grass et al. 2002; Kwon et al. 2005, Srivastava and Thakur 2006a, b). Different studies reported that many microbes are efficiently involved in the degradation of metals into their less harmful oxidation state like hexavalent chromium converted into tetravalent chromium. *Rhodobium marinum* and *Rhodobacter sphaeroides* have the potential to remove heavy metals like copper, zinc, cadmium, and lead from the contaminated environments either by biosorption or biotransformation. Therefore, bacteria have been designated for assessing pollution through their tolerance and biosorption of heavy metals (Kumar et al. 2011). Genetic modification of microbes enhances their tolerance towards heavy metals.

- **Mechanism of Microbial Detoxification of Heavy Metal**

Microorganisms use numerous techniques to interact with and survive in the presence of inorganic metals. Biotransformation is one of the approaches utilized by bacteria to survive metal toxicity. Besides this many other techniques are involved in the bioremediation process, electrostatic contact, ion exchange, precipitation, redox process, and surface complexation are all part of the mechanism (Soloman et al. 2009). Metal oxidation, methylation, enzymatic decrease, metal-organic complexion, metal decrease, metal-ligand degradation, metal flux pumps, demethylation, intracellular and extracellular metal sequestration, exclusion by permeability barrier, and production of metal chelators such as these enzymes and biosurfactants are the major mechanical means by which microorganisms resist heavy metals. Fungi have distinct detoxification processes than eukaryotic. Chelation, precipitation, and cell wall binding are examples of extracellular procedures. Binding to sulfur compounds, organic acids, peptides, polyphosphates, and transport into intracellular compartments are examples of intercellular techniques (Bellion et al. 2006; Kumaret al. 2016). Microorganisms actively convert metal ions into less toxic oxidation states by using electron acceptors and donors for the generation of energy (Srivastava and Thakur 2012). In oxidized form, metal serves as an acceptor for aerobic respiration. Metal ions behave as an important component and transport to the cell with essential nutrients (Green-Ruiz 2006; Gupta and Mohapatra 2003). Figure 1 shows a general outline of mechanism of bioremediation by microbes for heavy metals.

Active transport is also considered an important mechanism that is used by microbes to show resistance toward heavy metals. This system needs three families of protein RND (resistance, nodulation, cell division), CDF (cation diffusion

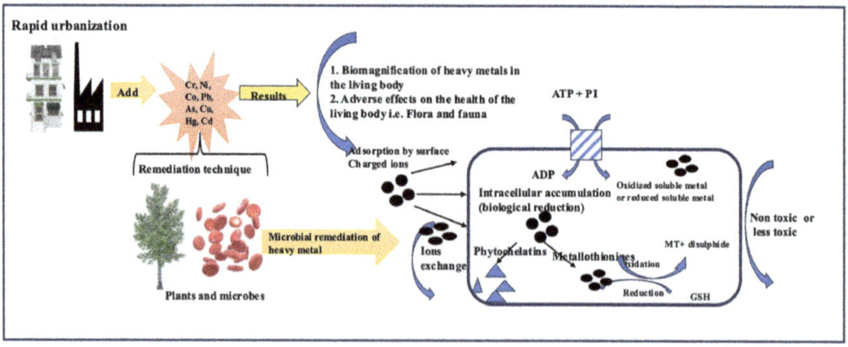

Fig. 1 Bioremediation and uptake mechanism of heavy metals by microbes

facilitator), and P-type ATPase. These proteins are specific for different types of metal ions (Niu et al. 2009). Discussed bioremediation and uptake mechanisms are outlined in Fig. 1.

4 Factors Affecting the Pathway of Microbes in the Bioremediation Process

Remediation through microbes basically involves six main processes these are degradation, removal, modification, immobilization or detoxification of various contamination through physical and chemical changes. Microbes undergo various enzymatic pathways that act as a biocatalyst and initiates biochemical reactions that cause the degradation of targeted pollutant (Abatenh et al. 2017). These microbes are only able to detoxify the contaminants when they are capable of using the different components of pollutants as a nutrient and able to generate energy. The efficacy of bioremediation through microbes depends on many factors which include chemical and concentration of contaminant or toxic compounds, physicochemical characteristics and nature of the environment, and availability of microbes (Fantroussi and Agathos 2005). The percentage of degradation by bacteria is also affected by to uneven spread of contaminants and microbes. Therefore, in situ, controlled, and static decontamination by microbes is not feasible and is a complex process. Table 1 shows showing involvement of different microbes in the degradation of different contaminants. Generally, according to the living and non-living nature, factors are divided into two categories that are biological and environmental factors. Biological factor involves the presence of microbes, their potentiality for degradation, and the presence of enzymes whereas environmental factors involve all physical aspects of the environment like temperature, humidity, pH type of soil, nutrient availability, presence of oxygen, electron acceptor pathways, etc. (Kumar et al. 2021).

- **Biological factors**

Microbes of different species act differently for the same type of pollutants. Different species may have competition because of limited carbon sources, their potentiality may be affected due to interaction between different microbes like pathogens, and macrophages. The potentiality may be affected because of the amount of catalyst. The expression of particular enzymes by microbes may enhance and decrease the process of degradation. Different biological factors that affect the decontamination process are mutation, horizontal gene transfer, enzymatic activity, interaction between different microbes as predator and prey, biomass, population size, and nature of microbes. Microbes also have a variety of intracellular and extracellular enzymes that degrade the pollutants and change them into sources of energy and carbon (Thakur et al. 2019). Microbes also have the capability to process required genetic changes which makes them capable of acquiring and adapting the new metabolic pathways for degrading the xenobiotics (Poirier et al. 2013; Igiri et al.

Table 1 List of toxic organic and inorganic contaminants and microbes involved in their degradation (Tegene and Tenkegna 2020)

Compound	Microbes	References
Azo dyes	*Exiguobacterium indicum, B. cereus, E. aurantiacums* and *A. baumanii*	Raquel et al. (2013)
Aromatic hydrocarbons	*Acinetobacter* sp., *Microbacterium* sp., *Pseudomonas* sp. and *Ralstonia* sp.	Basu et al. (2018)
Crude oil	*B. subtilis strain NAP1, NAP2, NAP4*	Phulpoto et al. (2016)
Acid dye	*T. versicolor, Coriolopsis polyzona; Perenniporia ochroleuca; Perenniporia tephropora, Pycnoporus sanguineus*	Vidali (2001)
Aldrin, lindane	*Bacillus* sp., *Exiguobacterium, aurantiacum, Pandoraea* sp., *Pseudomonas pseudoalcaligenes*	Iyer et al. (2005), Pararlioglu et al. (2005)
Carbofuran	*Novosphingobium sp*	Yang et al. (2005)
Carbamate Aldicarb	*Rotylenchulus reniformis*	Mai et al. (2000)
Diesel oil	*P. cepacia, B. coagulans, B. cereus, B. cereus A and Serratia ficaria*	Miri et al. (2022)
Chlorpyrifos	*Enterobacter sp. Enterobacter aerogenes, Escherichia coli, and Kleibsiella pneumonia*	Liang et al. (2005)
(Polycyclic aromatic hydrocarbons)	(*Burkholderia* sp., *Myceliophthorathermophila*)	Mohapatra et al. (2021)
Phenol	*Alcaligenes odorans, Corynebacterium propinquum, B. subtilis*, and *P. aeruginosa*	Gaur et al. (2018)
Methyl parathion and chlorpyrifos	*Acinetobactor* sp., *Pseudomonas* sp., *Enterobacter* sp. and *Photobacterium* sp.	Halak et al. (2006)
2,4-D	*Pseudomonas* sp.	Myslak and Bolt (1988)
Heavy metals		
Arsenic	*Bacillus* spp.	Tálos et al. (2009)
Cadmium	*A. versicolor, Paecilomyces* sp., *A. fumigatus, Paecilomyces* sp., *Terichoderma* sp. and *Cladosporium* sp.	Unuofin et al. (2021)
Lead	*Pseudomonasaeruginosa, Oedogonium rivulare, Saccharomyces cerevisiae*	Chatterjee et al. (2011), Xu (1996)
Chromium	*Pseudomonas mendoca Cellulosmicrobium cellula ns, Oedogoniumrivulare, Saccharomyces cerevisiae*	Jyoti and Harsh (2019), Kumaran et al. (2011)
Mercury	*Trichoderma Viride*	Xu (1996),
Copper, Nickel	*P. aeruginosa, Aeromonas* sp.	Gaur et al. (2022a, b)

2018). First and foremost, the strategy to utilize microbes for degradation is the isolation and purification of such capable native microbes which have the potential for degradation of pollutants (Chandran et al. 2020). Different studies found that the majority of microbes belonging to different species are not cultured in lab environment (Handelsman 2004). So, knowledge about these microbes is reliable on their information about nucleic acid, lipid formation and nature of the protein (Panigrahi et al. 2019). Although bioremediation through microbes is a better option but still extensive research is still required to understand the mechanism of microbes under different circumstances.

Microorganisms have the potential to degrade a wide number of organic and inorganic compounds and then convert them into less toxic or nontoxic compounds and sometimes utilize these as a source of energy and nutrients.

Molecular approaches like proteomics, genomics, transcriptomics, metabolomics, fluxomics, etc. These techniques give more insight into the population of microbes (De Sousa et al. 2018; Gupta et al. 2020; Rodríguez et al. 2020).

- **Environmental factors**

The success of the bioremediation technique by microbes depends on environmental factors and the interaction between the two. The growth and potentiality of microorganisms depend on pH, temperature, humidity, soil profile, availability of water and nutrients, redox potential and oxygen availability, and most importantly nature of pollutants (Madhavi and Mohini 2012; Adams et al. 2015). Degradation by microbes occurs in a wide range of pH of 6.5 to 8.5 and this pH value is optimal for degradation in both aquatic and territorial habitats. Moisture also influences the rate of degradation as the solubility of pollutant greatly depend on the availability of water as well as the osmotic pressure of the system. As microbes need enough amount of water to flourish in their growth and development but sometimes excess water may have an adverse effect on the growth of microbes. pH of soil also plays an important role and it is affected by the presence of pollutants as soil acidity, basicity, and alkalinity depend on it. pH of soil indicates the efficiency of microbes towards degradation, extreme higher and extremely low pH have a negative impact on the growth of microbes (Asira 2013; Wang et al. 2011).

Different species of microbes require different concentrations of oxygen few do not require high amounts of oxygen whereas few require good concentrations of oxygen. Degradation of pollutants by microbes can be aerobic and non-aerobic. As oxygen is important for degradation in most cases it increases the hydrocarbon metabolism (Macaulay 2014). Species of *Bacillus, Pseudomonas, Sphingomonas, Flavobacterium, Nocardia, Rhodococcus,* and *Mycobacterium* are aerobic in nature they can easily degrade a variety of compounds and utilize them as a source of energy in this case oxygen is becoming the limiting factor (Tegene and Tenkegna 2020). Along with pH temperature also is an important factor that monitors the growth and development of microbes. Temperature determines the components of hydrocarbon and the existence and capability of microbes to survive in extreme environments (Das and Chandran 2011). Enzymes that are required in the degradation process act only at particular temperatures, increase and decrease in the temperature may hinder the potentiality of microbes for degradation. Few microbes can survive at zero temperature whereas few can survive at 100 °C, moreover degradation process is very specific about the temperature its changes may slow and speed up the process as microbial activity increases with an increase in the temperature but at optimal temperature it will static but afterward increase in temperature may affect the activity of microbes (Yang et al. 2009).

The presence of nutrients that are required for the growth of microbes also has an important role in the process of degradation. Essential nutrients like carbon, nitrogen, and phosphorous may enhance the efficiency of degradation so by controlling the ratio of C:N:P, microbial activity can be controlled. Although hydrocarbon pollutants also serve as a source of nutrients the additional suitable quantity of nutrients may enhance the efficacy of microbes (Couto et al. 2014; Phulia et al. 2013). Supplements of nutrients in the cold habitat or environment may enhance the growth and population of microbes. Even in aquatic environments, nutrients are the major limiting factor for the microbes (Sharma et al. 2022a, b). High concentration of pollutants sometimes inhibits the growth of microbes as all microbes are not capable of surviving at high contamination which slow down the process of degradation. Mechanism of degradation may vary with the concentration of toxic compounds; sometimes high concentration speeds up the process but, in some cases, it degrades the process of degradation. Like metal contamination does not affect the growth of microbes if they are present in less quantity but if concentration is high then it may inhibit the growth of bacteria. Microbes that are capable of releasing chelating compounds and acids and also have redox potential can cause significant changes in the environment by enhancing the availability of heavy metals (Masindi and Muedi 2018). Bio absorption, ion complex formation, and adsorption are important processes for the interaction between metal and microbial communities (Gałwa-Widera et al. 2021).

5 Use of Advanced Techniques in Bioremediation

The use of microbes in bioremediation involves the degradation of pollutants either in less harmful products or in neutralizing the contaminants (Giri et al. 2021a, b). A different number of tools are developed to monitor and control the process. The implementation of the proteomic technique gives a clear study of microbes like gives detailed structural study of microbial protein (Vega-Páez et al. 2019).

- **Genomics**

In this, microbes' capability and potentiality towards bioremediation were extensively studied. The bioremediation technique uses a wide variety of species of microbes, to analyze in an appropriate way different tools are used like PCR, isotope distribution, DNA hybridization, molecular connectivity, metabolic footprinting, and metabolic engineering. For genotypic fingerprinting, different type of PCR-based techniques is available, which includes amplified fragment length polymorphisms (AFLP), amplified ribosomal DNA restriction analysis (ARDRA), automated ribosomal intergenic spacer analysis (ARISA), terminal-restriction fragment length polymorphism (T-RFLP), randomly amplified polymorphic DNA analysis (RAPD), single strand conformation polymorphism (SSCP), and length heterogeneity (Hakeem et al. 2020). Whereas for soil microbes, the Random Amplified Polymorphic DNA (RAPD) technique was widely used for their fingerprinting and structural analysis (Rodríguez et al. 2020). Microbial interaction with soil environmental factors can be studied by a combination of different fields like genetic fingerprinting, microradiography, FISH, stable isotope probing, and quantitative PCR (Gupta et al. 2020). A PCR-based quantitative analysis can be used to detect the abundance and presence of taxonomic and operational gene markers in the soil (Yunusa and Umar 2021). In order to analyze in a better way, the relationship between diverse microbial communities, cluster-assisted analysis, with fingerprints of different samples, could be used.

- **Transcriptomics and Metabolomics**

Transcriptomes are generally used to present all RNAs including coding and non-coding, these are the link between cellular phenotype, interactome, genome, and proteome and involves all genes that are transcribed ta specific site, time, and condition. Transcriptomes provide an overall view of the entire human genome. In transcriptomics, analysis through DNA microarray is a powerful tool to know about the expression levels of mRNA expression (Sharma et al. 2022a, b). Knowledge of Transcriptomics and Meta transcriptomics gives insight about the microbial structure and their efficiency towards environmental factors. These tools can be used by researchers for further studies.

Besides transcriptomics, metabolomics is mainly concerned with the total metabolites produced by organisms in a given time frame (Tripathi et al. 2018). These are more promising and efficient for functional analysis (Gaur et al. 2022a, b). Analysis of protein can be done by use of proteomics, as compared to genomics this tool is more promising. Profiling of metabolite, foot printing, and analysis of targets are a few important tools in the toolbox of microbial metabolomics which can be used to identify and quantify the myriad of cellular byproducts present in living organisms (Sanghvi et al. 2020). When a cell is exposed to environmental stress the cell releases primary and secondary metabolites which helps to understand the behaviour of microbes. The collective employment of advanced analytical methodologies, and bioinformatics software boosts up the amalgamation of metabolomics into systems biology.

- **Microbes and nanotechnology**

Nanometres are the smallest scale which is used in nanotechnology. A variety of toxic pollutants are removed by using this technology, and nanoparticles have promising abilities of adsorption and absorption. This technology is widely used in the treatment of wastewater. For the treatment of water, nanotechnology, and efficient microbes are collectively used. The use of nanotechnology expands the use of microbes in the remediation of a wide variety of pollutants like polyaromatic hydrocarbons with many rings which are mutagenic and non-biodegradable and can be treated with the use of nanoparticles with microbes. The use of nanotechnology with microbes for bioremediation opened a new frontier to enhance the rate of biodegradation. Nanoparticles are also prepared with the help of PMUA (poly-(ethylene) glycol-modified urethane acrylate). These PMUA nanoparticles efficiently work with a wide range of bacteria to remediate the environment.

- **Genetic and metabolic engineering**

Genetic engineering refers to the modification of genes through addition, deletion, and substitution of fragments of gene or DNA molecules. TALENs, ZFNs, and CRISPRs are prominently used transcription activators for the editing of genes (Dangi et al. 2019). Among all these CRISPR-Cas is the most common and efficient gene editing tool. ZENs also have 30 amino acid binding sites for DNA. The Folk1 causes breakdown of the DNA strand and is used as an important tool for genetic engineering (Phale et al. 2022). *Streptococcus pyogenes* has a special ability against the resistance to viruses. In model species like *Pseudomonas and Escherichia coli,* the CRISPR-Cas system has been widely accepted by researchers. In non-model species like *Rhodococcus ruber* TH, *Achromobacter* sp. HZ01, and *Comamonas testosterone* the area of bioremediation is also exploring new insights into CRISPR tool kits and the synthesis of gRNA for the production of remediation-specific genes.

Microbes which have resistance towards pollutant make them a better choice for genetic engineering as these are accustomed to tolerating and sorting a variety of toxic and non-degradable xenobiotic compounds under unfavourable conditions. Moreover, species of *Bacillus* like DG-02 release haloalkane dehalogenases from the degradation of harmful pollutants (Chen et al. 2019). With the use of genetic engineering, recombinant enzyme can modify and their effect can enhance significantly. In few studies it was found that some extracellular enzyme plays a significant role in bioremediation and such type of remediation is called enzymatic bioremediation. As studies showed that enzyme lignin peroxidase (LiPs) which is released by *Pseudomonas chrysosporium plays* a significant role in degrading polyaromatic hydrocarbon.

Although pollutants can be consumed by microbes during pathway as a substrate and release nontoxic, less toxic or simpler substances. Like LiP enzyme is able to decompose benzopyrene into three different simpler compound (1,6-quinol, 3,6-quinine, and 6,12-quinine) (Baker et al. 2019).

Genetically engineered microbes are more efficient in bioremediation as compare to non-engineered microbes like *Pseudomonas putida* KT2440 was used in organophosphate and pyrethroid bioremediation experiments (Gong et al. 2018; Jaain et al. 2019). After the development of genetic engineering, many persistent and harmful pollutant are degraded successfully by different micro-organism.

6 Advantages and Disadvantages of Microbial Bioremediation

As pollutants in the environment causes harm to both humans and other living creatures. Although many different techniques were developed to eliminate these pollutants from environment completely but it is very tough job. Use of different techniques to reduce this pollution may causes harm to the environment may be not cost effective. However, use bioremediation to eradicate the persistent pollutants may prove helpful for the environment. Bioremediations are the cost effective, environment friendly and economically sound techniques but still this technique has some disadvantages and advantages. Aim of bioremediation is to supply some important conditions which improves the process and enhances the degradation process. Bioremediation through microbes only effective when different environmental conditions are suitable for the growth of microbes (Kumar et al. 2011). Through bioremediation pollutants can be completely remove to some extent as microbial degradation convert the compounds into naturally occurring compounds like carbon dioxide, water other inorganic compounds which are safe for the human and environment (Pankaj and Vivek 2012).

7 The Advantages of Bioremediation

- It is a natural and self-regulated process. When contamination level is high then degradation rate is also high and when concentration of contaminants declines then the rate of degradation is also decreases. Remains of this process are generally harmless and store in environment as water vapours, carbon dioxide etc.
- It demands very little effort and is frequently performed on-site without causing substantial interruptions to usual activity.
- Transporting rubbish away from the site and the potential hazard to human health and the environment occur while transport.
- It works in a cost-effective process since it wastes less than other conventional ways (technologies) for hazardous waste cleanup. An important method for detoxifying oil-contaminated locations (Hussain et al. 2022).
- It also aids in the full disappearance of pollutants, as many dangerous compounds can be converted to harmless products, and this aspect reduces the prospect of future liability concerning the treatment and disposal of contaminated material.
- No hazardous chemicals are used. Nutrients, specifically fertilizers, are provided to promote active and rapid microbiological development. Usually utilized on lawns and gardens. Because bioremediation converts dangerous chemicals into innocuous gases and water, the harmful compounds are entirely eliminated (Giri et al. 2021a, b)

- As a result of their natural role in the environment, they are simple, labour-intensive, and inexpensive. It does not require any post-treatment of pollutants. Environmentally friendly and long-lasting (Tarekegn et al. 2020).
- Contaminants get eliminated rather than merely transported to other environmental mediums. Non-intrusive, potentially enabling continuous usage of the site. Relative simplicity of implementation (Bhavya et al. 2021).
- An efficient method of removing contaminants from ecosystems that are naturally occurring without being ecologically sound. Widely accepted by the public (Tegene 2020)

8 The Disadvantage of Bioremediation

- It is only relevant to biodegradable toxic substances. Not all chemicals have the potential of complete and rapid disintegration.
- There are some by product of biodegradations which are not degradable and toxic in nature.
- Bioremediation required some biological components and they are site specific, need some metabolically active compounds which degrade the contaminants.
- More extensive knowledge is required to deal with microbes and generally this process take more time as compare to some other techniques.
- Many microbes cannot culture in laboratory conditions and it also not possible to generalize the results of pilot scale study with large scale study.
- Ethical and Regulatory Concerns is required. Introduction of genetically modified bacteria for bioremediation raises concerns regarding unexpected repercussions and long-term implications.
- It is critical to develop accurate methods for monitoring the development of bioremediation and assessing its long-term effectiveness.

9 Conclusion

Environmental pollution produced by many sources such as industry, farming, and urbanization activities causes an ecological crisis, biodiversity loss, and catastrophic climate change. Various garbage elimination solutions have been used to address these worldwide issues. Bioremediation is a step-by-step technology that uses the efficiency of biological agents to eradicate environmental contaminants and detoxify hazardous toxic pollutants. Ex situ and in situ, bioremediation processes such as biopiling, composting, land farming, bio venting, biosparging, biostimulation, and bioaugmentation are used to treat a wide variety of pollutants like heavy metal waste, hydrocarbon petroleum waste, chlorinated compound and plastic waste, organic and volatile compound pollutant, organic and inorganic municipal solid waste, lignocellulose organic agriculturally based rubbish, industry effluent waste, nuclear waste, and various dyestuff. During the bioremediation process, oxidoreductases, transferases, hydrolases, lyases, isomerases, and ligases (synthetases) are used for degradation and mineralization process. Understanding the mechanism, mode of action, and role of microorganisms in the bioremediation process is critical for harnessing microbial potential and developing waste management strategies.

The emergence of influential molecular techniques, transformed and altered molecular and microbial ecology over the last few decades. Different advanced methods were employed which provided direct access to environmental microorganisms irrespective of their habitat. High-throughput technologies are the means through which information on the functionality and enormous characterization of microbial communities is produced. The improvement of methodologies that are helpful in the study of breakdown reactions and thus give significant information on the management and cleanup of contaminated sites and effluents in an environmentally friendly way is providing a boost to bioremediation research. However, due to economic and technological constraints, the use of high-throughput genomic techniques for bioremediation is still in its early stages.

Conflict of Interest The authors declare no conflict of interest.

References

Abatenh E, Gizaw B, Tsegaye Z, Wassie M (2017) The role of microorganisms in bioremediation-A review. *Open Journal of Environmental Biology*, *2*(1), 038-046.

Adams GO, Fufeyin PT, Okoro SE, Ehinomen I (2015) Bioremediation, Biostimulation and Bioaugmention: A Review. International Journal of Environmental Bioremediation & Biodegradation 3: 28–39. Link: https://goo.gl/9XY7ni

Asira, Enim Enim (2013) Factors that Determine Bioremediation of Organic Compounds in the Soil. Academic Journal of Interdisciplinary Studies 2: 125- 128. Link: https://goo.gl/aZ2ntP

Baker, S. E., Serafim, A. B., Morales-Agudelo, P., Vidal, M., Calafat, A. M., & Ospina, M. (2019). Quantification of DEET and neonicotinoid pesticide biomarkers in human urine by online solid-phase extraction high-performance liquid chromatography-tandem mass spectrometry. Analytical and bioanalytical chemistry, 411, 669-678.

Basu S, Rabara RC, Negi S, Shukla P (2018) Engineering PGPMOs through gene editing and systems biology: A solution for phytoremediation? Trends Biotechnol. 36, 499–510

Bellion M, Courbot M, Jacob C, Blaudez D, Chalot M (2006) Extracellular and cellular mechanisms sustaining metal tolerance in ectomycorrhizal fungi. FEMS Microbiol Lett 254: 173-181.

Bhavya G, Belorkar SA, Mythili R, Geetha N, Shetty HS, Udikeri SS, Jogaiah S (2021) Remediation of emerging environ-mental pollutants: A review based on advances in the uses of eco-friendly biofabricated nanomaterials. Chemosphere 275, 129975.

Chandran H, Meena M, Sharma K (2020) Microbial biodiversity and bioremediation assessment through omics approaches. *Frontiers in Environmental Chemistry*, *1*, 570326.

Chatterjee S, Gupta D, Roy P, Chatterjee NC, Saha P, Dutta (2011) Study of a lead tolerant yeast strain BUSCY1. Afr J Microbiol Res 5: 5362-5372

Chen, J., McIlroy, S. E., Archana, A., Baker, D. M., & Panagiotou, G. (2019). A pollution gradient contributes to the taxonomic, functional, and resistome diversity of microbial communities in marine sediments. Microbiome, 7, 1-12.

Couto N, Fritt-Rasmussen J, Jensen PE, Højrup M, Rodrigo AP, et al. (2014) Suitability of oil bioremediation in an Artic soil using surplus heating from an incineration facility. Environmental Science and Pollution Research 21: 6221- 6227. Link: https://goo.gl/eXKYxm

Dangi, A. K., Sharma, B., Hill, R. T., & Shukla, P. (2019). Bioremediation through microbes: systems biology and metabolic engineering approach. Critical reviews in biotechnology, 39(1), 79-98.

Das N, Chandran P (2011) Microbial degradation of petroleum hydrocarbon contaminants: An overview. Biotechnol Res Int 2011: 1–13. Link: https://goo.gl/jt4THZ

De Sousa CS, Hassan SS, Pinto AC, Silva WM, De Almeida SS, Soares SD (2018) "Microbial omics: applications in biotechnology," in Omics Technologies and Bio-Engineering, eds D. Barh and V. Azevedo (Cambridge, MA: Academic Press), 3–20. doi: https://doi.org/10.1016/B978-0-12-815870-8.00001-2

Delille D, Duval A, Pelletier E (2008) Highly efficient pilot bio piles for onsite fertilization treatment of diesel oil-contaminated sub-Antarctic soil. Cold Reg Sci Technol 54: 7-18

El Fantroussi S, Agathos SN (2005) Is bioaugmentation a feasible strategy for pollutant removal and site remediation? Current Opinion in Microbiology 8: 268–275. Link: https://goo.gl/y6kLsc

Emami S, Pourbabaee AA, Alikhani HA (2012) Bioremediation Principles and Techniques on Petroleum Hydrocarbon Contaminated Soil. Technical Journal Engineering and Applied Sciences 2: 320-3

Gałwa-Widera, M. (2021). Biochar–production, properties, and service to environmental protection against toxic metals. Handbook of Assisted and Amendment: Enhanced Sustainable Remediation Technology, 53-75.

Gaur N, Narasimhulu K, PydiSetty Y (2018) Recent advances in the bio-remediation of persistent organic pollutants and its effect on environment. J. Clean. Prod. 198, 1602–1631

Gaur V.K, Gautam K, Sharma P, Gupta P, Dwivedi S, Srivastava JK, Varjani, S, Ngo HH, Kim SH, Chang J.S (2022) Sustainable strategies for combating hydrocarbon pollution: Special emphasis on mobil oil bioremediation. Sci. Total Environ, 832, 155083.

Gaur VK, Tripathi V, Manickam N (2022b) Bacterial-and fungal-mediated biodegradation of petroleum hydrocarbons in soil. In Development in Wastewater Treatment Research and Processes; Elsevier: Amsterdam, The Netherlands, pp. 407–427.

Giri BS, Geed S, Vikrant K, Lee SS, Ki KH, Kailasa SK, Vithanage M, Chaturvedi P, Rai BN, Singh RS (2021) Progress in bioremediation of pesticide residues in the environment. Environ. Eng. Res. 26, 200446.

Giri, B.S.; Geed, S.; Vikrant, K.; Lee, S.S.; Kim, K.H.; Kailasa, S.K.; Vithanage, M.; Chaturvedi, P.; Rai, B.N.; Singh, R.S. Progress in bioremediation of pesticide residues in the environment. Environ. Eng. Res. 2021b, 26, 200446. [CrossRef]

Godlewska-Zyłkiewicz B (2006) Microorganisms in inorganic chemical analysis. Analytical and Bioanalytical Chemistry 589: 119-125.

Gong, T., Xu, X., Dang, Y., Kong, A., Wu, Y., Liang, P., ... & Yang, C. (2018). An engineered Pseudomonas putida can simultaneously degrade organophosphates, pyrethroids and carbamates. Science of The Total Environment, 628, 1258-1265.

Grass G, Wong MD, Rosen BP, Smith RL, Rensing C (2002) ZupT is a Zn (II) uptake system in Escherichia coli. J Bacteriol 184: 864-866

Green-Ruiz, C. (2006). Mercury (II) removal from aqueous solutions by nonviable Bacillus sp. from a tropical estuary. Bioresource technology, 97(15), 1907-1911.

Gupta, R., & Mohapatra, H. (2003). Microbial biomass: an economical alternative for removal of heavy metals from waste water.

Gupta, K., Biswas, R., and Sarkar, A. (2020). "Advancement of omics: prospects for bioremediation of contaminated soils," in Microbial Bioremediation & Biodegradation, ed M. Shah (Singapore: Springer), 113–42.

Hakeem, K.R.; Bhat, R.A.; Qadri, H. Bioremediation and Biotechnology; Springer: Cham, Switzerland, 2020.

Halak, S.; Basta, T.; Bürger, S.; Contzen, M.; Stolz, A. Characterization of the genes encoding the 3-carboxy-cis, cis-muconatelactonizing enzymes from the 4-sulfocatechol degradative pathways of Hydrogenophaga intermedia S1 and Agrobacterium radiobacter S2. Microbiology 2006, 152, 3207–3216

Handelsman, J. (2004). Metagenomics: application of genomics to uncultured microorganisms. Microbiol. Mol. Bio. Rev. 68, 669–685. doi: https://doi.org/10.1128/MMBR.68.4.669-685.2004

Hussain, A.; Rehman, F.; Rafeeq, H.; Waqas, M.; Asghar, A.; Afsheen, N.; Rahdar, A.; Bilal, M.; Iqbal, H.M. In-situ, Ex-situ, and nano-remediation strategies to treat polluted soil, water, and air—A review. Chemosphere 2022, 289, 133252

Igiri, B. E., Okoduwa, S. I. R., Idoko, G. O., Akabuogu, E. P., Adeyi, A. O., and Ejiogu, I. K. (2018). Toxicity and bioremediation of heavy metals contaminated ecosystem from tannery wastewater: a review. J. Toxicol. 2018:2568038. doi: https://doi.org/10.1155/2018/2568038

Iyer A, Mody K, Jha B (2005) Biosorption of heavy metals by a marine bacterium. Mar Pollut Bull 50: 340-343

Jaain, R.; Patel, A. Bioremediation of Gurugram–Faridabad Dumpsite at Bandhwari. In Waste Valorisation and Recycling; Springer: Singapore, 2019; pp. 433–4

Jyoti B, Harsh KSN (2019) Utilizing Aspergillus nigerfor bioremediation of tannery effluent. Octa Journal of Environmental Research 2: 77-81.

Kour D, Kaur T, Devi R, Yadav A (2021) Singh, M.; Joshi, D. Beneficial microbiomes for bioremediation of diverse contaminated environments for environmental sustainability: Present status and future challenges. Environ. Sci. Pollut. Res, 28, 24917–24939.

Krzmarzick, M.J.; Taylor, D.K.; Fu, X.; McCutchan, A.L. Diversity and niche of archaea in bioremediation. Archaea 2018, 2018, 3194108

Kumar A, Bisht B S, Joshi V D, Dhewa T (2011) Review on Bioremediation of Polluted Environment: A Management Tool. international journal of environmental sciences 1: 1079–1093. Link: https://goo.gl/P6Xeqc .

Kumar, A., & Chandra, R. (2020). Ligninolytic enzymes and its mechanisms for degradation of lignocellulosic waste in environment. Heliyon, 6(2).

Kumar S, Chaurasia P, Kumar A (2016) Isolation and Characterization of Microbial Strains from-Textile Industry Effluents of Bhilwara, India: Analysis with Bioremediation. Journal of Chemical and Pharmaceutical Research 8: 143-50.

Kumar, A, Gupta, A., and Srivastava, S. (2021) Degradation and Biotransformation of Pentachlorophenol by Microorganisms" In Book Climate Resilience and Environmental Sustainability Approaches.Springer Nature, pp 299–318.

Kumaran NS, Sundaramanicam A, Bragadeeswaran S (2011) Adsorption studies on heavy metals by isolated cyanobacterial strain (nostoc sp.) from uppanar estuarine water, southeast coast of India. Journal of Applied Sciences Research 7: 1609- 1615.

Kwon GS, Sohn HY, Shin KS, Kim E, Seo BI (2005) Biodegradation ofthe organochlorine insecticide, endosulfan, and the toxic metabolite, endosulfansulfate, by Klebsiella oxytoca KE–8. Appl. Microbiol. Biotechnol 67: 845-850.

Leahy JG, Colwell RR (1990) Microbial degradation of hydrocarbons in the environment. Microbiological Reviews 54: 305-315.

Lee IS, OK Kim, YY Chang, B Bae, HH Kim, et al. (2002) Heavy metal concentrations and enzyme activitiesin soil from a contaminated Korean shooting range. J Biosci Bioeng 94: 406-411.

Liu L, Zhang X, Zhong T (2015) Pollution and health risk assessment of heavy metals in urban soil in China. Hum Ecol Risk Assess An Int. 22: 424-434.

Lopez L, Pozo C, Rodelas B, Calvo C, Juarez B, et al. (2005) Identification of bacteria isolated from an oligotrophic lake with pesticide removal capacities. Ecotoxicology, 14: 299- 312.

Macaulay BM (2014) Understanding the behavior of oil-degrading microorganisms to enhance the microbial remediation of spilled petroleum. Appl Ecol Environ Res 13: 247–262. Link: https://goo.gl/JfFVWd

Madhavi GN, Mohini DD (2012) Review paper on – Parameters affecting bioremediation. International journal of life science and pharma research 2: 77–80. Link: https://goo.gl/tBP2C6

Mai C, Schormann W, Milstein O, Huttermann A (2000) Enhanced stability of laccase in the presence of phenolic compounds. Applied Microbiology and Biotechnology 54: 510-514.

Masindi, V.; Muedi, K.L. Environmental contamination by heavy metals. Heavy Met. 2018, 10, 115–132. [CrossRef]

Miri, S.; Rasooli, A.; Brar, S.K.; Rouissi, T.; Martel, R. Biodegradation of p-xylene—A comparison of three psychrophilic Pseudomonas strains through the lens of gene expression. Environ. Sci. Pollut. Res. 2022, 29, 21465–21479

Mohapatra, B.; Phale, P.S. Microbial degradation of naphthalene and substituted naphthalenes: Metabolic diversity and genomic insight for bioremediation. Front. Bioeng. Biotechnol. 2021, 9, 144

Myslak ZW, Bolt HM (1988) Berufliche Exposition Gegenüber Azofarbstoffen Harnblasenkarzinom-Risiko. Zbl. Arbeitsmed 38: 310-21

Neill C, Hawke FR, Hawkes DL, Lourenço ND, Pinheiro HM, et al. (1999) Colour in textile effluents sources, measurement, discharge consents and simulation: a review. Journal of Chemical Technology and Biotechnology 74: 1009-18.

Niu GL, Zhang JJ, Zhao S, Liu H, Boon N (2009) Bioaugmentation of a 4-chloronitrobenzene contaminated soil with *Pseudomonas putida* ZWL73. Environ Pollut. 157: 763- 771

Panigrahi, S., Velraj, P., & Rao, T. S. (2019). Functional microbial diversity in a contaminated environment and application in bioremediation. In *Microbial diversity in the genomic era* (pp. 359–385). Academic press.

Pankaj Kumar Jain, Vivek Bajpai (2012) Biotechnology of bioremediation review. International journal of environmental sciences 3: 535–549. Link: https://goo.gl/EhLwbz

Pararlioglu NK, Urek RO, Ergum F (2005) biodecolourization of direct blue15 immobilized *Phanerochate chrysosprium*. ProcessBiochem 40: 1923-1929.

Phale, P.S.; Mohapatra, B.; Malhotra, H.; Shah, B.A. Eco-physiological portrait of a novel *Pseudomonas* sp. CSV86: An ideal host/candidate for metabolic engineering and bioremediation. Environ. Microbiol. 2022, 24, 2797–2816.

Phulia V, Jamwal A, Saxena N, Chadha NK, Muralidhar, et al. (2013) Technologies in aquatic bioremediation. 65–91.

Phulpoto. A.H, Qazi. M.A, Mangi. S, Ahmed. S, Kanhar. N (2016) A. Biodegradation of oil-based paint by Bacillus species monocultures isolated from the paint warehouses. Int. J. Environ. Sci. Technol. 13, 125–134

Poirier, I., Hammann, P., Kuhn, L., and Bertrand, M. (2013). Strategies developed by the marine bacterium *Pseudomonas fluorescens* BA3SM1 to resist metals: a proteome analysis. Aquat. Toxicol. 128, 215–232. doi: https://doi.org/10.1016/j.aquatox.2012.12.006

Pongratz R, Heumann KG (1999) Production of mehylated mercury, lead and cadmium by marine bacteria as a significant natural source for atmospheric heavy metals in Polar Regions. Chemosphere 39: 89-102

Raquel S, Natalia G, Luis Fernando B, Maria Carmen M (2013) Biodegradation of high-molecular-weight polycyclic aromatic hydrocarbons by a wood-degrading consortium at low temperatures. FEMS Microbiol. Ecol, 83, 438–449

Rodríguez, A., Castrejón-Godínez, M. L., Salazar-Bustamante, E., GamaMartínez, Y., Sánchez-Salinas, E., Mussali-Galante, P., et al. (2020). Omics approaches to pesticide biodegradation. Curr. Microbiol. 77, 545–563. doi: https://doi.org/10.1007/s00284-020-01916-5

Sanghvi, G.; Thanki, A.; Pandey, S.; Singh, N.K. Engineered bacteria for bioremediation. In Bioremediation of Pollutants; Elsevier: Amsterdam, The Netherlands, 2020; pp. 359–374.

Sharma, P.; Singh, S.P.; Iqbal, H.M.; Tong, Y.W. Omics approaches in bioremediation of environmental contaminants: An integrated approach for environmental safety and sustainability. Environ. Res. 2022, 211, 113102.

Sharma, P.; Singh, S.P.; Parakh, S.K.; Tong, Y.W. Health hazards of hexavalent chromium (Cr (VI)) and its microbial reduction. Bioengineered 2022b.

Soloman PA, Basha CA, Velan M, Ramamurthi V, Koteeswaran K, et al. (2009) Electrochemical degradation of Remazol Black Bdye effluent. CLEAN–Soil Air Water 37: 889–900

Srivastava S. and Thakur I.S. (2006a). Isolation and Process Parameter Optimization of Aspergillus sp. for Removal of Chromium from Tannery Effluent. Bioresource Technology. 97, 1167-1173.
Srivastava S. and Thakur, I.S. (2006b). Evaluation of bioremediation and detoxification potentiality of Aspergillus niger for removal of hexavalent chromium in soil microcosm. Soil Biology and Biochemistry 38, 1904-1911.
Srivastava S. and Thakur IS. (2012) Biosorption and biotransformation of chromium by Serratia sp. isolated from tannery. Environmental Technology. 33 (1):113–122 ISSN: 0959–3330.
Stentiford EI (1996) Composting Control: Principles and Practice, Blackie Academic and Professional
Tálos K, Páger C, Tonk S, Majdik C, Kocsis B, Kilár F, Pernyeszi T (2009) Cadmium biosorption on native Saccharomyces cerevisiaecells in aqueous suspension. Agriculture and Environment 1: 20-30.
Tarekegn, M., Zewdu Salilih, F., & Ishetu, A. I. (2020). Microbes used as a tool for bioremediation of heavy metal from the environment. Cogent Food & Agriculture, 6(1), 1783174.
Tegene B G (2020). Mode of action, mechanism and role of microbes in bioremediation service for environmental pollution management. *Journal of Biotechnology & Bioinformatics Research*, *2*(3), 1-18.
Tegene BG, Tenkegna TA (2020) Mode of Action, Mechanism and Role of Microbes in Bioremediation Service for Environmental Pollution Management. J. Biotechnol. Bioinform. Res. 2, 1–18.
Thakur M, Medintz I L, Walper S.A (2019). Enzymatic bioremediation of organophosphate compounds—progress and remaining challenges. Front. Bioeng. Biotechnol. 7, 289. doi: https://doi.org/10.3389/fbioe.2019. 00289
Tripathi M, Singh DN, Prasad N, Gaur R (2021) Advanced Bioremediation Strategies for Mitigation of Chromium and Organics Pollution in Tannery. In Rhizobiont in Bioremediation of Hazardous Waste; Kumar, V., Prasad, R., Kumar, M., Eds.; Springer: Singapore, pp. 195–215.
Tripathi M, Singh DN, Vikram S, Singh VS, Kumar S (2018) Metagenomic approach towards bioprospection of novel biomolecule(s) and environmental bioremediation. Ann. Res. Rev. Biol, 22, 1–12.
Unuofin JO, Falade AO, Aladekoyi OJ (2021) Applications of microbial laccases in bioremediation of environmental pollutants: Potential issues, challenges, and prospects. In Bioremediation for Environmental Sustainability; Elsevier: Amsterdam, The Netherlands, pp. 519–540.
Vega-Páez JD, Rivas RE, Dussán-Garzón J (2019) High Efficiency Mercury Sorption by Dead Biomass of Lysinibacillussphaercus—New Insights into the Treatment of Contaminated Water. Materials , 12, 1296.
Vidali M (2001) Bioremediation. An overview. Pure Appl cm 73: 1163-1172
Wang Q, Zhang S, Li Y, Klassen W (2011) Potential Approaches to Improving Biodegradation of Hydrocarbons for Bioremediation of Crude Oil Pollution. Environ Protection J 2: 47–55. Link: https://goo.gl/pyGZ6a
Xu F (1996) Oxidation of phenols, anilines, and benzenethiols by fungal laccases: Correlation between activity and redox potentials as well as halide inhibition. Biochemistry, 35, 7608–7614.
Yang L, Zhao YH, Zhang BX, Yang CH, Zhang X (2005) Isolation and characterization of a chlorpyrifos and 3, 5, 6– trichloro–2–pyridinol degradingbacterium. FEMS Microbiol Lett 251: 67-73.
Yang, J., Kloepper, J. W., & Ryu, C. M. (2009). Rhizosphere bacteria help plants tolerate abiotic stress. Trends in plant science, 14(1), 1-4.
Yunusa YR, Umar ZD (2021) Effective microbial bioremediation via the multi-omics approach: An overview of trends, problems and prospects. UMYU J. Microbiol. Res. , 6, 127–145.

Bioaugmentation for Reducing Plastic Pollution

Manu Shukla, Sanskriti Thapa, and Juhi Gupta

1 Introduction

In an era characterized by escalating concerns over environmental degradation and the detrimental impacts of plastic pollution on ecosystems, the concept of bioaugmentation has emerged as a promising and innovative strategy. As plastic waste continues to accumulate in landfills, water bodies, and even remote natural habitats, the urgent need to address this crisis calls for novel approaches that can alleviate its pervasive effects. Bioaugmentation, a technique rooted in harnessing the power of microbial communities, offers a glimmer of hope in the fight against plastic pollution. Bioaugmentation involves the intentional introduction of specialized microorganisms into polluted environments to enhance the degradation of plastic materials. Unlike traditional methods that primarily focus on mechanical removal or recycling, bioaugmentation capitalizes on nature's inherent ability to break down complex polymers. Microbes with the capacity to enzymatically degrade plastics, such as polyethylene and polypropylene, have been identified and isolated from various sources including soil, marine sediments, and even the guts of certain organisms. By introducing these plastic-degrading microbes into polluted sites, it is possible to accelerate the decomposition process and transform persistent plastics into less harmful substances. This revolutionary approach offers several distinct advantages. Firstly, bioaugmentation presents a targeted and site-specific solution, allowing for the customization of microbial communities based on the specific plastic types present in an environment. Moreover, it operates in harmony with natural processes, minimizing disruptions to ecosystems. Additionally, bioaugmentation holds the potential to reduce the dependence on fossil-fuel-derived plastics, fostering a more sustainable and circular economy.

M. Shukla · S. Thapa · J. Gupta (✉)
Amity Institute of Environmental Sciences, Amity University, Noida 201303, India
e-mail: jgupta2@amity.edu

J. Gupta and A. Verma (eds.), *Microbiology-2.0 Update for a Sustainable Future*,
https://doi.org/10.1007/978-981-99-9617-9_6

However, despite its immense promise, the practical application of bioaugmentation for plastic pollution reduction also presents challenges. Ensuring the effective establishment and survival of introduced microbial communities, addressing potential unintended consequences, and scaling up the process from laboratory settings to real-world conditions require careful consideration. In this pursuit of mitigating plastic pollution through bioaugmentation, ongoing research, multidisciplinary collaboration, and a deep understanding of microbial interactions are paramount. As we navigate the complexities of our modern environmental crisis, embracing innovative strategies like bioaugmentation could pave the way for a cleaner, greener, and more harmonious coexistence with our planet.

2 Bioaugmentation

Bioaugmentation is defined as the technique of bioremediation which involves the addition of microbial cultures to a polluted environment to support the native microbial species in degrading the pollutants and to restoring the quality of the area (Adams et al. 2015). This approach comes into light whenever the native microbial species are unable to decompose the target pollutants. Bioremediation, in turn, refers to the decontamination of polluted soil and water through the usage of microbes (Gupta et al. 2022; Huang and Ye 2020). Since bioremediation deals with multiple forms of environmental degradation in comparison to bioaugmentation, which is mainly used for polluted sites, the former has a much broader scope than the latter. Bio-stimulation is another type of bioremediation in which the environment is physically or chemically modified to make the conditions more favourable which in turn induce the existing microbial population to begin breaking down the pollutants (Huang and Ye 2020). This can be done through the addition of limiting nutrients and electron acceptors such as phosphorus, nitrogen, oxygen, or carbon. Bio stimulation also requires the evaluation of both the natural degradation capacities of the native microbial population as well as the environmental parameters involved in the process (Adams et al. 2015). One significant advantage of this process is that bioremediation is performed by native microbes, which show high adaptability in the subsurface environment and are evenly distributed throughout the subsurface. However, the likelihood of these additives reaching the subsurface microbes is determined by the local geology of the area. If the subsurface is tight and impermeable, the likelihood of these additives spreading throughout the affected area will be quite low (Adams et al. 2015). The presence of fractures within the subsurface results in uneven distribution of the additives since they will only flow through the pathways created by the fractures. Further, the addition of nutrients may also facilitate the growth of heterotrophic bacteria, thus creating unnecessary competition for the native microbial species.

Depending on the type of microbes being used, bioaugmentation can be divided into two types: indigenous and exogenous bioaugmentation. If the microorganisms used are indigenous, i.e., those which can be found and isolated from the polluted site, the process is known as indigenous bioaugmentation and in case of non-availability of indigenous microbes or lack of metabolic capability of the indigenous variety, exogenous or non-native microbes are used to degrade the pollutants and the process is known as exogenous bioaugmentation. For bioaugmentation to be successful, the foreign microbes must be able to effectively compete with the native microbes, survive in an alien environment, and maintain their genetic stability and viability during storage. Some of the factors affecting microbial growth in bioaugmentation are: the chemical structure and concentration of the pollutant, the availability of the contaminant to the microbes, the size and nature of the microbial population and the physical environment.

3 Plastic

3.1 Introduction

The various kinds of synthetic or semi-synthetic polymers are known as plastics. Plastics are commonly obtained from non-renewable energy sources such as natural gas and petroleum (Brydson 1999). Between 1950 and 2017, it has been estimated that more than 9.2 billion tonnes of plastic have been produced (Brydson 1999).

3.2 Structure of Plastics

Plastics are essentially synthetic polymers, i.e., they have been created entirely by humans. As per the definition given by the IUPAC (International Union of Pure and Applied Chemistry), a polymer can be defined as a substance which is made up of macromolecules, which in turn consist of multiple repetition of units and possess very high molecular mass (Jansen 2016). However, for synthesizing a polymer in the first place, a monomer is required. It may be defined as a molecule which links with another monomer molecule by covalent bonds which are formed through the binding sites available on it. The monomer must be converted into a repeating unit to create the polymer chain (Jansen 2016). The part of the polymer which produces the entire polymer chain upon repetition (except for the end groups) is known as the repeating unit or repeat unit. During the process of polymerization, the repeat unit is formed when the carbon- carbon double bond in the monomer is converted into a carbon–carbon single bond, opening two new vacancies (Jansen 2016). These vacancies are filled up by the bonding of more repeat units which together make up the entire polymer chain. Several thousand repeating units constitute a polymer chain.

There are two types of chains in a polymer: the main chain and the side chain (Jansen 2016). As per the IUPAC, that straight chain to which all the other chains (long or short or both) may be regarded as hanging downwards is called the main chain or the backbone. The long or short branches of the main chain which influence the properties of the plastic are referred to as the side chains (Jansen 2016). The side chains are attached to the monomers before the process of polymerization is carried out.

3.3 Properties of Plastics

One of the main properties of plastic is its plasticity, i.e., the ability of a solid material to become permanently deformed once a certain force is applied upon it. Plastic objects of various shapes can be created by virtue of this property (Brydson 1999). They are in popular use mainly because of their low production cost, ability to last over a long time period, flexibility, and low weight (Gupta et al. 2021a).

4 Various Kinds of Plastics

S. no	Kind of plastic	Characteristics	Examples
1	Thermoplastics	Those plastics whose shape can be changed again and again since they don't decompose upon heating are called thermoplastics	Polyethylene (PE), Polypropylene (PP), Polystyrene (PS), and Polyvinyl Chloride (PVC)
2	Thermosets/thermosetting plastics	a. Those plastics whose shape cannot be changed through melting since they decompose upon heating are known as thermosetting plastics b. It is an irreversible chemical process	The vulcanization of rubber changes natural rubber(polyisoprene), which is initially a sticky material, into hard and dry rubber
3	Biodegradable plastics	a. Those plastics that decompose in the presence of sunlight, UV (Ultraviolet) rays, moisture, microbes such as bacteria, higher organisms such as waxworms and mealworms and by wind currents, are known as biodegradable plastics b. The decomposition of plastic in the presence of oxygen is called aerobic biodegradation, and in the absence of oxygen is termed anaerobic biodegradation	Poly Hydroxy Butyrate (PHB)

(continued)

(continued)

S. no	Kind of plastic	Characteristics	Examples
4	Bioplastics	a. Those plastics which are synthesized using renewable sources are known as bioplastics b. Vegetable fats and oils, corn starch, straw, recycled food waste, etc. are some examples of the renewable sources from where we can derive bioplastics. c. Bioplastics can be derived by directly processing natural biopolymers such as starch, cellulose, chitosan, and soy protein d. In the second method, oils and fats from plants or animals are also used in the synthesis of bioplastics. e.Anaerobic digestion of sugars and lipids also gives us bioplastics	Bio-PET(Polyethyleneterepthalate), Bio-polyethylene, etc.

4.1 *Types of Plastic*

S. no.	Type of plastic	Properties and uses	Examples
1	Commodity plastics	a. Those plastics which possess weak mechanical properties and are produced in very large quantities are known as commodity plastics b. The production cost for such plastics is quite low and these are mostly used for applications such as packaging, making disposable plates and cups, photographic and magnetic tape, clothing, medical trays, and seeding trays, etc	Polyethylene terephthalate (PET), High Density Polyethylene (HDPE), Polyvinyl chloride(PVC), Low Density Polyethylene(LDPE), Polypropylene(PP), Polystyrene(PS), Polyurethanes(PURs)

(continued)

(continued)

S. no.	Type of plastic	Properties and uses	Examples
2	Engineering plastics	a. Those plastics which possess superior characteristics compared to other plastic materials and can provide high performance are known as engineering plastics b. Since they can tolerate high mechanical stress, remain stable over a wide temperature range, and resist chemical or physical changes in the environment, they are mostly used for structural applications c. They are used to manufacture vehicle parts, building and construction material, and machine parts	Acrylonitrile Butadiene Styrene (ABS), Polycarbonate (PC), etc.
3	High performance plastics	These are those plastics which meet higher requirements than engineering and commodity plastics They are created through the combination of stabilizers and reinforcing materials such as glass and carbon fibres They can operate at temperatures up to 149 Degree Celsius These plastics also display resistance to a wide range of chemicals	Polyphenylene sulphide (PPS), Polyamide imide (PAI), Teflon or PTFE(Polytetrafluoroethylene), PEEK(Polyetheretherketone),etc.

Source Hudson and Hudson (1995); FOY (1969); Platt (2003)

5 Plastic—A Bane for the Environment

The most used raw material for manufacturing plastic materials is petroleum. When propylene, a component of petroleum, is heated in the presence of a catalyst, the monomers are joined together through the carbon–carbon single bond and the polymer polypropylene is formed (Tokiwa et al. 2009). Evolution has created organisms which are capable of breaking down only certain types of bonds which can be found very often in nature Such organisms are capable of breaking down polysaccharides (including sugar) and wood (Tokiwa et al. 2009). Since a large amount of energy is involved in creating the carbon–carbon single bonds in polypropylene, the naturally occurring microorganisms find it difficult to decompose polypropylene (Tokiwa et al. 2009). In contrast, peptide bonds, which are found in proteins, frequently occur in nature and are easier to break down since relatively lower amounts of energy are involved during their synthesis (Tokiwa et al. 2009). Thus, plastics take a very long time to decompose as compared to organic matter. Plastics cannot be synthesized using peptide bonds to enhance their biodegradability since the shelf life of such plastics is very short.

5.1 Causes of Plastic Pollution in the Marine Ecosystems

The build-up of plastic waste in the environment which damages environmental health and human health is known as plastic pollution. There are two main sources of plastic pollution: land- based sources and water-based sources. Land-based sources include freshwater from rivers, the domestic activities, tourism, and harbor operations (Thushari and Senevirathna 2020). The coastal zone has become a source of land based plastic pollution due to the large number of communities living here as well as the extensive industrialization and urbanization of this zone. Most of the plastic waste generated ends up in the rivers and streams or is leached into the groundwater since the wastewater treatment systems are incapable of removing a large volume of plastic waste from the sewage water. The rivers and streams, in turn, carry all the plastic waste into the oceans. Plastic waste may also build up in coastal ecosystems in case the lotic freshwater ecosystems carrying the plastic debris possess higher flow rates. For example, as many as 2 billion plastic fragments enter the ocean every three days due to the aggregation of plastic waste in the ocean system around California (Thushari and Senevirathna 2020).

Tourism and recreational activities are also one of the main sources of plastic pollution on beaches, contributing more than 60% of the total beach debris on the eastern coast of Thailand (Thushari and Senevirathna 2020). Careless disposal of items such as plastic bottles by many tourists leads to the buildup of a large volume of plastic waste on beaches. Upon entering the sea, the larger plastics break down into microplastics. These types of plastics have also been reported off the northeastern coast of Brazil (Thushari and Senevirathna 2020). Harbor operations also contribute to plastic pollution through the careless disposal of monofilament and nylon fishing nets.

Extreme climatic conditions such as storms and hurricanes increase the rate of buildup of plastic waste in the oceans (Cheung et al. 2019). This is because during a storm, most of the coastal plastic debris is swept into the sea due to the strong action of waves, which increases the concentration of microplastics in the ocean. These plastics are carried over vast distances by ocean currents, thus enabling them to become widely distributed. For instance, stormy weather increases the density of microplastics in the water from California by six times as compared to normal conditions (Thushari and Senevirathna 2020).

Ocean-based sources of plastic pollution include commercial fisheries, waste disposal at sea, navigation-related activities and fish culture or aquaculture related operations (Thushari and Senevirathna 2020). During fishing operations, the careless disposal of damaged fishing nets also leads to plastic pollution in the oceans. Marine vessels dispose of plastic waste into the ocean at the rate of 6.5 million tons per year, as reported in the early 1990s (Thushari and Senevirathna 2020).

Classification of the types of plastics found in the ocean based on the variations in size:

S. no	Type of plastic	Size range
1	Mega plastics	More than 1 m
2	Macroplastics	Less than 1 m
3	Mesoplastics	Less than 2.5 cm
4	Microplastics	Less than 5 mm

Classification of the types of plastics based on range of lengths:

S. no.	Type of plastic	Range of length
1	Megaplastics	More than 100 mm
2	Macroplastics	20–100
3	Mesoplastics	5–20 m
4	Microplastics	Less than 5 mm

When larger plastics are decomposed in the presence of physical, chemical, and biological processes such as UV radiation and wind, small particles called microplastics are formed which are generally less than 5 mm in size (Thushari and Senevirathna 2020). Microplastics are the most abundant type of plastic litter found in the oceans and their distribution in the water column is determined by their densities. The ones with lower densities will float on the surface and those with higher densities will sink to the bottom of the ocean (Thushari and Senevirathna 2020). Examples of low density microplastics are polypropylene and polyethylene and examples of high density microplastics are polystyrene, Polyvinyl Chloride (PVC), polyamide, and Polyethylene Terephthalate (PET), etc. The presence of microplastics is also determined by the salinity of water since there is a high likelihood of microplastic formation in coastal and marine ecosystems.

S. no.	Type of microplastic	Characteristics and examples
1	Primary microplastic	a. Those plastic particles which are manufactured to possess a microscopic size and are used for a particular purpose are known as primary microplastics b. These are utilized during the manufacture of cosmetics such as cleansers, shower gel, medicines, etc Ex: Polyethylene and polypropylene microplastics can be found in cosmetics
2	Secondary microplastic	a. When larger plastics decompose on land or water, the smaller particles formed are known as secondary microplastics b. larger plastics break down faster in beach ecosystems due to the higher amounts of sunlight and oxygen c. These plastics are further subjected to the action of waves which break them down into even finer particles and finally these microplastics are formed

Adapted and modified from: Abbasi et al. (2018); Thushari and Senevirathna (2020); Guo and Wang (2019a, b); Acot et al. (2022)

5.2 *Impacts of Plastic Pollution on the Marine Ecosystem*

S. no.	Mode of adverse effect	Examples
1	Entanglement	a. Entanglement occurs when individual organisms get intertwined with fishing nets or the plastic rope in fishing gears b. It has been reported that more than 30,000 individuals from close to 250 marine species suffered from entanglement related issues

(continued)

(continued)

S. no.	Mode of adverse effect	Examples
2	Ingestion	a. Birds can easily consume carelessly disposed fishing hooks which can build-up within their stomach, resulting in their deaths b. More than 80% of the individuals of the commercially valuable lobster species, *Nephrops norvegicus*, contained plastic filaments, which can lead to biomagnification c. The build-up of plastic particles of more than 80 micro meter in size within the epithelial cells of the digestive tubules of the invertebrates result in many inflammatory issues d. High-density microplastics lying on the ocean floor are consumed by the benthic organisms
3	Behave as vectors for toxic substances	a. Bisphenol-A(BPA), flame retardants, and antibiotics, Polychlorinated Biphenyls (PCBs), organochlorine pesticides, Dioxins/Polychlorinated Dibenzo-p-Dioxins (PCDDs), furans b. Flame retardants and phthalates are toxic to fish, mollusks, and other mammals c. Also adsorb antibiotics such as Sulfamethoxazole (SMX or SMZ) Sulfamethazine (SMT) on their surface d. The presence of antibiotics in water stimulates the rapid growth of Antibiotic Resistant Bacteria (ARB) which pose danger to humans and aquatic life e. Microplastic also show a high tendency for combining with metal compounds
4	Transport of invasive species	a. Certain organisms such as bivalves, crustaceans, annelids, echinoderms, gastropods, coelenterates, bryozoans, insects, sponges, polychaetes, seagrasses, seaweeds, etc. and microbes were reported to be growing on plastic waste b. Creates competition between the native species and invasive species and completely disturbs the native ecosystem's composition, structure, and equilibrium c. Can grow on plastic debris such as microplastics, fishing gear parts, nets, ropes, and packaging material

(continued)

(continued)

S. no.	Mode of adverse effect	Examples
5	Coral reefs	a. The coral reef and seagrass ecosystems in the benthic zone are damaged by the buildup of macro and mega plastic debris on the ocean floor b. It has been reported that abandoned fishing equipment negatively impacts nearly 70 percent of the coral reefs c. Biodiversity levels fall since coral reefs act as the support systems of many marine species
6	Reduced photosynthesis	The amount of light entering the water is significantly reduced by the plastic waste floating on the surface Since plants are unable to perform photosynthesis in the absence of adequate sunlight, the concentration of Dissolved Oxygen (DO) in the water body is drastically reduced This negatively impacts primary productivity and consequently entire marine food chains

Adapted and modified from: Hossain et al. (2020); Theil et al. (2018); Guo et al. (2019a, b); Guo and Wang (2019a)

5.3 *Demonstration of the Effectiveness of Bioaugmentation in Degrading Plastic Waste*

A field study was conducted by Naveen Mohan and R. Usha with the objective of determining the effectiveness of bioaugmentation in reducing plastic pollution in a particular area.

Methodology of the Field Study

The field area was divided into two equal parts as follows: the control land and the augmented land. The nature and texture of both these parts was also quite similar. For this study, the authors chose Actinomycetes, a bacterial strain which is incapable of movement but possesses many filaments (Mohan and Usha 2018) This strain was chosen since it can break down plastic waste very efficiently (Mohan and Usha 2018).

1. Low Density Polyethylene (LDPE) sheets were collected and each of them were cut into sizes of 5 × 2 cm (Mohan and Usha 2018)
2. Six small wells, each having a depth of 3–5 cm were dug in both control and augmented land.
3. The LDPE sheets were inserted into each of the wells.

4. No exogenous microbes are added to the control land whereas the Actinomycetes strain (DSR 2) is added equally to each of the six wells in augmented land with rice hull acting as the carrier for the foreign microbes (Mohan and Usha 2018)
5. The initial weight of all the LDPE strips before the addition of Actinomycetes strain to the augmented land is noted down.
6. Each of the wells are labelled as 15, 30, 45, 60, 75, and 90 in both control and augmented land (Mohan and Usha 2018)
7. The weight of the strips from the corresponding wells in both control and augmented land was recorded every fifteen days (up to 90 days) after being washed with 70% ethanol and being air-dried. The reduction in weight is calculated and compared.

   The formula for calculating weight loss is as follows:

   Weight loss % = Initial weight − final weight/initial weight × 100
8. The pH value of the soil near these wells was recorded after every 15 days and compared with its respective initial value for both control and augmented land (Mohan and Usha 2018).

   Weight loss in non-augmented land/reference land

Initial weight (mg)	After 15 days	after 30 days	After 45 days	after 60 days	after 75 days	After 90 days
25	25	24.9	24.6	24.3	24.1	24

Adapted and modified from: (Mohan and Usha 2018)

Weight loss on land containing Actinomycetes strain

Initial WT (mg)	After 15 days	After 30 days	After 45 days	After 60 days	After 75 days	After 90 days
25	25	24.6	24	23.7	23	22.8

Adapted and modified from: (Mohan and Usha 2018)

Variation In pH of non-augmented land/reference land

Initial pH	After 15 days	After 30 days	After 45 days	After 60 days	After 75 days	After 90 days
6.8	6.8	6.7	6.7	6.6	6.6	6.5

Adapted and modified from: (Mohan and Usha 2018)

Variation in pH of land containing Actinomycetes strain

Initial pH	After 15 days	After 30 days	After 45 days	After 60 days	After 75 days	After 90 days
6.8	6.8	6.7	6.5	6.5	6.2	5.7

Adapted and modified from: (Mohan and Usha 2018)

Percentage of degradation for the indigenous microbes (non-augmented land)

After 15 days	After 30 days	After 45 days	After 60 days	After 75 days	After 90 days
0	0.4	1.6	2.8	3.6	4

Adapted and modified from: (Mohan and Usha 2018)

Percentage of degradation for the Actinomycetes strain (augmented land)

After 15 days	After 30 days	After 45 days	After 60 days	After 75 days	After 90 days
0	1.6	4	5.2	8	8.8

Adapted and modified from: (Mohan and Usha 2018)

6 Results

It was observed that there was no significant reduction in weight of the LDPE strips for the control land since only native microbes were present in that land which possess lower efficiency. The highest percentage of degradation of the native microbes was found to be 4% (Mohan and Usha 2018). On the other hand, there was a significant reduction in the weight of the LDPE strips for the augmented land since the Actinomycetes strain possessed a higher capacity to break down the plastic strips as compared to the native microbes. The highest percentage of degradation of the Actinomycetes strain was found to be 8.8% (Mohan and Usha 2018). As per the results obtained for control land, there wasn't any noticeable change in the pH of the soil around the wells (pH 6.8–6.5) (Mohan and Usha 2018). On the other hand, there was a significant decrease in pH of the soil for augmented land (pH 6.8–5.7) which verifies that the Actinomycete strain utilizes the LDPE strips as its only carbon source (Mohan and Usha 2018).

7 Discussion

The results obtained clearly indicate that microbes decompose plastic by utilizing it as a food source. For the degradation process to occur in a optimal manner, the plastic surface must be first colonized by the microbes. Colonization of plastic surface occurs through the formation of biofilms by the microbes (Mohan and Usha 2018). Since the Actinomycetes strain possessed higher efficiency as compared to the native microbial species, they were added to the soil to support the native microbes and accelerate the rate of decomposition of plastic waste. Thus, this study clearly proves that bioaugmentation can be utilized for decomposing plastic waste in an effective manner.

8 How Microbes Degrade the Plastic Waste

8.1 Biological Aspect

There are different steps involved in the biodegradation of plastic polymers by bacteria which include: Bio-deterioration, Bio-fragmentation, Assimilation and Mineralization.

a. Bio-deterioration:

   It is the degradation that occurs at the surface of the plastic and alters its mechanical, physical, and chemical properties (Dussud and Ghiglione 2014). Abiotic factors also play a significant role in this step as they work in coordination with the microbes in breaking down the plastic and cause the plastic to begin decomposing. Bio-deterioration is initiated when a microbial biofilm starts growing on the plastic surface (Jain et al., 2023). A biofilm is defined as a thick layer of microbes that have come together to create a colony (Dussud and Ghiglione 2014). The composition and structure of plastic, as well as the environmental conditions will determine how the biofilm will develop on the plastic surface. If the plastic being decomposed is water repelling in nature (such as Polyethylene and Polystyrene), then the bacterial surface must also be water-repelling in nature to form a biofilm having high stability. For instance, *Rhodococcus ruber* formed a highly stable biofilm on polyethylene that remained adhered to the surface of the plastic even after 60 days in incubation (Dussud and Ghiglione 2014).

The formation of the microbial biofilm causes physical and chemical deterioration which are as follows:

Physical deterioration: Extracellular Polymeric Substances (EPS) are secreted during the formation of a biofilm that strengthens its cohesion and its adhesion to the plastic surface. Microbes begin growing inside once EPS enters the pores of the plastic, which results in enlargement of the hole and the formation of cracks which weaken the plastic's physical properties (Gupta et al. 2019a, b, 2021b; Dussud and Ghiglione 2014).

Chemical deterioration: Microbial communities of various kinds develop on the plastic surface and biofilms of chemolithotrophic bacteria secrete various inorganic acids to degrade the plastic. For example, *Nitrosomonas* spp. secretes nitrous acid,

Nitrobacter releases nitric acid, and *Thiobacillus* spp. secretes sulphuric acid. On the other hand, biofilms of chemoorganotrophic bacteria secrete many organic acids such as oxalic acid, citric acid, fumaric acid, gluconic acid, glutaric acid, to decompose plastic. The release of acids results in a decrease in pH inside the pores of the plastic, causing the break down to occur in a progressive manner, changing the microstructure of the plastic (Dussud and Ghiglione 2014).

b. Bio-fragmentation:

Plastic polymers cannot be taken through the cell wall as such because of their high molecular weight. Reactions occur at the surface of the polymer due to the secretion of extracellular enzymes or exoenzymes by the microbes. To initiate the break down of plastic, the microbes require an imbalance in electric charge within the polymer chain. In a normal situation, there is no imbalance of electric charge in the structure of the polymer. Therefore, oxygenase enzymes are released by bacteria to destabilize this electric charge by adding oxygen atoms to the carbon chains (Dussud and Ghiglione 2014). For instance, Mono-oxygenase enzyme adds one oxygen atom to form the alcohol group and the Dioxygenase enzyme adds two oxygen atoms to form the peroxyl group. The decomposition of plastic is thus accelerated since both functional groups are easier to biodegrade. Once the carboxylic group is formed, lipases and esterases catalyze the rest of the chemical reactions. Below are some of the bacterial species which can break down plastic polymers:

Pseudomonas is capable of breaking down polyethylene, Polyvinyl Chloride (PVC), Polyhydroxy Butyrate (PHB), poly(3-hydroxy-co-3-mercaptopropionate) and poly(3-droxypropionate). Bacillus brevis is capable of breaking down only polycaprolactone (PCL). Streptomyces can decompose PHB, poly(3-hydoxybutyarate-co-3-hydroxyvalerate), and starch or polyester (Dussud and Ghiglione 2014). *Ochrobactrum* TD can also break down PVC. PHB, PCL, and PLA can be decomposed by 39 bacterial strains belonging to the classes Firmicutes and Proteobacteria. Other bacterial species with high plastic degrading capacities include *Staphylococcus* sp., *Streptococcus* sp., *Diplococcus* sp., *Micrococcus* sp., and *Moraxella* sp.

c. Assimilation:

The intake of digested food molecules by an organism into its cells is known as assimilation. Specific carriers are utilized by monomers to cross the cell wall and cytoplasmic membrane and enter the microbial cell. Oxidation of monomers takes place inside the cell through catabolism to generate energy, cell structure and new biomass. Based on the ability of microbes to grow in aerobic or anaerobic conditions, there are three pathways available to perform catabolism of the monomer: aerobic respiration, anaerobic respiration, and fermentation (Dussud and Ghiglione 2014) Many secondary metabolites (SMs) are produced, but if the microbes are unable to decompose these compounds or do not need to store it, they are excreted from the cell. Another cell may assimilate and break down the secondary metabolites or they may remain in the environment as such.

d. Mineralization:

This process involves the elimination of completely oxidized metabolites by microbes which are generated from the complete breakdown of primary and secondary metabolites. Some examples of completely oxidized metabolites include carbon dioxide (CO_2), nitrogen gas (N_2), methane (CH_4), and water (H_2O) (Dussud and Ghiglione 2014).

8.2 Use of Biotechnology in Solving Plastic Problem

In 2016, *Ideonella sakaiensis,* a bacterium which consumes plastic for food, was discovered in Sakai, Japan among the plastic bottles outside a recycling facility (Yoshida et al. 2016). It was found to be capable of decomposing Polyethylene Terephthalate (PET) plastic. It possessed two digestive enzymes, hydrolyzing PET and PETase. The plastic is degraded by these enzymes into two monomers, i.e., terephthalic acid and ethylene glycol (Yoshida et al. 2016). Energy is released through the further degradation of monomers which are utilized for bacterial growth. Many genetic scientists worked with *Ideonella sakaiensis* to improve its efficiency. One such project involved the conversion of the *E. coli* bacterium into a PETase factory, the steps for which are given below:

- The gene responsible for the synthesis of PETase enzyme is extracted from *Ideonella sakaiensis* (Cui et al. 2021)
- The gene for PETase enzyme is then inserted into a plasmid backbone, which is a circular piece of DNA.
- Heat or electricity is used to insert this plasmid backbone into *E. coli* bacteria.
- Collection and purification of PETase enzymes is done after they are produced by genetically engineered *E. coli* bacteria (Cui et al. 2021).

9 Types of Microbes that Degrade the Plastic

Plastic pollution has emerged as one of the most pressing environmental challenges of our time. Traditional plastic disposal methods, such as landfilling and incineration, contribute significantly to environmental degradation and are unsustainable in the long run. However, recent advances in biotechnology have introduced a promising solution known as bioaugmentation, wherein microorganisms are utilized to degrade and break down plastics (Gupta et al. 2020). This report delves into the concept of bioaugmentation by various microbes that degrade them and explores its potential as a sustainable approach for plastics waste management and environmental protection.

Plastics are synthetic polymers widely used in various industries due to their versatility, durability, and low cost. Unfortunately, their long-lasting nature has led to the accumulation of plastic waste in landfills, oceans, and other natural ecosystems, causing detrimental impacts on biodiversity and ecosystem health. Bioaugmentation, a biotechnological strategy, involves introducing specialized microorganisms into the environment to enhance the natural breakdown of plastics.

9.1 Mechanisms of Plastic Degradation by Microorganisms

Microorganisms play a crucial role in the degradation of plastics through enzymatic actions. Some microorganisms possess unique enzymes, such as esterase, lipases, and proteases, which enable them to break down various types of plastics. The most notable plastic-degrading microorganisms belong to the bacterial genera Pseudomonas, Bacillus, and Ide Onella.

Types of Plastics Subject to Bioaugmentation

Bioaugmentation techniques are versatile and can be applied to various types of plastics, including polyethylene (PE), polypropylene (PP), polyethylene terephthalate (PET), and polystyrene (PS). Each plastic type requires specific microbial strains, enzymes, and environmental conditions to promote efficient degradation.

Biodegradation Pathways of Plastics

The biodegradation pathways of plastics vary depending on the microbial species involved and the plastic type. Generally, microorganisms produce extracellular enzymes that depolymerize the plastics into smaller fragments. These fragments are then assimilated and utilized as a carbon source by the microorganisms, leading to the complete breakdown of plastics.

Factors Influencing Bioaugmentation Efficiency

The efficiency of bioaugmentation in degrading plastics is influenced by several factors, including environmental conditions (e.g., temperature, pH, and oxygen availability), microbial population dynamics, plastic surface area, and the presence of other pollutants. Optimizing these factors is essential to enhance the overall effectiveness of the bioaugmentation process.

9.2 Various Types of Bioplastics

- Photodegradable bioplastics

In this type of plastic, the groups that are connected to the polymer backbone are very light sensitive. By giving, long time exposure of the UV radiation, the polymeric structure can be fragmented. In case the radiation supply, stops then fragmentation is not possible. Landfills they lack sunlight, plastics in landfill are non-degradable,

Artificial photo-degradation and can lead to the release of toxic volatile organic compounds (VOCs), these are potentially hazardous and are associated with the environmental degradation of plastic remains (Elahi et al. 2021).

- Bio-based bioplastics

Types of plastics from which we can obtain 100% of carbon using renewable resources, like forest, resources from agricultural sector are known as bio-based plastics. Corn, Starch, soybean and are the examples of these resources (Elahi et al. 2021).

- Compostable bioplastics

In composting process, the bioplastic requires a specific order to break down, while in whereas biodegradable products it breaks down naturally, the plastics are decomposed, biologically and leave no toxic material. The rate of composting, of the bioplastic is similar to the other compostable material (Elahi et al. 2021).

10 Microorganisms that Degrade Plastic

Bacterial microorganisms	Fungal microorganisms	Marine microorganisms
Pseudomonas spp. Pseudomonas species are among the most extensively studied bacteria used in bioaugmentation. They secrete a range of enzymes, including lipases and esterase, which effectively degrade various plastics like polyethylene and polypropylene. These microorganisms thrive in both terrestrial and marine environments, making them versatile candidates for plastic waste management	Aspergillus spp. Several species of Aspergillus fungi have demonstrated plastic-degrading capabilities. These fungi produce extracellular enzymes, such as esterases and laccases, which play a pivotal role in degrading plastics like polyethylene and polystyrene. Aspergillus species are well-suited for bioaugmentation in terrestrial habitats and composting systems	Marine Bacteria Marine environments are particularly susceptible to plastic pollution, making the bioaugmentation of plastics in these ecosystems essential. Some marine bacteria, such as Alteromonas, Vibrio, and Halomonas species, have demonstrated the capability to degrade marine plastics like polyethylene. These microorganisms have adapted to the harsh marine conditions, making them ideal candidates for plastic waste management in marine habitats

(continued)

(continued)

Bacterial microorganisms	Fungal microorganisms	Marine microorganisms
Bacillus spp. Bacillus species are Gram-positive bacteria known for their robust survival capabilities in challenging environments. Their capacity to produce a diverse array of enzymes, such as cutinases and proteases, enables them to break down different plastic types. Bacillus strains are particularly efficient in degrading polyethylene terephthalate (PET) plastics, commonly used in beverage containers	Trichoderma spp. Trichoderma fungi possess a diverse set of enzymes, including cellulases and hemicellulases, which contribute to their plastic-degrading abilities. They have shown potential in degrading agricultural films and other biodegradable plastics, making them relevant for bioaugmentation in specific applications	Marine Fungi Marine fungi also play a role in plastic degradation in marine environments. Certain marine fungal species, such as Zalerion maritimum, produce enzymes that assist in breaking down plastics like polyethylene. Bioaugmentation with marine fungi holds promise for combating plastic pollution in oceans and coastal regions (Mohan and Usha 2018)
Ideonella sakaiensis Ideonella sakaiensis is a newly discovered bacterium that has garnered significant attention due to its unique ability to break down PET plastics. This microorganism produces a PET-degrading enzyme, PETase, capable of efficiently cleaving PET into its constituent monomers. I. sakaiensis offers promising prospects for the bioaugmentation of PET contaminated environments	Algae: Certain marine algae, such as Chlorella vulgaris and Ulva sp., have demonstrated the potential to aid in plastic bioaugmentation. Through a process known as enzymatic-assisted biodegradation, these algae secrete enzymes capable of breaking down certain types of plastics found in marine environments	

10.1 Synergistic Microbial Consortia

In some cases, the combined action of different microorganisms in a consortium can enhance plastic degradation. Synergistic microbial consortia involving bacteria and fungi have been investigated for their potential in bioaugmentation strategies. These consortia can utilize a broader range of plastic types and offer better adaptability to diverse environmental conditions.

In a review paper on study "Plastics degradation by microbes: A sustainable approach summarizes that Various types of degradation processes can lead to the breakdown of plastics:

- Photo-oxidative degradation: Plastics can be damaged by UV radiation, leading to the formation of reactive groups, and breaking of C–C bonds (Elahi et al. 2021)
- Thermal degradation: High temperatures can cause accidental or depolymerization reactions in plastics, resulting in the breakdown of polymers into smaller monomers (Elahi et al. 2021)
- Ozone degradation: Ozone present in the atmosphere can react with polymers, leading to the formation of carbonyl and unsaturated carbonyl products.
- Mechanochemical degradation: Polymer chains can break down under mechanical stress or ultrasonic irradiation, leading to changes in molecular weight and double bond concentration.
- Catalytic degradation: Plastics can be transformed into hydrocarbons through catalytic processes, providing an opportunity to achieve desired products (Elahi et al. 2021)

Biodegradation of plastics occurs through the action of microorganisms, which attach to the polymer surface and secrete enzymes to break down the polymers into smaller molecules. This process involves depolymerization and mineralization, leading to the formation of inorganic end products. Analytical procedures, such as gas chromatography/mass spectrometry and micro-Fourier transform infrared spectroscopy, are used to assess the biodegradability of polymeric materials (Elahi et al. 2021).

Overall, understanding the different degradation processes and biodegradation mechanisms is crucial for addressing plastic pollution and finding effective solutions to tackle this environmental issue. Plastics are widely used materials and their accumulation in the environment poses a serious threat to ecosystems and human health. Various degradation processes can lead to the breakdown of plastics, such as photo-oxidative degradation, thermal degradation, ozone degradation, mechanochemical degradation, and catalytic degradation. Microbial degradation of plastics involves several steps, including biodeterioration, bio-fragmentation, mineralization, and assimilation. Microorganisms can attach to plastic surfaces and secrete enzymes to break down the polymers into smaller molecules, which are then utilized for energy production and biomass growth. Polyethylene (PE) is one of the most common plastics, and microbial degradation of PE has been studied. Certain microorganisms, such as bacteria and fungi, have been found to have the ability to degrade PE, although the process is relatively slow. Researchers have identified various enzymes involved in PE degradation, and the biodegradation efficiency of different organisms has been assessed.

Biodegradable plastics, including oxo-biodegradable and hydro-biodegradable plastics, are being explored as potential solutions to the plastic pollution problem. Microbes play a role in the degradation of these biodegradable plastics as well. To enhance the effectiveness of microbial plastic degradation, it is essential to investigate potential microbial strains, maintain optimum conditions for microbial activity, and consider the use of microbial consortia and enzymes. The successful implementation of microbial degradation strategies can offer a sustainable solution to plastic pollution and contribute to environmental preservation.

11 Advantages of Reducing Plastic Waste Using Bioaugmentation

Bioaugmentation of plastics using microorganisms offers several significant advantages that make it a promising approach for plastic waste management and environmental protection:

(a) Biodegradation of Diverse Plastics:

One of the primary advantages of bioaugmentation is its ability to degrade a wide range of plastic types. Different microorganisms possess specific enzymes capable of breaking down various plastics, such as polyethylene, polypropylene, polystyrene, and PET. This versatility allows bioaugmentation to address the diverse plastic pollution problem.

(b) Environmentally Friendly:

Compared to traditional plastic disposal methods like incineration or landfilling, bioaugmentation is an eco-friendlier approach. It utilizes natural processes of microbial degradation, reducing the need for energy-intensive and polluting practices. As a result, it minimizes the carbon footprint associated with plastic waste management.

(c) Reduction of Plastic Accumulation:

By enhancing the biodegradation of plastics, bioaugmentation helps to reduce the accumulation of plastic waste in landfills, oceans, and other ecosystems. This, in turn, mitigates the harmful impacts of plastic pollution on wildlife, marine life, and the overall environment.

(d) Circular Economy Promotion:

Bioaugmentation aligns with the principles of a circular economy, where waste materials are reused or recycled back into the production cycle. By breaking down plastics into biodegradable by-products, it allows the converted materials to be utilized in various applications, reducing the demand for new plastic production.

(e) Versatility in Environmental Applications:

Bioaugmentation can be applied in various environmental settings, including landfills, contaminated soil, wastewater treatment plants, and marine environments. This adaptability makes it a versatile tool for managing plastic pollution in diverse ecosystems.

(f) Enhanced Efficiency:

Microorganisms employed in bioaugmentation are specifically selected for their plastic-degrading capabilities. As a result, they often exhibit higher efficiency in breaking down plastics compared to naturally occurring microbial populations. This improved efficiency accelerates the biodegradation process, reducing the persistence of plastic waste in the environment.

(g) Potential for Synergistic Interactions:

In some cases, bioaugmentation involves the use of microbial consortia, which are combinations of different microorganisms. These consortia can exhibit synergistic interactions, where the combined action of multiple microorganisms enhances plastic degradation. Synergy can lead to faster and more complete degradation of plastics than individual microorganisms alone.

(h) Low Cost and Scalability:

The implementation of bioaugmentation is relatively cost-effective compared to other advanced waste management technologies. Microorganisms can be cultured and propagated on a large scale at a reasonable cost. Moreover, the scalability of the process allows it to be applied in both small-scale and large-scale plastic waste management initiatives.

(i) Potential for Genetic Modification:

Advancements in biotechnology offer the potential to genetically modify microorganisms to enhance their plastic-degrading abilities further. Through genetic engineering, scientists can optimize existing plastic-degrading enzymes or introduce new ones, making bioaugmentation even more effective in tackling plastic pollution.

(j) Non-Toxic and Safe:

Bioaugmentation does not involve the use of harmful chemicals or pollutants, making it a non-toxic and safe method for plastic waste management. Unlike some conventional waste treatment methods, bioaugmentation does not produce harmful by-products that could negatively impact the environment or human health.

12 Disadvantages of Reducing Plastic Waste Using Bioaugmentation

The use of plastic waste augmentation, also known as plastic waste recycling or upcycling, does have some disadvantages despite its positive intentions, which involves using microorganisms to break down plastic waste, is a promising approach to reducing plastic pollution. However, it also comes with several disadvantages and challenges. Here are some of the main disadvantages associated with plastic waste augmentation.

(a) Limited Range of Plastics: Bioaugmentation is typically effective on certain types of plastics, particularly those that are more easily biodegradable, like some polyesters. However, it may not be effective on a wide range of plastic types, including those commonly used in packaging, like polyethylene and polypropylene.

(b) Slow Decomposition: While bioaugmentation can accelerate the natural degradation of plastics, the process is still relatively slow compared to the rate at which plastic waste is generated. This could lead to accumulation of plastic waste in the environment before it can be adequately broken down.
(c) Environmental Conditions: The effectiveness of bioaugmentation depends on various environmental conditions, such as temperature, moisture, and pH levels. Maintaining these conditions at optimal levels for microbial activity can be challenging, especially in real-world outdoor settings.
(d) Unintended Consequences: Introducing new microorganisms into an ecosystem through bioaugmentation can have unintended consequences. These microorganisms might interact with native species in unpredictable ways, potentially disrupting the balance of the ecosystem.
(e) Resource Requirements: The process of bioaugmentation requires a consistent supply of suitable microorganisms and nutrients to support their growth and activity. This demand for resources could potentially strain other aspects of resource management.
(f) Regulatory Concerns: The release of genetically modified organisms or non-native microorganisms into the environment raises regulatory concerns. There might be uncertainty and debate about the long-term effects and potential risks associated with these interventions.
(g) Cost and Scale: Implementing bioaugmentation on a large scale can be costly due to the need for research, development, production, distribution, and monitoring of microorganisms. Scaling up this approach to address the global plastic waste problem would require significant investment.
(h) Lack of Infrastructure: Bioaugmentation requires a certain level of infrastructure to be in place, such as facilities for cultivating and dispersing microorganisms. This could be a barrier in regions with limited resources or less developed waste management systems.
(i) Public Perception and Acceptance: Introducing genetically modified or non-native microorganisms into the environment might face resistance from the public due to concerns about safety, ethics, and potential unintended consequences.
(j) Complexity of Plastic Mixtures: Plastic waste often consists of mixtures of different polymers and additives, which can make it challenging for microorganisms to effectively degrade them. Bioaugmentation might not provide a comprehensive solution for such complex plastic waste streams (Wierckx et al. 2018).

13 Conclusion

Bioaugmentation has emerged as a promising new technique for reducing the problem of plastic pollution. It is the technique of introducing microbes into a contaminated site which, in turn, decompose plastic by utilizing it as a source of carbon and energy. If the native microbes are involved in the degradation of plastic, then it is known as indigenous bioaugmentation, and if exogenous microbes are used, then it is known as exogenous bioaugmentation. There are four steps involved in the complete decomposition of plastic which are bio-deterioration, bio- fragmentation, assimilation, and mineralization. During the process of bio-fragmentation, oxygenase enzymes are released by the bacteria which speed up the decomposition of plastic through the addition of oxygen atoms which make the plastic polymer more susceptible towards biodegradation. Biotechnology can also contribute towards solving the plastic problem through the creation of genetically engineered bacteria which can secrete enzymes capable of decomposing plastic. For instance, the creation of genetically engineered *E. coli* bacteria to synthesize PETase enzyme for breaking down Polyethylene Terephthalate (PET) plastic. Advantages of bioaugmentation such being environmentally friendly in nature and producing biodegradable products of plastic waste which can be reused in the circular economy highlight the need for promoting this type of approach towards plastic waste management. However, the disadvantages associated with bioaugmentation, such as the dangers of introducing invasive species into a new ecosystem, as well as the high cost involved, need to be addressed at the earliest for bioaugmentation to emerge as the best solution for solving the plastic crisis.

Conflict of Interest Statement The authors declare that there is no conflict of interest.

References

Abbasi, S., Soltani, N., Keshavarzi, B., Moore, F., Turner, A. and Hassanaghaei, M., 2018. Microplastics in different tissues of fish and prawn from the Musa Estuary, Persian Gulf. *Chemosphere*, *205*, pp.80–87. (Abbasi, S et al., 2018)

Acot Jr, F.T., Sajorne, R.E., Omar, N.A.K., Suson, P.D., Rallos, L.E.E. and Bacosa, H.P., 2022. Unraveling macroplastic pollution in rural and urban beaches in Sarangani Bay Protected Seascape, Mindanao, Philippines. *Journal of Marine Science and Engineering*, *10*(10), p.1532. (Acot Jr, F.T., 2022).

Adams, G.O., Fufeyin, P.T., Okoro, S.E. and Ehinomen, I., 2015. Bioremediation, biostimulation and bioaugmention: a review. International Journal of Environmental Bioremediation & Biodegradation, 3(1), pp.28-39.

Brydson, J.A., 1999. Plastics materials. Elsevier.

Cheung, P.K., Hung, P.L. and Fok, L., 2019. River microplastic contamination and dynamics upon a rainfall event in Hong Kong, China. *Environmental Processes*, *6*, pp.253–264. (Cheung, P.K. et al., 2019).

Cui, L., Qiu, Y., Liang, Y., Du, C., Dong, W., Cheng, C. and He, B., 2021. Excretory expression of IsPETase in E. coli by an enhancer of signal peptides and enhanced PET hydrolysis. International Journal of Biological Macromolecules, 188, pp.568-575.

Dussud, C. and Ghiglione, J.F., 2014. Bacterial degradation of synthetic plastics. In CIESM workshop monogr (Vol. 46, pp. 49–54).

Elahi, A., Bukhari, D.A., Shamim, S. and Rehman, A., 2021. Plastics degradation by microbes: A sustainable approach. Journal of King Saud University-Science, 33(6), p.101538. 7. Filiciotto, L. and Rothenberg, G., 2021. Biodegradable plastics: Standards, policies, and impacts. ChemSusChem, 14(1), pp.56-72

FOY, G.F., 1969. Summary of Engineering Plastics.

Guo, X. and Wang, J., 2019a. The chemical behaviors of microplastics in marine environment: A review. *Marine pollution bulletin, 142*, pp.1–14. (Guo, X. and Wang, J., 2019)

Guo, X. and Wang, J., 2019b. The phenomenological mass transfer kinetics model for Sr2+ sorption onto spheroids primary microplastics. *Environmental Pollution, 250*, pp.737–745. (Guo, X. and Wang, J., 2019)

Guo, X., Chen, C. and Wang, J., 2019a. Sorption of sulfamethoxazole onto six types of microplastics. *Chemosphere, 228*, pp.300–308. (Guo, X, et al., 2019)

Guo, X., Liu, Y. and Wang, J., 2019b. Sorption of sulfamethazine onto different types of microplastics: a combined experimental and molecular dynamics simulation study. *Marine Pollution Bulletin, 145*, pp.547–554. (Guo, X et al.,2019)

Gupta, J., Rathour, R., Singh, R. and Thakur, I.S., 2019b. Production and characterization of extracellular polymeric substances (EPS) generated by a carbofuran degrading strain Cupriavidus sp. ISTL7. Bioresource Technology, 282, pp.417-424.

Gupta, J., Rathour, R., Dupont, C., Mishra, A. and Thakur, I.S., 2022. Biogeochemical profiling and taxonomic characterization of municipal landfill site by metagenomic sequencing. Bioresource Technology, 351, p.126936.

Gupta, J., Tyagi, B., Rathour, R. and Thakur, I.S., 2019a. Microbial treatment of waste by culture-dependent and culture-independent approaches: Opportunities and challenges. Microbial Diversity in Ecosystem Sustainability and Biotechnological Applications: Volume 1. Microbial Diversity in Normal & Extreme Environments, pp.415–446.

Gupta, J., Rathour, R., Medhi, K., Tyagi, B. and Thakur, I.S., 2020. Microbial-derived natural bioproducts for a sustainable environment: a bioprospective for waste to wealth. In Refining biomass residues for sustainable energy and bioproducts (pp. 51–85). Academic Press.

Gupta, J., Rathour, R., Maheshwari, N. and Thakur, I.S., 2021a. Integrated analysis of Whole genome sequencing and life cycle assessment for polyhydroxyalkanoates production by Cupriavidus sp. ISTL7. Bioresource Technology, 337, p.125418.

Gupta, J., Rathour, R., Dupont, C.L., Kaul, D. and Thakur, I.S., 2021b. Genomic insights into waste valorized extracellular polymeric substances (EPS) produced by Bacillus sp. ISTL8. Environmental Research, 192, p.110277.

Hossain, M.S., Rahman, M.S., Uddin, M.N., Sharifuzzaman, S.M., Chowdhury, S.R., Sarker, S. and Chowdhury, M.S.N., 2020. Microplastic contamination in Penaeid shrimp from the Northern Bay of Bengal. *Chemosphere, 238*, p.124688. (Hossain, M.S. et al., 2020)

Huang, H. and Ye, L., 2020. Biological technologies for cHRPs and risk control. In High-risk pollutants in wastewater (pp. 209–236). Elsevier.

Hudson, R. and Hudson, R., 1995. Commodity Plastics-as Engineering Materials?: A Report from Rapra's Industry Analysis and Publishing Group. iSmithers Rapra Publishing.

Jain, N., Sharma, P., Verma, A. and Gupta, J., 2023. Enhancement of thermo-mechanical, creep-recovery, and anti-microbial properties in PVA-based biodegradable films through cross-linking with oxalic acid: implications for packaging application. Biomass Conversion and Biorefinery, pp.1–10. DOI: https://doi.org/10.1007/s13399-023-04748-y

Jansen, J.A., 2016. Plastics–It's All About Molecular Structure. Plast. Eng, 72(8), pp.44-49.

Mohan, N. and Usha, R., 2018. Bio-augmentation-Effective Method of Treating Plastic Waste-A field Study. Journal of Pure and Applied Microbiology.

Platt, D.K., 2003. Engineering and high performance plastics market report: a Rapra market report. iSmithers Rapra Publishing.

Thiel, M., Luna-Jorquera, G., Álvarez-Varas, R., Gallardo, C., Hinojosa, I.A., Luna, N., Miranda-Urbina, D., Morales, N., Ory, N., Pacheco, A.S. and Portflitt-Toro, M., 2018. Impacts of

marine plastic pollution from continental coasts to subtropical gyres—fish, seabirds, and other vertebrates in the SE Pacific. *Frontiers in Marine Science*, p.238. (Theil, M et al., 2018)
Thushari, G.G.N. and Senevirathna, J.D.M., 2020. Plastic pollution in the marine environment. Heliyon, 6(8). (Thushari, G.G.N. and Senevirathna, J.D.M., 2020)
Tokiwa, Y., Calabia, B.P., Ugwu, C.U. and Aiba, S., 2009. Biodegradability of plastics. International journal of molecular sciences, 10(9), pp.3722-3742.
Wierckx, N., Narancic, T., Eberlein, C., Wei, R., Drzyzga, O., Magnin, A., Ballerstedt, H., Kenny, S.T., Pollet, E., Averous, L. and O'Connor, K.E., 2018. Plastic biodegradation: Challenges and opportunities. *Consequences of microbial interactions with hydrocarbons, oils, and lipids: biodegradation and bioremediation*, pp.1–29.
Yoshida, S., Hiraga, K., Takehana, T., Taniguchi, I., Yamaji, H., Maeda, Y., Toyohara, K., Miyamoto, K., Kimura, Y. and Oda, K., 2016. A bacterium that degrades and assimilates poly (ethylene terephthalate). Science, 351(6278), pp.1196-1199.

Unlocking the Therapeutic Potential of Microorganisms in Revolutionizing Pharmaceuticals

Shreya, Juhi Gupta, and Akarsh Verma

1 The Historical Use of Microorganisms in Medicine

Since the beginning of human civilization microbes have been a crucial part in the development of medicines and drugs. People from various civilizations Observed that certain fermented products hard ability to treat certain infections and had healing properties. In various civilizations various fermented foods where used in order to maintain the health of people (Gasbarrini et al. 2016).

Louis Pasteur found out that are microbes responsible for the process of fermentation at the beginning of the twentieth century, while E. Metchnikoff attributed the longer life expectancy of Bulgarian farmers to their regular use of fermented dairy products like yogurt. E. Metchnikoff proposed that lactobacilli have the potential remedy for the putrefactive effects of gastrointestinal metabolism, which he claimed contributed to diseases and aging. Metchnikoff regarded lactobacilli as probiotics (also known as "probios," which are beneficial to the host's life rather than antibiotics); probiotics may improve our health and slow the aging process (Gasbarrini et al. 2016).

Nowadays these fermented foods are sold in the market in the name of probiotics. Nearly every civilization in history has developed some kind of fermented food for maintaining the health of people. In the Chinese Neolithic settlement of Jiahu and in

Shreya
Indian Institute of Forest Management, Bhopal, Madhya Pradesh, India

J. Gupta (✉)
Amity Institute of Environmental Sciences, Amity University, Noida 201303, India
e-mail: jgupta2@amity.edu

A. Verma
Department of Mechanical Engineering, University of Petroleum and Energy Studies, Dehradun 248007, India

Department of Mechanical Science and Bioengineering, Osaka University, Osaka 560-8531, Japan

J. Gupta and A. Verma (eds.), *Microbiology-2.0 Update for a Sustainable Future*,
https://doi.org/10.1007/978-981-99-9617-9_7

Mesopotamia as early as 5000 BC, archaeologists have discovered evidence of the production of a fermented beverage. Rice was the primary ingredient in fermented beverages in Asia, whereas ancient Egypt and Mesopotamia manufactured wine, mead, and beer from of fruits, honey, and malted grains, respectively (Gasbarrini et al. 2016).

History and origins of some fermented foods

Food origin	Approx. year of introduction	Location/region
Fermented milk	10,000 BC	Middle East
Fermented milk products	7000–5000 BC	Egypt, Greece, Italy
Dahi (yogurt)	6000–4000 BC	India
Cheese	7000 BC	Iraq
Cheese	2000 BC	Middle East, China
Mushroom	4000 BC	China
Soy sauce	3000 BC	China, Korea, Japan
Wine	3000 BC	North Africa, Middle East, Europe
Fermented rice	2000 BC	China, Asia
Fermented honey (mead)	2000 BC	North Africa, Middle East
Fermented barley to beer	2000 BC	North Africa, Middle East, China
Bread	1500 BC	Egypt, Europe
Fermented meat	1500 BC	Middle East
Sourdough bread	1000 BC	Europe
Fish sauce	1000 BC	South-East Asia, North Africa
Garum (made from the fish gut)	400 BC	Greece, Italy (Rome)
Pickled vegetable	1000 BC	China, Europe
Tea	200 BC	China
Mouldy soybean curd	500 BC	China

2 Discovery of Vaccines

In the late eighteenth century, the world witnessed one of the earliest successes in using microorganisms for medical purposes. As a child, Edward Jenner was infected with smallpox and he remembered the horrible experiences of isolation and fevers (Stewart and Devlin 2006). Despite his own experience, he variolated his patients with smallpox as that was the most effective preventive measure against smallpox mortality. He noticed that the milkmaids who were infected with cowpox from their cows were immune to smallpox (Stewart and Devlin 2006). As a result, in 1796, he gave a local milkmaid's cowpox (a milder version of smallpox) to James Phipps, the son of his gardener. He then gave Phipps smallpox, to which he, fortunately, developed immunity. In 1798, Jenner published the results of his investigation. Cow inoculum faced strong criticism from both the medical establishment and the general public (Stewart and Devlin 2006). Edward Jenner toured worldwide to demonstrate his technique and to distribute cowpox serum. Modern-day vaccines are also developed on this basic fundamental where the weaker or milder versions of a microorganism is used to develop immunity against the other stronger and more harmful microbes.

Later in the twentieth century, the world witnessed the development of various other vaccines for diseases like polio, measles, and influenza. These developments helped in the prevention of epidemics and saved millions of lives.

3 Discovery of Antibiotics

Alexander Fleming was a Scottish physician and scientist best known for discovering penicillin. This life-saving antibiotic earned Fleming, together with Howard Florey and Ernst Chain, the Nobel Prize in Physiology/Medicine in 1945. Fleming left a Petri plate unprotected near an open window while testing with staphylococcal bacteria in 1928, resulting in mould spore contamination. He recognised the mould as Penicillium (Tan and Tatsumura 2015) after noticing that bacteria surrounding the mould were dying and the agar gel around it was clearing, and he realised that a chemical it generated was responsible for killing Gram-positive pathogens. This 'mould juice' was given the name penicillin (Tan and Tatsumura 2015). Despite the fact that Fleming reported his discoveries in 1929, early interest was lacking, and mass manufacture of penicillin continued to be difficult (Tan and Tatsumura 2015). Florey and Chain's research failed to enable large-scale manufacture of penicillin until 1940. When mass production of penicillin started it became a crucial instrument during the Second World War and revolutionised medicine.

4 Importance of Microorganisms in Modern Pharmaceuticals

Microorganisms play a significant role in the creation of modern medications, particularly in pharmaceutical research, development, and manufacturing (Gupta et al. 2022). Using the unique skills of microorganisms, researchers employ bioprospecting to discover new chemicals with medical promise (Verma and Ogata 2023). Microbes, such as bacteria and fungi, produce a large amount of bioactive compounds, which serve as the basis for many drugs, including antibiotics, antivirals, and anticancer treatments (Gupta et al. 2021a; Demain and Sanchez 2009). Furthermore, genetic engineering approaches enable bacteria to be manipulated to produce complex therapeutic compounds using biotechnological operations. Microbial fermentation has revolutionized pharmaceutical production, allowing for the cost-effective and scalable production of biologics such as insulin and vaccines (Demain and Sanchez 2009). Furthermore, bacteria contribute in the understanding of illness causes and pharmaceutical interactions, which benefits in preclinical testing and the evaluation of treatment efficacy. In essence, microorganisms have evolved into critical tools for advancing scientific advancement. Microbes, in essence, have become critical tools in pharmaceutical research, from discovery through final delivery of life-saving medications (Gupta et al. 2021b; Demain and Sanchez 2009).

4.1 Bacteria, Fungi, and Viruses as Sources of Natural Products

Dubos, who had just demonstrated the bactericidal activity of an antibiotic obtained from a soil bacterium, motivated Waksman and Woodruff to build on Fleming's findings. When clearings were seen, soil sample suspensions were cultivated with bacteria, and colonies were isolated for pure growth. A piece of broth culture filtrate from a discovered 'Actinomyces sp.' shown strong antibacterial action in an early experiment. Actinomycin was one of the first actinomycete-derived antibacterial medicines to be isolated in crystalline form. Using the same approach, Waksman and coworkers discovered the medically important antibiotic streptomycin.

These successes gave rise to the 'Waksman platform': the most successful antibacterial drug research platform and the start of the 'golden period' (Lewis 2013). Our most significant antibiotic classes, such as tetracycline, chloramphenicol, neomycin, erythromycin, vancomycin, kanamycin, rifamycin, and gentamicin, Mahajan and Balachandran (2012) were developed on this platform. This platform was separated from Fleming's work by the systematic nature of screening, and for around 20 years researchers were able to work their way through organisms (primarily actinomycetes), mining new antibacterial drugs, until old hits began to re-emerge, and the platform became outdated.

While natural products were thriving, only a few synthetic medications (e.g., isoniazid) were found, largely through species-specific screening against pathogens such as Mycobacterium TB. While the Waksman platform began to fade in the 1960s, expectations were revived when Lesher et al. (1962) synthesised synthetic 'nalidixic acid'. This optimism was short-lived, since it produced relatively few effective synthetics and no new broad-spectrum antibacterial medications following its optimization to produce successful fluoroquinolones.

While natural products thrived, only a few synthetic drugs (e.g., isoniazid) were discovered, mostly through species-specific screening against diseases like Mycobacterium tuberculosis (Lewis 2013). While the Waksman platform began to atrophy in the 1960s, expectations were reawakened when Lesher et al. (1962) synthesised synthetic 'nalidixic acid'. Following its optimisation to develop viable fluoroquinolones, it produced very few effective synthetics and no new broad-spectrum antibacterial drugs.

Only Cyanophyta (blue-green algae) and Pyrrophyta (dinoflagellates) have been examined for biomedical natural products; cyanobacteria produced the bulk of the metabolites. Secondary metabolites having biotechnological significance in the pharmaceutical business were initially assumed to be abundant in cyanobacteria. Because of the creation of bioactive compounds with commercial and therapeutic consequences, there has recently been a spike in interest in studying these species (Demain and Sanchez 2009). In actuality, cyanobacteria produce a number of compounds with antifungal, antibiotic, and anticancer effects in addition to powerful poisons. Furthermore, antimicrobial, immunosuppressive, antibacterial, anticoagulant, anticancer, anti-HIV (human immunodeficiency virus), antifungal, anti-inflammatory, antimalarial, antiprotozoal, antituberculosis, antiviral, and antitumor characteristics are found in cyanobacterial metabolites (Demain and Sanchez 2009).

5 Various Culture Mediums for the Growth of Microorganisms

Culture media are carefully prepared substances or materials that offer a nourishing environment for the development, culture, and preservation of microorganisms such as bacteria, fungi, viruses, and other microscopic creatures in the discipline of microbiology. These media are precisely constructed to offer the nutrients required for the optimal development of various types of microorganisms, including carbon sources, nitrogen sources, minerals, and vitamins. Culture media can be solid, liquid, or semisolid in consistency, and they can be enriched, selective, differential, or specialised for certain microorganisms or conditions for growth (Bonnet et al. 2019).

Microbiological research, clinical diagnostics, and a variety of industrial uses all rely on culture medium. They allow scientists to isolate, examine, and alter microorganisms, allowing them to better understand their traits, behaviour, and interactions (Bonnet et al. 2019). The numerous types of culture medium available adapt to the individual demands of different microorganisms, allowing for the cultivation and research of a varied spectrum of microbial species in controlled laboratory conditions (see Fig. 1).

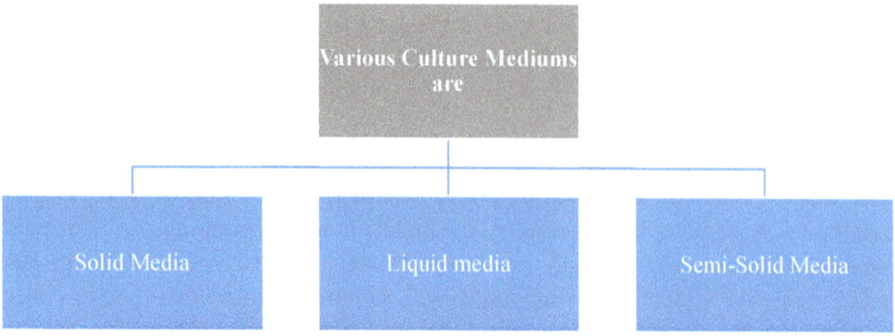

Fig. 1 Different types of culture medium

5.1 Solid Media

Agar, a gelatinous material produced from seaweed, is the primary component of solid medium, which is an essential component of microbiological practice. These media, infused with nutrients, salts, and growth-promoting components, provide a supportive environment for microbial development (Bonnet et al. 2019). Their value stems from the ability to cultivate and isolate microorganisms like as bacteria and fungus into separate colonies, allowing for the examination of colony properties such as shape, size, and colour. This facilitates the separation of pure cultures and allows for the investigation of individual microbial strains. Solid media have a number of advantages, including the ability to isolate colonies for individual research, facilitate pure culture production, and observe morphological features important for identification.

However, there are certain disadvantages to using solid media. When compared to liquid media, the preparation, sterilisation, and chilling processes take longer (Bonnet et al. 2019). Furthermore, the agar's restricted dispersion of nutrients and gases can stifle the growth of picky organisms. Solid media are versatile because they may be found in both continuous solid states (e.g., potato slices, coagulated blood serum, coagulated egg) and liquefiable forms (e.g., nutrient-agar medium, nutrient-gelatin medium, potato dextrose-agar medium). When heated, the latter change from solid to liquid (or vice versa), allowing for manipulation for a variety of purposes.

Agar, a complex polysaccharide generated from diverse red algae and distinguished by its unique qualities of liquefaction upon heating and gelling upon chilling, serves as the foundation of these media. In a nutshell despite their inherent constraints solid media serve as a vital tool, providing insights into microbial growth, variety, and behavior (Bonnet et al. 2019).

5.2 *Liquid Media*

Liquid media, an essential tool in microbiological research, are nutrient-rich solutions that serve as the foundation for microbial development (Bonnet et al. 2019). They permit a variety of uses, including large-scale microbial culture, biochemical testing, and research into microbial physiology. Notably, liquid medium promote fast development due to unfettered movement of nutrients and gases, as well as their simple inoculation technique. Furthermore, liquid medium are easily scaled up for large-scale fermentation operations aiming at creating useful chemicals. Their lack of isolation capacity, on the other hand, impedes the acquisition of pure cultures, posing a serious constraint (Bonnet et al. 2019). Furthermore, the liquid environment's vulnerability to contamination endangers culture quality.

Liquid media, often known as broth, differs from solid media in that it is fluid and is best suited for bacterial cultivation (Bonnet et al. 2019). Such media include nutrient broth, glucose broth, beef extract, skimmed milk, and peptone solutions. Beef extract and peptone solution, both semi-digested sources of protein, are the most widely used in bacteriological laboratories. By adding specialised chemicals or materials, these basic liquid media may be further adjusted to suit certain microbe species or responses. In conclusion, liquid media offer as a dynamic platform for the investigation of microbial growth and behaviour, but with some constraints that researchers traverse deftly in order to get significant insights (Bonnet et al. 2019).

5.3 *Semisolid Media*

Semisolid media are a distinct category within microbiological research, distinguished by a gel-like texture caused by a lower agar content than solid media (Hiss 1897). Their function is simple but critical: they are typically used for motility assays to determine the movement of microorganisms. This distinctive use stems from their inherent benefits, the most notable of which is their value in aiding the observation of microbial movement, allowing the distinction of motile and non-motile strains (Hiss 1897). The structural integrity of motility patterns is preserved by slower nutrient and gas transport in comparison to liquid mediums.

Semisolid media, remain somewhat limited, centred mostly on motility experiments, making them less adaptable than other media types. One disadvantage is that they have a lower nutritional content, which may result in slower microbial growth when compared to more nutrient-rich liquid equivalents. This special media class highlights its specific function in microbiological research, capitalising on its distinct consistency to uncover the enthralling complexities of microbial motion.

5.4 Enrichment Culture

Enrichment culture is a critical method in microbiology that selectively promotes the development of a specific bacterium among a diverse population by precisely altering nutritional and environmental parameters (Singh 2022). This approach is useful for culturing microorganisms that may exist in trace amounts in complicated combinations. The required nutrients are given to encourage the growth of the target organisms while simultaneously limiting the expansion of undesired counterparts through the painstaking preparation of enrichment medium (Singh 2022). This optimised growth environment promotes the establishment of a microbial population that is specifically tailored to the medium's composition. Enrichment culture is critical in the early phases of isolating and characterising certain microorganisms from materials with mixed populations, such as clinical specimens and environmental samples.

The use of Selenite F Broth, developed for the enrichment of Salmonella species from faecal material, is an example of this approach. This medium successfully restricts the growth of normal gut flora by utilising selective agents, allowing for the enrichment and subsequent detection of Salmonella. Enrichment culture, in essence, is a necessary technique for uncovering the hidden complexities of microbial communities and gaining insights into the broad diversity of microorganisms found in complicated samples (Singh 2022).

5.5 Selective Media

Selective media are important in microbiological research because they provide a customised environment that promotes the development of select bacteria while inhibiting the growth of undesirable equivalents (Bonnet et al. 2019). Their complex composition is made up of carefully selected components that make the environment unfavourable to non-target species, generally by including inhibitors such as antibiotics, dyes, or controlling pH levels. When it comes to isolating or identifying certain groups of microorganisms among a diverse population, researchers use selective media expertly (Bonnet et al. 2019). Mannitol Salt is a classic example of such a medium. Agar acts as an exemplary paradigm, exemplifying Staphylococcus species preference. By infusing the medium with a high concentration of salt, an environment unfriendly to many bacteria is created, with the exception of staphylococci, which thrive under these conditions (Bonnet et al. 2019). Similarly, MacConkey agar, which is selective for Gram-negative bacteria, aids in the identification of enteric organisms such as E. coli and Klebsiella, while also distinguishing lactose fermenters from non-fermenters. This demonstrates how selective media cleverly modify growth circumstances to elucidate the complexities of individual bacteria, enabling insights into microbial diversity throughout complicated communities. In a larger sense, the use of selective media is an artistic method in microbiology, revealing the nuances of microbial communities and allowing for specialised inquiry into desirable microbial populations (Singh 2022).

5.6 *Transport Media*

Transport media play an important role in the medical realm, catering to a wide range of applications. Their importance grows when patients are geographically distant from pathology facilities. They are specifically designed to ensure the preservation of pathogenic germs during transportation from collecting locations to laboratories (Singh 2022). Delicate pathogens, such as Neisseria gonorrhoeae, which causes gonorrhoea, risk succumbing to unfavourable conditions en route, while normal microorganisms, such as Escherichia coli, may outcompete pathogenic counterparts like Salmonella, Shigella, and Vibrio cholerae before reaching the laboratory. Transport mediums have been cleverly created to bypass these obstacles. Buffers to regulate pH, nutrients to support microbial life, and occasionally antimicrobial chemicals to inhibit contaminant development are all included into their design (Singh 2022). These media, by limiting overgrowth and conserving microbial features, serve as crucial protections for maintaining clinical specimen integrity, particularly in circumstances involving delayed analysis. Stuart's Transport Medium, for example, is well-known for its ability to preserve bacterial swabs throughout transport while also providing a favourable habitat for a variety of microbes. Overall, transport medium play a vital role in preserving microbial viability during the important shift from patient to laboratory (Singh 2022).

5.7 *Differential Media*

Differential media are important in microbiology because they allow bacteria to be distinguished based on their distinct metabolic processes or other visible characteristics. These media, which include chemicals that react with certain microbes, produce observable changes such as changing colours or precipitate formation, allowing them to be identified (Singh 2022). Eosin Methylene Blue Agar (EMB) is an example of a method used to detect enteric bacteria that is capable of separating lactose fermenters (with black colonies) from non-fermenters (with colourless colonies). Furthermore, differential media act as markers, allowing for preliminary microbe identification. Blood-Agar medium, MacConkey-Agar medium, and Christensen's media are notable examples. Blood-Agar distinguishes between hemolytic and non-hemolytic bacteria by the presence or absence of distinct zones surrounding colonies, which indicate the destruction of red blood cells (Singh 2022). The colour of MacConkey-Agar, which contains lactose and neutral red, is determined by the pH variations generated by lactose fermentation. Christensen's medium contains urease and phenol red, which cause a colour change when urease-producing bacteria generate alkaline pH (Singh 2022). These distinct media, functioning as interpreters of microbial features, highlight their importance in unravelling the complexities of microorganism identities.

5.8 *Enriched Media*

Enriched media are important in microbiology because they provide a fostering environment rich in nutrients, growth factors, and even blood or serum components. These medium promote robust development by being tailored to the specific nutritional demands of many bacteria, even fastidious ones (Singh 2022). Blood Agar, for example, is an example of enriched media, promoting the development of a variety of bacteria, particularly those that require extra nutrients such as haemoglobin (Singh 2022). These media are cleverly designed to culture a diverse range of bacteria seen in complicated clinical cases. Enriched media are painstakingly made by supplementing basal mediums with specialized growth components such as blood, serum, and egg, ensuring the nutritional needs of metabolically complex microbes are satisfied (Singh 2022). Blood Agar for Streptococcus isolation, chocolate agar for Neisseria and Haemophilus isolation, Bordet-Gengou medium for Bordetella, and Loefler's serum slope for Corynebacterium diphtheriae isolation are all notable examples. In essence, enriched media serve as an important instrument, promoting the growth of a variety of microorganisms in various samples and so assisting in the numerous goals of microbial study and identification.

Advantages and disadvantages

Microbial sources, such as bacteria, fungus, and algae, have emerged as major contributors to the pharmaceutical industry, providing a varied array of bioactive chemicals with potential therapeutic uses. These sources have provided transformational benefits, but they also present certain problems that must be overcome in order to fully realise their promise.

Advantages:

(a) **Biological Diversity and Compound Diversity**: Microbes are a tremendous and mostly unexplored source of biodiversity. The microbial kingdom has a vast variety of species, each having its own genetic make-up, enzymatic capabilities, and metabolic processes. This variety results in a diverse set of bioactive chemicals, including antibiotics, antivirals, anticancer medicines, immunomodulators, and enzymes (Demain and Sanchez 2009). Access to such a wide pool of compounds is critical for combating the increasing difficulties posed by drug-resistant pathogens and emerging diseases.

(b) **Drug Discovery and Development**: Microbial sources have sped up the drug discovery and development process dramatically. This is shown by the long history of antibiotic discovery from Streptomyces, as pioneered by Waksman and Schatz (1943). Microbes provide a quick and low-cost way to isolate and discover new chemicals with potential pharmacological action (Demain and Sanchez 2009). Furthermore, genetic engineering techniques allow microorganisms to be manipulated to increase the synthesis of certain chemicals, allowing the creation of novel medicines and therapeutic agents.

(c) **Biotechnological Production**: Microbial fermentation has revolutionised pharmaceutical manufacture. Microbial cultures' scalability and repeatability, together with advances in bioprocess engineering, enable efficient large-scale manufacturing of complex compounds such as insulin, growth hormones, and vaccines (Bonnet et al. 2019). This has decreased not just manufacturing costs but also the reliance on sophisticated extraction procedures from precious natural resources.

(d) **Targeted Engineering**: Microbes can be genetically modified to create high-value substances that would otherwise be difficult to get from traditional sources. This comprises recombinant proteins, monoclonal antibodies, and personalised pharmaceuticals. The capacity to edit microbial genomes using tools such as CRISPR-Cas9 has opened up possibilities for adapting microbial strains to specific pharmacological demands (Tekamp-Olson and Valenzuela 1990).

Challenges:

(a) **Selectivity and Bioactivity**: While microbial sources provide a wealth of bioactive chemicals, not all are acceptable for medicinal use. To isolate molecules with high therapeutic effectiveness and minimal toxicity, thorough screening and testing are required (Demain and Sanchez 2009). Many substances exhibit promising results in preliminary studies but fail to fulfil the demanding safety and efficacy standards during clinical trials. This difficulty emphasises the importance of extensive preclinical and clinical testing.

(b) **Regulatory Hurdles**: To bring microbial-derived medications to market, numerous regulatory routes must be navigated. To assure repeatability and safety, strain characterisation, manufacturing uniformity, and quality control are necessary. Regulatory bodies such as the FDA and EMA demand extensive data on the production process, product stability, and pharmacokinetics.

(c) **Fermentation Optimization**: While microbial fermentation has significant production benefits, it also has limitations in terms of optimising growing conditions, yield, and scale-up (Demain and Sanchez 2009). To ensure constant and high-yield production, variables like as nutrient availability, oxygen levels, and fermentation kinetics must be accurately managed.

(d) **Unpredictable Genetic Variation**: The genetic makeup of microorganisms can be unstable and vulnerable to mutations, which can impact the synthesis of desirable chemicals (Tekamp-Olson and Valenzuela 1990). It is critical to maintain genetic stability through generations, especially for strains employed in industrial production. Continuous monitoring, strain improvement, and selection procedures are required.

(e) **Competition and Sourcing**: Because of the expanding interest in microbial sources for medicines, competition for novel strains and bioactive chemicals is developing. Some bacteria may be rare or difficult to culture, making it difficult to access and preserve vital microbial resources (Demain and Sanchez 2009).

To summarise, using microbial sources in pharmaceuticals is a viable path for drug discovery, development, and manufacture. Microbial diversity, efficient drug discovery, biotechnological manufacturing, and focused engineering are all important drivers of innovation (Demain and Sanchez 2009). However, problems such as bioactivity screening, control, fermentation optimisation, genetic stability, and competition must be successfully overcome. Overcoming these issues will enable the continuous study and use of microbial resources for the progress of pharmaceutical science through multidisciplinary research, technical improvements, and regulatory collaboration.

6 The Role of Microbial Fermentation in Drug Discovery and Development

6.1 Antibiotics

Microbial fermentation is a critical step in the manufacture of antibiotics such as penicillin and cephalosporins (Zinner 2007). These antibiotics occur because of the complex metabolic activity of microbes, notably fungus and bacteria. The amazing discovery of penicillin, for example, demonstrated the revolutionary power of microbial fermentation. Penicillin, generated from the Penicillium fungus, transformed the treatment of bacterial illnesses by preventing the growth of a wide variety of pathogenic bacteria (Tan and Tatsumura 2015). Microbial fermentation is used to produce penicillin by cultivating the Penicillium fungus in controlled surroundings and optimising variables such as temperature, pH, and nutrient availability (Tan and Tatsumura 2015). Penicillin is synthesised by the fungus as a defence mechanism against competing bacteria as it develops. After harvesting, the penicillin can be removed and purified for medicinal use (see Fig. 2).

Similarly, cephalosporins, another powerful antibiotic family, are produced by microbial fermentation (Zinner 2007). These antibiotics, which have a broad spectrum of effectiveness against many bacterial illnesses, are derived from the fungus Cephalosporium. The fungus is grown under regulated settings in this method,

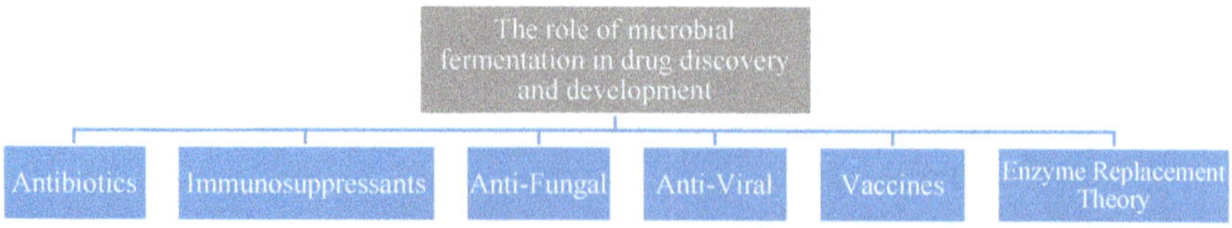

Fig. 2 Functions served by Microbial fermentation in drug discovery and development

allowing it to manufacture cephalosporins as part of its natural defence mechanism. Following that, the antibiotic component is separated and purified from the fermentation broth. The capacity to harness these bacteria to create antibiotics on a bigger scale has been a game changer in medicine, allowing for the bulk manufacturing of these life-saving medications.

Microbial fermentation has various advantages in the manufacture of antibiotics. For starters, because microorganisms can be cultivated repeatedly in controlled settings, it provides a renewable and sustainable source of antibiotics.

6.2 *Immunosuppressants*

Microbial fermentation is essential in the manufacturing of immunosuppressants, such as the commonly used medication cyclosporine. This immunosuppressant is produced by the fermentation of the fungus Tolypocladium inflatum (Survase et al. 2011). Cyclosporine, by reducing the immune response, is an important therapeutic strategy used to avoid organ rejection in transplantation and to control autoimmune illnesses.

To produce cyclosporine by microbial fermentation, the fungus Tolypocladium inflatum is grown in a controlled setting, such as a bioreactor (Survase et al. 2011). This regulated atmosphere provides ideal conditions for fungal growth for cyclosporine synthesis. As the fungus grows, it produces cyclosporine as a secondary metabolite, most likely as a defence mechanism in its natural environment.

Following the completion of the fermentation process, the fungal biomass is harvested and the cyclosporine component is isolated (Survase et al. 2011). Purification of the extracted cyclosporine results in a highly concentrated and pharmaceutically pure version of the immunosuppressant. This refined cyclosporine can subsequently be synthesised into different pharmaceutical dosage forms, such as oral capsules or intravenous solutions, for delivery.

6.3 *Antifungal*

Microbial fermentation is crucial in the manufacturing of antifungal medications, with the example of amphotericin B demonstrating the importance of this process. Amphotericin B is a highly effective antifungal agent produced from the bacteria Streptomyces nodosus (Jakubiec-Krzesniak et al. 2018). This medication is used to treat serious fungal infections caused by Candida and Aspergillus species, which can be fatal in immunocompromised people.

The culture of Streptomyces nodosus in a controlled setting, such as a fermentation tank or bioreactor, is the first step in producing amphotericin B by microbial fermentation. These regulated circumstances promote bacterial growth and stimulate the formation of amphotericin B as a secondary metabolite (Jakubiec-Krzesniak et al. 2018). As part of their natural survival tactics, Streptomyces bacteria create a variety of bioactive chemicals, including antibiotics and antifungals.

The bacteria synthesises amphotericin B during fermentation, which is subsequently secreted into the culture medium (Jakubiec-Krzesniak et al. 2018). Following the completion of the fermentation process, the culture medium containing amphotericin B is collected. The drug is then extracted and purified from the growth medium, resulting in a concentrated and pharmaceutical-grade version of amphotericin B.

6.4 Antiviral

Microbial fermentation is a critical step in the production of antiviral medicines, with ribavirin acting as an example. Ribavirin, an antiviral medicine used to treat viral illnesses such as hepatitis C and respiratory syncytial virus (RSV), is manufactured by a variety of microorganisms, including bacteria and yeast. This procedure emphasises the critical importance of biotechnology in developing antiviral treatments (Kausar et al. 2021).

Microbial fermentation begins with the growth of chosen microorganisms, such as bacteria or yeast, in controlled settings such as fermentation tanks or bioreactors (Kausar et al. 2021). These settings provide the ideal circumstances for microbial development and subsequent ribavirin production. These microbes metabolise precursors during the fermentation process, producing ribavirin as a useful secondary product.

Following the completion of fermentation, the next stages entail extracting and purifying ribavirin from the culture medium. Various separation and purification processes are used to guarantee that the antiviral component is concentrated and of good quality (Kausar et al. 2021). Purified ribavirin can then be synthesised into pharmaceutical formulations for dosing, assuring successful viral infection therapy.

6.5 Vaccines

Microbial fermentation is a critical step in the manufacture of vaccines, as demonstrated by the hepatitis B vaccine. The vaccine production process comprises complicated biotechnological processes that use microorganisms' ability to generate vaccine components. For example, the hepatitis B vaccine uses recombinant DNA technology in collaboration with the yeast Saccharomyces cerevisiae to manufacture the hepatitis B surface antigen (HBsAg), a critical component of the vaccination (Canouï and Launay 2019).

The introduction of the genetic material encoding the HBsAg protein into yeast cells initiates the manufacture of the hepatitis B vaccine via microbial fermentation. This is accomplished by the use of recombinant DNA technology, in which the gene encoding HBsAg is introduced into the yeast genome. The genetically engineered yeast cells are then cultivated in controlled fermentation conditions to optimise their growth and protein expression (Canouï and Launay 2019). The HBsAg protein is naturally produced by yeast cells as part of their biological activities as they grow.

The yeast cells secrete the HBsAg protein into the culture medium during fermentation. Following the completion of the fermentation process, the culture medium is harvested and the HBsAg protein is isolated (Canouï and Launay 2019). Purification techniques are subsequently used to create a highly concentrated and physiologically active version of HBsAg. Purified HBsAg is the foundation of the hepatitis B vaccine, as it stimulates the immune response and gives protection against the hepatitis B virus upon immunisation.

6.6 *Enzyme Replacement Therapy*

Microbial fermentation is a critical technology in the manufacture of enzymes for enzyme replacement therapy, as demonstrated by recombinant enzymes used to cure a variety of hereditary illnesses (Mitragotri et al. 2014). This method includes using microbes, specifically modified bacteria or yeast, to synthesise medicinal enzymes. For example, in the context of diabetes treatment, microbial fermentation is used to make recombinant insulin, whereas recombinant factor VIII is used to treat haemophilia.

The introduction of the gene encoding the therapeutic enzyme into the genome of the microorganism is the first step in the manufacture of recombinant enzymes via microbial fermentation (Mitragotri et al. 2014). This genetic alteration is accomplished by the use of recombinant DNA technology, which allows the microorganisms to express the required enzyme (Mitragotri et al. 2014). The altered microbes are then grown in controlled fermentation settings where temperature, pH, and nutrient availability are optimised to boost enzyme synthesis.

The recombinant enzyme is naturally produced by bacteria as part of their biological activities as they develop and proliferate. During fermentation, the enzyme is normally secreted into the culture media. Following the completion of the fermentation process, the culture medium is harvested and the recombinant enzyme is isolated. The enzyme is subsequently purified to produce a highly concentrated and physiologically active version of the therapeutic enzyme.

7 Numerous Important Drugs Have Been Developed Using Microbial Fermentation. Here Are a Few Notable Examples

(a) **Penicillin**

Penicillin, one of the first and most well-known examples, revolutionised the treatment of bacterial diseases (Demain and Sanchez 2009). Penicillin is produced by the fungus Penicillium during fermentation (Tan and Tatsumura 2015). This discovery ushered in the antibiotic era, which saved many lives by efficiently treating bacterial illnesses.

(b) **Cyclosporin A**

Cyclosporin A was found in the 1970s as a narrow-spectrum antifungal peptide generated in aerobic fermentation by the mould Tolypocladium nivenum (formerly Tolypocladium inflatum) (Survase et al. 2011). The discovery of this secondary metabolite's immunosuppressive effect led to its usage in heart, liver, and kidney transplants, as well as the tremendous success of the organ transplant sector.

(c) **Streptomycin**

Streptomycin, another ground-breaking antibiotic, is manufactured by fermentation by the bacteria Streptomyces griseus (Tan and Tatsumura 2015). It was the first successful TB therapy and had an important role in limiting the spread of this fatal illness.

(d) **Erythromycin**

This antibiotic, which is used to treat a variety of bacterial illnesses, is generated by fermentation by the bacterium Saccharopolyspora erythraea (Tan and Tatsumura 2015). Erythromycin and its derivatives have shown to be invaluable in the treatment of respiratory, skin, and other infections.

(e) **Insulin**

Synthetic insulin is produced using recombinant DNA technology and microbial fermentation. Engineered bacteria or yeast are used to create human insulin, revolutionising diabetes therapy and offering a safer and more consistent supply of this vital hormone.

(f) **Vaccination against Hepatitis B**

The hepatitis B surface antigen (HBsAg) for the hepatitis B vaccination is produced using recombinant DNA technology employing yeast (Saccharomyces cerevisiae) (Demain and Sanchez 2009). This vaccination has proven to be effective in avoiding hepatitis B infections and consequences.

(g) **Human Growth Hormone**

Recombinant hGH, which is used to treat growth problems, is manufactured by fermentation using genetically altered bacteria. Individuals with growth hormone shortages have reported an improvement in their quality of life as a result of this medication.

(h) **Recombinant Factor VIII**

Recombinant factor VIII, a blood clotting protein manufactured by microbial fermentation, is frequently used in the treatment of haemophilia. This medicine has revolutionised haemophilia care by averting potentially fatal bleeding episodes.

(i) **Interferons**

Interferons are antiviral and immune-regulating proteins that are created by microbial fermentation. They're used to treat things like hepatitis, multiple sclerosis, and certain malignancies (Demain and Sanchez 2009).

(j) **Statins**

These cholesterol-lowering medicines are made by fermenting fungus such as Aspergillus terreus (Demain and Sanchez 2009). They have made major contributions to the management of cardiovascular disorders and the reduction of the risk of heart attacks and strokes (Lovastatin, Simvastatin, and others).

(k) **Monoclonal Antibodies**

Some monoclonal antibodies used in cancer therapy and autoimmune illnesses are created in fermentation techniques employing mammalian cell cultures. These biologic medications target certain illnesses in a highly targeted and effective manner (Demain and Sanchez 2009).

(l) **Agents that block enzyme activity**

Enzyme inhibitor drugs may be useful not only in treating human disease, but also in agriculture, enzyme structure elucidation, and reaction processes. Microbes have produced a number of enzyme inhibitors with a variety of industrial applications (Demain and Sanchez 2009). The statins and hypocholesterolemic medications listed here are among the most essential. Fungal compounds are also utilised to block enzymes that cause cancer, diabetes, toxicity, and Alzheimer's disease. Acetylcholinesterase, protein kinase, tyrosine kinase, glycosidases, and other enzymes are blocked.

(m) **Mycotoxins**

Mycotoxins, which are fungi-produced poisons, have proven to be effective therapeutic agents for a range of medical illnesses and disorders (Demain and Sanchez

2009). Consumption of bread manufactured from grain infected with Claviceps species had resulted in the death of humans and animals for ages. Mycotoxins, on the other hand, were subsequently discovered to be effective for angina pectoris, hypertonia, serotonin-related disorders, suppression of protein release in agalactorrhea, decrease in postpartum haemorrhage, and prevention of implantation in early pregnancy. Their physiological functions include the suppression of adrenalin, noradrenaline, and serotonin activity, as well as the contraction of uterine smooth muscles (Demain and Sanchez 2009).

(n) **Antineoplastic drugs**

Over 12 million new cases of cancer were detected worldwide in 2008, resulting in 7.6 million deaths. The most common cancers were lung (12.7%), breast (10.9%), and colorectal (9.8%). Taxol and camptothecin, secondary metabolites produced from plants and fungus, are two anticancer medicines in clinical use (Demain and Sanchez 2009).

These examples demonstrate the broad influence of microbial fermentation on medication discovery, which has altered medical therapies and greatly improved patient outcomes (Demain and Sanchez 2009).

8 Overview of Biopharmaceuticals and Their Production Using Microbial Expression Systems

The usage of biopharmaceuticals dates back to the nineteenth century, and in the next 5–10 years, a significant fraction of medications under research, up to 50%, are expected to be biopharmaceuticals. During the 1980s, the biopharmaceutical sector saw significant expansion, which was characterised by the prolific synthesis and approval of recombinant proteins such as interferons (IFN, and) and growth hormones. These biopharmaceuticals, also known as biologics, are medicinal treatments that are generated from living cells or creatures, typically via the use of recombinant DNA technology (Mitragotri et al. 2014). Vaccines, monoclonal antibodies, hormones, enzymes, and growth factors are all included in this broad group. The use of microbial expression systems, notably bacteria and yeast, has been critical in their production. These microbes have greatly contributed to the manufacture of biopharmaceuticals due to their powerful genetic modification techniques, high growth rates, and ease of cultivation (Tekamp-Olson and Valenzuela 1990).

Biopharmaceuticals

Biopharmaceuticals are huge, complex compounds created from live creatures or cells, as opposed to standard small-molecule medications. Proteins, peptides, nucleic acids, and complex carbohydrates are among them. Because of their biological origins, biopharmaceuticals frequently have great specificity and effectiveness, making them excellent tools for treating a wide range of illnesses such as cancer, autoimmune disorders, and genetic abnormalities (Mitragotri et al. 2014).

Microbial Expression Systems

Microorganisms such as bacteria (e.g., Escherichia coli) or yeast (e.g., Saccharomyces cerevisiae) are used in microbial expression systems to create biopharmaceuticals (Mitragotri et al. 2014). Because of their rapid growth, simplicity of genetic modification, and cost-effective fermentation techniques, these systems are desirable. Biopharmaceutical manufacture employing microbial expression systems is a complicated and painstakingly organised set of procedures that combine scientific acumen and technology developments to create therapeutic proteins with great accuracy and efficacy (Mitragotri et al. 2014). This method is critical in modern medicine because it allows for the large-scale production of biologically derived medications, which have revolutionised illness treatment and management.

8.1 Gene Cloning

The manufacturing process begins with gene cloning, a critical step in which the gene encoding the therapeutic protein of interest is obtained and placed into the genetic composition of a chosen microorganism. This innovation is made possible by the clever use of recombinant DNA technology, which allows scientists to precisely control and modify genetic material (Tekamp-Olson and Valenzuela 1990). The modified gene is smoothly incorporated into the genome of the microbe, essentially providing it with the genetic blueprint for manufacturing the required protein. The transformed microbe transforms into a cellular factory, ready to carry out the complex process of protein synthesis.

8.2 Cultivation

Following gene editing, the altered microbes are grown in specialised bioreactors, which provide a controlled environment optimised for growth and protein synthesis (Tekamp-Olson and Valenzuela 1990). Temperature, pH, oxygen levels, and nutrition availability are all carefully monitored and controlled within these bioreactors. These settings are fine-tuned to promote optimum microorganism growth and protein synthesis, creating a favourable environment for therapeutic protein production.

8.3 Protein Expression

As the microbes grow in the bioreactor, they begin to produce proteins in line with the genetic instructions contained by the injected gene. The changed genetic code is read by the cellular machinery, which then orchestrates the creation of the required protein molecule (Tekamp-Olson and Valenzuela 1990). Depending on the therapeutic protein's unique properties, this protein may be generated and maintained within the cells or released into the culture media.

8.4 Harvesting

The culture is harvested after the necessary amount of protein synthesis is reached. In certain cases, this means extracting the protein from the cellular biomass or the culture media, which necessitates meticulous separation processes. The recovered material serves as the foundation for further purifying operations.

8.5 Purification

Although enriched with the therapeutic protein, the collected material contains impurities and pollutants that must be rigorously removed to assure the final product's safety and efficacy. A range of purification methods are used, including chromatography, filtering, and centrifugation. These techniques isolate and concentrate the therapeutic protein while removing undesirable contaminants.

8.6 Formulation

The purified therapeutic protein is subsequently put through formulation methods that are specific to the mode of delivery. The protein can be converted into injectable solutions, lyophilized powders, or tablets depending on the purpose. This procedure improves the drug's stability, bioavailability, and usability, all of which contribute to its therapeutic efficacy.

8.7 Quality Control

The end of this lengthy production process is recognised by stringent quality control inspections. These tests include a battery of studies to confirm the biopharmaceutical product's authenticity, purity, potency, and safety. Before being licenced for clinical usage, each batch is thoroughly analysed against preset parameters to ensure that it fulfils the highest quality requirements.

To summarise, the manufacturing of biopharmaceuticals employing microbial expression systems is a multifaceted endeavour that incorporates genetic modification, culture, protein synthesis, purification, and quality control. This approach demonstrates the symbiotic link between science and medicine, providing creative ways for treating a wide range of ailments by precisely synthesising therapeutic proteins.

Advantages and Challenges of Using Microbial Expression Systems in Pharmaceuticals

Microbial expression systems, which employ microorganisms to create recombinant proteins and bioactive compounds, have transformed the pharmaceutical industry. These systems have various benefits, but they also bring certain problems that must be carefully considered for successful adoption (McMillan and Kuehn 2021).

Advantages:

(a) **Cost Effectiveness**: Microbial expression systems, specifically bacteria and yeast, offer a low-cost platform for protein synthesis (Tekamp-Olson and Valenzuela 1990). They can be grown on a huge scale with relatively cheap growing medium. This cost-effectiveness is critical, particularly in the manufacture of therapeutic proteins and enzymes used in medications.
(b) **High Yields and Rapid Expression**: Microbial systems are recognised for their fast growth rates, which allows for the rapid synthesis of target proteins (Tekamp-Olson and Valenzuela 1990). Many microorganisms can also produce large amounts of protein, resulting in high yields. This is especially beneficial for satisfying the demands of the pharmaceutical business.
(c) **Ease of Genetic Manipulation**: Microbes can be genetically manipulated, allowing for the insertion, deletion, or mutation of desired genes. Techniques such as recombinant DNA technology and CRISPR-Cas9 have enabled microbial strains to be engineered for optimum protein production and secretion. This level of control is critical for customizing expression systems to meet certain pharmaceutical requirements (Akbari Kordkheyli et al. 2022).
(d) **Scalability**: Microbial fermentation techniques are extremely scalable, allowing the move from laboratory-scale to industrial-scale manufacture possible. This scalability is critical for fulfilling the needs of large-scale pharmaceutical production while maintaining product quality consistency.
(e) **Safety**: Many of the microbial hosts utilised in expression systems, such as E. coli and Saccharomyces cerevisiae, have a long history of being employed safely in a variety of sectors. This decreases the possibility of impurities or undesirable substances being introduced into medicinal goods.

Challenges:

(a) **Protein Folding and Post-Translational Modifications**: Microbes frequently lack the sophisticated machinery necessary for complicated post-translational modifications (PTMs), which are required for appropriate protein folding and functioning. As a result, misfolded or non-functional proteins may be produced (Hogwood et al. 2013). To overcome this difficulty, eukaryotic expression systems or microbial host modification to accommodate specific PTMs are frequently required.

(b) **Formation of an Inclusion Body**: Microbial systems, despite high expression levels, can result in the production of solid protein clumps known as inclusion bodies. It can be difficult and time-consuming to refold these proteins into their active forms (Hogwood et al. 2013).
(c) **Limited Glycosylation**: Glycosylation, a vital PTM, plays an important role in the bioactivity and pharmacokinetics of many therapeutic proteins (Hogwood et al. 2013). When compared to mammalian cells, microbial hosts frequently synthesise proteins with distinct glycosylation patterns, which may impact protein function and immunogenicity.
(d) **Endotoxin Contamination**: Endotoxins, which can be harmful to human health, are naturally present in certain microbial systems, such as E. coli. Endotoxins must be removed from the finished product by rigorous purification stages, which complicates the manufacturing process.
(e) **Regulatory Approval**: Regulatory authorities such as the FDA need microbial expression systems to be well characterised and validated. Demonstrating the consistency, stability, and quality of goods created from these systems may be a time-consuming and labor-intensive procedure that requires substantial documentation.
(f) **Host Cells Protein Contamination**: During protein expression, host cells can leak their own proteins into the culture medium, potentially contaminating the end product. Effective purification procedures are required to ensure that these host-cell proteins are removed (Hogwood et al. 2013).

Microbial expression methods provide a potent and adaptable platform for the generation of therapeutic proteins. Their low cost, fast expression, scalability, and simplicity of genetic modification make them very appealing. However, problems with protein folding, PTMs, glycosylation, contamination, and regulatory compliance must be handled carefully (Hogwood et al. 2013). Advances in genetic engineering, fermentation technology, and downstream processing are constantly enhancing microbial expression systems' capabilities and performance. As these obstacles are handled with creative solutions, interdisciplinary collaboration, microbial expression systems will continue to play a pivotal role in shaping the future of pharmaceuticals.

9 Microbial Resistance

Microbial resistance, defined as the ability of germs to adapt and withstand the effects of antimicrobial treatments, has emerged as a worldwide healthcare concern with far-reaching consequences for medication development, patient care, and public health.

Microbial Resistance:

Microbial resistance refers to bacteria, viruses, fungi, and parasites' ability to survive the effects of medications meant to kill them (Demain and Sanchez 2009). This adaptive response has the potential to render previously successful therapies ineffective, resulting in chronic infections, increased morbidity and death, and higher healthcare expenditures **(Frontiers in Microbiology 2019) 'Sec**. Antimicrobial resistance offers a serious challenge to modern medicine, threatening the efficacy of a wide variety of medical procedures such as surgery, cancer therapy, and common infections (Morrison and Zembower 2020).

9.1 Causes and Mechanisms

Natural selection is primarily responsible for the evolution of microbial resistance. Antimicrobial medication exposure creates selection pressure, promoting the survival and growth of bacteria with resistance-inducing genetic changes or characteristics (Morrison and Zembower 2020). Horizontal gene transfer, in which resistant genes are transferred among microorganisms, hastens the spread of resistance. Overuse and abuse of antibiotics, ineffective infection prevention and control practises, and the availability of suboptimal drug concentrations in the environment all contribute to the fast formation and spread of resistance (Demain and Sanchez 2009).

Microbial resistance has a tremendous influence on the landscape of drug development, posing complicated difficulties that echo across healthcare systems, economies, and patient outcomes (Morrison and Zembower 2020).This section expands on the varied consequences of microbial resistance on drug development, including restricted treatment options, increased healthcare costs, longer drug development timeframes, worse return on investment, and increasing clinical trial complexity.

(a) **Limited Treatment Options**

Microbial resistance depletes effective antibiotic alternatives, leaving healthcare providers with a limited toolset to tackle infectious infections (Christaki et al. 2020). The emergence of multidrug-resistant (MDR) and extensively drug-resistant (XDR) bacteria, which render previously curable illnesses resistant to frequently used medications, is especially worrying. This constraint can result in poor patient outcomes, extended illnesses, and higher death rates, particularly in disadvantaged groups.

(b) **Elevated Healthcare Costs**

The rise of resistance demands the investigation of innovative treatment techniques, raising R&D expenses. The ineffectiveness of current therapies adds to longer hospital stays, increased healthcare use, and increased expenses for treating resistant infections (Christaki et al. 2020). Patients, healthcare professionals, and healthcare systems all bear a financial cost, putting a pressure on resource allocation and impeding efficient illness management.

(c) **Prolonged Drug Development**

The time-consuming and resource-intensive process of generating new antimicrobial drugs is inherently time-consuming (McEwen and Collignon 2018). Because of the dynamic nature of microbial resistance, medication research tactics must be constantly innovated and adapted (Christaki et al. 2020). As researchers work to battle emerging resistance mechanisms, the development process becomes more time-consuming, necessitating ongoing investment and expertise.

(d) **Diminished Return on Investment**

Because of the short duration of treatment regimens and worries about the quick evolution of resistance, antimicrobial medication development offers particular economic hurdles. This distinct feature limits pharmaceutical firms' ability to generate significant financial gains. As a result, the financial incentives for investing in new antibiotic research and development reduce, resulting in a smaller pipeline of novel antimicrobial drugs (McEwen and Collignon 2018).

(e) **Clinical Trial Complexity**

The emergence of microbial resistance complicates the planning and execution of clinical trials for novel antimicrobial medicines. Trials require specialised patient groups with resistant illnesses, which frequently causes difficulties in patient recruitment and enrolment (McEwen and Collignon 2018). Patients with few remaining treatment alternatives may be unwilling to engage in studies that contain possible dangers, which raises ethical concerns.

9.2 *Mitigation Strategies*

To address the rising threat of microbial resistance, a broad and integrative strategy spanning healthcare, research, legislation, and public awareness is required. The following explication emphasises the significance of important mitigation techniques for mitigating the impact of microbial resistance.

(a) **Judicious Use of Antibiotics**

The cautious and focused use of antibiotics is a critical step in preventing germ resistance. Antibiotics should only be prescribed when absolutely required, according to evidence-based standards (Demain and Sanchez 2009). Overuse, improper prescriptions, and the use of broad-spectrum antibiotics when narrower choices are effective can all contribute to reduce the selection pressure that promotes resistance (Christaki et al. 2020).

(b) **Infection Prevention and Control Measures**

To minimise the transmission of resistant bacteria in hospital settings, strict infection prevention and control practises are required (McEwen and Collignon 2018).

Implementing adequate hygiene practises, isolating sick patients, and encouraging healthcare staff and patients to comply with infection control measures are critical to avoiding the spread of resistance.

(c) **Development of New Drugs and Therapies**

It is vital to invest in research and development for novel antimicrobial drugs. Novel medications, alternative treatments, and novel treatment techniques must be sought in order to avoid resistance mechanisms (Waclaw 2016). Governments, academia, and pharmaceutical corporations should work together to encourage the development of antibiotics and other antimicrobial medications, taking into account regulatory routes that speed approvals for critically needed treatments (McEwen and Collignon 2018).

(d) **Investment in Research**

Funding for research into novel drug targets, diagnostics, and therapeutic alternatives is critical to staying ahead of developing microbial resistance (Waclaw 2016). Investment in fundamental and translational studies fosters development and the discovery of innovative interventions.

(e) **Public Awareness Campaigns**

It is critical to educate the general population, healthcare practitioners, and legislators about the consequences of microbial resistance. Effective communication efforts may encourage ethical antibiotic use, stress the significance of finishing recommended treatments, and raise awareness about the global consequences of uncontrolled resistance.

(f) **International Collaboration**

Microbial resistance crosses national boundaries, demanding global collaboration. Collaboration across nations, international organisations, and stakeholders is vital for developing consistent policies, disseminating research findings, and implementing collective initiatives. To oppose the growth of resistance, collaborative projects can encourage the flow of knowledge, skills, and resources (Christaki et al. 2020).

(g) **Policy Interventions**

Through policy interventions, governments and regulatory agencies play a critical role in combating microbial resistance. Implementing rules that encourage responsible antibiotic use, support research and development of novel medicines, and limit antibiotic abuse in agriculture are critical to reducing resistance (Christaki et al. 2020).

To summarise, microbial resistance is a crucial problem that has a significant impact on medication development. The growing problem threatens the efficacy of present therapies, prompting quick innovation and worldwide collaboration to solve the crisis and ensure the long-term viability of our medical interventions.

10 Strategies for Developing Drugs that Overcome Microbial Resistance

Developing medications that efficiently overcome microbial resistance is a multidimensional task that necessitates novel approaches to keep up with the developing resistance pathways used by bacteria (Demain and Sanchez 2009). This section expands on many comprehensive techniques being investigated by researchers and pharmaceutical corporations in order to produce medications capable of combating microbial resistance.

10.1 Targeting Novel Pathways

Targeting new biological pathways appears to be a potential method in the fight against germ resistance. Researchers can pave the road for the creation of novel drug candidates by discovering critical survival pathways or activities within microbes that vary from those addressed by current antibiotics (Waclaw 2016). This strategy seeks to target pathogen weaknesses while avoiding existing resistance mechanisms. Targeting particular enzymes or interrupting cellular processes that are not generally addressed by present medications, for example, can result new molecules that are less prone to resistance. To identify novel therapeutic targets, this technique necessitates extensive research in microbial biology, genetics, and proteomics (Waclaw 2016). Furthermore, new computational approaches and high-throughput screening technologies are critical in finding promising candidates for further development. Pursuing this path might result in a new generation of medications that avoid resistance mechanisms, assuring ongoing efficacy in the face of changing microbial threats.

10.2 Combination Therapies

Combination medicines are a deliberate method to increasing antimicrobial efficacy while combating resistance development. Researchers create a harsh environment for microbes by using various medications with complimentary modes of action, making it difficult for them to simultaneously build resistance to multiple treatments (Coates et al. 2020). This strategy may include medications that target different parts of microbial physiology or even numerous locations within a single pathway. Combining medications has a synergistic impact that can improve efficacy while potentially lowering individual drug dosages, minimising negative effects (Coates et al. 2020). While developing combination medicines necessitates careful consideration of drug interactions, toxicity, and dosage regimes, their potential to extend the life of current antibiotics and prevent resistance development is significant. Computational modelling and systems biology advances make it easier to identify appropriate medication combinations, directing researchers towards the creation of effective and long-lasting treatment regimens.

10.3 Repurposing Existing Drugs

Repurposing existing medications for new antibacterial uses is a practical way to combat microbial resistance quickly (Hu et al. 2018). Because of common molecular targets or processes, many medicines developed for non-microbial purposes may demonstrate unexpected antibacterial action. This method makes use of existing safety profiles, pharmacokinetic data, and regulatory insights to accelerate the development process. Researchers can avoid the obstacles of de novo drug discovery, such as long development timetables and expensive costs, by repurposing pharmaceuticals (Hu et al. 2018). Screening libraries of authorised medications or research compounds against a variety of microbial diseases might reveal candidates with previously unknown antibacterial properties. This technique demonstrates the benefits of cross-disciplinary collaboration across domains such as infectious illnesses and pharmacology, allowing for the faster development of new therapeutic alternatives.

10.4 Immune System Modulation

Boosting the host's immune response to battle drug-resistant bacteria is a novel approach that complements standard antibiotic therapies. Immunomodulatory medications boost immune system function, bolstering the body's natural defences against infections. These medications help the body fight infections by enhancing immune cell functioning, cytokine synthesis, and phagocytic activity (Chen and Musa 2021). This strategy is especially important when antibiotic effectiveness is hampered by microbial resistance. Immunomodulators, such as interferons or cytokine agonists, can be given alongside antibiotics to improve treatment results. Modulation of the immune system is a critical part of anticancer therapies including immunotherapy, chemotherapy, and radiotherapy (Chen and Musa 2021). Furthermore, progress in studies into the microbiome's effects on immune response opens the door to more targeted medicines that better harness the body's defences.

10.5 Disrupting Biofilm Formation

The biofilm matrix surrounding bacteria makes them tolerant to harsh conditions and resistant to antibacterial treatments (Roy et al. 2018). Drug development that directly targets biofilm production or weakens biofilm matrix is a specialised technique for combating microbial resistance in chronic illnesses. Antibiotic resistance is increased in biofilms, which are complex microbial populations wrapped in protective matrices. Researchers can sensitise microbes to traditional antimicrobial treatments by interfering with biofilm development, stability, or communication activities (Roy et al. 2018). This technique has the potential to address infections caused by medical devices, wound healing difficulties, and chronic illnesses such as cystic fibrosis. Strategies include blocking matrix-producing enzymes and interfering with quorum sensing, a communication mechanism required for biofilm growth. To successfully enter and destroy the protective matrix, the development of biofilm-targeting therapeutics necessitates a thorough understanding of microbial biofilm biology, biomaterial interactions, and improved drug delivery technologies.

10.6 Antibiotic Adjuvants

Antibiotic adjuvants are a promising technique for enhancing the efficacy of current antibiotics and combating resistance mechanisms (Douafer et al. 2019; Wright 2016). These chemicals work in tandem with antibiotics, either increasing their effectiveness or assisting in the removal of resistance barriers. Adjuvants can alter the microenvironment around the microbe, resulting in circumstances favourable to antibiotic activity. Furthermore, they may disrupt resistance mechanisms such as efflux pumps or enzymatic inactivation, making the bacterium more vulnerable to antibiotics. Some adjuvants have the capacity to weaken microbial cell walls, allowing antibiotics to get through (Douafer et al., 2019; Wright 2016). Adjuvant research entails discovering chemicals that improve the antibacterial efficacy of current medications without sacrificing safety. Antibiotic adjuvants provide a strategic approach to restore or increase the efficacy of antibiotics by targeting processes that lead to antibiotic resistance.

10.7 Precision Medicine Approaches

Precision medicine develops as a highly personalised technique for combating microbial resistance by tailoring therapies to the unique features of infections and patients. The selection of appropriate medications or combination treatments is guided by genetic information about the microorganism's resistance mechanisms (Ho et al. 2020). Precision medicine improves treatment results by identifying a pathogen's specific weaknesses. This method not only improves treatment efficacy but also minimises the possibility of resistance development. The combination of genetics, molecular diagnostics, and bioinformatics allows for the quick detection of resistance determinants, allowing healthcare practitioners to make educated treatment regimen selections. Precision medicine is transforming the way antimicrobial medicines are prescribed, shifting away from a one-size-fits-all approach and towards personalised interventions that increase the likelihood of success (Ho et al. 2020).

10.8 Therapies Targeting Virulence Factors

The development of medicines that target microbial virulence factors rather than killing microorganisms is a unique technique for reducing resistance. These medicines diminish the microorganism's potential to cause harm without necessarily eliminating it by inhibiting the production or activity of virulence factors that contribute to its pathogenicity (Dehbanipour and Ghalavand 2022). Because microbes that retain virulence components but have lower pathogenicity are still disadvantaged, this method reduces the selective pressure for resistance development. Virulence factors can be targeted to disrupt host–pathogen interactions, prevent biofilm development, and alter the immunological response of the host (Dehbanipour and Ghalavand 2022). Designing medicines that precisely target virulence factors involves a thorough understanding of microbial pathogenicity, host colonisation processes, and methods to reduce virulence without generating resistance.

10.9 Nanotechnology and Drug Delivery

Nanotechnology provides a paradigm change in medicine delivery and targeting that holds considerable promise in combating microbial resistance. Recently, there are a number of outstanding applications of nanomedicine (chemotherapeutic agents, biological agents, immunotherapeutic agents etc.) in the treatment of various diseases (Patra et al. 2018). Antimicrobial medications can be encapsulated in nanoparticles and nanomaterials, enhancing their stability, solubility, and bioavailability. These nanocarriers allow for regulated medication release at infection sites, resulting in greater concentrations locally while minimising systemic exposure and associated negative effects (Patra et al. 2018). Nanoparticles can be functionalized to precisely interact with microbial surfaces or biofilms, hence improving medication penetration and effectiveness (Amreddy et al. 2018). Furthermore, nanotechnology allows for synergistic medication and therapeutic agent combinations, allowing for various treatment methods. Creating novel nanocarriers necessitates sophisticated material design, optimisation of drug loading and release characteristics, and thorough safety testing (Patra et al. 2018). Researchers can use nanotechnology to improve the performance of current antibiotics, reduce resistance, and perhaps introduce novel medicines to combat microbial infection.

Finally, the multiple tactics discussed, such as antibiotic adjuvants, precision medicine, virulence factor targeting, and nanotechnology-driven drug delivery, highlight the novel ways being investigated to combat microbial resistance (Amreddy et al. 2018). These tactics highlight the need of staying ahead of resistance mechanisms and providing effective ways to combat increasingly complicated illnesses. By adopting these tactics, researchers and pharmaceutical firms hope to usher in a new age of antimicrobial medicines that are both effective and long-lasting in the face of emerging microbial threats.

11 Prospects and Challenges

Certainly, the integration of microbes into the pharmaceutical industry is poised to bring about transformative changes in drug discovery, production, and therapeutic strategies.

11.1 Precision Therapeutics

The capacity to use the human microbiome for precision medicine offers enormous promise. We should anticipate to uncover distinct microbial fingerprints linked with certain health issues as our understanding of microbial populations within the human body increases. This understanding might lead to the creation of medicines that are personalised to a person's microbiome composition (Ho et al. 2020). Pharmaceuticals might be engineered to perform more effectively and with fewer side effects by leveraging the symbiotic connections between the microbiota and the human host. This level of personalisation has the potential to transform treatment techniques, making them more focused and efficient (Ho et al. 2020).

11.2 Microbiome-Based Therapies

As our understanding of the microbiome's significance in human health grows, we should expect the appearance of novel medicines that alter the microbiome to treat a wide range of illnesses (Bajaj et al. 2022). Obesity, inflammatory illnesses, and even neurological issues have all been related to the microbiome. Future therapeutics might include faecal microbiota transplantation (FMT), which involves transferring a healthy donor's microbiome to a patient in order to restore microbial balance and ease symptoms (Bajaj et al. 2022). Furthermore, synthetic biology technologies may enable the development of designer bacteria capable of producing medicinal chemicals directly within the body.

11.3 Advanced Drug Delivery Systems

Microbes may play an important role in improving medication delivery strategies. Engineered microorganisms might act as medicine transporters, directing them to precise target locations within the body. Nanoparticles and modified microbes have the ability to carry medications to previously inaccessible areas, such as tumours or particular cells (Amreddy et al. 2018). Targeted medication distribution might reduce systemic adverse effects while improving overall treatment success. Furthermore, bacteria might be programmed to respond to precise triggers within the body, releasing medications just when they are required.

11.4 Synthetic Biology and Genetic Engineering

Synthetic biology and genetic engineering advances are set to transform the capacities of microorganisms in the pharmaceutical business. The capacity to precisely edit microbial genomes and pathways allows for the creation of organisms with tailored functions (Cubillos-Ruiz et al. 2021). Engineers can design organisms that are optimised for certain activities, such as manufacturing complex therapeutic chemicals, metabolising harmful substances, or monitoring and responding to illness indicators within the body. This potential for personalised microbial designs broadens the field of drug discovery and production, enabling for the development of previously unthinkable medications, diagnostic tools, and cures (Ho et al. 2020). The combination of cutting-edge methods with microbial systems has the potential to open up totally new avenues of pharmaceutical discovery.

12 Challenges for the Future

12.1 Microbial Resistance Evolution

As bacteria become more integrated into pharmaceutical processes, the threat of microbial resistance development grows (Demain and Sanchez 2009). The fast adaptability of microorganisms to novel treatment drugs emphasises the importance of maintaining ongoing surveillance and calls for new preventive measures to slow down this process (Waclaw 2016). Researchers and pharmaceutical developers must remain ahead of resistance mechanisms by developing novel drugs, combination treatments, and techniques to limit the establishment and dissemination of resistance genes. Continuous monitoring and knowledge of microbial behaviour are essential for retaining the efficacy of microbial-based therapy (Jampilek 2022).

12.2 Regulatory and Safety Complexities

The use of microorganisms into pharmaceutical practises poses complex regulatory and safety issues. To strike a balance between stimulating innovation and maintaining patient safety, comprehensive regulatory frameworks are required. These structures must address the distinct features of microbial-based therapeutics, such as genetic changes and innovative delivery modalities (Jampilek 2022). Harmonising international rules, shortening approval procedures, and encouraging cooperation among regulatory agencies, researchers, and industry will be critical to successfully navigating these difficulties.

12.3 Ethical Considerations and Public Perception

The ethical implications of using bacteria, particularly genetically engineered organisms, must be carefully considered (Jampilek 2022). To balance the potential benefits of microbial-based medicines with concerns about biosecurity, environmental impact, and the unintended effects of genetic tinkering, open debate and ethical frameworks are required. Because public perception influences acceptance and adoption of these unique techniques, education and open communication are critical to gaining trust and support (Jampilek 2022).

12.4 Complex System Interactions

Microbes interact with their environment, including the human body, in complicated and unknown ways. Understanding these complicated relationships is critical for properly utilising microorganisms for medicinal reasons. To understand the different effects of microbial manipulations on multiple systems, multidisciplinary collaboration incorporating microbiology, immunology, ecology, and other sciences is required.

12.5 Scale-Up and Commercialization

While bacteria have shown promise in small-scale investigations, the leap from lab-scale trials to large-scale manufacturing is difficult. Maintaining consistency, scalability, and cost-effectiveness while complying to regulatory norms necessitates rigorous microbiological process engineering and optimisation. It will be critical to develop strong fermentation techniques and manufacturing pipelines in order to provide a consistent supply of microbial-based medications (Jampilek 2022).

The use of microorganisms in the pharmaceutical business has enormous promise for transformational improvements, but it also poses a number of complicated issues that must be addressed. Researchers and stakeholders may usher in a new age of creative pharmaceutical solutions by capitalising on the opportunities provided by precision medicines, microbiome-based treatments, enhanced drug delivery technologies, and synthetic biology (Jampilek 2022). To harness the full power of microbes for the betterment of healthcare, however, navigating the obstacles of resistance evolution, regulatory complications, ethical considerations, multifaceted interactions, and commercialization will require coordinated efforts, responsible practices, and ongoing studies.

13 Conclusion

To summarise, the historical and present importance of microbes in medicine and pharmaceuticals is indisputable, changing the landscape of drug discovery, manufacturing, and treatment techniques (Gupta et al. 2019a; Gasbarrini et al. 2016). Human interactions with microbes have changed over time, from ancient practises to current biotechnology achievements with far-reaching ramifications for healthcare. The path from utilising mouldy bread to design microbes for targeted medicine delivery systems highlights the incredible advances made in harnessing the potential of microorganisms. Microorganisms' historical usage in medicine laid the path for our understanding of their numerous uses. The early understanding of microbes' therapeutic abilities, while restricted in scope, laid the groundwork for the creation of antibiotics and vaccines that have revolutionised medical practise. With the introduction of modern medicines, there has been a greater respect for microbes as natural product sources and agents for biopharmaceutical manufacturing (Mahajan and Balachandran 2012). Microorganisms' importance in current medications cannot be emphasized. They are significant sources of biologically active chemicals, allowing for the development of innovative medications that target a variety of ailments. Bacteria, fungi, and viruses have produced a wealth of natural compounds with medicinal promise, extending pharmaceutical researchers' toolset (Demain and Sanchez 2009). The use of various culture media for the development of microbes has opened up new possibilities, providing specialised settings that boost productivity and output (Bonnet et al. 2019).

However, including microbes into therapeutic research is not without difficulties. While the benefits are obvious, problems like as optimising culture conditions, assuring repeatability, and resolving regulatory issues remain. The delicate balance between increasing microbial production and reducing contamination risk highlights the complexities of microbial-based pharmaceutical development (Bonnet et al. 2019). The fermentation of microbes is important in drug discovery and development because it produces antibiotics, enzymes, and other physiologically active molecules. The capacity to alter microbial metabolism using fermentation methods has resulted in the discovery of life-saving medications that have transformed medical practise.

The process of microbial fermentation exemplifies the interaction of scientific innovation, engineering, and biomanufacturing. The capacity to design microbes to create complex proteins and medicinal compounds has allowed for the development of novel treatment techniques. However, the issues of optimising expression levels, preserving product integrity, and dealing with post-translational changes demand ongoing study and innovation. Microbial resistance is becoming an increasingly serious problem for antibiotic development. The capacity of microorganisms to adapt and develop in response to antimicrobial drugs necessitates creative techniques to avoid resistance mechanisms (Jampilek 2022). Therapeutic development that targets numerous areas of vulnerability, as well as the investigation of synergistic therapeutic combinations, are ways for mitigating the establishment of resistance. Finally, the path from historical uses to cutting-edge biotechnology breakthroughs emphasises microorganisms' critical role in influencing the pharmaceutical sector (Gupta et al. 2019b). As science and technology advance, the symbiotic interaction between microbes and drugs has the potential to transform healthcare, but the intricacies inherent in this integration need an ongoing dedication to research, innovation, and ethical concerns. The incorporation of microorganisms into medicines will surely continue to influence the future development of medicine via unwavering determination to overcome hurdles and capitalizing on possibilities.

Acknowledgements Author "Akarsh Verma" is grateful to the monetary support provided by the University of Petroleum and Energy Studies (UPES)-SEED Grant program.

Declarations

Conflicts of Interest The authors declare that there is no conflict of interest. Their primary aim is the dissemination of information.

Ethical Approval The authors hereby state that the present work is in compliance with the ethical standards.

Competing Interests The authors declare no competing interests.

Availability of Data and Materials Not applicable.

References

Akbari Kordkheyli V, Rashidi M, Shokri Y, Fallahpour S, Variji A, Nabipour Ghara E, Hosseini SM. (2022) CRISPER/CAS System, a Novel Tool of Targeted Therapy of Drug-Resistant Lung Cancer. Adv Pharm Bull, 12(2), 262–273. https://doi.org/10.34172/apb.2022.027.

Amreddy N, Babu A, Muralidharan R, Panneerselvam J, Srivastava A, Ahmed R, Mehta M, Munshi A, Ramesh R. Recent Advances in Nanoparticle-Based Cancer Drug and Gene Delivery. Adv Cancer Res. 2018; 137:115–170. https://doi.org/10.1016/bs.acr.2017.11.003. PMID: 29405974.

Bajaj JS, Ng SC, and Schnabl B. (2022) 'Promises of microbiome-based therapies', Journal of Hepatology, 76(6), pp. 1379–1391. https://doi.org/10.1016/j.jhep.2021.12.003. Available at: https://www.sciencedirect.com/science/article/pii/S0168827821022431.

Bonnet M, Lagier JC, Raoult D, Khelaifia S. Bacterial culture through selective and non-selective conditions: the evolution of culture media in clinical microbiology. New Microbes New Infect. 2019 Nov 30;34:100622. https://doi.org/10.1016/j.nmni.2019.100622. PMID: 31956419.

Canouï E, Launay O. (2019) Histoire et principes de la vaccination [History and principles of vaccination]. Rev Mal Respir, 36(1), 74–81. https://doi.org/10.1016/j.rmr.2018.02.015.

Chen L, Musa AE. Boosting the immune system against cancer by resveratrol. Phytother Res. 2021 Oct;35(10):5514–5526. https://doi.org/10.1002/ptr.7189. PMID: 34101276.

Christaki E, Marcou M, Tofarides A. (2020) Antimicrobial Resistance. Gastrointest Endosc Clin N Am, 30(4), 619–635. https://doi.org/10.1016/j.giec.2020.06.004.

Coates ARM, Hu Y, Holt J, Yeh P. Antibiotic combination therapy against resistant bacterial infections: synergy, rejuvenation and resistance reduction. Expert Rev Anti Infect Ther. 2020 Jan;18(1):5–15. https://doi.org/10.1080/14787210.2020.1705155. PMID: 31847614.

Cubillos-Ruiz A, Guo T, Sokolovska A, Miller PF, Collins JJ, Lu TK, Lora JM. Engineering living therapeutics with synthetic biology. Nat Rev Drug Discov. 2021 Dec;20(12):941–960. https://doi.org/10.1038/s41573-021-00285-3. PMID: 34616030.

Dehbanipour R, Ghalavand Z. Anti-virulence therapeutic strategies against bacterial infections: recent advances. Germs. 2022 Jun 30;12(2):262–275. https://doi.org/10.18683/germs.2022.1328. PMID: 36504617.

Demain AL, Sanchez S. (2009) Microbial drug discovery: 80 years of progress. J Antibiot (Tokyo), 62(1), 5–16. https://doi.org/10.1038/ja.2008.16.

Douafer H, Andrieu V, Phanstiel O 4th, Brunel JM. Antibiotic Adjuvants: Make Antibiotics Great Again! J Med Chem. 2019 Oct 10;62(19):8665–8681. https://doi.org/10.1021/acs.jmedchem.8b01781. PMID: 31063379.

Gasbarrini G, Bonvicini F, and Gramenzi A. (2016) 'Probiotics History', Journal of Clinical Gastroenterology, 50(Suppl), pp. S116–S119. https://doi.org/10.1097/MCG.0000000000000697.

Gupta J, Rathour R, Dupont C, Mishra A and Thakur IS. 2022. Biogeochemical profiling and taxonomic characterization of municipal landfill site by metagenomic sequencing. Bioresource Technology, 351, p. 126936.

Gupta J, Rathour R, Dupont CL, Kaul D and Thakur IS. 2021b. Genomic insights into waste valorized extracellular polymeric substances (EPS) produced by Bacillus sp. ISTL8. Environmental Research, 192, p. 110277.

Gupta J, Rathour R, Maheshwari N and Thakur IS. 2021a. Integrated analysis of Whole genome sequencing and life cycle assessment for polyhydroxyalkanoates production by Cupriavidus sp. ISTL7. Bioresource Technology, 337, p. 125418.

Gupta J, Rathour R, Medhi K, Tyagi B and Thakur IS. 2020. Microbial derived natural bioproducts for a sustainable environment: a bioprospective for waste to wealth. In Refining biomass residues for sustainable energy and bioproducts (pp. 51–85). Academic Press.

Gupta J, Rathour R, Singh R and Thakur IS. 2019b. Production and characterization of extracellular polymeric substances (EPS) generated by a carbofuran degrading strain Cupriavidus sp. ISTL7. Bioresource Technology, 282, pp. 417–424.

Gupta J, Tyagi B, Rathour R and Thakur IS. 2019a. Microbial treatment of waste by culture-dependent and culture-independent approaches: Opportunities and challenges. Microbial Diversity in Ecosystem Sustainability and Biotechnological Applications: Volume 1. Microbial Diversity in Normal & Extreme Environments, pp. 415–446.

Hiss PH. (1897) On a method of isolating and identifying bacillus typhosus, based on a study of bacillus typhosus and members of the colon group in semi-solid culture media. J Exp Med, 2(6), 677–700. https://doi.org/10.1084/jem.2.6.677

Ho D, Quake SR, McCabe ERB, Chng WJ, Chow EK, Ding X, Gelb BD, Ginsburg GS, Hassenstab J, Ho CM, Mobley WC, Nolan GP, Rosen ST, Tan P, Yen Y, Zarrinpar A. Enabling Technologies for Personalized and Precision Medicine. Trends Biotechnol. 2020 May;38(5):497–518. https://doi.org/10.1016/j.tibtech.2019.12.021. PMID: 31980301.

Hogwood CE, Bracewell DG, Smales CM. Host cell protein dynamics in recombinant CHO cells: impacts from harvest to purification and beyond. Bioengineered. 2013 Sep-Oct;4(5):288–91. https://doi.org/10.4161/bioe.23382. PMID: 23328085.

Hu Y, Liu Y, Coates A. Azidothymidine Produces Synergistic Activity in Combination with Colistin against Antibiotic-Resistant Enterobacteriaceae. Antimicrob Agents Chemother. 2018 Dec 21;63(1):e01630-18. https://doi.org/10.1128/AAC.01630-18. PMID: 30373798.

Jakubiec-Krzesniak K, Rajnisz-Mateusiak A, Guspiel A, Ziemska J, Solecka J.Secondary Metabolites of Actinomycetes and their Antibacterial, Antifungal and Antiviral Properties. Pol J Microbiol. 2018;67(3):259–272. https://doi.org/10.21307/pjm-2018-048. PMID: 30451442; PMCID: PMC7256786.

Jampilek J. (2022) 'Drug repurposing to overcome microbial resistance', Drug Discovery Today, 27(7), 2028–2041. https://doi.org/10.1016/j.drudis.2022.05.006. Available at: https://www.sciencedirect.com/science/article/pii/S1359644622001933.

Kausar S, Said Khan F, Ishaq Mujeeb Ur Rehman M, Akram M, Riaz M, Rasool G, Hamid Khan A, Saleem I, Shamim S, Malik A. (2021) A review: Mechanism of action of antiviral drugs. Int J Immunopathol Pharmacol, 35, 20587384211002621. https://doi.org/10.1177/20587384211002621.

Lesher GY, Froelich EJ, Gruett MD, Bailey JH, Brundage RP (1962) 1,8-naphthyridine derivatives, a new class of chemotherapy agents. Journal of Medicinal Chemistry, 5, 1063-1065.

Lewis K (2013) Platforms for antibiotic discovery. Nature Reviews Drug Discovery, 12(5), 371-387. https://doi.org/10.1038/nrd3975. PMID: 23629505.

Mahajan GB, Balachandran L (2012) Antibacterial agents from actinomycetes - a review. Frontiers in Bioscience (Elite Edition), 4(1), 240-253. https://doi.org/10.2741/373. PMID: 22201868.

McEwen SA, Collignon PJ. (2018) Antimicrobial Resistance. Gastrointest Endosc Clin N Am, 30(4), 619–635. https://doi.org/10.1016/j.giec.2020.06.004.

McMillan HM, Kuehn MJ. The extracellular vesicle generation paradox: a bacterial point of view. EMBO J. 2021 Nov 2;40(21):e108174. https://doi.org/10.15252/embj.2021108174. PMID: 34636061.

Mitragotri S, Burke PA, Langer R. Overcoming the challenges in administering biopharmaceuticals: formulation and delivery strategies. Nat Rev Drug Discov. 2014 Sep;13(9):655–72. https://doi.org/10.1038/nrd4363. PMID: 25103255.

Morrison L, Zembower TR. (2020) Antimicrobial Resistance. Gastrointest Endosc Clin N Am, 30(4), 619–635. https://doi.org/10.1016/j.giec.2020.06.004.

Patra JK, Das G, Fraceto LF, Campos EVR, Rodriguez-Torres MDP, Acosta-Torres LS, Diaz-Torres LA, Grillo R, Swamy MK, Sharma S, Habtemariam S, Shin HS. Nano based drug delivery systems: recent developments and future prospects. J Nanobiotechnology. 2018 Sep 19;16(1):71. https://doi.org/10.1186/s12951-018-0392-8. PMID: 30231877.

Roy R, Tiwari M, Donelli G, Tiwari V. Strategies for combating bacterial biofilms: A focus on anti-biofilm agents and their mechanisms of action. Virulence. 2018 Jan 1;9(1):522–554. https://doi.org/10.1080/21505594.2017.1313372. PMID: 28362216.

Singh A. (2022, February 24). Title of the Webpage. Conduct Science. https://conductscience.com/culture-media/.

Stewart AJ & Devlin PM. (2006) The history of the smallpox vaccine. Journal of Infection, 52(5), 329–334. https://doi.org/10.1016/j.jinf.2005.07.021

Survase SA, Kagliwal LD, Annapure US, Singhal RS. (2011) Cyclosporin A–a review on fermentative production, downstream processing and pharmacological applications. Biotechnol Adv, 29(4), 418–435. https://doi.org/10.1016/j.biotechadv.2011.03.004.

Tan SY and Tatsumura Y. (2015) 'Alexander Fleming (1881-1955): Discoverer of penicillin', Singapore Medical Journal, 56(7), pp. 366-367. https://doi.org/10.11622/smedj.2015105.

Tekamp-Olson P, Valenzuela P. (1990) Gene expression and engineering in yeast and other fungi. Curr Opin Biotechnol, 1(1), 28–35. https://doi.org/10.1016/0958-1669(90)90007-8.

Verma A and Ogata S. 2023. Magnesium based alloys for reinforcing biopolymer composites and coatings: a critical overview on biomedical materials. Advanced Industrial and Engineering Polymer Research. https://doi.org/10.1016/j.aiepr.2023.01.002

Waclaw B. Evolution of Drug Resistance in Bacteria. Adv Exp Med Biol. 2016;915:49-67. https://doi.org/10.1007/978-3-319-32189-9_5. PMID: 27193537.

Waksman SA, Schatz A (1943) Strain specificity and production of antibiotic substances. Proceedings of the National Academy of Sciences of the United States of America, 29(2), 74–79. https://doi.org/10.1073/pnas.29.2.74. PMID: 16588605; PMCID: PMC1078562.

Zinner SH. Antibiotic use: present and future. New Microbiol. 2007 Jul;30(3):321–5. PMID: 17802919.

Microbial Marvels: Transforming the Way We Diagnose

Surbhi Gupta, Anushka Bhattacharyya, and Juhi Gupta

1 Introduction

Diagnostic microbiology is an emerging branch of microbiology that arises in the wake of increasing infectious diseases worldwide. Infectious disease has been spreading rapidly over the years due to environmental pollution, and globalization beyond their simultaneous detection. Globally, infectious diseases place a heavy strain on public health and society because of their unpredictability, transmissibility, and preventability (Jing et al. 2023), with the recent pandemic of COVID-19 serving as an example. 540 million cases of SARS-COV-2 infections have been reported and over 6.4 million deaths were reported worldwide (Solante et al. 2023). These global and local pandemics hampers the civilization process and aid the collective collaboration of all nations to understand the most diverse group of organisms i.e., microorganisms. Despite our reliance on the microbial world's inhabitants, we know very little about their demographics or nature. Less than 0.01% of the anticipated total number of microorganisms have been cultured and characterized (Parija et al. 2023). The study of microbiology has made significant improvements and researchers now fully comprehend the role of microbes in various fields ranging from agriculture, industries, medicine, and the environment.

One of the most fascinating applications of microbes is diagnostics. Microbes have been employed as diagnostic tools in medicine, agriculture, and environmental studies because of their unique physiological and metabolic variations (Gupta et al.

Surbhi Gupta and Anushka Bhattacharyya these authors have equally contributed to the current work.

S. Gupta · A. Bhattacharyya
Dr. B.R. Ambedkar Center of Biomedical Research, University of Delhi, Delhi 110007, India

J. Gupta (✉)
Amity Institute of Environmental Sciences, Amity University, Noida 201303, India
e-mail: jgupta2@amity.edu

J. Gupta and A. Verma (eds.), *Microbiology-2.0 Update for a Sustainable Future*,
https://doi.org/10.1007/978-981-99-9617-9_8

2019a, b). Microbes are employed in agriculture to detect plant disease and evaluate the condition of the soil (Gupta et al. 2022a). Additionally, microbes can be utilized as bioindicators to track environmental contamination levels. Because some bacteria and fungi are sensitive to particular contaminants, their presence or absence can reveal how contaminated an area is, as *Escherichia coli* presence indicates fecal contamination of water bodies which causes diarrhea. Microbes are used as diagnostic tools to aid in timely identification and diagnosis of infections. Pathogens like bacteria, viruses, fungi, and parasites are different and distinct from each other thus requiring an individual diagnostic approach.

The invention of microscopes was essential for identifying and understanding microorganisms, which helped establish the famous germ theory of disease. Microbes were unknown to scientists before the invention of microscopes; their existence was only speculated about (Gupta et al. 2022b). The germ theory of disease, which was developed in the mid-nineteenth century, laid the foundation principle of modern medicine and the objective of clinical microbiology laboratories to identify bacterial, fungal, viral, and parasitic pathogens and the diseases they elicit (Schmitz et al. 2022). This idea has profoundly changed healthcare and our understanding of infectious diseases. The era known as the "Golden Age of Microbiology" which is mostly credited to Louis Pasteur and Robert Koch, was a turning point in the growth of the discipline. Louis Pasteur, a French scientist who is often regarded as the founder of microbiology, was one of the significant pioneers in the development of germ theory (Scavone et al. 2019). Robert Koch, a German physician, is another pioneer in the microbial world. His renowned work in identifying various microbes as the etiological agents of different diseases. Koch's postulates, a set of standards for determining if a microbe and a disease are causally related, were a key technique for discovering pathogenic microbes (Grimes et al. 2006). Based on their discovery various techniques were developed which helped in the detection of microbial pathogens. Microscopy invention led to the development of more advanced detection tools. Understanding the molecular and genetic makeup of microbes allowed researchers to culture them and do molecular analysis. Viral infections are often detected by viral culturing methods and serological assays that look for antibodies against the viral antigens, whereas bacterial detection uses gram staining and culturing methods. Shortcomings of past diagnostic approaches have been rectified by modern diagnostic techniques.

Understanding microbiomes led to the widespread use of these microbes as diagnostic tools with advanced techniques like MALDI, ELISA, PCR, and whole genome sequencing and improved our knowledge of the diverse array of diseases and environmental circumstances. CRISPR/Cas system is another potential diagnostic tool that can be applied for the detection of many DNA/RNA pathogens. Detection of SARS-CoV-2 and its variants such as the real-time reverse transcriptase-polymerase chain reaction (RT-qPCR) and the Clustered Regularly Interspaced Short Palindromic Repeats (CRISPR)-based approaches detecting the nucleic acids of the virus (Aouida et al. 2023). We can now use the information-processing capabilities of live cells to identify the disease and then treat it in a controlled manner owing to innovations in synthetic biology. Living sensors may be developed to accurately detect disease indicators, such as inflammation-related byproducts, and to react by administering tailored therapies in the region that is affected (Inda and Lu 2020). Highly sensitive, selective, and stable microbial biosensors have been incorporated for detecting chemical and biological substances which are produced during specific infections in the body (Karman et al. 2023). The gut microbiome has also been related to a number of diseases, including obesity, diabetes, and inflammatory bowel disease, and might one day be used as a diagnostic tool. The potential benefits of microorganisms in diagnosis are projected to increase as our understanding of microbiology grows, resulting in more precise and effective diagnostic procedures (Verma and Ogata 2023).

2 Past Techniques

Microbes have always been a part of speculation in scientific history. The first identification of microbial importance happened due to the invention of the microscope which provides the ability to view and characterize diverse microbes in detail, leading to significant advancements in microbiology. The visualization and characterization of bacteria, protozoa, fungi, viruses, and other microorganisms allowed researchers to elicit their role in various diseases, forwarding the understanding of their pathologies.

Microscopy was the first direct detection method. As microbiology progressed different techniques emerged such as microbial culturing, gram staining, serological testing, antibiotic sensitivity test, and Sanger sequencing. Traditionally, the presence of bacteria (*Staphylococcus*, *Enterococcus*, *Serratia*, and *Pseudomonas)* in or on humans during health or disease has been determined by growing the bacteria in culture (Rhoads et al. 2012). Gram staining indicates bacterial infection and easily distinguishes between gram-positive bacteria (e.g., *Staphylococcus* species, and *Streptococcus* species) and gram-negative bacteria (e.g. *Enterobacteriaceae* and *E. coli* species). The most typical bacterial disease is urinary tract infection (UTI). UTIs are the second most common cause of bacterial infections in hospitalized patients and the most prevalent hospital-acquired infection, accounting large number of nosocomial infections. Pseudomonas, Proteus species, and enteric bacteria have been and continue to be the primary causes of UTI. Gram staining of centrifuged urine specimens, Gram staining of uncentrifuged urine specimens, or direct observation of bacteria in urine specimens can all be used to identify bacteriuria microscopically (Burillo et al. 2014). The need for precise viral diagnosis has risen as a result of antiviral medications. Dedicated virology laboratories generally carry out viral isolation using cell culture. The identification of viruses is made possible through viral culture (Enders et al. 1949). The most common method for identifying viral growth in cell culture is to observe morphological alterations in the cells, or CPE (cytopathic effect). The CPE's properties are frequently distinct enough for the lab to suspect a certain virus as the reason for it (Storch et al. 2000). Culturing of adenovirus, enterovirus, rhinovirus etc. with various cell lines in traditional cell culture approaches help in the detection of various viral diseases (Leland and Ginocchio

2007). The therapeutic efficacy of the conventional approach for identifying viral infections by looking for specific antiviral antibodies is constrained by the requirement to compare acute and convalescent antibody titers. EBV (IgM antibodies to the viral capsid antigen), CMV, hepatitis A virus, and other viruses are among those for which the detection of virus-specific IgM antibodies is advantageous (Storch 2000). Epstein-Barr virus (EBV) antigens are often used to define infection level and for the differential diagnosis of other pathogens generating mononucleosis syndrome in serological diagnosis of EBV infection (De Paschale and Clerici 2012). When common microorganisms present with uncommon phenotypes, reliance on phenotype can compromise accurate identification (Petti et al. 2005). Identification by gene sequencing is more reliable, does not require ideal growth or even a live microorganism, and has the added advantage of establishing taxonomic relationships among microorganisms. Amplification of genes and sequencing of wide range of gene targets using chain termination methods developed by Sanger for microorganisms like bacteria and fungi have emerged as important methods of diagnosis of pathologies (Reller et al. 2007) (Table 1 and Fig. 1).

These traditional approaches to diagnosis showed significant advancement in their time but they have various limitations. Microbial and viral culturing has poor sensitivity with limited microorganism culturing availability. Serological testing results are generally retrospective since an immunosuppressed host may be unable to mount a response. Other molecular diagnostics techniques like Sanger sequencing, and antibiotic sensitivity tests require facility, sophisticated research due to high levels of contamination chances. These techniques are time-consuming and labor-intensive. However, they laid the foundation stone for the development of more advanced and rapid diagnostic methods we use today, such as polymerase chain reaction (PCR), nucleic acid sequencing, and immunological assays, which have revolutionized disease diagnosis by offering improved sensitivity, specificity, and speed.

3 Present Diagnostic Techniques

Infectious diseases are a major cause of deaths worldwide accounting for approximately 40% of annual deaths globally (Hassan et al. 2020). Timely detection of such diseases prevents further spread of the infection and thus prevents any further comorbidities associated with the progression of the disease. Various methods have been utilized for clinically detecting the pathogens causing infections and can be conveniently categorized according to the flow chart in Fig. 2.

(a) Rapid-Antigen tests help in identifying the pathogen either by using antibody-antigen conjugates coupled with fluorescence or by label-free methods like MALDI-TOF.
(b) Many commercially biochemical microbial identification assays are available that help in profiling the disease-causing organism such as gonochek II, and hippurate. Disk tests are also available for antibiotic susceptibility testing

Table 1 Description of past techniques for microbial diagnosis

S. no.	Techniques	Description	Turnaround time	History	References
1	Microbial culturing	• Patient samples (blood, urine, and sputum) are cultured to promote the growth of potential pathogens on nutrient-rich agar plates • The resulting colonies were then examined macroscopically and microscopically to identify the causative microorganism responsible for the disease	2–14 days	Louis Pasteur developed the first liquid artificial culture medium in 1860	Mostowy (2022)
2	Gram staining	• Staining technique to differentiate bacteria into Gram-positive and Gram-negative bacteria based on their cell wall composition • Bacteria show distinct staining which aids in identification • Gram positive bacteria retain the purple stain of crystal violet whereas gram negative bacteria show pink stain of safranin	30–40 min	Danish bacteriologist Hans Christian Gram who first introduced it in 1882	Panicker et al. (2023)
3	Serological testing	• Detection of specific antibodies or antigens produced in response to an infection, aiding in the diagnosis of infectious diseases • Complement fixation test-Detection of antibodies that fix complement, helping diagnose various infectious diseases • Agglutination test-Detection of antibody-antigen reactions by causing clumping or agglutination of specific particles, aiding in the diagnosis of infections	4–6 h	Bordet's research on the destruction of bacteria and red corpuscles in blood serum, conducted at the Pasteur Institute, Paris (1894–1901)	Mahajan (2023)
4	Viral culture	• Isolation and identification of viruses by infecting susceptible cell lines with patient samples and observing viral cytopathic effects	5–10 days	Weller and Enders in 1948	Laqueur et al. (1991) Enders et al. (1949)

(continued)

Table 1 (continued)

S. no.	Techniques	Description	Turnaround time	History	References
5	Antibiotic sensitivity	• Testing the susceptibility of microbes to different antibiotics using methods like the Kirby-Bauer disk diffusion technique • The zone of inhibition surrounding each disc revealed whether the microbe was sensitive to the antibiotic or resistant to it, assisting in the decision of the best course of action	18–24 h	Bolmstron and Eriksson in 1980 developed etest	Morris et al. (2005)
6	Sanger sequencing	• Using the chain termination approach for DNA sequencing, which relies on the DNA polymerase enzyme's selective incorporation of chain-terminating dideoxy nucleotides, nucleotides in single-stranded DNA molecules are identified by complementary synthesis of polynucleotide chains	7–14 days	Fredrick Sanger developed the first major breakthrough in sequencing techniques in 1977	Sanger et al. (1977)

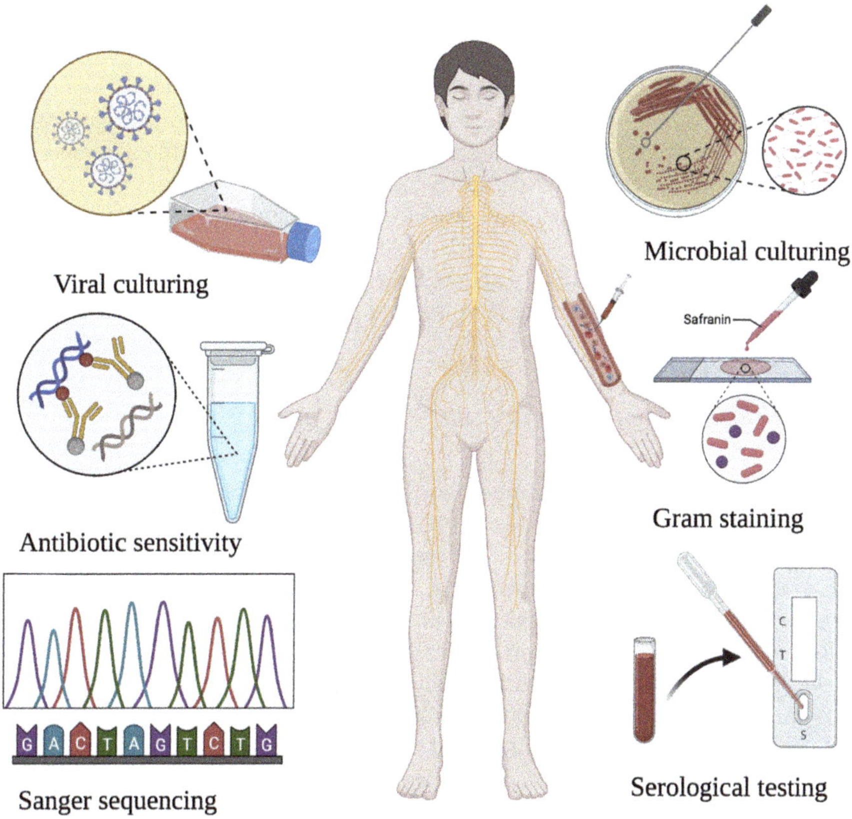

Fig. 1 Illustration of past techniques used to diagnose microbial diseases

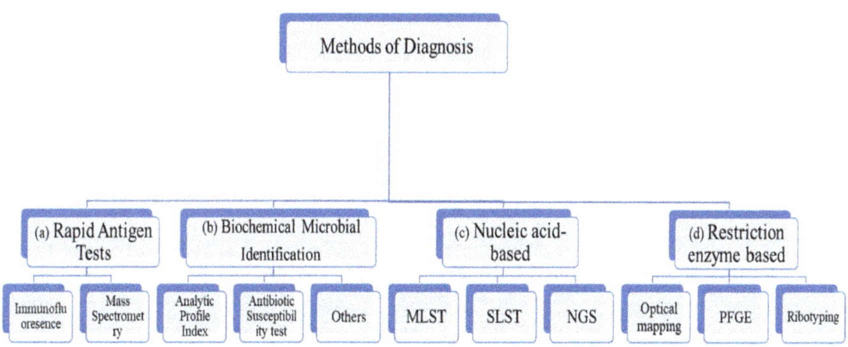

Fig. 2 Flowchart of all the current methods available for clinical diagnosis

or butyrate disks for confirming the presence of gram-negative diplococcus *Moraxella catarrhalis* etc.

(c) With the advances made in the extraction of nucleic acids, sequencing of nucleic acids either multi-locus or single-locus or different types of next-generation sequencing helps in creating libraries for referencing the different strains and their properties, thus, enabling faster identification.

(d) All the methods listed under restriction enzyme based have the standard first step of fragmenting the DNA or rRNA which is further used as either a 'barcode' to reference back to a library and identify the unknown pathogen as in Optical mapping or run on a gel-like in Pulse-field gel electrophoresis (PFGE).

3.1 Automated Culture Systems

Any infection is clinically confirmed by isolating and detecting the correct specimen causing the infection. Antibiotic susceptibility tests (AST) are further required to help guide physicians in choosing effective treatment strategies (Tang et al. 2013). With the ever-increasing workload, increasing demand for faster results, and declining number of microbiological technicians, automated cultures for the identification of infectious agent has become apparent and in use. (Williams and Trotman 1969). Lower bacterial load along with high chances of contamination are some factors that are circumvented with automated culture systems (Islam et al. 2022). These help in maintaining sterile conditions and along with maintaining large volumes and flows, many systems come with photography apparatus which help in studying the morphology and physiology of the colonies or plaques, thus determining the strain of the pathogen.

Automated blood culture is a great example of automation helping in reducing the mortality rates caused due to sepsis. Bloodstream invasion is a condition in which the whole body can undergo an inflammatory state which can turn deadly. With unclear symptoms, antibiotics can't be administered, thus delaying treatment. Such delays can turn worse, particularly for children (Ahmad et al. 2017). Early diagnosis and initiation of proper antibiotic therapy of bloodstream infections (BSI) initially is based on isolation, identification, and performance of susceptibility testing on microorganisms isolated from blood cultures.

Two decades have already passed since the launch of the first automated blood culture systems, and with time, better versions have become commercially available. While initially, the major advantage stemmed from automatic monitoring of microbial growth based on their CO_2 production levels along with agitation of the culture to facilitate aeration, newer systems come with the advantage of shorter times and integrating culture system (incubator) with measurement system (microplate reader) (Horinouchi et al. 2014).

Blood cultures are useful to detect bacteremia, especially in immunocompromised patients. Patients suffering from AIDS are often infected with *Mycobacterium avium, Mycobacterium tuberculosis.* Sepsis can be caused by infectious strains

Table 2 Commercial continuously monitoring blood culturing systems (CMBCSs)

S. no.	Manufacturer	Microbial detection mechanism	System Names	References
1	BioMerieux	Colorimetric for CO_2 production	BacT/ALERT	Thorpe et al. (1990)
2	Becton–Dickinson	Fluorescent for CO_2 production	BACTEC	Ryan and Murray (1993)
3	Thermo scientific	Manometric for CO_2 production	Versa TREK	Nanua et al. (2009)

of *Escherichia coli, Staphylococcus aureus, Pseudomonas aeruginosa*, and yeast. Systems are available to monitor and detect septicemia caused by pathogens (bacteria, viruses, fungi, and protozoa) using colorimetric, manometric, and fluorescent mechanisms to detect the production of CO_2 (Tang et al. 2013) (Table 2).

3.2 *Rapid Antigen Testing*

Molecular analysis of microbial pathogens by antigens, antibodies, or nucleic acids is an important rapid complement to culture diagnosis. The rapid analysis represents a major advantage over culture, which generally takes 1–2 days or, in the case of fastidious microorganisms, up to several weeks. Rapid antigen tests are tools that reduce the time it takes to detect the microorganism, from the receipt of the sample to the final results, these tests don't take much time. Rapid antigen tests can be of various types as explained in Fig. 2. Immunofluorescence, chemifluorescence, different types of Polymerase chain reactions (PCR), ELISA, and many others. Those which are most in use are explained further.

3.2.1 Real-Time PCR

The polymerase chain reaction has evolved in many different forms, RT-PCR, qPCR, multiplex PCR, and touchdown PCR. With the SARS-CoV-2 pandemic, RT-PCR has redefined the gold standard test for accurate diagnosis (Saeed and Ahmad 2013) (Table 3).

As with other PCR methods discussed in the above table, Real time-PCR also uses double-stranded DNA and denatures it at high temperatures of 95 °C, then cools it at around 50–60 °C which helps in primer annealing at their specific sites, and then the temperature is again raised to 72 °C along with the addition of *Taq* polymerase, thus extending the primers and creating a cDNA. However, RT-PCR differs from conventional PCR due to the use of fluorophore labels along with quenchers in a probe design which allows the monitoring of the cDNA production (Saeed and Ahmad 2013).

Table 3 Different types of modifications of Polymerase chain reaction (PCR)

S. no	Types of PCR	Description	References
1	Real-time PCR (qPCR)	• Similar to conventional PCR • Quantitation is visible due to the presence of a fluorescent probe	Kralik and Ricchi (2017)
2	Nested PCR	• Increases specificity by using two sets of primers • First set of primers, generate specific and non-specific DNA products • Second set of primers is used on the products generated by the first set of primers • This helps amplify only the targeted sequence	Green and Sambrook (2019)
3	Reverse transcriptase PCR (RT-PCR)	• Used to amplify a known RNA sequence	Huggett et al. (2005)
4	Multiplex PCR	• Used to amplify multiple targets in a single PCR reaction • Used to identify exonic/intronic sequences in specific genes	Elnifro et al. (2000)
5	Touchdown PCR	• Increases specificity in a complex template DNA amplification • Achieved by increasing the annealing temperature, allowing only perfectly compatible primer-template hybrids are formed	Korbie and Mattick (2008)

Real time-PCR has become useful in the clinical setting due to its software providing quantifiable data i.e., the melting curve. A duplex RT-PCR, helps in determining the differential expression of a gene, by comparing it with the copy numbers of a housekeeping gene that will not change.

Multiplex, duplex RT-PCR have been used in many assay kits for the detection of microbes such as *Helicobacter pylori* in gastric biopsies and stools (Redondo et al. 2018), skin and nail dermatophyte infections due to fungi such as *Trichophyton* and *Epidermophyton* (Lin et al. 2021) or even for the detection of respiratory viruses (Cho et al. 2014). Other examples of Real Time-PCR-based devices detecting parasites, such as *Plasmodium falciparum* and *Plasmodium vivax*, or bacteria, like multidrug-resistant tuberculosis (MDR-TB), *E coli* O157 mono: H7, *Salmonella typhimurium*, and *Vibrio parahaemolyticus citogeni, or* viruses such as Ebola and HIV, are available nowadays (Veron et al. 2009).

Advances in lab-on-a-chip technology have led to the development of some Point-of-Care (POC) devices using Real-Time PCR and approved by the United States Food and Drug Administration (Garzarelli et al. 2022). The miniaturized version of RT-PCR has helped in the rapid diagnosis of dengue serotypes (Moser et al. 2022), to onsite rapid detection of SARS-CoV-2 patients.

3.2.2 ELISA

ELISA is a type of enzyme immunoassay (EIA) that detects immunological reactions and quantifies them using the catalytic property inherently present in the enzymes (Aydin 2015). It can be performed in a microwell, tube, or bead and can be categorized into four categories depending on the order of the sample and reagents added and what is used for quantification (antibody or antigen). Direct ELISA uses antigen-coated plates to screen antibodies, Indirect ELISA uses antibody-coated plates to screen for antigens/antibodies, Sandwich ELISA uses an antibody-coated plate to screen antigens while the last one, Competitive ELISA is used to screen antibodies.

The immunological reaction of antibody-antigen has helped in directing the use of microbes for clinical diagnosis of infections using ELISA. The disease could vary from bacterial infections such as leprosy caused by *Mycobacterium leprae* (Duthie et al. 2014) to viral infections of dengue (Kadkhoda et al. 2017) or chikungunya (Shukla et al. 2009).

Further coupling of ELISA with other molecular techniques such as RT-PCR has helped in better evaluation because of the increased sensitivity of such tests. The samples are first screened using the antibodies in ELISA against the sample biomarkers and then a confirmatory test is done using RT-PCR or q-PCR, such as for the diagnosis of *Flavivirus* causing acute dengue virus infections (Iqbal et al. 2023) and to detect mycobacterial antigens for pulmonary tuberculosis (Sharma et al. 2023).

3.3 MALDI-TOF

Matrix-associated laser desorption/ionization time-of-flight mass spectrometry (MALDI-TOF) is a useful technique for rapidly characterizing and identifying bacteria, fungi, or viral infections (Kocaman et al. 2019). As the name suggests this technique utilizes the energy of a laser associated with matrix desorption for the breakdown of large molecules into their associated ions which are then detected and further evaluated for a better understanding of the molecule and its structure. This technology's attributes include rapid identification, accuracy, low cost, and waste minimization (Hou et al. 2019).

3.3.1 Rapid Bacterial Identification

In 1975, Catherine Fenselau and John Anhalt proposed the use of mass spectrometry to identify microorganisms specifically bacterial identification (Anhalt and Fenselau 1975). Kochi Tanaka's demonstration of MALDI-TOF use for biological macromolecules in 1985, rewarded his discovery with a Nobel Prize in Chemistry (Tanaka 2003). MALDI-TOF has become an indispensable tool for bacterial identification. It has become useful for bacterial analysis as direct patient samples along with the whole bacterial cells from environmental samples (Avanzi et al. 2017) can be used for analysis. Thus, the samples can be whole-bacterial cell cultures, or they can be further processed.

- **DNA/RNA Based**:

  DNA/RNA based methods, especially small subunit ribosomal rRNA methods have been useful for MALDI-TOF/MS studies because of their applicability across bacterial and archeal domains (Kim et al. 2012). The process can be briefly described such that, initially the DNA sequence from an unknown bacterium is prepared in the same way as that in any other techniques, and the extracted nucleic acid is then conjugated with a chosen primer and the targeted region is amplified using PCR. The cDNA is further transcribed to produce RNA such as in the case of 16S rRNA. Then sequencing analysis is conducted using MALDI-TO and the results are then compared against a database of known sequences, for example, GenBank and the closest type of strain match is used for potential species identification (Kim et al. 2014).

- **Protein Based**:

  Bacterial proteins can be obtained for analysis by directly depositing the cell lysis on the target plate. This is often accomplished by mixing, washing, depositing, and co-crystallizing the lysate onto the MALDI target plate after adding a highly acidic MALDI matrix solution to bacterial cells (Soro-Yao et al. 2014). Using this method, both the external surface proteins and intracellular proteins can be analyzed (Alizadeh et al. 2021).

MALDI-TOF has been useful to identify both gram-negative and gram-positive strains of bacteria and has been helpful to identify various mycobacteria and acid-fast bacteria, such as *Mycobacterium tuberculosis*, *Mycobacterium intracellulare*, *Mycobacterium avium*, *Mycobacterium bovis*, *Mycobacterium kansasii*, and *Mycobacterium fortuitum*. Rapid and precise detection of this bacteria is crucial due to the development of tuberculosis by *M. tuberculosis*, which is one of the deadliest global illnesses and health concerns (Deng et al. 2011).

Identification of anaerobic bacteria is another crucial application of MALDI-TOF MS that faces many challenges because of their difficult growth on solid media, the inefficiency of routine biochemical tests for their identification, and the requirement for particular environmental conditions that develop the need for new and efficient diagnostic procedures (Nagy et al. 2012).

Apart from the clinical significance, MALDI-TOF has been utilized for veterinary studies of associated diseases such as in the case of fish pathogens. Several serotypes of *Vibrio anguillarum* and *Yersinia ruckeri*, and *Aeromonas salmonicida* have been successfully identified upto their different biotypes using MALDI-TOF, and thus have been useful to prevent monetary loss in the business (Jansson et al. 2020).

3.3.2 Mycological Analysis: Identification of Fungi

The majority of yeast identification in the clinical laboratory has relied on traditional techniques, such as microscopy and biochemical testing, which can take up to several days, are expensive, and demand highly qualified lab employees (Bader 2013).

Just as in bacterial analysis, fungal analysis can also be done by direct transfer on MALDI-TOF/MS target plate, commonly used for yeasts or by using the off-plate method with protein extraction step used for filamentous fungi (Patel 2019).

For the identification of the peaks generated, databases are curated with large datasets for speedy identification. Different MALDI systems have been compared and according to their datasets and species level cutoffs, have been useful for the identification of different types of fungi. Vulvovaginal candidiasis caused by *Candida albicans* (Pereira et al. 2021) is an important clinical application, but using MALDI biotyping presence of filamentous fungi has also been established for *Aspergillus, Fusarium,* and *Mucorales* (De Carolis et al. 2012). *Aspergillus* infections are known to cause respiratory distress with pulmonary aspergillosis and rhinosinusitis being the most prominent examples (Vuong and Waymack 2023). An infection known as mucormycosis is brought on by a fungus of the Mucorales order. Lungs, sinuses, soft tissue, skin, gastrointestinal tract, and bloodstream are among the infection sites (Anaissie et al. 2003). The *Fusarium spp.* has collectively been known to cause a range of infections in a variety of species of humans, animals as well as plants (Gomes et al. 2015). *Fusarium solani* and *F. oxysporum* are globally distributed and represent the most common complexes (Batista et al. 2020).

3.3.3 MALDI-TOF and Virology

Studies have been performed using the multiplex MALDI-TOF MS method which demonstrated that successfully can detect human herpesviruses (HHVs) from a wide variety of archival biological specimens (Sjöholm et al. 2008). The detection limits of the MS approaches were equivalent to earlier studies of multiplex herpesvirus detection utilizing an oligonucleotide microarray and multiplex PCR techniques, and the concordance rate between the MALDI-TOF MS methodology and reference methods was high. The few results that differ from reference methods may be due to low virus loads close to the method's detection limit and mild cross-reactivity. Therefore, this multiplex technique may prove to be very helpful for multiplex clinical diagnostic tests as well as for large-scale epidemiological research projects requiring broad virus detection (Cobo 2013).

In addition to gastroenteritis, hand, foot, and mouth disease, and serious neurological consequences such as aseptic meningitis, encephalitis, and acute neurogenic pulmonary edema, enteroviruses are a significant cause of sickness in children. The conventional techniques for detecting enteric viruses include indirect immunofluorescence assays, neutralizing tests, cell culture, and immunohistochemistry detection. The MALDI-TOF MS technology has shown a great capacity to diagnose these viruses with high throughput and quick turnaround time (Camarasa and Cobo 2018).

3.4 *Next-Generation Sequencing Techniques*

DNA sequencing has been a growing field of research which has helped for many years the basic understanding of cells and their organelles along with their function. Knowledge of the DNA sequence has helped to understand and localize various genes which are responsible for producing proteins and toxins that are responsible for the pathogenic activity of different pathogens (bacteria/fungi/virus/protozoa).

Next-generation sequencing techniques are considered the third generation of sequencing techniques with Sanger sequencing and the Maxam–Gilbert method of sequencing as pioneers, followed by pyrosequencing, ion-torrent sequencing and bridge amplification sequencing more commonly known as Illumina, as the second generation that helped automate sequencing (Hu et al. 2021).

Next-generation sequencing (NGS) technologies have become increasingly available for use in the clinical microbiology diagnostic environment. There are three main applications of these technologies in the clinical microbiology laboratory: whole genome sequencing (WGS), targeted metagenomics sequencing and shotgun metagenomics sequencing. These applications are being utilized for the initial identification of pathogenic organisms, the detection of antimicrobial resistance mechanisms and for epidemiologic tracking of organisms within and outside hospital systems.

3.4.1 Whole Genome Sequencing (WGS)

Whole genome sequencing is the process of comprehensively analyzing the entirety or nearly the entirety of an organism's genome at a single time. For confirmatory tests of infections, it takes a considerable amount of time to identify and process antibiotic susceptibility tests (Gupta et al. 2020). With the advent of WGS, there is a rapid fall in cost and total turnaround time thus becoming a valuable addition in diagnostic and reference clinical laboratories. The combined reduction of time and costs ultimately helps bridge the gap between research and diagnosis, which can further help in a better understanding of pathogens and their virulent life cycle and would help in managing the foreseeable trend of global antimicrobial resistance, as identified by the World Health Organisation (WHO) (Infectious Diseases Society of America 2010; Köser et al. 2012).

For performing WGS, the first step is to extract DNA from the organism (often cultured and taken from culture plates, which then undergo due process for extraction), this extracted DNA is then sheared into fragments and then given adaptors with unique barcodes that enable tracking of the different DNAs collected from different bacteria. This bar-coded mixture is then sequenced and pooled in a library. There are multiple ways to assemble the genome, one of them is de novo in which fragments are assembled in contigs, this process however is not very productive as high-quality genomes are difficult to obtain. Another process is using a reference library and aligning the fragments in accordance with the reference genome to get a consensus sequence. The last method is a hybrid of previously discussed methods with first generating the contigs and then aligning with the reference library. This way genomic data is collected. Now, for its application in diagnosis.

Since 2019, WGS has been used to detect foodborne illness and disease outbreaks in the US using the PulseNet method for bacteria such as *Salmonella*, *Escherichia coli* (O157 and other *Shiga* toxin-producing *E. coli*), *Campylobacter*, *Listeria monocytogenes*, *Shigella*, *Vibrio cholerae*, *Vibrio parahaemolyticus* and *Cronobacter* (Aladhadh 2023).

Foodborne diseases caused due these above-mentioned bacteria need identification of their specific serotypes, to identify the correct treatment for enteric fevers. The already available methods are not helpful for serotype identification and WGS in comparison to the previously used method of pulse-field gel electrophoresis (PFGE), has demonstrated better results (Stanton et al. 2022; Kersh et al. 2020).

Apart from serotype identification, WGS has been further helpful to track vaccine-preventable such as *Neisseria meningitidis* serogroups, *Streptococcus pneumoniae* serotypes (Tsang 2021), and antimicrobial-resistant pathogens such as multi-drug resistant *Mycobacterium tuberculosis* and prevent outbreaks caused by fastidious organisms such as *Mycobacterium chimaera* or nosocomial infections such as methicillin-resistant *Staphylococcus aureus,* carbapenem-resistant *Klebsiella pneumoniae,* and multi drug resistant *Acinetobacter baumannii* (Hilt and Ferrieri 2022).

While WGS provides many advantages for diagnosis and tracking and provides in-depth knowledge with the help of genomic analysis, the problem remains at the initial stage of WGS for decreasing turnaround time and reducing cost i.e. culturing of the pathogens (Gupta et al. 2021a). If the culture process is entirely eliminated from the overall process, there will be a drastic decrease in time and increased efficiency. There are methods available in the form of targeted sequencing and shotgun metagenomics, both of which came under direct sample metagenomic sequencing (Gupta et al. 2021b).

3.4.2 Targeted Sequencing

Targeted sequencing is a method by which a group of organisms is selected and further amplified either by PCR amplification or probe hybridization. The advantage of using these methods is low sample DNA needed for detection and higher sensitivity with low human DNA interference.

PCR amplification or deep amplicon sequencing has been helpful for the identification of bacteria mostly with the help of their 16S ribosomal RNA and 28S rRNA or ribosomal ITS genes for fungal identification. The deep amplicon sequencing of 16S RNA has been useful clinically to identify difficult-to-grow organisms that are generally unidentifiable by bacterial cultures. 16S RNA amplification is possible in a variety of samples such as blood, cerebrospinal fluid (CSF) or joint fluids (Köser et al. 2012).

While direct deep amplicon sequencing has limited success in fungal infections, it has shown considerable progress in clinical virology diagnosis specifically with drug-resistant cytomegalovirus (CMV) and human immunodeficiency virus (HIV).

The probe hybridization method has great potential for tracking and detection of global viral pathogens, especially seen in the 2010 Zika virus outbreak and the recent pandemic with the severe acute respiratory coronavirus 2 (SARS-CoV-2).

3.4.3 Shotgun Metagenomics

Shotgun metagenomic sequencing, as opposed to targeted sequencing, casts a larger net because all the nucleic acids in a sample are sequenced. It is possible to identify almost all pathogens, such as bacteria, fungi, viruses, and parasites, with a single test by sequencing all of the nucleic acids (Naccache et al. 2014). This method of sequencing has been successful in detecting an infection from many different specimen types including sources that are normally sterile such as CSF.

Shotgun metagenomics has been used commercially to create assays such as microbial cell-free DNA tests or a non-invasive liquid biopsy created by Karius laboratory, Explify™ Respiratory, which is performed primarily on BALs and was launched by IDbyDNA and ARUP Laboratories and Explify™ Respiratory to the diagnosis of urinary tract infections with antimicrobial resistance markers (Azar et al. 2021) (Fig. 3).

3.5 Microbial Biomarkers and Biosensors

Biosensors are nowadays used as key point-of-care analytical devices that are used for monitoring the progression and treatment of various diseases (Nikhil et al. 2016). The analytes for biosensors differ from whole cell viruses or bacteria, specific proteins to small ions and nucleic acids such as DNA or RNA (Rushworth and Hirst 2013). Most biosensors can be categorized according to their method of signal transduction

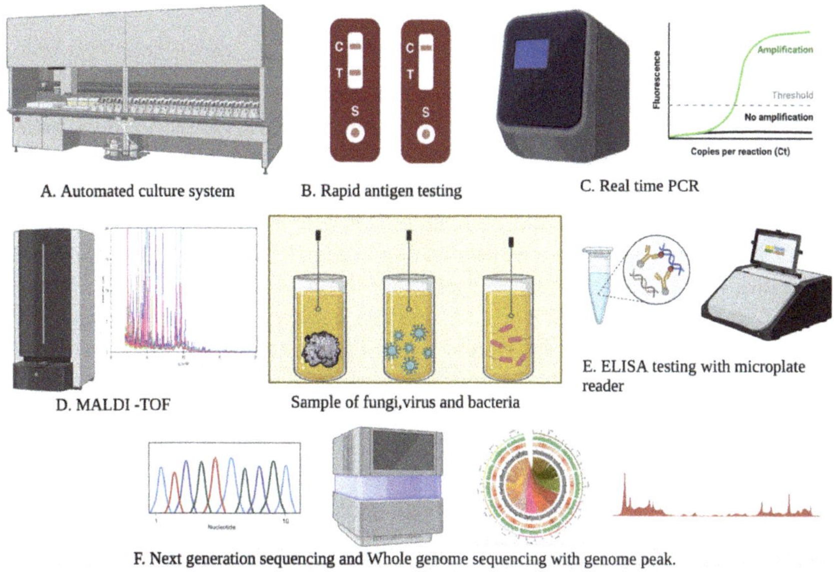

Fig. 3 Representation of various modern microbial diagnostic techniques used to detect the causative organisms

and by the type of bioreceptor employed i.e., catalytic or affinity-based. The affinity-based biosensors are much more utilized with respect to catalytic types due to their high sensitivity and specificity (Ahmed et al. 2014). The affinity-based biosensors utilize antibodies, bacteriophages, etc. and thus with the help of microbial epitopes and markers can further serve to prevent the massive spread of harmful diseases such as the recent case of COVID-19 infections that ravaged globally.

Lectins and bacteriophages share the similar characteristics of selective binding to different structural components of bacterial cell surfaces (Serra et al. 2008; Tolba et al. 2012). These have been employed for specific binding to bacterial receptor proteins for detection in various infections (Tolba et al. 2012; Ahmed et al. 2014). All the biosensors that employ lectins and viruses have to immobilize the lectin or bacteriophage cell wall binding domain (CBD) to act as a bioreceptor and make them used for selectivity and specificity. Bacteriophage biosensors have been designed to be utilized for early diagnosis of nosocomial infections caused by *Escherichia coli* (*E. coli*) O157:H7 and methicillin-resistant *Staphylococcus aureus* (MRSA) and thus useful in preventing life-threatening illness such as hemorrhagic colitis, septicemia, etc. (Tawil et al. 2012).

Molecular markers are another way that is utilized to detect and monitor various diseases like cancer, tuberculosis (TB), human immunodeficiency virus (HIV), microbial infections, and pregnancy screening (Li 2018). While lectins, and bacteriophages employing biosensors use these as bioreceptors, protein biomarkers using peptide biosensors utilize these markers with electrochemical transducer systems and thus can detect even 10^{-7} to 10^{-9} M (Hassan et al. 2020).

4 CRISPR-Cas9

The clustered regularly interspaced short palindromic repeats associated with protein 9 (CRISPR-Cas9) is a powerful gene-editing tool that has revolutionized the genetic engineering sector (Tavakoli et al. 2021) and has a high potential for therapeutic applications.

CRISPR-Cas9 is a heritable immunological defense system first identified in *Escherichia coli* and later discovered in a range of bacteria and archaea, which helps in defense against bacteriophage attacks (Strich and Chertow 2019). The mechanism of action of the CRISPR-Cas system can be described in three major steps: adaptation, maturation, and interference. During adaptation, the foreign substance is recognized either by PAM i.e., protospacer-adjacent motif, or by the help of integrated host factor (IHF) bends the foreign DNA so it can be recognized by Cas 1-Cas 2 complex. Once the correct sequence is recognized, the spacer integration step occurs which is important to generate mature crRNA as the orientation of the spacer sequence determines the correct reading of the leader sequence preceding the CRISPR array.

The mature crRNA is generated by cleaving the pre-crRNA and a variety of enzymes have been identified for this purpose. According to the subtype of the system, these molecules differ for example, in most subtypes Cas6 protein is used for cleaving the pre-crRNA and acts like a dimer, while in subtype II-A and B systems, Cas9 binds to a crRNA-tracrRNA complex and recruits RNase III, a host protein, to cleave the pre-crRNA repeat segment (Deltcheva et al. 2011).

There are two types of interference machinery i.e., class I and class II. The major difference between the mechanisms of interference is the complex of proteins involved. Class I machinery is termed a CRISPR-associated complex for antiviral defense (Cascade), where Cas5 and Cas6 bind to the 5′ and 3′ portion of crRNA, and a Cas8 subunit unwinds the foreign DNA. The Cas11 protein further forms and stabilizes the generated R-loop. Once there is a conformational change in Cas8 and Cas11 subunits, Cas3 is recruited to cleave the targeted unwinded DNA (Xiao et al. 2017). Class II machinery consists of a characteristic bilobed Cas9 protein that acts as a nuclease. This nuclease is guided to its target site by tracrRNA along with crRNA. The tracrRNA binds to the complementary repetitive regions on pre-crRNA recruiting Cas9, which then recognizes the PAM sequence on the foreign DNA. Recognition of the PAM sequence along with the presence of the tracrRNA—crRNA complex, results in a double-stranded cleavage of the targeted foreign DNA (Deltcheva et al. 2011).

The realm of infectious illnesses has benefited from broader applications due to a better knowledge of CRISPR-Cas biology. Tools made possible by CRISPR technology hold the promise of elucidating fundamental host-microbe interactions, assisting in developing quick and precise diagnostics, and advancing the detection and management of infectious diseases (Strich and Chertow 2019).

The identification of nucleic acids, such as DNA or RNA, linked to particular disorders is one use of CRISPR-Cas9 in disease diagnostics. The Cas9 enzyme is used by researchers to target and cleave specific DNA or RNA sequences in CRISPR-based diagnostic testing. The presence of a target sequence can be determined by combining Cas9 with a fluorescent or colorimetric signal, demonstrating the presence of the genetic material linked to the disease.

This method, often referred to as CRISPR diagnostics or CRISPR-based nucleic acid detection, has the power to completely change the way diseases are discovered. It has a high specificity and sensitivity, making it possible to identify target nucleic acids at extremely low quantities. Additionally, CRISPR diagnostics can be developed to detect numerous targets concurrently, making them helpful for identifying distinct strains or subtypes of an organism or for detecting complicated disorders.

Cancer, genetic problems, and infectious diseases like COVID-19 are all being studied in relation to CRISPR-Cas9-based diagnostics. As an instance, CRISPR-based tests have been created for the quick and precise identification of SARS-CoV-2, the virus that causes COVID-19 (Bailey and Diamond 2021) Compared to more conventional techniques like the polymerase chain reaction (PCR), these assays target and identify viral genetic material using CRISPR-Cas9.

The detection of genetic alterations linked to inherited disorders is another application for CRISPR-based diagnostics that has the potential to be exploited. Researchers can detect the existence of these mutations in patient samples by building CRISPR systems to target particular genomic sequences known to be associated with particular illnesses. Antibiotic resistance has become a rising cause of concern and researchers believe that the CRISPR-Cas system can be used against them. Researchers have established the presence of the I-F CRISPR system in *E. coli,* which is associated with antibiotic sensitivity. It has been proposed to use CRISPR technology to create particularly titratable antimicrobials to get rid of diseases. A subtype I-E CRISPR-Cas system was employed in these investigations to kill individual strains of *Salmonella enterica* and *E. coli* in pure and mixed culture tests (Javed et al. 2023).

While CRISPR-based diagnostics appear promising, it's vital to remember that they are still in the research and development stage. Optimizing sensitivity, lowering false positives or negatives, and assuring dependable and cheap implementation are just a few of the issues that need to be resolved. Nevertheless, CRISPR-Cas9's potential for disease diagnostics remains a topic of ongoing research, and future developments might result in the creation of novel diagnostic instruments (Table 4).

5 Limitations

The transformation of diagnostics techniques has revolutionized the way of identifying the causative organisms of an infectious disease. It is an important field which helps health care professionals make informed decisions about patient care. However, like any other technique, they too have limitations. Some of the key limitations are listed below:

Table 4 Description of microbes as biomarkers

Microbes	Virus 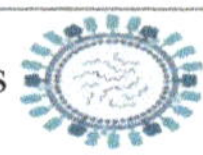	Bacteria 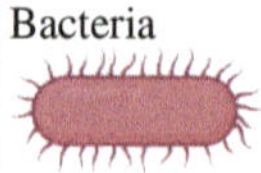	Fungi 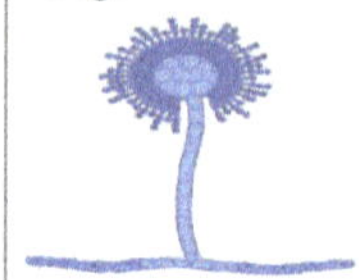
Size	1–2.5 Mb	2–8 μm long and 0.2–0.5 μm in diameter	2–10 μm in diameter and several cm in length
Major characteristics	• Capsid-coated genome • Host dependent replication	• Unicellular prokaryotes • Autonomous metabolism and replication	• Osmotrophic eukaryotes • Environmental or animal infection
Biomarkers	• Tumor necrosis factor-related apoptosis-inducing ligand (TRAIL) • Interferon (IFN)-γ-induced protein-10 (IP-10) • Myxovirus resistance protein A (MxA)	• 16S, 23S rRNA • 16S-23S rDNA ISR and IS256 • rpoB, gyrB, dnaK	• Nuclear ribosomal internal transcribed spacer (ITS) region • 18S rDNA
References	Tsao et al. (2020)	Li et al. (2023)	Oliveira and Azevedo (2022)

- **Culture-dependent techniques**: In traditional diagnostic microbiology, culture-dependent techniques are often used. In these techniques, bacteria are cultivated on culture media to aid identification and susceptibility testing. However, certain infections are difficult to propagate such as siderophores bacteria and many anaerobic bacteria. because they need particular growth conditions or lengthy incubation times. This causes a delay in the diagnosis, which might lead to missing or delayed treatment.
- **Slow turnaround**: Culture-based techniques and sequencing techniques like sanger sequencing and nucleic acid amplification techniques may take several days to get conclusive findings. The postponement in diagnosis may affect patient care, especially when prompt action is essential. Quick identification, for instance, is essential for putting in place the proper infection control measures in the case of serious diseases or outbreaks.
- **Limited scope**: The study of microbial pathogens detection and characterization is the main emphasis of diagnostic microbiology. The microbial world is so diverse i.e., a lot of infectious diseases are brought on by pathogens such as viruses, fungi, bacteria, parasites, or atypical pathogens that may be difficult to find using present techniques due to a lack of information about microbes. This restriction may make

diagnosing the cause more difficult and take longer. Sequencing techniques have tried to shorten this limitation but a holistic study of the genetic makeup of these microorganisms is needed.

- **False negatives and false positives**: Diagnostic microbiology techniques can sometimes produce false-negative or false-positive results. False negatives occur when the test fails to detect the presence of a pathogen, leading to missed diagnoses. False positives, on the other hand, can occur due to contamination or cross-reactivity, leading to unnecessary treatments and costs, as well as potential patient anxiety. The main limitation of ELISA, once the microbe has been eliminated, these antibodies may continue to exist for months or even years, making it important to carefully interpret test findings. The difficulty to judge specimen quality, they have to establish occasionally arbitrary cutoffs, hook effects, interfering elements such as rheumatoid factors and heterophile antibodies, and the requirement for cautious and thorough washing to prevent false-positive findings are all disadvantages.
- **Lack of sensitivity and specificity**: The sensitivity and specificity of diagnostic tests determine their accuracy in detecting and identifying pathogens. Some diagnostic methods may have limitations in terms of sensitivity (ability to detect true positive cases) or specificity (ability to rule out false positives). Improving the sensitivity and specificity of diagnostic tests is crucial for reducing false results and ensuring accurate diagnoses. The downside of MALDI is that because the co-crystallized sample is heterogeneous, there may be variations in the intensity of the spectral peaks from various locations on the target plate. Additionally, since the matrix was chosen empirically, the mass spectrum is protocol-dependent. Due to the different methods used, comparison of the spectrum patterns of bacteria described in the literature is difficult due to the influence of laser energy and sample preparation.
- **Limited point-of-care testing**: Point-of-care testing (POCT) refers to diagnostic tests performed at or near the site of patient care, providing rapid results for immediate decision-making. While POCT has been beneficial in various medical fields, its application in diagnostic microbiology is limited. Many microbiological tests still require specialized laboratory equipment and trained personnel, hindering their implementation at the point of care.

Most of the detection methods are time-consuming and thus, continuous progress is being made to better meet the rising demand of decreasing the turnaround time for pathogen diagnosis. Through the creation and application of innovative diagnostic technologies, initiatives are being made to get beyond these constraints. Cost-effective and easily accessible detection techniques need to be developed for reducing the financial burden on the nation. Future developments aim to improve precision, efficiency, and the ability to detect a broader spectrum of pathogens.

6 Conclusion and Prospects

Diagnostic microbiology has been using novel methods for accurate identification of disease using modern technologies with slow turnaround time and high efficacy. Over the past two decades, the discipline of diagnostic microbiology has seen perpetual and significant developments in a variety of disciplines, including bacteriology, mycology, mycobacteriology, parasitology, and virology. Microbes remain an integral part of human life since the beginning. Evolved molecular study of microorganisms enables them to utilize them as diagnostic tools also. In the past few years cases of viral disease (like H1N1, monkey virus, zika virus, Covid-19), bacterial disease (such as tuberculosis, gonorrhea, etc.) protozoan diseases (malaria), and fungal diseases (like candidiasis and pneumocystis) are few current examples that show the microbial diseases are more severe than we perceived. Therefore, efforts require a better understanding of the microbiome, genetic makeup, and how they affect humans at the molecular level. The unrestricted spread of infectious disease globally affects job, lifestyle, reduced lifespan, increase stress, and comorbidities globally, so genomic libraries need to be created not only to understand the lifecycle and interactions of microbes with host at molecular level but also helps to combat the superbugs, multiple resistant strains and hospital acquired diseases-nosocomial infections. The developments in technology have had a significant impact on the physical design of laboratories, personnel patterns, job flow, and turnaround time. Implementing molecular methods based on nucleic acid amplification, next-generation sequencing, CRISPR/Cas and other techniques offers supplementary, rapid, cost-effective, and on-demand diagnosis services. Utilization of machine learning in early detection of disease can be taken as a future goal. As an outcome of these advancements, diagnostic microbiology will undoubtedly become an emerging discipline capable of surpassing any challenges that arise in the years ahead.

Conflict of Interest The authors of this book declare no conflict of interest. Our primary goal remains the dissemination of accurate and unbiased information.

References

Ahmad, A., Iram, S., Hussain, S. and Yusuf, N.W., 2017. Diagnosis of paediatric sepsis by automated blood culture system and conventional blood culture. *Yeast*, *5*, p.3.

Ahmed, A., Rushworth, J.V., Hirst, N.A. and Millner, P.A., 2014. Biosensors for whole-cell bacterial detection. *Clinical microbiology reviews*, *27*(3), pp.631-646

Aladhadh, M., 2023. A Review of Modern Methods for the Detection of Foodborne Pathogens. Microorganisms, 11(5), p.1111.

Alizadeh, M., Yousefi, L., Pakdel, F., Ghotaslou, R., Rezaee, M.A., Khodadadi, E., Oskouei, M.A., Soroush Barhaghi, M.H. and Kafil, H.S., 2021. MALDI-TOF mass spectroscopy applications in clinical microbiology. *Advances in Pharmacological and Pharmaceutical Sciences, 2021*, pp.1-8.

Anaissie, E.J., McGinnis, M.R. and Pfaller, M.A., 2003. Clinical mycology. *Annals of Internal Medicine*, *138*(9), p.776., Heitman, J. ed., 2006. *Molecular principles of fungal pathogenesis.* Asm Press.Mendoza, L., Vilela, R., Voelz, K., Ibrahim, A.S., Voigt, K. and Lee, S.C., 2014. Human fungal pathogens of Mucorales and Entomophthorales. *Cold Spring Harbor perspectives in medicine*, p.a019562.

Anhalt, J.P. and Fenselau, C., 1975. Identification of bacteria using mass spectrometry. *Analytical chemistry*, *47*(2), pp.219-225

Aouida, M., Saifaldeen, M., Al-Ansari, D.E., Taleb, S., Hssain, A.A. and Ramotar, D., 2023. A CRISPR-based approach using dead Cas9-sgRNA to detect SARS-CoV-2. *Frontiers in Molecular Biosciences*, *10*, p.1201347.

Avanzi, I.R., Gracioso, L.H., Baltazar, M.D.P.G., Karolski, B., Perpetuo, E.A. and do Nascimento, C.A.O., 2017. Rapid bacteria identification from environmental mining samples using MALDI-TOF MS analysis. *Environmental Science and Pollution Research*, *24*, pp. 3717–3726.

Aydin, S., 2015. A short history, principles, and types of ELISA, and our laboratory experience with peptide/protein analyses using ELISA. *Peptides*, *72*, pp.4-15.

Azar, M.M., Schlaberg, R., Malinis, M.F., Bermejo, S., Schwarz, T., Xie, H. and Cruz, C.S.D., 2021. Added diagnostic utility of clinical metagenomics for the diagnosis of pneumonia in immunocompromised adults. *Chest*, *159*(4), pp.1356-1371.

Bader, O., 2013. MALDI-TOF-MS-based species identification and typing approaches in medical mycology. *Proteomics*, *13*(5), pp.788-799.

Bailey, A.L. and Diamond, M.S., 2021. A Crisp (r) new perspective on SARS-CoV-2 biology. *Cell*, *184*(1), pp.15-17.

Batista, B.G., Chaves, M.A.D., Reginatto, P., Saraiva, O.J. and Fuentefria, A.M., 2020. Human fusariosis: An emerging infection that is difficult to treat. *Revista da Sociedade Brasileira de Medicina Tropical*, *53*.

Burillo, A., Rodríguez-Sánchez, B., Ramiro, A., Cercenado, E., Rodríguez-Créixems, M. and Bouza, E., 2014. Gram-stain plus MALDI-TOF MS (matrix-assisted laser desorption ionization-time of flight mass spectrometry) for a rapid diagnosis of urinary tract infection. *PloS one*, *9*(1), p.e86915.

Camarasa, C.G. and Cobo, F., 2018. Application of MALDI-TOF mass spectrometry in clinical virology. In *The Use of Mass Spectrometry Technology (MALDI-TOF) in Clinical Microbiology* (pp. 167–180). Academic Press.

Cho, C.H., Lee, C.K., Nam, M.H., Yoon, S.Y., Lim, C.S., Cho, Y. and Kim, Y.K., 2014. Evaluation of the AdvanSure™ real-time RT-PCR compared with culture and Seeplex RV15 for simultaneous detection of respiratory viruses. *Diagnostic microbiology and infectious disease*, *79*(1), pp.14-18.

Cobo, F., 2013. Application of maldi-tof mass spectrometry in clinical virology: a review. *The open virology journal*, *7*, p.84.

De Paschale, M. and Clerici, P., 2012. Serological diagnosis of Epstein-Barr virus infection: Problems and solutions. *World journal of virology*, *1*(1), p.31.

De Carolis, E., Posteraro, B., Lass-Flörl, C., Vella, A., Florio, A.R., Torelli, R., Girmenia, C., Colozza, C., Tortorano, A.M., Sanguinetti, M. and Fadda, G., 2012. Species identification of Aspergillus, Fusarium and Mucorales with direct surface analysis by matrix-assisted laser desorption ionization time-of-flight mass spectrometry. *Clinical microbiology and infection*, *18*(5), pp.475-484.

Deltcheva, E., Chylinski, K., Sharma, C.M., Gonzales, K., Chao, Y., Pirzada, Z.A., Eckert, M.R., Vogel, J. and Charpentier, E., 2011. CRISPR RNA maturation by trans-encoded small RNA and host factor RNase III. *Nature*, *471*(7340), pp.602-607.

Deng, C., Lin, M., Hu, C., Li, Y., Gao, Y., Cheng, X., Zhang, F., Dong, M. and Li, Y., 2011. Exploring Serological Classification Tree Model of Active Pulmonary Tuberculosis by Magnetic Beads Pretreatment and MALDI-TOF MS Analysis. *Scandinavian Journal of Immunology*, *74*(4), pp.397-405.

Duthie, M.S., Raychaudhuri, R., Tutterrow, Y.L., Misquith, A., Bowman, J., Casey, A., Balagon, M.F., Maghanoy, A., Beltran-Alzate, J.C., Romero-Alzate, M. and Cardona-Castro, N., 2014. A rapid ELISA for the diagnosis of MB leprosy based on complementary detection of antibodies against a novel protein-glycolipid conjugate. *Diagnostic microbiology and infectious disease, 79*(2), pp.233-239.

Elnifro, E.M., Ashshi, A.M., Cooper, R.J. and Klapper, P.E., 2000. Multiplex PCR: optimization and application in diagnostic virology. *Clinical microbiology reviews, 13*(4), pp.559-570.

Enders JF, Weller TH, Robbins FC,1949. Cultivation of the Lansing strain of poliomyelitis virus in cultures of various human embryonic tissues, *Science*, vol. 109

Garzarelli, V., Chiriacò, M.S., Cereda, M., Autuori, I. and Ferrara, F., 2022. Miniaturized Real-Time PCR systems for SARS-CoV-2 detection at the Point-of-Care. *Clinica Chimica Acta.*

Gomes, L.B., Ward, T.J., Badiale-Furlong, E. and Del Ponte, E.M., 2015. Species composition, toxigenic potential and pathogenicity of F usarium graminearum species complex isolates from southern B razilian rice. *Plant Pathology, 64*(4), pp.980-987.

Green, M.R. and Sambrook, J., 2019. Nested polymerase chain reaction (PCR). *Cold Spring Harbor Protocols, 2019*(2).

Grimes, D.J., 2006. Koch's postulates-then and now. *Microbe-American Society for Microbiology, 1*(5), p.223.

Gupta, J., Rathour, R., Singh, R. and Thakur, I.S., 2019b. Production and characterization of extracellular polymeric substances (EPS) generated by a carbofuran degrading strain Cupriavidus sp. ISTL7. Bioresource Technology, 282, pp.417-424.

Gupta, A., Singh, U.B., Sahu, P.K., Paul, S., Kumar, A., Malviya, D., Singh, S., Kuppusamy, P., Singh, P., Paul, D. and Rai, J.P., 2022a. Linking soil microbial diversity to modern agriculture practices: A review. *International Journal of Environmental Research and Public Health, 19*(5), p.3141.

Gupta, J., Rathour, R., Dupont, C., Mishra, A. and Thakur, I.S., 2022. Biogeochemical profiling and taxonomic characterization of municipal landfill site by metagenomic sequencing. Bioresource Technology, 351, p.126936.

Gupta, J., Tyagi, B., Rathour, R. and Thakur, I.S., 2019a. Microbial treatment of waste by culture-dependent and culture-independent approaches: Opportunities and challenges. Microbial Diversity in Ecosystem Sustainability and Biotechnological Applications: Volume 1. Microbial Diversity in Normal & Extreme Environments, pp.415–446.

Gupta, J., Rathour, R., Medhi, K., Tyagi, B. and Thakur, I.S., 2020. Microbial-derived natural bioproducts for a sustainable environment: a bioprospective for waste to wealth. In Refining biomass residues for sustainable energy and bioproducts (pp. 51–85). Academic Press.

Gupta, J., Rathour, R., Maheshwari, N. and Thakur, I.S., 2021a. Integrated analysis of Whole genome sequencing and life cycle assessment for polyhydroxyalkanoates production by Cupriavidus sp. ISTL7. Bioresource Technology, 337, p.125418.

Gupta, J., Rathour, R., Dupont, C.L., Kaul, D. and Thakur, I.S., 2021b. Genomic insights into waste valorized extracellular polymeric substances (EPS) produced by Bacillus sp. ISTL8. Environmental Research, 192, p.110277.

Hassan, S.U., Donia, A., Sial, U., Zhang, X. and Bokhari, H., 2020. Glycoprotein-and lectin-based approaches for detection of pathogens. Pathogens, 9(9), p.694.

Hilt, E.E. and Ferrieri, P., 2022. Next generation and other sequencing technologies in diagnostic microbiology and infectious diseases. *Genes, 13*(9), p.1566.

Horinouchi, T., Minamoto, T., Suzuki, S., Shimizu, H. and Furusawa, C., 2014. Development of an automated culture system for laboratory evolution. *Journal of Laboratory Automation, 19*(5), pp.478-482.

Hou, T.Y., Chiang-Ni, C. and Teng, S.H., 2019. Current status of MALDI-TOF mass spectrometry in clinical microbiology. *Journal of food and drug analysis, 27*(2), pp.404-414.

Hu, T., Chitnis, N., Monos, D. and Dinh, A., 2021. Next-generation sequencing technologies: An overview. *Human Immunology, 82*(11), pp.801-811

Huggett, J., Dheda, K., Bustin, S. and Zumla, A., 2005. Real-time RT-PCR normalisation; strategies and considerations. *Genes & Immunity*, *6*(4), pp.279-284.

Inda, M.E. and Lu, T.K., 2020. Microbes as biosensors. *Annual Review of Microbiology*, *74*, pp.337-359.

Infectious Diseases Society of America, 2010. The 10×'20 Initiative: pursuing a global commitment to develop 10 new antibacterial drugs by 2020. *Clinical Infectious Diseases*, *50*(8), pp.1081-1083.,

Iqbal, G., Javed, H., Raza, F.A., Gohar, U.F., Fatima, W. and Khurshid, M., 2023. Diagnosis of Acute Dengue Virus Infection Using Enzyme-Linked Immunosorbent Assay and Real-Time PCR. *Canadian Journal of Infectious Diseases and Medical Microbiology, 2023.*

Islam, M.B., Yusuf, M.A., Khan, M.S. and Islam, S., 2022. Rapid Detection of Microorganisms by Automated Blood Culture System: Experience of 3220 cases of Blood Culture. *Bangladesh Journal of Infectious Diseases*, *9*(1), p.3.

Jansson, E., Haenen, O., Nonnemann, B., Madsen, L., Van Gelderen, E., Aspán, A., Saker, E., Boonstra, M., Gulla, S., Colquhoun, D.J. and Roozenburg-Hengst, I., 2020. MALDI-TOF MS: A diagnostic tool for identification of bacterial fish pathogens. *Bulletin of the European Association of Fish Pathologists*, *40*(6), pp.240-248.

Javed, M.U., Hayat, M.T., Mukhtar, H. and Imre, K., 2023. CRISPR-Cas9 System: A Prospective Pathway toward Combatting Antibiotic Resistance. *Antibiotics, 12*(6), p.1075.

Jing, F., Li, Z., Qiao, S., Zhang, J., Olatosi, B. and Li, X., 2023. Using geospatial social media data for infectious disease studies: a systematic review. *International Journal of Digital Earth*, *16*(1), pp.130-157.

Kadkhoda, K., Gretchen, A. and Racano, A., 2017. Evaluation of a commercially available Zika virus IgM ELISA: specificity in focus. *Diagnostic Microbiology and Infectious Disease*, *88*(3), pp.233-235.

Karman, S.B., Ibrahim, S.B. and Khuen, C.C., 2023. Engineered Microbial Sensing Element-Based Biosensor for Sustainable Biomedical Engineering Application. In *Sustainable Material for Biomedical Engineering Application* (pp. 183–205). Singapore: Springer Nature Singapore.

Kersh, E.N., Pham, C.D., Papp, J.R., Myers, R., Steece, R., Kubin, G., Gautom, R., Nash, E.E., Sharpe, S., Gernert, K.M. and Schmerer, M., 2020. Expanding US Laboratory capacity for Neisseria gonorrhoeae antimicrobial susceptibility testing and whole-genome sequencing through the CDC's antibiotic resistance laboratory network. *Journal of Clinical Microbiology*, *58*(4), pp.10-1128

Kim, O.S., Cho, Y.J., Lee, K., Yoon, S.H., Kim, M., Na, H., Park, S.C., Jeon, Y.S., Lee, J.H., Yi, H. and Won, S., 2012. Introducing EzTaxon-e: a prokaryotic 16S rRNA gene sequence database with phylotypes that represent uncultured species. *International journal of systematic and evolutionary microbiology*, *62*(Pt_3), pp.716–721.

Kim, M., Oh, H.S., Park, S.C. and Chun, J., 2014. Towards a taxonomic coherence between average nucleotide identity and 16S rRNA gene sequence similarity for species demarcation of prokaryotes. *International journal of systematic and evolutionary microbiology*, *64*(Pt_2), pp.346–351.

Kocaman, E. and Sumer, A.R.A.S., 2019. General overview of the studies with MALDI-TOF. *Communications Faculty of Sciences University of Ankara Series C Biology*, *28*(1), pp.1–9.

Korbie, D.J. and Mattick, J.S., 2008. Touchdown PCR for increased specificity and sensitivity in PCR amplification. *Nature protocols*, *3*(9), pp.1452-1456.

Köser, C.U., Ellington, M.J., Cartwright, E.J., Gillespie, S.H., Brown, N.M., Farrington, M., Holden, M.T., Dougan, G., Bentley, S.D., Parkhill, J. and Peacock, S.J., 2012. Routine use of microbial whole genome sequencing in diagnostic and public health microbiology.

Kralik, P. and Ricchi, M., 2017. A basic guide to real-time PCR in microbial diagnostics: definitions, parameters, and everything. *Frontiers in microbiology*, *8*, p.108.

Laqueur, T.W., 1991. Viral cultures. *New Republic*, *205*(2), pp.36-41.

Leland, D.S. and Ginocchio, C.C., 2007. Role of cell culture for virus detection in the age of technology. *Clinical microbiology reviews*, *20*(1), pp.49-78.

Li, G., 2018. *Nano-inspired biosensors for protein assay with clinical applications.* Elsevier.

Li, Y., Tan, Z., Wang, X. and Hou, L., 2023. Metabolic changes and potential biomarkers in" Candidatus Liberibacter solanacearum"-infected potato psyllids: implications for psyllid-pathogen interactions. *Frontiers in Plant Science, 14.*

Lin, B.B., Pattle, N., Kelley, P. and Jaksic, A.S., 2021. Multiplex RT-PCR provides improved diagnosis of skin and nail dermatophyte infections compared to microscopy and culture: A laboratory study and review of the literature. *Diagnostic Microbiology and Infectious Disease, 101*(3), p.115413.

Mahajan, V.K., 2023. Lyme Disease: An Overview. *Indian Dermatology Online Journal.*

Morris, R.P., Nguyen, L., Gatfield, J., Visconti, K., Nguyen, K., Schnappinger, D., Ehrt, S., Liu, Y., Heifets, L., Pieters, J. and Schoolnik, G., 2005. Ancestral antibiotic resistance in Mycobacterium tuberculosis. *Proceedings of the National Academy of Sciences, 102*(34), pp.12200-12205.

Moser, N., Yu, L.S., Rodriguez Manzano, J., Malpartida-Cardenas, K., Au, A., Arkell, P., Cicatiello, C., Moniri, A., Miglietta, L., Wang, W.H. and Wang, S.F., 2022. Quantitative detection of dengue serotypes using a smartphone-connected handheld lab-on-chip platform. *Frontiers in Bioengineering and Biotechnology, 10,* p.892853.

Mostowy, S., 2022. Louis Pasteur continues to shape the future of microbiology. *Disease Models & Mechanisms, 15*(12), p.dmm050011.

Naccache, S.N., Federman, S., Veeraraghavan, N., Zaharia, M., Lee, D., Samayoa, E., Bouquet, J., Greninger, A.L., Luk, K.C., Enge, B. and Wadford, D.A., 2014. A cloud-compatible bioinformatics pipeline for ultrarapid pathogen identification from next-generation sequencing of clinical samples. *Genome research, 24*(7), pp.1180-1192.

Nagy, E., Becker, S., Kostrzewa, M., Barta, N. and Urbán, E., 2012. The value of MALDI-TOF MS for the identification of clinically relevant anaerobic bacteria in routine laboratories. *Journal of medical microbiology, 61*(10), pp.1393-1400.

Nanua, S., Weber, C., Isgriggs, L. and Dunne Jr, W.M., 2009. Performance evaluation of the VersaTREK blood culture system for quality control testing of platelet units. *Journal of clinical microbiology, 47*(3), pp.817-818.

Nikhil, B., Pawan, J., Nello, F. and Pedro, E., 2016. Introduction to biosensors. *Essays Biochem, 60*(1), pp.1-8.

Oliveira, M. and Azevedo, L., 2022. Molecular markers: An overview of data published for fungi over the last ten years. *Journal of Fungi, 8*(8), p.803.

Panicker, V., Nayak, P., Krishna, R., Sreenivaasan, N., Thomas, J., Sreedevan, V., Anjaneyan, G., Jagadeesan, S. and Lekshmi, S., 2023. Gram stain. *Journal of Skin and Sexually Transmitted Diseases, 5*(1), pp.60-61.

Parija, S.C., 2023. *Textbook of microbiology and immunology.* Springer.

Patel, R., 2019. A moldy application of MALDI: MALDI-ToF mass spectrometry for fungal identification. *Journal of Fungi, 5*(1), p.4.

Pereira, L.C., Correia, A.F., da Silva, Z.D.L., de Resende, C.N., Brandão, F., Almeida, R.M. and de Medeiros Nóbrega, Y.K., 2021. Vulvovaginal candidiasis and current perspectives: new risk factors and laboratory diagnosis by using MALDI TOF for identifying species in primary infection and recurrence. *European Journal of Clinical Microbiology & Infectious Diseases, 40,* pp.1681-1693.

Petti CA, Polage CR, Schreckenberger P., 2005.Misidentification of microorganisms by conventional methods: role of 16S rRNA gene sequencing, *J Clin Microbiol*,vol. 43 (pg. 6123–5)

Redondo, J.J., Keller, P.M., Zbinden, R. and Wagner, K., 2018. A novel RT-PCR for the detection of Helicobacter pylori and identification of clarithromycin resistance mediated by mutations in the 23S rRNA gene. *Diagnostic microbiology and infectious disease, 90*(1), pp.1-6.

Reller, L.B., Weinstein, M.P. and Petti, C.A., 2007. Detection and identification of microorganisms by gene amplification and sequencing. *Clinical infectious diseases, 44*(8), pp.1108-1114.

Rhoads, D.D., Wolcott, R.D., Sun, Y. and Dowd, S.E., 2012. Comparison of culture and molecular identification of bacteria in chronic wounds. *International journal of molecular sciences, 13*(3), pp.2535-2550.

Rushworth, J.V. and Hirst, N.A., 2013. Impedimetric biosensors for medical applications: current progress and challenges.

Ryan, M.R. and Murray, P.R., 1993. Historical evolution of automated blood culture systems. *Clinical Microbiology Newsletter*, *15*(14), pp.105-108.

Saeed, K. and Ahmad, N.S., 2013. Real-time polymerase chain reaction: applications in diagnostic microbiology. *International Journal of Medical Students*, *1*(1), pp.28-36.

Sanger, F., Nicklen, S. and Coulson, A.R., 1977. DNA sequencing with chain-terminating inhibitors. *Proceedings of the national academy of sciences*, *74*(12), pp.5463-5467.

Scavone, P., Carrasco, V., Umpiérrez, A., Morel, M., Arredondo, D. and Amarelle, V., 2019. Microbiology can be comic. *FEMS Microbiology Letters*, *366*(14), p.fnz171.

Schmitz, J.E., Stratton, C.W., Persing, D.H. and Tang, Y.W., 2022. Forty years of molecular diagnostics for infectious diseases. *Journal of Clinical Microbiology*, *60*(10), pp.e02446-21

Serra, B., Gamella, M., Reviejo, A.J. and Pingarron, J.M., 2008. Lectin-modified piezoelectric biosensors for bacteria recognition and quantification. *Analytical and bioanalytical chemistry*, *391*, pp.1853-1860.

Sharma, S., Suri, D., Aggarwal, A.N., Yadav, R., Sethi, S., Laal, S. and Verma, I., 2023. Evaluation of immunodominant peptides of in vivo expressed mycobacterial antigens in an ELISA-based diagnostic assay for pulmonary tuberculosis. *Brazilian Journal of Microbiology*, pp.1–9.

Shukla, J., Khan, M., Tiwari, M., Sannarangaiah, S., Sharma, S., Rao, P.V.L. and Parida, M., 2009. Development and evaluation of antigen capture ELISA for early clinical diagnosis of chikungunya. *Diagnostic Microbiology and Infectious Disease, 65*(2), pp.142-149.

Sjöholm, M.I., Dillner, J. and Carlson, J., 2008. Multiplex detection of human herpesviruses from archival specimens by using matrix-assisted laser desorption ionization-time of flight mass spectrometry. *Journal of Clinical Microbiology, 46*(2), pp.540-545.

Solante, R., Alvarez-Moreno, C., Burhan, E., Chariyalertsak, S., Chiu, N.C., Chuenkitmongkol, S., Dung, D.V., Hwang, K.P., Ortiz Ibarra, J., Kiertiburanakul, S. and Kulkarni, P.S., 2023. Expert review of global real-world data on COVID-19 vaccine booster effectiveness and safety during the omicron-dominant phase of the pandemic. *Expert review of vaccines*, *22*(1), pp.1-16.

Soro-Yao, A.A., Schumann, P., Thonart, P., Djè, K.M. and Pukall, R., 2014. The use of MALDI-TOF mass spectrometry, ribotyping and phenotypic tests to identify lactic acid bacteria from fermented cereal foods in Abidjan (Côte d'Ivoire). *The Open Microbiology Journal*, *8*, p.78.

Stanton, R.A., Campbell, D., McAllister, G.A., Breaker, E., Adamczyk, M., Daniels, J.B., Lutgring, J.D., Karlsson, M., Schutz, K., Jacob, J.T. and Wilson, L.E., 2022. Whole-genome sequencing reveals diversity of carbapenem-resistant Pseudomonas aeruginosa collected through CDC's emerging infections program, United States, 2016–2018. *Antimicrobial Agents and Chemotherapy*, *66*(9), pp.e00496-22

Storch, G.A., 2000. Diagnostic virology. *Clinical infectious diseases*, *31*(3), pp.739-751.

Strich, J.R. and Chertow, D.S., 2019. CRISPR-Cas biology and its application to infectious diseases. *Journal of clinical microbiology*, *57*(4), pp.10-1128.

Tanaka, K., 2003. The origin of macromolecule ionization by laser irradiation (Nobel lecture). *Angewandte Chemie International Edition*, *42*(33), pp.3860-3870.

Tang, Y.W., Stratton, C.W. and Tang, Y.W., 2013. *Advanced techniques in diagnostic microbiology*. New York: Springer.

Tavakoli, K., Pour-Aboughadareh, A., Kianersi, F., Poczai, P., Etminan, A. and Shooshtari, L., 2021. Applications of CRISPR-Cas9 as an advanced genome editing system in life sciences. *BioTech*, *10*(3), p.14.

Tawil, N., Sacher, E., Mandeville, R. and Meunier, M., 2012. Surface plasmon resonance detection of E. coli and methicillin-resistant S. aureus using bacteriophages. *Biosensors and Bioelectronics*, *37*(1), pp.24–29.

Thorpe, T.C., Wilson, M.L., Turner, J.E., DiGuiseppi, J.L., Willert, M., Mirrett, S. and Reller, L., 1990. BacT/Alert: an automated colorimetric microbial detection system. *Journal of clinical microbiology*, *28*(7), pp.1608-1612.

Tolba, M., Ahmed, M.U., Tlili, C., Eichenseher, F., Loessner, M.J. and Zourob, M., 2012. A bacteriophage endolysin-based electrochemical impedance biosensor for the rapid detection of Listeria cells. *Analyst*, *137*(24), pp.5749-5756.

Tsang, R.S., 2021. A narrative review of the molecular epidemiology and laboratory surveillance of vaccine-preventable bacterial meningitis agents: Streptococcus pneumoniae, Neisseria meningitidis, Haemophilus influenzae and Streptococcus agalactiae. *Microorganisms*, *9*(2), p.449.

Tsao, Y.T., Tsai, Y.H., Liao, W.T., Shen, C.J., Shen, C.F. and Cheng, C.M., 2020. Differential markers of bacterial and viral infections in children for point-of-care testing. *Trends in molecular medicine*, *26*(12), pp.1118-1132.

Verma, A. and Ogata, S., 2023. Magnesium based alloys for reinforcing biopolymer composites and coatings: a critical overview on biomedical materials. Advanced Industrial and Engineering Polymer Research. DOI: https://doi.org/https://doi.org/10.1016/j.aiepr.2023.01.002

Veron, V., Simon, S. and Carme, B., 2009. Multiplex real-time PCR detection of P. falciparum, P. vivax and P. malariae in human blood samples. *Experimental Parasitology, 121*(4), pp.346-351.

Vuong, M.F. and Waymack, J.R., Continuing Education Activity 2023.

Williams, R.E. and Trotman, R.E., 1969. Automation in diagnostic bacteriology. *Journal of Clinical Pathology. Supplement (College of Pathologists).*, *3*, p.8.

Xiao, Y., Luo, M., Hayes, R.P., Kim, J., Ng, S., Ding, F., Liao, M. and Ke, A., 2017. Structure basis for directional R-loop formation and substrate handover mechanisms in type I CRISPR-Cas system. *Cell*, *170*(1), pp.48-60.

Role of Biochar as a Sustainable Sorbent for Fipronil Removal from Aqueous and Soil Environments

Rajat Kumar Sharma, T. P. Singh, Deepshikha Azad, Himanshu Adhikari, and Akarsh Verma

1 Introduction

Chemical pesticides have made substantial contributions to increasing agricultural productivity. Of the frequently employed pesticides, around 40% exhibit chirality, including widely used pyrethroid insecticides, phenoxypropionic acid herbicides, and organophosphorus insecticides. Intriguingly, a significant number of these chiral pesticides are applied in their racemic forms within the agricultural setting. One such agriculture chemical is fipronil Fipronil, an n-phenylpyrazole insecticide with broad-spectrum properties, finds extensive application in rice fields for the eradication of pest insects around the world. Fipronil, scientifically known as (±)-5-amino-1-(2,6-dichloro-α,α,α-trifluoro-p-tolyl)-4-trifluoromethylsulfinylpyrazole-3-carbonitrile ($C_{12}H_4C_{12}F_6N_4OS$), belongs to the phenylpyrazole chemical family. This pesticide is non-systemic and can be found in a range of products under various trade names. It demonstrates broad-spectrum efficacy against insect pests present in agricultural and horticultural crops, as well as on golf courses and domestic animals (Gunasekara et al. 2007). It has low volatility, moderate hydrophobicity, low-to-moderate water solubility and persistence in the environment. Degradation of fipronil occurs via photolysis, oxidation, pH-dependent hydrolysis, and reduction. The main stable and persistent degradation products of fipronil are disulfonyl fipronil, fipronil sulfone, fipronil amide, and

R. K. Sharma (✉) · T. P. Singh · D. Azad · H. Adhikari
Department of Farm Machinery and Power Engineering, Govind Ballabh Pant University of Agriculture & Technology, Pantnagar 263145, India
e-mail: rajat_sharma@hotmail.com

A. Verma
Department of Mechanical Engineering, University of Petroleum and Energy Studies, Dehradun 248007, India

Department of Mechanical Science and Bioengineering, Osaka University, Osaka 560-8531, Japan

J. Gupta and A. Verma (eds.), *Microbiology-2.0 Update for a Sustainable Future*,
https://doi.org/10.1007/978-981-99-9617-9_9

fipronil sulfide, which are more toxic than their parent pesticide (Qu et al. 2018). Fipronil can enter aquatic ecosystems with runoff or leaching. It generally binds to soil and hence has low soil mobility (Miller et al. 2020; Fent 2014).

Fipronil destroys the central nervous system by disrupting the gamma amino butyric acid receptors in insects (Bhatti et al. 2019; Bhatt et al. 2021). It is a chiral compound and has fipronil and fipronil that have different biological activities and environmental fate (Gong et al. 2021). It is also an endocrine disrupting compound (EDC) which mimics natural hormones. It adversely affects nervous system and reproductive organs of human being as well as other aquatic organisms (Rathi et al. 2019; Liu et al. 2009). According to United States Environmental Protection Agency, the efficiency of the existing wastewater treatment plants are not up to the mark and it can remove only 30% of these compounds and the rest persists and accumulate in the sludge (Environmental Protection Agency, 2017).

Some of the treatment processes used to remove EDC are nano-filtration, advance oxidation process, biodegradation, ozonation, electrochemical oxidation and adsorption. Adsorption is a successful treatment method for pollutant removal for its easy operation, less capital input, and almost no possibility of secondary harmful products generation. The adsorption efficiency of different adsorbents is the crucial point to be considered for better treatment effect.

Waste is any substance which is discarded after primary use and is considered to be of no use. However, most of the waste entities are refuge which can be further used as potential resources. Turning waste into resource is the key to achieving sustainable development, including environmental and economical sustainability. The modern biomass conversion technologies, most importantly pyrolysis convert a wide range of biomass (mostly agriculture refuge) to different products such as bio-oil, biochar, and syngas (Azad et al. 2023; Mandal and Sharma 2023).

Biochar is the finely grained, highly porous, black carbonaceous product prepared from a wide spectrum of solid wastes. The common environmental wastes employed for preparation of biochar include agro-forestry waste, animal waste, fresh/marine algae, biological digestates, and weed plants. There are wide possibilities in terms of selection of feedstock, pyrolysis methods, modification methods and applications of biochar. The idea is to make use of enormous tons of waste as an economical, viable resource for further application. Biochar has different characteristics which are function of the pyrolysis conditions and feedstock (Azad et al. 2023; Cantrell et al. 2012). It is a multi-faceted product with intent application in environmental remediation, agronomy, carbon sequestration and geo-engineering (Wani et al. 2021). It is used as an efficient adsorbent for fipronil removal from aqueous as well as soil medium (Boehm et al. 2020a; Li et al. 2018). Biochar has also been successfully used as an amendment for increasing biomass production and improving soil quality (Mensah and Frimpong 2018). The in-situ application of biochar is cost effective, eco-friendly and sustainable way of dealing with waste right from conversion of different biomass for its preparation to its final application for on-site removal of pesticide and other pollutants from soil. The cost of biochar varies depending upon the type of raw material, pyrolysis operation implied and transport. In comparison to construction sand which costs about \$8.6 per metric, the cost of biochar is at higher side that range from \$350 to \$18 000 per metric ton (Alhashimi and Aktas 2017; Shackley et al. 2015). However, as the biochar market is becoming more competitive, it is estimated that it will become cheaper. The efficient performance of biochar augmented biofilters to remove contaminants would require lesser filters with longer lifetimes in comparison to the conventional filters to meet the water quality standards which will also reduce the overall price of construction. For example, Ulrich et al. (2017) have estimated the exhaustion time to be about five months to seven years for wood-based biochar pyrolyzed at 600 °C and under gasification respectively (Ulrich et al. 2017). However, the knowledge about the actual long-term effects such as extreme flow condition and environmental conditions on long term performance of degradation needs further research. In case of pesticide remediation, addition of biochar in soil facilitates better results primarily because it improves physico-chemical properties and microbial activity of soil. Earlier studies have clearly demonstrated that it increased the adsorption capacity for pesticides and decreased desorption, mobility of pesticide and bioavailability in soil pore water (Safaei Khorram et al. 2016).

2 Effect of Pyrolysis Temperature on Properties of Biochar

Biochar can be produced through the pyrolysis process, which involves subjecting diverse feedstock sources to varying pyrolysis temperatures, heating rates, durations, and using different ancillary inputs (such as nitrogen, oxygen, steam, etc.). Additionally, partial pressures, reaction vessels, and pre-treatment/post-treatment modifications can also influence the preparation of biochar (Yargicoglu et al. 2014; Yavari et al. 2017; Mandal et al. 2022). Among these factors, the physico-chemical properties of the raw material and the pyrolysis temperature significantly influence both the quantity and quality of biochar. To achieve the most favorable functional properties, it is essential to optimize the pyrolysis conditions through rigorous testing. Liu et al. (2018a), (2018b) highlighted that the absence of an optimized pyrolysis temperature can lead to varying effects on pesticide removal when biochar is prepared at either high or low temperatures (Liu et al. 2018a). In the former case, biochar exhibits a larger surface area, while in the latter case, it contains a higher number of functional groups. Moreover, the pyrolysis temperature also plays a critical role in determining the longevity of biochar, as biochar produced at temperatures above 500 °C generally exhibits longer half-lives (exceeding 1000 years) (Ippolito et al. 2020).

2.1 *Effect of Temperature on Porosity and Surface Area*

The primary characteristics of biochar that influence pesticide sorption are its porosity and surface area, achieved through pore filling and surface adsorption. The porosity and surface area of biochar exhibit considerable variation based on the pyrolysis temperature employed. An optimal higher pyrolysis temperature is observed, beyond which the surface area starts to decrease along with other textural properties. Numerous previous studies have demonstrated that higher temperatures during pyrolysis lead to greater pore volume and surface area (Cantrell et al. 2012; Liu et al. 2017). According to Chen et al. (2014), as the temperature during biochar production increases from 500 to 900 °C, the pore volume of the biochar rises from 0.056 to 0.099 $cm^3\ g^{-1}$, and the surface area increases from 25.4 to 67.6 $m^2\ g^{-1}$. However, it has been observed in some instances that when the biochar is produced at temperatures exceeding 550 °C, it exhibits lower surface area and porosity (Chen et al. 2014). Leng et al. (2021) conducted a study plotting the surface area of various biochar samples against pyrolysis temperature. The plot illustrates that the surface area of the different biochar samples increases with rising temperature until it reaches a threshold point, often between 800 and 900 °C, beyond which the surface area starts to decrease due to the distortion of the biochar's structure (Leng et al. 2021). The elevated temperatures increase the likelihood of CO_2 being released from the biochar, which triggers a secondary reaction between the released CO_2 and the carbon present in the biochar matrix. As a consequence, previously closed pores open up, leading to increased porosity. This enhanced porosity enhances the biochar's ability to adsorb substances (Thines et al. 2017). However, at very high temperature higher volatilization of organic compounds occur that got re-condensed on surface and resulted in blocked pores and thus decreased the surface area (Atkinson et al. 2010).

2.2 *Effect of Temperature on pH of Biochar*

Much like porosity and surface area, the pH of biochar is influenced by both the pyrolysis temperature and the type of feedstock used. Typically, biochar produced at higher pyrolysis temperatures tends to exhibit an alkaline pH. However, there are exceptions to this trend. For instance, biochar derived from oak wood at 200 °C (Zhang et al. 2015a), as well as from corn stover, wheat straw, and wastewater sludge at low temperatures like 350 and 600 °C (Nguyen et al. 2010), display an acidic pH (Nguyen et al. 2010; Hossain et al. 2011; Zhang et al. 2015b). The elevated alkalinity of biochar at higher temperatures results from the depletion of acidic functional groups and the formation of mineral phases containing Ca, Mg, Na, and K (Cao and Harris 2010; Novak et al. 2009). Generally, biochar prepared at lower temperatures contains higher levels of volatile matter, consisting of easily decomposable substances. These types of biochar support plant growth and are well-suited for agronomic purposes (Robertson et al. 2012; Mukherjee and Zimmerman 2013). On the other hand, higher temperatures during pyrolysis lead to the development of an aromatic structure and a greater surface area in the biochar. This enhances the adsorption capacity and recalcitrant nature, making such biochar highly effective for environmental remediation and carbon sequestration (Lehmann 2007). Moreover, biochar with an alkaline pH can accelerate the hydrolysis of organophosphorus and carbamate pesticides through an alkali catalysis mechanism (Zhang et al. 2012).

2.3 Effect of Temperature on Nutrients and Aromatic Structure

The efficiency of biochar for pesticide adsorption is significantly influenced by both its carbon content and aromatic structure (Liu et al. 2018a). Pyrolysis induces more substantial changes in the elemental composition of biochar compared to the original feedstock material. At higher pyrolysis temperatures, biochar with condensed and aromatic structures is produced, making it more biologically resistant (Bruun et al. 2014). During pyrolysis and carbonization of biomass, the higher temperatures lead to enhancements in C, P, K, Ca, ash content, pH, and specific surface area, while the contents of N, H, and O tend to decrease. This suggests the occurrence of decarboxylation, decarbonylation, and dehydration processes (Ahmad et al. 2012; Ghanim et al. 2016). In general, slow pyrolysis, which involves low to moderate temperatures ranging from 300 to 500 °C, longer residence times of 5 to 30 min, and heat transfer rates between 5 to 20 °C/min, tends to yield biochars with higher levels of N, S, available P, Ca, Mg, surface area, and cation exchange capacity (CEC) compared to fast pyrolysis (Sharma et al. 2022). As the pyrolysis temperature increases, there is a noticeable decrease in the H/C and O/C as well as (O + N)/C ratios. The H/C atomic ratio is considered an indicator of carbonization, aromaticity, and the resistance of biochar to microbial and chemical degradation. A decline in this ratio suggests the formation of aromatic condensed carbon structures at higher pyrolysis temperatures. An increase in aromaticity, in turn, indicates ongoing dehydration and decarboxylation reactions (Mohan et al. 2014).

The (O + N)/C and O/C atomic ratios serve as indicators of the polarity of the biochar. These ratios also change with increasing temperature as the elemental concentration changes. The oxygen to carbon (O/C) ratio is particularly associated with the surface hydrophilicity of biochar, reflecting the content of polar groups. A lower O/C ratio at higher temperatures indicates increased carbonization and a reduction in polar functional groups, making the biochar more aromatic and less hydrophilic. Consequently, the polarity decreases, and the number of oxygen-containing functional groups reduces due to the removal of hydroxyl groups through dehydration, the loss of carboxyl and carbonyl groups via decarboxylation, and cleavage of ester and ether bonds through hydrolysis (Funke and Ziegler 2010; Smith et al. 2016).

2.4 Effect of Temperature on Surface Functional Groups

Surface functional groups located on the biochar's surface, including carboxylic, hydroxyl, lactonic, amide, and amine groups, have a significant impact on its sorption capacity (Antón-Herrero et al. 2018; Li et al. 2017). The quantities of these functional groups on the biochar surface are mainly influenced by two crucial factors: the pyrolysis temperature and the raw material used (Chen et al. 2015). However, in contrast to properties such as surface area, porosity, and pH that tend to increase at higher temperatures, the functional groups experience a reduction at those elevated temperatures (Liu et al. 2017). Table 1 shows the effect of carbonization temperature on elemental and physical properties of biochar.

Table 1 Effect of carbonization temperature on properties of biochar

Temperature (°C)	Properties of biochar								References
	C (%)	N (%)	SSA (m^2/g)	pH	Pore volume (cm^3/g)	CEC (cmol/kg)	Ash (wt %)	H/C Ratio	
400	59.8	2.08	4.75	7.61	0.02	–	21.7	0.05	Břendová et al. (2017)
450	58.8	1.88	40.1	8	0.04	–	23	0.04	Břendová et al. (2017)
500	54.9	1.95	64.3	9.82	0.06	–	27.3	0.04	Břendová et al. (2017)
550	44.2	1.79	105.0	9.96	0.09	–	35.7	0.04	Břendová et al. (2017)
230	60.0	–	4 0.0	4.1	–	30.0	–	1.2	Budai et al. (2016)
600	83.0	–	39 0.0	9.5	–	13.0	–	0.3	Budai et al. (2016)
369	79.0	–	2 0.0	9.2	–	21.0	–	0.7	Budai et al. (2016)
416	87.0	–	4 0.0	10.1	–	16.0	–	0.6	Budai et al. (2016)
562	74.0	–	45 0.0	9.4	–	13.0	–	0.4	Budai et al. (2016)
796	92.0	–	92.0	9.4	–	5.0	–	0.1	Budai et al. (2016)
500	48.8	2.18	0.17	10.3	–	–	30	–	Wang et al. (2013)
500	55.0	2.81	0.36	10.1	–	–	25.6	–	Wang et al. (2013)
500	50.5	2.64	0.2	10.2	–	–	28.8	–	Wang et al. (2013)
700	51.1	1.97	20.5	10.7	–	–	29.4	–	Wang et al. (2013)
700	51.8	1.83	112.0	11	–	–	31.1	–	Wang et al. (2013)
700	46.6	1.66	160.0	10.8	–	–	34.1	–	Wang et al. (2013)

3 Effect of Type of Raw Material on Biochar Properties

The plant biomass used as feedstock for biochar production mainly consists of hemicellulose, cellulose, and lignin. The proportion of these compounds in the raw biomass determines the extent of chemical, physical, and structural changes that occur during the pyrolysis process. At lower temperatures, the weight loss is primarily due to the evaporation of water. As the temperature rises, these components undergo a gradual pyrolysis process. The primary decomposition process occurs within the temperature range of 200 to 500 °C. Hemicellulose degradation occurs at the lower end of this range, specifically between 200 and 260 °C. On the other hand, cellulose degradation takes place within the temperature range of 240 to 350 °C. Lignin, which is more resistant to decomposition, undergoes degradation between 280 and 500 °C during the pyrolysis process (Wani et al. 2021; Mandal et al. 2022).

3.1 Effect of Raw Material on Surface Area and Porosity

The characteristics of biochar are equally impacted by the composition of the raw material, much like the influence of pyrolysis temperature. Ippolito et al. (2020) emphasized that the selection of the raw material plays the most crucial role in determining the final biochar properties. Biochar derived from woody biomass, which contains a significant amount of lignin, is known for its porous nature and large surface area (Ippolito et al. 2020). This type of biochar contains high levels of aromatic organic components and has a low ash content. During pyrolysis, thermal cracking frequently occurs in wood-based biochar due to the unequal shrinkage between the interior parts and the exterior surfaces of the rigid wood masses. This leads to the loss of micro-molecule organic compounds, creating voids within the biochar matrix and subsequently increasing the overall surface area and pore volume (Guo and Chong Lua 1998; Weber and Quicker 2018). In contrast, biochar produced from manures or biosolids tends to have a low surface area due to deformation, structural cracking, or micro-pore blockage (Ahmad et al. 2014). The feedstock itself in these cases exhibits less distinct porous structures compared to wood-based biochar (Ippolito et al. 2020). Herbaceous feedstocks, containing a high percentage of silica components, often produce porous biochar with a high ash composition. While animal manures, initially being not very porous, undergo improvements in porosity during pyrolysis (Yavari et al. 2017; Kloss et al. 2012).

In general, for biochar production, biomass with high lignin content results in a macroporous structure, while biomass with high cellulose content leads to microporous structures (Cantrell et al. 2012; Liu et al. 2018a, 2017; Li et al. 2017; Joseph 2007). Another crucial factor that can negatively impact the surface area and pore volume is prolonged pyrolysis time, which can cause melting, shrinkage, and elemental realignment. This destruction may be more pronounced for softwood biomass, resulting in reduced adsorption capacity (Yavari et al. 2017).

3.2 Effect of Raw Material on CEC and pH

Unlike biochar produced from woody biomass, biochar obtained from non-woody biomass, such as crop waste, grasses, and manures, demonstrates higher soil cation exchange capacity (CEC) and pH. The process of pyrolysis leads to the formation of oxygenated surface and inorganic functional groups, contributing to the enhanced CEC (Briggs et al. 2012). When woodchip biochar is applied to soil, it results in higher saturated hydraulic conductivities compared to manure-based biochar (Lei and Zhang 2013). For example, the biochemical composition and structure of seaweeds differ significantly from ligno-cellulosic biomasses. Seaweed-derived biochar is typically characterized by a high yield, nutrient content, and CEC, but it tends to have a relatively low organic carbon content and surface area (Roberts et al. 2015).

3.3 Effect of Raw Material on Nutrients Biochar

The mineral composition of biochar also plays a significant role in pesticide sorption from soil, as it can reduce pesticide availability through mechanisms such as surface chelation and/or surface acidity (Wei et al. 2001; Dias et al. 2010). In general, biochars derived from wood tend to have higher carbon content and lower plant-available nutrients, while biochars from manures have lower carbon content and higher plant-available nutrients. Biochars produced from grasses typically fall somewhere in between woody and manure biochars in terms of their mineral composition (Ippolito et al. 2020). For example, hardwood-based biochar supplies less than 0.002% of the total nitrogen, 2.2% of the total phosphorus, and 17% of the total potassium present. On the other hand, biochar from softwood can supply 27% of the total phosphorus and 6% of the total potassium present (Ippolito et al. 2020; Dieguez-Alonso et al. 2018). Table 2 describe the effect of feedstocks composition on elemental and physical properties of biochar.

4 Role of Biochar in Fipronil Removal from Soil and Aqueous Medium

Biochar has been employed as an adsorbent to eliminate fipronil from both aqueous solutions and soil mediums (Li et al. 2018; Boehm et al. 2020b; Yang et al. 2010). The use of biochar in stormwater management is particularly exciting due to its renewable origin from energy systems and its ability to remediate various contaminants. Both laboratory and field tests have demonstrated that biochar can be utilized in filtration media, either for new construction or as a retrofit, to enhance existing stormwater management systems (Miles 2016). Biochar-augmented biofilters (BAB)

Table 2 Effect of feedstocks composition on different properties of biochar

Feedstock	Feedstock Composition			Biochar properties				Refs
	Lignin	Cellulose	Hemicellulose	C (%)	N (%)	SSA (m^2/g)	pH	
Corn stover	22.23	28.78	23.27	78.24	0.66	–	9.03	(Intani et al. 2018)
Corn stover	22.39	36.04	13.36	61.76	1.73	–	9.6	(Liu et al. 2018b)
Corn stover	3	30	38	77.6	0.85	0	7.816	(Mullen et al. 2010)
Corn stover	6	48	29	57.29	1.47	3.10	7.157	(Mullen et al. 2010)
Wood	25.0	45 ± 5	28	79.49	0.17	372	8.77	(Carrier et al. 2013)
Wood	24.22	45.75	9.57	86.8	3.68	0.26	–	(Oginni and Singh 2020)
Switchgrass	8.84	34.44	28.84	82.75	1.20	519.49	10.29	(Oginni et al. 2017)
Pine needles	25.91	24.51	19.09	71.87	1.75	15.59	6.91	(Mandal et al. 2019)

have shown promising results in reducing trace organic contaminants at the laboratory scale (Boehm et al. 2020b). The pollution caused by unregulated pesticides that reach urban stormwater and eventually urban water sources is a significant environmental concern (Masoner et al. 2019; Wolfand et al. 2019; Weston et al. 2015). Wolfand et al. (2019) demonstrated the effectiveness of BAB at the watershed scale in reducing pesticide concentrations in the San Diego River. The BAB exhibited a reduction of 93% for bifenthrin and 78% for fipronil, as compared to only 55% reduction for fipronil by a conventional biofilter without biochar (Wolfand et al. 2019). The study showed that the incorporation of biochar in biofilters was particularly successful for hydrophilic fipronil, which is poorly removed by conventional biofilters. Moreover, the conventional biofilter was unable to reduce the concentrations of bifenthrin and fipronil below concentration-based aquatic toxicity benchmarks. However, the installation of BAB in the watershed led to a reduction in concentrations of bifenthrin and fipronil below toxicity benchmarks for 46% and 97% of the time, respectively (Boehm et al. 2020b).

The dissipation and bioavailability of chlorpyrifos and fipronil were investigated in Chinese chives (Allium tuberosum) grown in different clay loam soils amended with biochar. The biochar was obtained by pyrolyzing cotton straw chips at temperatures of 450 °C and 850 °C, resulting in BC450 and BC850, respectively. Various concentrations of biochar (0, 0.1, 0.5, and 1% by soil dry weight) were mixed with 50 mg kg^{-1} of each pesticide. After 35 days of incubation, the control soil showed higher pesticide losses (58–68%), while the addition of 1% BC850 resulted in lower losses of chlorpyrifos (34%) and fipronil (32%). The presence of biochar reduced the uptake of both pesticides by the Chinese chives. This reduction was more pronounced with increasing doses of biochar prepared at higher temperatures (BC850) in the amended soil. BC850 also enhanced the biomass of Chinese chives compared to the control and BC450-amended soils. These findings demonstrate the positive role of temperature in regulating the functional properties of biochar, enabling it to sequester more pesticides and support better plant growth (Yang et al. 2010). The increased removal of pesticides may be attributed to the changes in the textural properties of biochar, particularly the increase in surface area and its more porous nature at the optimal higher temperature (Mandal et al. 2022; Sharma et al. 2022; Holanda et al. 2023).

The results obtained from biochar derived from Magnolia wood (Magnolia denudata) prepared at different pyrolysis temperatures of 300, 500, and 700 °C for 4 h indicated a consistent increase in the adsorption of the pollutants fipronil and ethiprole. Particularly noteworthy was the biochar prepared at 700 °C, which exhibited the highest adsorption capacity, characterized by a low polarity index. Moreover, extending the contact time between sediment-biochar from 7 to 30 days resulted in a further increase in the adsorption of fipronil and ethiprole. This enhanced removal can be attributed to the augmented total surface area in biochar prepared at higher pyrolysis temperatures. Additionally, another contributing factor to the increased adsorption was the heightened π-π interaction between the many electron-withdrawing groups of fipronil/ethiprole and the aromatic, heterocyclic rings present in the biochar, acting as electron donors. The biochar prepared at higher temperatures also exhibited a higher level of aromaticity, as evidenced by the decreased H/C ratio (Li et al. 2018). Another significant property that is influenced by pyrolysis temperature is the pH of biochar. Numerous studies have demonstrated that increasing the temperature during pyrolysis leads to an increase in the pH of biochar (Azad et al. 2023). This higher pH of biochar can be advantageous in the degradation of fipronil and ethiprole, as previous research has shown that their degradation tends to be faster at higher pH levels (Bobé et al. 1998; Liu et al. 2014).

In sediments, the presence of biochar prepared at higher temperatures resulted in a decrease in the freely dissolved concentration of fipronil and ethiprole. The maximum removal observed was 81.1% for ethiprole by biochar prepared at 700 °C, while the minimum removal was 34.9% for fipronil by biochar prepared at 300 °C. Moreover, as the dose of biochar in the sediments increased, the concentration of both fipronil and ethiprole decreased (e.g., fipronil decreased from 10.9 to 0.8 μg/L and ethiprole from 99.4 to 2.0 μg/L for a 10% biochar dose). This reduction in concentration indicates that the distribution, transport, and bioaccumulation of both chemicals in the sediments decreased due to the presence of biochar. However, it should be noted that the reduction in fipronil was found to be less significant compared to that of ethiprole. One possible reason for this difference could be the influence of organic matter, which may cause blockage of pores within the biochar or compete for active sites on the biochar surface, thereby interfering with the adsorption of target compounds (Li et al. 2018).

Fipronil has demonstrated enantioselective toxicity towards loach (Misgurnus anguillicaudatus), where it exhibited greater sensitivity to the S-enantiomer. Fipronil exists in two enantiomers, S-(+)- and R-(−)-fipronil. Among these, the S-(+) enantiomer has been found to be more toxic and has been associated with adverse effects, such as DNA methylation in zebrafish embryos (Qian et al., 2017), toxicity in loach liver (Qu et al. 2018), as well as in Danio rerio and ceriodaphnia dubia, and Procambarus clarkia (Qu et al. 2018; Konwick et al. 2006; Overmyer et al. 2007). Conversely, for Chlorella pyrenoidosa, the R-enantiomer was found to be more toxic than the S-enantiomer (Gong et al. 2021).

Loach (Misgurnus anguillicaudatus) is a prevalent fish species distributed widely in rice fields and freshwater ecosystems, making it a valuable biological model for water pollutant studies. Three different sets of experiments were conducted, where loaches were exposed to S-fipronil, R-fipronil, or rac-fipronil alone, as well as combined with biochar derived from maize straw. After 72 h of exposure, the LC50 value was found to be lowest for rac-fipronil (0.99 mg L−1) and highest for S-fipronil (approximately 0.49 mg L−1). The introduction of biochar alleviated the LC50 values after the same exposure time. For instance, the LC50 value of S-fipronil increased fourfold from 0.49 mg L−1 to 2.17 mg L−1, and its metabolites, fipronil sulfide and sulfone, increased up to six to seven times, indicating a positive effect of biochar in reducing the toxicity of S-fipronil. This improvement in toxicity mitigation is likely due to the adsorption of these contaminants by biochar from the solution, leading to a significant decrease in their freely dissolved concentration. The toxicity of fipronil and its degradates on liver of D.rerio were analysed by measuring superoxide dismutase (SOD) activity by Gong et al. (2021) . SOD activity induced by S-fipronil was stronger than that induced by rac- and R-fipronil, which indicated that S-fipronil is more toxic to D.rerio. Levels of SOD were higher in groups treated in the presence of biochar. This demonstrated the positive role of biochar to mitigate the fipronil.

The addition of biochar also resulted in a significant reduction of fipronil sulfide, desulfinyl, and sulfone in the liver of loach. Similar findings were reported by Qu et al. (2018), where biochar reduced the bioavailability of hexabromocyclododecanes (HBCDs) to earthworms in soil (Qu et al. 2018; Li et al. 2017). In contrast, only fipronil was detected in the loach brain at a concentration lower than 15 ng $g-^1$, and there was no significant enantiomeric accumulation. This observation might be attributed to the presence of the blood–brain-barrier (BBB) system, which prevents the transport of hydrophobic compounds into the brain and acts as a protective barrier (Qu et al. 2018; Cravedi et al. 2013). Additionally, the introduction of biochar was found to significantly decrease the levels of fipronil in the loach brain to 1.5 ng g^{-1} on the 5th day.

Gong et al. (2021) conducted a study to investigate the effectiveness of biochar in remediation of fipronil from water and leachate. The results indicated that fipronil and its metabolites, namely fipronil desulfinyl, fipronil sulfide, and fipronil sulfone, were highly toxic to Chlorella pyrenoidosa and Danio rerio (Gong et al. 2021). Among the metabolites, fipronil sulfide and sulfone showed higher toxicity and persistence compared to rac-fipronil. However, the introduction of biochar significantly mitigated the toxicity of fipronil and its metabolites to non-target organisms and reduced the residues by 90% for fipronil, 72% for sulfone, 65% for sulfide, and 60% for desulfinyl. The study also investigated the impact of fipronil and its degradants on the content of chlorophyll a pigment in Chlorella pyrenoidosa, which serves as a sensitive biomarker for contamination threats to plants. Without biochar treatment, the content of chlorophyll a noticeably decreased, with the maximum reduction observed in the R-fipronil treatment group, indicating significant inhibition of photosynthesis in algae by R-fipronil. However, the addition of biochar resulted in a significant increase in chlorophyll a content. Interestingly, the presence of biochar also influenced the enantioselectivity of fipronil toxicity. For instance, at 0.05 mg L^{-1} treatment concentration, rac-fipronil showed the least toxicity, whereas S-enantiomer demonstrated the least toxicity at 0.5 mg L^{-1}. In both cases, R-fipronil exhibited the highest toxicity to algae and caused the most damage to chlorophyll a. This indicates that biochar effectively reduced the toxicity of fipronil and its metabolites through adsorption. Furthermore, biochar at a concentration of 0.05 mg L^{-1} successfully decreased the toxicity level of the synergistic treatment combination of leachate with fipronil and its metabolites on algae. Lastly, the authors have a vast experience in the field of bio composites and computational mechanics (Verma and Singh 2016; Verma 2022; Verma et al. 2022a, 2022b, 2020a, 2019, 2020b, 2023; Raja et al. 2022; Prabhakaran et al. 2022; Sethi et al. 2022; Chaudhary et al. 2022; Verma et al. 2022; Lila et al. 2022; Kumar et al. 2022; Verma and Sharma 2022; Chaturvedi et al. 2022a, 2022b, 2022c; Shankar et al. 2022a, 2022b; Verma and Ogata 2022, 2023; Homer et al. 2022; Arpitha et al. 2022, 2023a, 2023b; Thimmaiah et al. 2022; Kataria et al. 2022, 2023; Kataria et al. 2022; Bharath et al. 2021; Kataria et al. 2023; Chaturvedi et al. 2023; Deji et al. 2022a, 2022b, 2022c, 2022d; Verma and Parashar 2020; Dogra et al. 2023a, 2023b; TG, Y.G. et al. 2023; Sati et al. 2023; Agrawal et al. 2023; Gupta et al. 2023a, 2023b; Tongco et al. 2023; Manasa et al. 2023; Singh et al. 2023; Zafar and Verma 2023; Robinson et al. 2023).

5 Conclusion

Fipronil, a modern pesticide, presents significant environmental risks due to the higher toxicity of its degradation products compared to the original compound. Conventional wastewater treatment methods have limited efficacy in completely removing fipronil, resulting in its persistence as a pollutant. Adopting biochar for fipronil removal offers an environmentally friendly and cost-effective solution, with low-cost inputs and minimal generation of secondary by-products. The successful application of biochar for fipronil removal depends on carefully selecting pyrolysis temperature and feedstock, as these factors influence essential properties such as porosity, surface area, pH, nutrients, aromatic structure, surface functional groups, and cation exchange capacity (CEC). Biochar has proven effective in removing fipronil from both water and soil, and it can even modify the enantioselectivity of the pesticide. However, the precise mechanism of fipronil removal by biochar remains unclear due to limited research. Future investigations should focus on gaining a deeper understanding of how biochar properties, altered through varying pyrolysis conditions and raw materials, can enhance the removal of fipronil.

Declarations

Conflicts of Interest "There are no conflicts of interest to declare by the authors."

Ethical Approval The authors hereby state that the present work is in compliance with the ethical standards.

Competing Interests The authors declare no competing interests.

Availability of Data and Materials Not applicable.

References

Act CW, United States Environmental Protection Agency. obtained from https://www.epa.gov/sites/default/files/2018-08/documents/epa-2018-science_annualreport_508compressed.pdf accessed on 12.08.2023

Agrawal PK, Sharma P, Verma A, Singh VK, Chaudhary AK, Chauhan S. Impact of graphite particle surface modification on the strengthening of cross-linked polyvinyl alcohol composites: A comprehensive investigation. *Proceedings of the Institution of Mechanical Engineers, Part L: Journal of Materials: Design and Applications.* 2023. DOI:https://doi.org/10.1177/14644207231188601

Ahmad M, Lee SS, Dou X, et al (2012) Effects of pyrolysis temperature on soybean stover- and peanut shell-derived biochar properties and TCE adsorption in water. Bioresour Technol 118:536–544. https://doi.org/10.1016/j.biortech.2012.05.042

Ahmad M, Lee SS, Lim JE, et al (2014) Speciation and phytoavailability of lead and antimony in a small arms range soil amended with mussel shell, cow bone and biochar: EXAFS spectroscopy and chemical extractions. Chemosphere 95:433–441. https://doi.org/10.1016/j.chemosphere.2013.09.077

Alhashimi HA, Aktas CB (2017) Life cycle environmental and economic performance of biochar compared with activated carbon: A meta-analysis. Resour Conserv Recycl 118:13–26. https://doi.org/10.1016/j.resconrec.2016.11.016

Antón-Herrero R, García-Delgado C, Alonso-Izquierdo M, et al (2018) Comparative adsorption of tetracyclines on biochars and stevensite: Looking for the most effective adsorbent. Appl Clay Sci 160:162–172. https://doi.org/10.1016/j.clay.2017.12.023

Arpitha, G.R., Verma, A., MR, S., Gorbatyuk, S., Khan, A., Sobahi, T.R., Asiri, A.M. and Siengchin, S., 2022. Bio-composite film from corn starch based vetiver cellulose. *Journal of Natural Fibers*, 19 (16), pp.14634–14644.

Arpitha, G.R., Mohit, H., Madhu, P. and Akarsh Verma. (2023) "Effect of sugarcane bagasse and alumina reinforcements on physical, mechanical, and thermal characteristics of epoxy composites using artificial neural networks and response surface methodology", *Biomass Conversion and Biorefinery.* https://doi.org/10.1007/s13399-023-03886-7

Arpitha, G.R., Jain, N. and Verma, A., 2023. Banana biofiber and glass fiber reinforced hybrid composite for lightweight structural applications: Mechanical, thermal, and microstructural characterization. *Biomass Conversion and Biorefinery*, pp.1–10. https://doi.org/10.1007/s13399-023-04300-y

Atkinson CJ, Fitzgerald JD, Hipps NA (2010) Potential mechanisms for achieving agricultural benefits from biochar application to temperate soils: a review. Plant Soil 337:1–18. https://doi.org/https://doi.org/10.1007/s11104-010-0464-5

Azad D, Pateriya RN, Sharma RK (2023) Chemical activation of pine needle and coconut shell biochar: production, characterization and process optimization. International Journal of Environmental Science and Technology. https://doi.org/10.1007/s13762-023-04913-w

Bharath, K.N., Madhu, P., Gowda, T.Y., Verma, A., Sanjay, M.R. and Siengchin, S., 2021. Mechanical and chemical properties evaluation of sheep wool fiber–reinforced vinylester and polyester composites. *Materials Performance and Characterization, 10*(1), pp.99–109.

Bhatt P, Rene ER, Kumar AJ, et al (2021) Fipronil degradation kinetics and resource recovery potential of Bacillus sp. strain FA4 isolated from a contaminated agricultural field in Uttarakhand, India. Chemosphere 276:130156. https://doi.org/10.1016/j.chemosphere.2021.130156

Bhatti S, Satyanarayana GNV, Patel DK, Satish A (2019) Bioaccumulation, biotransformation and toxic effect of fipronil in Escherichia coli. Chemosphere 231:207–215. https://doi.org/10.1016/j.chemosphere.2019.05.124

Bobé A, Meallier P, Cooper J-F, Coste CM (1998) Kinetics and Mechanisms of Abiotic Degradation of Fipronil (Hydrolysis and Photolysis). J Agric Food Chem 46:2834–2839. https://doi.org/10.1021/jf970874d

Boehm AB, Bell CD, Fitzgerald NJM, et al (2020) Biochar-augmented biofilters to improve pollutant removal from stormwater-can they improve receiving water quality? Environ Sci (Camb) 6:1520–1537. https://doi.org/10.1039/d0ew00027b

Boehm AB, Bell CD, Fitzgerald NJM, et al (2020) Biochar-augmented biofilters to improve pollutant removal from stormwater–can they improve receiving water quality? Environ Sci (Camb) 6:1520–1537. https://doi.org/10.1039/D0EW00027B

Břendová K, Száková J, Lhotka M, et al (2017) Biochar physicochemical parameters as a result of feedstock material and pyrolysis temperature: predictable for the fate of biochar in soil? Environ Geochem Health 39:1381–1395. https://doi.org/10.1007/s10653-017-0004-9

Briggs C, Breiner JM, Graham RC (2012) Physical and Chemical Properties of Pinus ponderosa Charcoal. Soil Sci 177:263–268. https://doi.org/10.1097/SS.0b013e3182482784

Bruun S, Clauson-Kaas S, Bobuľská L, Thomsen IK (2014) Carbon dioxide emissions from biochar in soil: role of clay, microorganisms and carbonates. Eur J Soil Sci 65:52–59. https://doi.org/10.1111/ejss.12073

Budai A, Rasse DP, Lagomarsino A, et al (2016) Biochar persistence, priming and microbial responses to pyrolysis temperature series. Biol Fertil Soils 52:749–761. https://doi.org/10.1007/s00374-016-1116-6

Cantrell KB, Hunt PG, Uchimiya M, et al (2012) Impact of pyrolysis temperature and manure source on physicochemical characteristics of biochar. Bioresour Technol 107:419–428. https://doi.org/10.1016/j.biortech.2011.11.084

Cao X, Harris W (2010) Properties of dairy-manure-derived biochar pertinent to its potential use in remediation. Bioresour Technol 101:5222–5228. https://doi.org/10.1016/j.biortech.2010.02.052

Carrier M, Joubert J-E, Danje S, et al (2013) Impact of the lignocellulosic material on fast pyrolysis yields and product quality. Bioresour Technol 150:129–138. https://doi.org/10.1016/j.biortech.2013.09.134

Chaturvedi, S., Verma, A., Sethi, S.K. and Ogata, S., 2022. Defect Energy Calculations of Nickel, Copper and Aluminium (and Their Alloys): Molecular Dynamics Approach. In *Forcefields for Atomistic-Scale Simulations: Materials and Applications* (pp. 157–186). Springer, Singapore.

Chaturvedi, S., Verma, A., Sethi, S.K., Rangappa, S.M. and Siengchin, S., 2022. Stalk fibers (rice, wheat, barley, etc.) composites and applications. In *Plant Fibers, their Composites, and Applications* (pp. 347–362). Woodhead Publishing.

Chaturvedi, S., Verma, A., Singh, S.K. and Ogata, S., 2022. EAM Inter-Atomic Potential—Its Implication on Nickel, Copper, and Aluminum (and Their Alloys). In *Forcefields for Atomistic-Scale Simulations: Materials and Applications* (pp. 133–156). Springer, Singapore.

Chaturvedi, S., Kataria, A., Chaudhary, V., Verma, A., Naman Jain Mavinkere Rangappa Sanjay, Suchart Siengchin, 2023. Bionanocomposites reinforced with cellulose fibers and agro-industrial wastes. In *Cellulose Fibre Reinforced Composites* (pp. 1–22). Woodhead Publishing. https://doi.org/10.1016/B978-0-323-90125-3.00017-3

Chaudhary, A., Sharma, S. and Verma, A., 2022. Optimization of WEDM process parameters for machining of heat treated ASSAB'88 tool steel using Response surface methodology (RSM). *Materials Today: Proceedings*, *50*, pp.917-922.

Chen T, Zhang Y, Wang H, et al (2014) Influence of pyrolysis temperature on characteristics and heavy metal adsorptive performance of biochar derived from municipal sewage sludge. Bioresour Technol 164:47–54. https://doi.org/10.1016/j.biortech.2014.04.048

Chen G, Yao J, Liu J, et al (2015) Biomass to hydrogen-rich syngas via catalytic steam gasification of bio-oil/biochar slurry. Bioresour Technol 198:108–114. https://doi.org/10.1016/j.biortech.2015.09.009

Cravedi JP, Delous G, Zalko D, et al (2013) Disposition of fipronil in rats. Chemosphere 93:2276–2283. https://doi.org/10.1016/j.chemosphere.2013.07.083

Deji, R., Verma, A., Kaur, N., Choudhary, B.C. and Sharma, R.K., 2022. Density functional theory study of carbon monoxide adsorption on transition metal doped armchair graphene nanoribbon. *Materials Today: Proceedings*, *54*, pp.771–776.

Deji, R., Verma, A., Choudhary, B.C. and Sharma, R.K., 2022. New insights into NO adsorption on alkali metal and transition metal doped graphene nanoribbon surface: A DFT approach. *Journal of Molecular Graphics and Modelling*, *111*, p.108109.

Deji, R., Jyoti, R., Verma, A., Choudhary, B.C. and Sharma, R.K., 2022. A theoretical study of HCN adsorption and width effect on co-doped armchair graphene nanoribbon. *Computational and Theoretical Chemistry*, *1209*, p.113592.

Deji, R., Verma, A., Kaur, N., Choudhary, B.C. and Sharma, R.K., 2022. Adsorption chemistry of co-doped graphene nanoribbon and its derivatives towards carbon based gases for gas sensing applications: Quantum DFT investigation. *Materials Science in Semiconductor Processing*, *146*, p.106670.

Dias BO, Silva CA, Higashikawa FS, et al (2010) Use of biochar as bulking agent for the composting of poultry manure: Effect on organic matter degradation and humification. Bioresour Technol 101:1239–1246. https://doi.org/10.1016/j.biortech.2009.09.024

Dieguez-Alonso A, Funke A, Anca-Couce A, et al (2018) Towards Biochar and Hydrochar Engineering—Influence of Process Conditions on Surface Physical and Chemical Properties, Thermal Stability, Nutrient Availability, Toxicity and Wettability. Energies (Basel) 11:496. https://doi.org/10.3390/en11030496

Dogra, V., Kishore, C., Mishra, A., Gaur, A. and Verma, A., 2023. Sol Gel preparation and wetting behaviour analysis of hydrophobic Zirconium based nano-coating: Implications for solar panel coating. *Chemical Engineering Journal Advances*, *15*, p.100507. https://doi.org/10.1016/j.ceja.2023.100507

Dogra, V., Kishore, C., Mishra, A., Verma, A. and Gaur, A., 2023. Coatings: Types and Synthesis Techniques. In *Coating Materials: Computational Aspects, Applications and Challenges* (pp. 17–31). Singapore: Springer Nature Singapore. DOI: https://doi.org/10.1007/978-981-99-3549-9_2

Fent GM (2014) Fipronil. Encyclopedia of toxicology 2:596–597

Funke A, Ziegler F (2010) Hydrothermal carbonization of biomass: A summary and discussion of chemical mechanisms for process engineering. Biofuels, Bioproducts and Biorefining 4:160–177. https://doi.org/10.1002/bbb.198

Ghanim BM, Pandey DS, Kwapinski W, Leahy JJ (2016) Hydrothermal carbonisation of poultry litter: Effects of treatment temperature and residence time on yields and chemical properties of hydrochars. Bioresour Technol 216:373–380. https://doi.org/10.1016/j.biortech.2016.05.087

Gong W, Barrett H, Hu Y, et al (2021) Application of biochar: An approach to attenuate the pollution of the chiral pesticide fipronil and its metabolites in leachate from activated sludge. Process Safety and Environmental Protection 149:936–945. https://doi.org/10.1016/j.psep.2021.03.044

Gunasekara AS, Truong T, Goh KS, et al (2007) Environmental fate and toxicology of fipronil. J Pestic Sci 32:189–199. https://doi.org/10.1584/jpestics.R07-02

Guo J, Chong Lua A (1998) Characterization of chars pyrolyzed from oil palm stones for the preparation of activated carbons. J Anal Appl Pyrolysis 46:113–125. https://doi.org/10.1016/S0165-2370(98)00074-6

Gupta, H.S., Shankar, U., Verma, A., Gogoi, R. and Sethi, S.K., 2023. Computational Aspects: Self-clean Coatings, Plastics and Polymers in Coatings. In *Coating Materials: Computational Aspects, Applications and Challenges* (pp. 167–184). Singapore: Springer Nature Singapore. DOI: https://doi.org/10.1007/978-981-99-3549-9_8

Gupta, H.S., Sethi, S.K. and Verma, A., 2023. Applications of Coating Materials: A Critical Overview. *Coating Materials: Computational Aspects, Applications and Challenges*, pp.81–109. DOI: https://doi.org/10.1007/978-981-99-3549-9_4

Holanda MAS, Coelho Menezes JM, Coutinho HDM, Teixeira RNP (2023) Effectiveness of biochar as an adsorbent for pesticides: Systematic review and meta-analysis. J Environ Manage 345:118719. https://doi.org/10.1016/j.jenvman.2023.118719

Homer, E.R., Verma, A., Britton, D., Johnson, O.K. and Thompson, G.B., 2022, July. Simulated migration behavior of metastable Σ3 (11 8 5) incoherent twin grain boundaries. In *IOP Conference Series: Materials Science and Engineering* (Vol. 1249, No. 1, p. 012019).

Hossain MK, Strezov V, Chan KY, et al (2011) Influence of pyrolysis temperature on production and nutrient properties of wastewater sludge biochar. J Environ Manage 92:223–228. https://doi.org/10.1016/j.jenvman.2010.09.008

Intani K, Latif S, Cao Z, Müller J (2018) Characterisation of biochar from maize residues produced in a self-purging pyrolysis reactor. Bioresour Technol 265:224–235. https://doi.org/10.1016/j.biortech.2018.05.103

Ippolito JA, Cui L, Kammann C, et al (2020) Feedstock choice, pyrolysis temperature and type influence biochar characteristics: a comprehensive meta-data analysis review. Biochar 2:421–438. https://doi.org/10.1007/s42773-020-00067-x

Joseph S DAMPCALJ (2007) Biochar for carbon sequestration, reduction of greenhouse gas emissions and enhancement of soil fertility; A review of the materials science. In: Proceedings of the Australian Combustion Symposium. pp 130–133.

Kataria, A. Verma, A., Sanjay, M.R. and Siengchin, S., 2022. "Molecular modeling of 2D graphene grain boundaries: Mechanical and fracture aspects," *Mater Today Proc*, vol. 52, pp. 2404–2408, Jan.

Kataria, A., Verma, A., Sethi, S.K. and Ogata, S., 2022. Introduction to Interatomic Potentials/Forcefields. In *Forcefields for Atomistic-Scale Simulations: Materials and Applications* (pp. 21–49). Springer, Singapore.

Kataria, A., Zafar, S., Verma, A. and Ogata, S., 2023. Molecular Dynamics Simulations in Coatings. In *Coating Materials: Computational Aspects, Applications and Challenges* (pp. 125–156). Singapore: Springer Nature Singapore. DOI: https://doi.org/10.1007/978-981-99-3549-9_6

Kataria, A., Chaturvedi, S., Chaudhary, V., Verma, A., Naman Jain Mavinkere Rangappa Sanjay, Suchart Siengchin, 2023. Cellulose fiber-reinforced composites—History of evolution, chemistry, and structure. In *Cellulose Fibre Reinforced Composites* (pp. 1–22). Woodhead Publishing. https://doi.org/10.1016/B978-0-323-90125-3.00012-4

Kloss S, Zehetner F, Dellantonio A, et al (2012) Characterization of Slow Pyrolysis Biochars: Effects of Feedstocks and Pyrolysis Temperature on Biochar Properties. J Environ Qual 41:990–1000. https://doi.org/10.2134/jeq2011.0070

Konwick BJ, Garrison AW, Black MC, et al (2006) Bioaccumulation, Biotransformation, and Metabolite Formation of Fipronil and Chiral Legacy Pesticides in Rainbow Trout. Environ Sci Technol 40:2930–2936. https://doi.org/10.1021/es0600678

Kumar, G., Mishra, R.R., Verma, A. (2022). Introduction to Molecular Dynamics Simulations. In *Forcefields for Atomistic-Scale Simulations: Materials and Applications* (pp. 1–19). Springer, Singapore. https://doi.org/10.1007/978-981-19-3092-8_1

Lehmann J (2007) Bio-energy in the black. Frontiers Ecol Environ 5:381–387

Lei O, Zhang R (2013) Effects of biochars derived from different feedstocks and pyrolysis temperatures on soil physical and hydraulic properties. J Soils Sediments 13:1561–1572. https://doi.org/10.1007/s11368-013-0738-7
Leng L, Xiong Q, Yang L, et al (2021) An overview on engineering the surface area and porosity of biochar. Science of The Total Environment 763:144204. https://doi.org/10.1016/j.scitotenv.2020.144204
Li H, Dong X, da Silva EB, et al (2017) Mechanisms of metal sorption by biochars: Biochar characteristics and modifications. Chemosphere 178:466–478. https://doi.org/10.1016/j.chemosphere.2017.03.072
Li S, Wang P, Liu X, et al (2018) Polyoxymethylene passive samplers to assess the effectiveness of biochar by reducing the content of freely dissolved fipronil and ethiprole. Science of the Total Environment 630:960–966. https://doi.org/10.1016/j.scitotenv.2018.02.221
Lila, M.K., Verma, A. and Bhurat, S.S., 2022. Impact Behaviors of Epoxy/Synthetic Fiber Composites. In *Handbook of Epoxy/Fiber Composites* (pp. 1–18). Singapore: Springer Singapore.
Liu X, Dong F, Xu J, et al (2014) Dissipation and Adsorption Behavior of the Insecticide Ethiprole on Various Cultivated Soils in China. J Integr Agric 13:2471–2478. https://doi.org/10.1016/S2095-3119(13)60685-1
Liu Y, Yao S, Wang Y, et al (2017) Bio- and hydrochars from rice straw and pig manure: Intercomparison. Bioresour Technol 235:332–337. https://doi.org/10.1016/j.biortech.2017.03.103
Liu Y, Lonappan L, Brar SK, Yang S (2018) Impact of biochar amendment in agricultural soils on the sorption, desorption, and degradation of pesticides: A review. Science of The Total Environment 645:60–70. https://doi.org/10.1016/j.scitotenv.2018.07.099
Liu Z, Niu W, Chu H, et al (2018) Effect of the Carbonization Temperature on the Properties of Biochar Produced from the Pyrolysis of Crop Residues. Bioresources 13:. https://doi.org/10.15376/biores.13.2.3429-3446
Liu Z hua, Kanjo Y, Mizutani S (2009) Removal mechanisms for endocrine disrupting compounds (EDCs) in wastewater treatment - physical means, biodegradation, and chemical advanced oxidation: A review. Science of the Total Environment 407:731–748. https://doi.org/10.1016/j.scitotenv.2008.08.039
Manasa, C.M., Prince, B., Arpitha, G.R. and Verma, A., 2023. Hybrid Feature Selection Techniques to Improve the Accuracy of Rice Yield Prediction: A Machine Learning Approach. In *Coating Materials: Computational Aspects, Applications and Challenges* (pp. 409–421). Singapore: Springer Nature Singapore. DOI: https://doi.org/10.1007/978-981-99-3549-9_19
Mandal S, Haydary J, Bhattacharya TK, et al (2019) Valorization of Pine Needles by Thermal Conversion to Solid, Liquid and Gaseous Fuels in a Screw Reactor. Waste Biomass Valorization 10:3587–3599. https://doi.org/10.1007/s12649-018-0386-7
Mandal S, Sharma RK, Bhattacharya TK, et al (2022) Charring of pine needles using a portable drum reactor. Chemical Papers 76:1239–1252. https://doi.org/10.1007/s11696-021-01893-4
Mandal S, Sharma RK (2023) Application of Gasification Technology in Agriculture for Power Generation. In: Handbook of Energy Management in Agriculture. Springer Nature Singapore, Singapore, pp 1–22.
Masoner JR, Kolpin DW, Cozzarelli IM, et al (2019) Urban Stormwater: An Overlooked Pathway of Extensive Mixed Contaminants to Surface and Groundwaters in the United States. Environ Sci Technol 53:10070–10081. https://doi.org/10.1021/acs.est.9b02867
Mensah AK, Frimpong KA (2018) Biochar and/or Compost Applications Improve Soil Properties, Growth, and Yield of Maize Grown in Acidic Rainforest and Coastal Savannah Soils in Ghana. International Journal of Agronomy 2018:1–8. https://doi.org/10.1155/2018/6837404
Miles TR REGM (2016) Aqueous contaminant removal and stormwater treatment using biochar. In: Guo M, He Z, Uchimiya SM (eds) Agricultural and environmental applications of biochar: advances and barriers. Soil Science Society of America Special Publication 63, Madison, pp 341–376.

Miller JL, Schmidt TS, Van Metre PC, et al (2020) Common insecticide disrupts aquatic communities: A mesocosm-to-field ecological risk assessment of fipronil and its degradates in U.S. streams. Sci Adv 6:. https://doi.org/10.1126/sciadv.abc1299

Mohan D, Sarswat A, Ok YS, Pittman CU (2014) Organic and inorganic contaminants removal from water with biochar, a renewable, low cost and sustainable adsorbent – A critical review. Bioresour Technol 160:191–202. https://doi.org/10.1016/j.biortech.2014.01.120

Mukherjee A, Zimmerman AR (2013) Organic carbon and nutrient release from a range of laboratory-produced biochars and biochar–soil mixtures. Geoderma 193–194:122–130. https://doi.org/10.1016/j.geoderma.2012.10.002

Mullen CA, Boateng AA, Goldberg NM, et al (2010) Bio-oil and bio-char production from corn cobs and stover by fast pyrolysis. Biomass Bioenergy 34:67–74. https://doi.org/10.1016/j.biombioe.2009.09.012

Nguyen BT, Lehmann J, Hockaday WC, et al (2010) Temperature Sensitivity of Black Carbon Decomposition and Oxidation. Environ Sci Technol 44:3324–3331. https://doi.org/10.1021/es903016y

Novak JM, Busscher WJ, Laird DL, et al (2009) Impact of Biochar Amendment on Fertility of a Southeastern Coastal Plain Soil. Soil Sci 174:105–112. https://doi.org/10.1097/SS.0b013e3181981d9a

Oginni O, Singh K (2020) Influence of high carbonization temperatures on microstructural and physicochemical characteristics of herbaceous biomass derived biochars. J Environ Chem Eng 8:104169. https://doi.org/10.1016/j.jece.2020.104169

Oginni O, Singh K, Zondlo JW (2017) Pyrolysis of dedicated bioenergy crops grown on reclaimed mine land in West Virginia. J Anal Appl Pyrolysis 123:319–329. https://doi.org/10.1016/j.jaap.2016.11.013

Overmyer JP, Rouse DR, Avants JK, et al (2007) Toxicity of fipronil and its enantiomers to marine and freshwater non-targets. Journal of Environmental Science and Health, Part B 42:471–480. https://doi.org/10.1080/03601230701391823

Prabhakaran, S., Sharma, S., Verma, A., Rangappa, S.M. and Siengchin, S., 2022. Mechanical, thermal, and acoustical studies on natural alternative material for partition walls: A novel experimental investigation. *Polymer Composites*, 43 (7), 4711–4720.

Qian Y, Wang C, Wang J, Zhang X, Zhou Z, Zhao M, Lu C (2017) Fipronil-induced enantioselective developmental toxicity to zebrafish embryo-larvae involves changes in DNA methylation. Sci Rep 7(1):2284. https://doi.org/10.1038/s41598-017-02255-5

Qu H, Ma R, Wang F, et al (2018) The effect of biochar on the mitigation of the chiral insecticide fipronil and its metabolites burden on loach (Misgurnus.anguillicaudatus). J Hazard Mater 360:214–222. https://doi.org/10.1016/j.jhazmat.2018.07.078

Raja, S., Verma, A., Rangappa, S.M. and Siengchin, S., 2022. Development and experimental analysis of polymer based composite bipolar plate using Aquila Taguchi optimization: Design of experiments. *Polymer Composites*, *43*(8), pp.5522–5533.

Rathi A, Basu S, Barman S (2019) Adsorptive removal of fipronil from its aqueous solution by modified zeolite HZSM-5: Equilibrium, kinetic and thermodynamic study. J Mol Liq 283:867–878. https://doi.org/10.1016/j.molliq.2019.02.140

Roberts DA, Paul NA, Dworjanyn SA, et al (2015) Biochar from commercially cultivated seaweed for soil amelioration. Sci Rep 5:9665. https://doi.org/10.1038/srep09665

Robertson SJ, Rutherford PM, López-Gutiérrez JC, Massicotte HB (2012) Biochar enhances seedling growth and alters root symbioses and properties of sub-boreal forest soils. Can J Soil Sci 92:329–340. https://doi.org/10.4141/cjss2011-066

Robinson, J., Verma, A., Homer, E.R. and Thompson, G.B., 2023. Nanotwin stability in alloyed copper under ambient and cryo-temperature dependent deformation states. *Materials Science and Engineering: A*, *871*, p.144866. DOI: https://doi.org/10.1016/j.msea.2023.144866

Safaei Khorram M, Zhang Q, Lin D, et al (2016) Biochar: A review of its impact on pesticide behavior in soil environments and its potential applications. Journal of Environmental Sciences 44:269–279. https://doi.org/10.1016/j.jes.2015.12.027

Sati, P., Verma, A., Zindal, A., Chauhan, S. and Singh, V.K., 2023. PVA biopolymer-acidic functionalized graphene hybrid nano composite for vibration isolation application: An experimental approach with variable reflux and vacuum timings. *Chemical Physics Impact*, *6*, p.100212. DOI: https://doi.org/10.1016/j.chphi.2023.100212

Sethi, S.K., Gogoi, R., Verma, A. and Manik, G., 2022. How can the geometry of a rough surface affect its wettability?-A coarse-grained simulation analysis. *Progress in Organic Coatings*, *172*, p.107062.

Shackley S, Clare A, Joseph S, et al (2015) Economic evaluation of biochar systems: current evidence and challenges. In: Biochar for Environmental Management. Routledge, pp 845–884.

Shankar, U., Gogoi, R., Sethi, S.K. and Verma, A., 2022. Introduction to Materials Studio Software for the Atomistic-Scale Simulations. In *Forcefields for Atomistic-Scale Simulations: Materials and Applications* (pp. 299–313). Springer, Singapore.

Shankar, U., Sethi, S.K. and Verma, A., 2022. Forcefields and Modeling of Polymer Coatings and Nanocomposites. In *Forcefields for Atomistic-Scale Simulations: Materials and Applications* (pp. 81–98). Springer, Singapore.

Sharma RK, Singh TP, Mandal S, et al (2022) Chemical Treatments for Biochar Modification: Opportunities, Limitations and Advantages. In: Engineered Biochar. Springer Nature Singapore, Singapore, pp 65–84.

Singh, S.K., Chaurasia, A. and Verma, A., 2023. Basics of Density Functional Theory, Molecular Dynamics, and Monte Carlo Simulation Techniques in Materials Science. In *Coating Materials: Computational Aspects, Applications and Challenges* (pp. 111–124). Singapore: Springer Nature Singapore. DOI: https://doi.org/10.1007/978-981-99-3549-9_5

Smith AM, Singh S, Ross AB (2016) Fate of inorganic material during hydrothermal carbonisation of biomass: Influence of feedstock on combustion behaviour of hydrochar. Fuel 169:135–145. https://doi.org/10.1016/j.fuel.2015.12.006

TG, Y.G., Ballupete Nagaraju, S., Puttegowda, M., Verma, A., Rangappa, S.M. and Siengchin, S., 2023. Biopolymer-Based Composites: An Eco-Friendly Alternative from Agricultural Waste Biomass. *Journal of Composites Science*, *7*(6), p.242. https://doi.org/10.3390/jcs7060242

Thimmaiah, S.H., Narayanappa, K., Thyavihalli Girijappa, Y., Gulihonenahali Rajakumara, A., Hemath, M., Thiagamani, S.M.K. and Verma, A., 2022. An artificial neural network and Taguchi prediction on wear characteristics of Kenaf–Kevlar fabric reinforced hybrid polyester composites. *Polymer Composites*, 44 (1), pp.261–273.

Thines KR, Abdullah EC, Mubarak NM (2017) Effect of process parameters for production of microporous magnetic biochar derived from agriculture waste biomass. Microporous and Mesoporous Materials 253:29–39. https://doi.org/10.1016/j.micromeso.2017.06.031

Tongco, J.V., Sethi, S.K., Kumar, A., Verma, A. and Shankar, U., 2023. Introduction to Coatings: Types and Their Synthesis. In *Coating Materials: Computational Aspects, Applications and Challenges* (pp. 1–15). Singapore: Springer Nature Singapore. https://doi.org/10.1007/978-981-99-3549-9_1

Ulrich BA, Loehnert M, Higgins CP (2017) Improved contaminant removal in vegetated stormwater biofilters amended with biochar. Environ Sci (Camb) 3:726–734. https://doi.org/10.1039/C7EW00070G

Verma, A., 2022. A Perspective on the Potential Material Candidate for Railway Sector Applications: PVA Based Functionalized Graphene Reinforced Composite. *Applied Science and Engineering Progress*, *15*(2), pp.5727–5727.

Verma, A. and Ogata, S., 2023. Magnesium based alloys for reinforcing biopolymer composites and coatings: A critical overview on biomedical materials. *Advanced Industrial and Engineering Polymer Research*. DOI: https://doi.org/10.1016/j.aiepr.2023.01.002

Verma, A., Parashar, A. and Packirisamy, M., 2019. Effect of grain boundaries on the interfacial behaviour of graphene-polyethylene nanocomposite. *Applied Surface Science*, *470*, pp.1085–1092.

Verma, A., Oliver K. Johnson, Gregory B. Thompson, Ian Chesser, Shigenobu Ogata and Eric R. Homer. (2023) "Insights into factors that affect non-Arrhenius migration of a simulated

incoherent Σ3 grain boundary", Acta Materialia, 258, 119210. DOI: https://doi.org/10.1016/j.actamat.2023.119210
Verma, A. and Ogata, S., 2022. Computational Modelling of Deformation and Failure of Bone at Molecular Scale. In *Forcefields for Atomistic-Scale Simulations: Materials and Applications* (pp. 253–268). Springer, Singapore.
Verma, A. and Parashar, A., 2020. Characterization of 2D nanomaterials for energy storage. In *Recent Advances in Theoretical, Applied, Computational and Experimental Mechanics* (pp. 221–226). Springer, Singapore.
Verma, A. and Sharma, S., 2022. Atomistic Simulations to Study Thermal Effects and Strain Rate on Mechanical and Fracture Properties of Graphene like BC3. In *Forcefields for Atomistic-Scale Simulations: Materials and Applications* (pp. 237–252). Springer, Singapore.
Verma, A. and Singh, V.K., 2016. Experimental characterization of modified epoxy resin assorted with almond shell particles. *ESSENCE-International Journal for Environmental Rehabilitation and Conservation*, 7 (1), pp.36–44.
Verma, A., Jain, N., Parashar, A., Singh, V.K., Sanjay, M.R. and Siengchin, S., 2020. Lightweight Graphene Composite Materials. In *Lightweight Polymer Composite Structures* (pp. 1–20). CRC Press.
Verma, A., Parashar, A., Singh, S.K., Jain, N., Sanjay, S.M. and Siengchin, S., 2020. Modeling and Simulation in Polymer Coatings. In *Polymer Coatings* (pp. 309–324). CRC Press.
Verma, A., Jain, N., Singh, K., Singh, V.K., Rangappa, S.M. and Siengchin, S., 2022. PVA-based blends and composites. In *Biodegradable Polymers, Blends and Composites* (pp. 309–326). Woodhead Publishing.
Verma, A., Jain, N. and Mishra, R.R., 2022. Applications and Drawbacks of Epoxy/Natural Fiber Composites. In *Handbook of Epoxy/Fiber Composites* (pp. 1–15). Singapore: Springer Singapore.
Verma, A., Jain, N., Sanjay, M.R. and Siengchin, S., Viscoelastic Properties of Completely Biodegradable Polymer-Based Composites. In *Vibration and Damping Behavior of Biocomposites* (pp. 173–188). CRC Press.
Wang Y, Hu Y, Zhao X, et al (2013) Comparisons of Biochar Properties from Wood Material and Crop Residues at Different Temperatures and Residence Times. Energy & Fuels 27:5890–5899. https://doi.org/10.1021/ef400972z
Wani I, Ramola S, Garg A, Kushvaha V (2021) Critical review of biochar applications in geoengineering infrastructure: moving beyond agricultural and environmental perspectives. Biomass Convers Biorefin.https://doi.org/10.1007/s13399-021-01346-8
Weber K, Quicker P (2018) Properties of biochar. Fuel 217:240–261. https://doi.org/10.1016/j.fuel.2017.12.054
Wei J, Furrer G, Kaufmann S, Schulin R (2001) Influence of Clay Minerals on the Hydrolysis of Carbamate Pesticides. Environ Sci Technol 35:2226–2232. https://doi.org/10.1021/es000179d
Weston DP, Chen D, Lydy MJ (2015) Stormwater-related transport of the insecticides bifenthrin, fipronil, imidacloprid, and chlorpyrifos into a tidal wetland, San Francisco Bay, California. Science of The Total Environment 527–528:18–25. https://doi.org/10.1016/j.scitotenv.2015.04.095
Wolfand JM, Seller C, Bell CD, et al (2019) Occurrence of Urban-Use Pesticides and Management with Enhanced Stormwater Control Measures at the Watershed Scale. Environ Sci Technol 53:3634–3644. https://doi.org/10.1021/acs.est.8b05833
Yang X-B, Ying G-G, Peng P-A, et al (2010) Influence of Biochars on Plant Uptake and Dissipation of Two Pesticides in an Agricultural Soil. J Agric Food Chem 58:7915–7921. https://doi.org/10.1021/jf1011352
Yargicoglu EN, Reddy KR Evaluation of PAH and metal contents of different biochars for use in climate change mitigation systems. In: ICSI 2014. American Society of Civil Engineers, Reston, VA.

Yavari S, Malakahmad A, Sapari NB, Yavari S (2017) Sorption properties optimization of agricultural wastes-derived biochars using response surface methodology. Process Safety and Environmental Protection 109:509–519. https://doi.org/10.1016/j.psep.2017.05.002
Zafar, S. and Verma, A., 2023. Continuum Mechanics-Based Simulations in Coatings. In *Coating Materials: Computational Aspects, Applications and Challenges* (pp. 185–216). Singapore: Springer Nature Singapore. DOI: https://doi.org/10.1007/978-981-99-3549-9_9
Zhang M, Gao B, Yao Y, et al (2012) Synthesis of porous MgO-biochar nanocomposites for removal of phosphate and nitrate from aqueous solutions. Chemical Engineering Journal 210:26–32. https://doi.org/10.1016/j.cej.2012.08.052
Zhang J, Liu J, Liu R (2015) Effects of pyrolysis temperature and heating time on biochar obtained from the pyrolysis of straw and lignosulfonate. Bioresour Technol 176:288–291. https://doi.org/10.1016/j.biortech.2014.11.011
Zhang H, Voroney RP, Price GW (2015) Effects of temperature and processing conditions on biochar chemical properties and their influence on soil C and N transformations. Soil Biol Biochem 83:19–28. https://doi.org/10.1016/j.soilbio.2015.01.006

Molecular Mechanisms in Drug Delivery

Vishnu A. Gite, Raj K. Verma, and Ratna S. Katiyar

1 Introduction

Recent years have seen vital research in the improvement of drug delivery systems based on organic, inorganic, and hybrid nanoparticles as drug carriers (Yang et al. 2018). These systems are constructed from biocompatible, biodegradable nanomaterials or miniaturized devices, characterized by multifunctionality, high viscoelasticity, and long circulating half-lives (Chen et al. 2021; Li et al. 2016; Ilhan-Ayisigi and Yesil-Celiktas 2018). Their design aims to target locations, enhancing therapeutic efficacy while minimizing off-target accumulation within the body. Drugs can enter the body through several different pathways such as oral route of administration, buccal and sublingual routes of administration, nasal and ophthalmic, transdermal, subcutaneous, anal, transvaginal, and intravesical (Benson and Roberts 2021; Larrañeta et al. 2022). Furthermore, drug delivery systems represent technological frameworks designed to formulate and preserve drug molecules in suitable formats for administration, such as tablets or liquids (Tozaki et al. 1997; Viseras et al. 2010). Therefore, each drug delivery system possesses distinct characteristics that significantly influence the rate and mechanism of action within the body. In addition, the comprehensive review by Vargason et al. highlights recent strides in drug delivery systems. Their analysis culminated in the observation that the

V. A. Gite · R. S. Katiyar (✉)
Department of Chemical Engineering, Jawaharlal Nehru Engineering College, MGM University, Chhatrapati Sambhajinagar 431003, India
e-mail: rkatiyar@mgmu.ac.in

V. A. Gite
e-mail: vgite@mgmu.ac.in

R. K. Verma
Department of Chemical Engineering, Chaitanya Bharathi Institute of Technology, Hyderabad 500075, India
e-mail: rajkumar_chem@cbit.ac.in

J. Gupta and A. Verma (eds.), *Microbiology-2.0 Update for a Sustainable Future*,
https://doi.org/10.1007/978-981-99-9617-9_10

adaptation of controlled-release and sustained-release systems, originally designed for small molecules, have been extended throughout the therapeutic landscape. This exemplifies the integration of well-established drug delivery methodologies to improve the translation of emerging therapeutic modalities (Vargason et al. 2021). Conversely, technologies and delivery methods for novel treatment approaches have been strategically adapted to enhance the administration and conventional medications (Salunke et al. 2022; Miller et al. 2006). This adaptation primarily stems from the inherent disparities in morphological, chemical, and physical attributes of these substances, subsequently influencing their interactions with diverse pharmacological compounds (Mitcheson et al.1998). Truly these studies highlighted the importance of diffusion, chemical reaction, solvent reaction, and stimuli control as molecular mechanisms (Pillay et al. 2014; Abu-Thabit and Makhlouf 2018). In the upcoming section, we will comprehensively explore the diverse drug delivery mechanisms that form the fundamental basis for the design and operation of any drug delivery device.

2 Drug Delivery Mechanism

Selecting the appropriate mechanism constitutes a pivotal stage in the effective design and advancement of the drug delivery system. The choice of a specific mechanism hinges upon a multitude of parameters. Within this section, our focus is directed toward an exploration of the diverse mechanism at hand for drug delivery systems.

2.1 Swelling Mechanism

The swelling mechanism entails the release of the drug through the swelling of excipients, where the swelling step serves as a pivotal rate-controlling phenomenon. This process is heavily reliant on the hydrogel matrix of polymers such as hydroxypropylmethylcellulose (HPMC) (Siepmann et al. 1999a; Joshi 1861; Levina and Rajabi-Siahboomi 2004; Tundisi et al. 2021; Gao et al. 1996), sodium carboxymethyl cellulose, scleroglucan, and alginates. Hydrogel, the swelling agent, constitutes a three-dimensional network of hydrophilic polymer chains that are either physically or chemically cross-linked (Omidian and Park 2008; Peppas 1997; He et al. 2004). This network can absorb moisture or liquid water. Within the swelling process, equilibrium is maintained between electrostatic, osmotic, and entropy-driven forces, facilitating the dissolution of the polymer in water. Hydrogel can manifest in two distinct forms: physical or chemical. In the case of physical hydrogels, a precursor hydrophilic polymer is employed, it can be treated with agents that exert physical interactions, often involving ions or other functional groups. This enables effective interaction with these components. On the other hand, chemical hydrogels involve polymer chains that are cross-linked through covalent bonds. The cross-linked density is an important parameter in the swelling process. As cross-links obstruct swollen behavior where the polymer chains from collapsing to each other by intermolecular forces. This factor supports the to pause of water into hydrogel structure (Omidian and Park 2008; Pillai and Panchagnula 2001). The extent of cross-linking, hydrophilic composition, and ionic presence are pivotal factors governing the swelling mechanism. Upon exposure to water, the process of swelling starts, facilitated by the diffusion of water into the chain structure of the hydrogel matrix. As a consequence of this water ingress, the polymer matrix undergoes expansion, leading to swelling. The water will transfer to the interior of the hydrogel here Fick's law is applicable which relates the absorption of water to the time. Within the polymer matrix, the hydrogel exhibits higher water content at the surface, gradually decreasing as water penetrates deeper into the polymer matrix forming three regions as shown in Fig. 1. Two primary mechanisms are at play: firstly, water ingress into the polymer matrix, and secondly, the subsequent drug releases from the hydrogel matrix. The penetration of water into the hydrogel matrix is governed by diffusion, while the drug's liberation from the matrix hinges on both the water content within distinct regions of the hydrogel matrix and the drug's solubility in water. Despite the solubility of the drug, the release of the drug particles is controlled by the mechanical strength exhibited by various layers within the hydrogel matrix. This interplay of factors dictates the rate and efficacy of drug release within the system.

Polyelectrolytes exhibiting a mild charge hold remarkable promise as stimuli-responsive carriers, serving as potential aids in accommodating the neutral pH fluctuations within the body. Their responsiveness to pH variations can be harnessed for effective drug delivery. Notably with oral drug delivery, these polyelectrolytes prove advantageous for administering poorly soluble drugs disposed to aggregation in the stomach. This aggregation often leads to reduced intestinal absorption and diminished bioavailability of the drug (Kocak et al. 2017; Manallack et al. 2013; Mura et al. 2013; Schmaljohann 2006). In an ideal scenario, these carriers facilitate the encapsulation of drug molecules within a gastric environment, minimizing or preventing aggregation. Subsequently, the controlled release of these molecules occurs in the

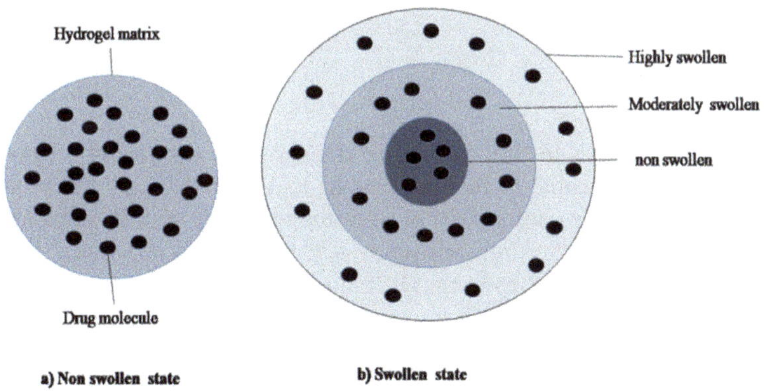

Fig. 1 Non swollen and swollen state in hydrogel matrix with drug molecules

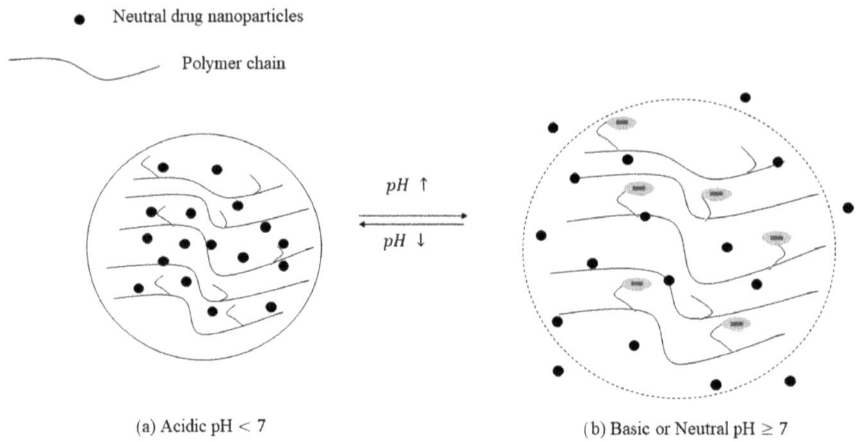

Fig. 2 Swelling mechanism of weak acidic pH-responsive polyelectrolytes with neutral drug nanoparticles

intestinal milieu (The process is represented in Fig. 2). Furthermore, intravenous drug delivery benefits from the capacity of these carriers to adapt to lower extracellular pH levels present in affected tissues, in contrast to the higher pH environment of the bloodstream. The pH responsiveness behavior of polyelectrolytes contributes significantly to enhancing drug delivery precision and efficiency (Kanamala et al. 2016). These studies that employ molecular simulation provide an in-depth understanding of how carriers and drugs respond to different pH environments, surrounding media, and their interactions. For instance, research has explored the impact of bile salts on the digestive process, shedding light on how carriers and drugs behave within such environments (Birru et al. 2017a, b). Similarly, the release of drugs through concentrated polymer solutions and gels has been studied, giving us valuable insights into the complex mechanism involved (Kocak et al. 2017; Jha et al. 1414).

2.2 *Diffusion*

Diffusion is the process in which molecule transfer from one region to another to equalize chemical potential or thermodynamic activity. Fick's law can be used to represent the quantitative description of some drug delivery systems. Fick's second law Eq. (1) is used in the drug release system if we know the initial boundary conditions (Siepmann and Siepmann 2012; Siepmann et al. 2012).

$$\frac{\partial c}{\partial t} = \frac{\partial}{\partial x}\left(D\frac{\partial c}{\partial x}\right) + \frac{\partial}{\partial y}\left(D\frac{\partial c}{\partial y}\right) + \frac{\partial}{\partial z}\left(D\frac{\partial c}{\partial z}\right) \tag{1}$$

where D is the diffusion coefficient, c is the concentration and t is the time.

The classifications of diffusion-controlled drug delivery systems are based on initial drug loading and its physical structure (Langer and Peppas 1981; Ranade 1991) i.e. reservoir system and monolithic system. In the reservoir system, the membrane is surrounded by a core–shell structure, and the drug is located in the centre. The drug release rate is controlled by the membrane. In the case of a monolithic system, the drug is homogeneously distributed in the matrix. Further, the reservoir system is classified into constant and non-constant activity sources. In the case of a constant activity source more amount of the drug was supplied into the depot and the concentration of the drug remain constant inside the membrane surface whereas in the case of a non-constant activity source the drug concentration was below solubility and it is not replaced. The Monolithic systems are classified into monolithic solution and monolithic dispersion which is based on drug loading and drug solubility ratio. In the first case, the initial drug loading is below solubility, and the drug is dissolved in the matrix. Whereas in the second case the initial drug loading is above solubility and the drug is partially dissolved in the matrix. Siepmann, J et al., have discussed the influence of factors such as device type, plasticizer quantity, and polymer type on diffusion-controlled drug delivery systems, contributing to a deeper understanding of their behavior and performance (Siepmann et al. 1999b).

2.2.1 Nonconstant Activity Sources

The drug delivered via membrane can be calculated according to Fick's law as follows Eq. (2). Here as the membrane is very thin compared to other dimensions therefore film approximation is useful (Siepmann and Siepmann 2012).

$$\frac{M_t}{M_\infty} = 1 - \exp\left(-\frac{ADKt}{VL}\right) \tag{2}$$

where M_t is the amount of drug released at time t and M_∞ is the amount of drug released at time infinity. A = Total surface area of the device, D = Diffusion coefficient, V is the volume of the reservoir, K is the partition coefficient of the drug between the membrane and the reservoir, and L is the thickness of the membrane. Thus, drug release follows first-order kinetics.

For a spherical reservoir surrounded by a thin membrane,

$$\frac{M_t}{M_\infty} = 1 - \exp\left(-\frac{3DKR_0t}{(R_0 - R_i) * R_i^2}\right) \tag{3}$$

where R_i and R_o are the inner and outer radii of the device. This expression results from substituting $L = R_o - R_i$, $A = 4\pi R_i^2$ and $V = (4/3)\pi R_i^3$ into above equation

2.2.2 Constant Activity Sources

As concentration is not changing inside the membrane it means the drug is delivered at constant rates which means that the equations follow zero order kinetics as follows for different geometry (Siepmann and Siepmann 2012).

Slab:

$$\frac{dM_t}{dt} = \frac{ADKc_s}{L} \tag{4}$$

Sphere:

$$\frac{dM_t}{dt} = \frac{4\pi DKc_s R_0 R_i}{R_0 - R_i} \tag{5}$$

Cylinder:

$$\frac{dM_t}{dt} = \frac{4\pi HDKc_s}{\ln\frac{R_0}{R_i}} \tag{6}$$

where c_s= concentration of saturated drug solution in the reservoir. R_i and R_o are the inner and outer radii, respectively, and H is the length of the cylinder.

2.2.3 Monolithic Devices

Similarly, Fick's second law of diffusion can be solved considering these initial and boundary conditions, leading to the following expressions for the fraction of drug released at time t for different geometry (Siepmann and Siepmann 2012),

Slab

$$\frac{M_t}{M_\infty} = 1 - \frac{8}{\pi^2}\sum_{n=0}^{\infty}\frac{1}{(2n+1)^2}\exp\left(-\frac{D(2n+1)^2\pi^2 t}{L^2}\right) \tag{7}$$

Sphere

$$\frac{M_t}{M_\infty} = 1 - \frac{6}{\pi^2}\sum_{n=0}^{\infty}\frac{1}{n^2}\exp\left(-\frac{Dn^2\pi^2 t}{R^2}\right) \tag{8}$$

Cylinder

$$\frac{M_t}{M_\infty} = 1 - \frac{32}{\pi^2}\sum_{n=0}^{\infty}\frac{1}{q_n^2}\exp\left(-\frac{q_n^2 Dt}{R^2}\right) * \sum_{p=0}^{\infty}\frac{1}{(2p+1)^2}\exp\left(-\frac{D(2p+1)^2\pi^2 t}{H^2}\right) \tag{9}$$

where R = radius of the cylinder and H = height of the cylinder, n, and p are summation indices, and q_n is the n's root of the zero-order Bessel function of the first kind. The last expression is valid when release is permitted in both the radial direction and through the ends of the cylinder.

2.3 Erosion

The loss of materials: oligomers, monomers, and parts of polymers from polymers are defined as erosion. Those are classified as surface and bulk erosion. Surface erosion is also called heterogeneous erosion in which the materials lose from the surface only. Their size reduces but their shape remains the same. In the case of the bulk erosion process, the size remains constant (Sackett and Narasimhan 2011; Göpferich and Williams 1996). Both processes are represented in Fig. 3. Polymer erosion is depending on various parameters such as swelling, degradation, dissolution, and diffusion of monomers and oligomers (Zuleger and Lippold 2001). The erosion mechanism is largely controlled by degradation and type of polymers. The polymer surface morphology also changes due to erosion. The surface of poly (anhydrides) changed after contact with buffer was studied by Achim and Robert. They observed that the surface roughness increases with proceeding erosion (Göpferich and Langer 1993). In the process of degradation, the oligomers and monomers need to be released from the polymer chain. The release of oligomers from microspheres of poly (D, L-lactic acid) has been studied in which its effect on drug release was also discussed (Park 1994). The release of monomer from the copolymer may be erosion-controlled or diffusion controlled. It has been assumed in erosion-controlled mechanisms that the bonds present in the copolymer backbone are cleaved due to which it has been released at different rates. Whereas the difference is solubility affects the diffusion release rates.

3 Influencing Factors in Molecular Mechanism

Certainly, within this section, we will outline the key factors that exert influence over molecular mechanisms. These factors involve:

3.1 Drug Dose

Certainly, the drug dose is a pivotal parameter that significantly shapes the design and development of effective, dependable, and sustainable drug delivery system. A recent study by Li et al. emphasized the crucial importance of addressing two major concerns in a mature drug delivery device: efficient actuation stimuli and a

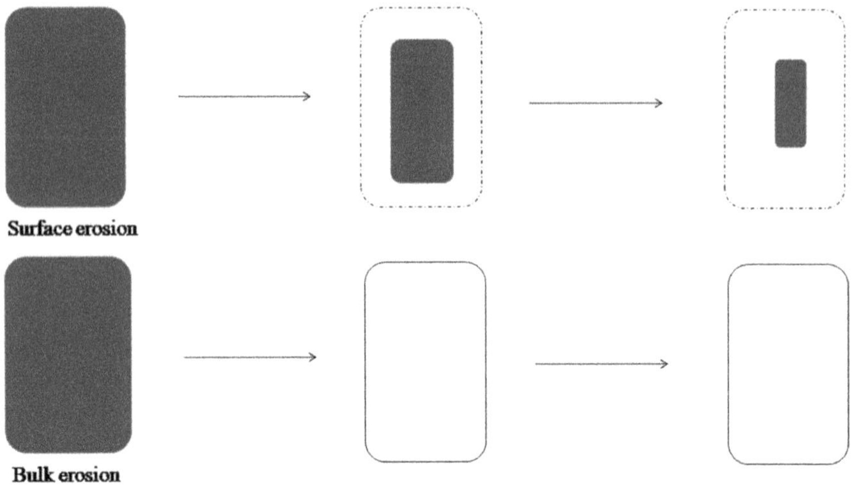

Fig. 3 Schematic explanation of surface and bulk erosion

controlled dosage release mechanism (Li et al. 2017). The drug dose is influenced by a range of factors including the specific drugs, mode of administration, and the body's physiological response (Yang et al. 2016). When determining drug doses, the risk of increased toxicity and potential side effects must be carefully considered, especially when the therapeutic window in narrow (the range between the lowest effective dose and the dose that leads to harmful effects) (Viscusi 2019; Wittmann and Helliwell 2013). The potential for dose-dependent toxicity is particularly relevant, where higher medication may escalate the occurrence and severity of adverse effects (Yang et al. 2018). When a medicine is used over an extended period, tolerance can develop, requiring greater doses to produce the same therapeutic benefit. Contrarily, sensitization can take place, whereby repeated exposure to a substance causes the body to become more receptive to lesser dosages of the drug (Berger and Whistler 2010; Löscher and Schmidt 2006; Winkelman and Johnston 2004).

3.2 Drug Type

The molecular mechanism of action of medication is heavily influenced by its specific characteristics. Various classes of medications exert distinct pharmacological effects due to their interaction with diverse molecular targets within the body (Chiotis et al. 2018; Bais et al. 2017). These interactions play a crucial role in determining how a drug function and produces its therapeutic effects. The intricate relationship between a drug's chemical structure, its mode of action, and the molecular pathways it engages with underscores the complexity of pharmaceutical interventions. Understanding these nuances is essential for tailoring drug treatment to specific medical conditions and optimizing their effectiveness (Muro 2012). Exactly, drugs that function as enzyme inhibitors play a significant role in modulating biochemical process within the body. These inhibitors work by binding to the active site of the specific enzymes, thereby impeding their normal functioning. This interference prevents the enzymes from catalyzing their usual biochemical reactions, disrupting the normal pathways that these enzymes would participate in. Enzyme inhibitors are utilized in various therapeutic contexts, including the treatment of diseases where excessive enzyme activity needs to be controlled (Chou and Talalay 1984). Examples include statins, which stop the production of cholesterol by blocking the HMG-CoA reductase enzyme, and angiotensin-converting enzyme (ACE) inhibitors, which stop the production of blood pressure by blocking the ACE enzyme. DNA interactions: Some medications can bind to DNA or intercalate between DNA base pairs to affect DNA. These medications' therapeutic benefits result from their ability to block DNA replication, transcription, or repair mechanisms. Examples include antibiotics like fluoroquinolones, which block the DNA gyrase and topoisomerase enzymes necessary for bacterial DNA replication, and anticancer medications like cisplatin, which produces cross-links with DNA strands (Chłopicki and Gryglewski 2005).

4 Effects of Various Parameters on the Drug Delivery Process

Positively, in the current section, we will discuss the influential parameters that shape molecular mechanisms in drug delivery. These parameters play a crucial role in determining the behavior of drug delivery systems and their interactions with the surrounding environment. The influencing parameters encompass.

4.1 Temperature

Indeed, polymers exhibiting a lower critical solution temperature (LCST) serve as key components for introducing temperature responsiveness into the polymer. Among these, poly(N-isopropyl acrylamide) (PNIPAM) stands as a traditional and widely used temperature-responsive material in drug delivery systems. PNIPAM demonstrates a unique behavior where, in an aqueous solution, it experiences a transition from a hydrophobic to a hydrophilic state when the temperature drops below its LCST, which is typically around 32 °C. This transition is primarily driven by the formation of hydrogen bonds between PNIPAM's amide groups and water molecules, inducing a phase shift in its structure (Göpferich and Williams 1996). Conversely, as the ambient temperature surpasses the LCST, the hydrogen bonds between PNIPAM and water molecules break, leading to the collapse of PNIPAM chains into compact globules. This collapse prompts the aggregation of the gel network, causing it to become hydrophobic. This distinctive temperature-responsive behavior of PNIPAM contributes to its utility in drug delivery systems, where temperature changes can be harnessed to modulate the release of therapeutic agents (Shen et al. 2011; Fu et al. 2018). Marizza et al. successfully used tuning temperature, pressure, time, and drug concentration in the fluid phase to develop a particular drug delivery system. An efficient drug delivery system was designed by incorporating ketoprofen into the poly(vinylpyrrolidone) (PVP) K10 matrix and impregnating it with supercritical CO_2. Then investigated that drug loading increases with time at each tested temperature at any tested pressure. Afterward, showed that even after an hour the equilibrium was not achieved between the polymer and CO_2 phases (Marizza et al. 2016). Bigham et al. studied the effect of calcination temperature on the drug delivery behavior of ordered mesoporous magnesium silicate (OMMS). They calcined the samples at 350 and 550 °C and observed that at this temperature it has a narrow small pore size distribution around 1.2 and 3.6 nm respectively. However, samples with a calcination temperature of 750 °C had wide pore size distribution. This found that the high specific surface area of the OMMS is 386–504 m^2/g due to the wide pore size. Further observed that OMMS had a quicker release rate than the others at a calcination temperature of 750 ºC and tested the toxicity of all samples at the calcination temperatures and found that there is no toxicity in the MTT experiment. Finally concluded that the temperature used to calcine OMMS samples may influence its drug delivery behavior by affecting other parameters such as pore size, surface area, and pore volume (Bigham et al. 2016).

4.2 pH

The consideration of pH is of paramount importance in the design of drug delivery systems, particularly for applications that involve targeted delivery or controlled release. pH plays a critical role in modulating drug release and interactions with the surrounding environment. Over the past few decades, researchers have made significant steps in developing pH-responsive carriers to enhance the precision and effectiveness of drug delivery systems (Mu et al. 2021; He et al. 2011). These carriers are designed to undergo specific changes in response to variations in pH, which can be exploited to achieve on-demand drug release at targeted sites within the body. By harnessing the pH gradient across different physiological compartments, these carriers enable controlled and site-specific drug delivery, minimizing off-target effects and enhancing therapeutic outcomes (Nazir et al. 2022; Zhao et al. 2020). Li et al. introduced an innovative approach to creating multi-stimuli responsive drug delivery systems, incorporating pH, temperature, and reduction stimuli. They employed the distillation precipitation polymerization technique, a novel methodology, to achieve this multifunctional design. Subsequently, they conducted in-vitro studies to assess the drug delivery response in the context of tumor diseases. Notably, their research focused on MAA/PNIPAM yolk/shell microspheres, which exhibited enhanced drug-loading capacity and superior responsiveness to the tumor microenvironment in controlled release experiments (Li et al. 2017; Dong and Hoffman 1991). Katiyar and Jha provide insights into the molecular level interaction between a model pH-responsive carrier polyacrylic acid (PAA) and cationic anticancer drug, doxorubicin (DOX) under different pH condition representative of various physiological conditions. Model oligomer at low pH environment, the carboxylic acid COOH groups of the PAA chains are protonated and available for hydrogen bonding interactions. This leads to the formation of aggregated, solid-like structure. At that condition, diffusion of the DOX molecules is a much faster lower pH value. Further, at a high pH environment, the COOH group on the PAA chains becomes ionized and negatively charged (COO^-) which results in repulsion between these groups of the polymer chain, causing the polymer to adopt a swollen, gel-like structure. The cationic drug molecules are likely to attach to the negatively charge carboxylic group (COO^-) on the PAA, leading to decreased diffusion of the drug. Collectively, their research provides valuable insights into the intricate drug release processes within pH-responsive polymer-drug formulations, thereby contributing to a deeper understanding of controlled drug delivery mechanisms (Katiyar and Jha 2018, 2017).

4.3 Light

Indeed, light can exert profound effects on drug delivery systems, particularly when they incorporate light-sensitive or photosensitive components. The interaction between light and these systems can result in various responses, either through stimulation or activation. A prime example is photodynamic therapy, a technique involving the administration of photosensitizers-medications sensitive to light (Perkins et al. 2020; Seggie et al. 1989). These photosensitizers remain inactive until exposed to light of a specific wavelength. Once activated by light, they release reactive oxygen species or other cytotoxic agents, selectively destroying diseased cells or tissues. Light-based treatments, such as photothermal therapy (PTT) and photodynamic therapy (PDT), have gained significant prominence in recent times (Hou et al. 2020). In photothermal therapy, light-absorbing agents generate heat upon exposure to specific light wavelengths, leading to localized hyperthermia and tumor destruction. Similarly, in photodynamic therapy, the interaction between light and photosensitizers triggers a cascade of cytotoxic reactions that target and eliminate malignant cells. The incorporation of light-sensitive elements into drug delivery systems represents a dynamic approach to achieving targeted and controlled therapeutic interventions, offering promising avenues for advancing the field of precision drugs (Fu et al. 2018). Fluorescence imaging and bioluminescence imaging stand out as remarkable light-based techniques that hold immense potential for visualizing and continuously monitoring drug delivery systems. By incorporating fluorescent dyes or biomarkers within drug carriers, these systems emit detectable light signals, facilitating non-invasive imaging of distribution, accumulation, and release dynamics. Another intriguing application lies in the light-induced activation of prodrugs. Prodrugs can transform into their active pharmacological forms upon light exposure. This innovative approach grants greater therapeutic selectivity and reduced off-target effects, enabling localized and controlled drug activation precisely at the intended site. Collectively, these light-based techniques revolutionize the way drug delivery systems are visualized, controlled, and activated. They offer unprecedented opportunities for advancing precision medicine, optimizing therapeutic efficacy, and enhancing patient outcomes (Alvarez-Lorenzo et al. 2009; Costa et al. 2019; Pan et al. 2021).

4.4 Pressure

Pressure plays a crucial role in determining the effectiveness of drug administration to specific body parts. In the realm of drug delivery systems, pressure directly influences two essential mechanisms. First Infusion, Pressure is a pivotal factor in delivering medications directly into the bloodstream or targeted tissues. Intravenous injections, for instance, rely on pressure to propel the medicinal solution into veins. Similarly, intramuscular or subcutaneous injections utilize pressure to facilitate the penetration of medication into a muscle or subcutaneous tissue, ensuring effective delivery to the desired location. Second Pressure-Driven Systems, Certain drug delivery systems leverage mechanical or pressure differentials to regulate the controlled release of medication (Yuan et al. 2022; Sebastian et al. 2022). Osmotic medication delivery devices exemplify this principle. These devices incorporate semipermeable membranes that allow water to enter when an osmotic pressure gradient is established. This influx of water generates internal pressure, prompting the system to release the drugs in a controlled manner. Moreover, pressure-based pumps or equipment, such as insulin pumps, are employed to administer medications at predetermined rates, ensuring precise dosing and consistent therapeutic effects. The utilization of pressure-related mechanisms underscores the sophistication and diversity of drug delivery strategies, enabling targeted and regulated administration of medications to enhance therapeutic outcomes (Boateng et al. 2008; Bae and Park 2011). Certainly, researchers worldwide are actively engaged in studying and developing strategies to address the effects of pressure on drug delivery systems. Here are a few notable examples of such endeavors: the system developed by Marizza et al. in which ketoprofen incorporated into the poly(vinylpyrrolidone) (PVP) K10 matrix and impregnating it with supercritical CO_2 for addressing the effects of pressure on drug dissolution (Marizza et al. 2016).

4.5 Ionic Strength

The term "ionic strength" states to the concentration of ions in a solution, and it plays a key role in influencing drug delivery systems in various ways. Solubility Impact:

High ionic strength, caused by an elevated concentration of salts or ions, can affect the solubility of certain drugs. This can lead to changes in the drug's ability to dissolve in the solution, potentially influencing its availability for absorption and distribution within the body. Stability Considerations: The presence of ions in a solution can impact the stability of pharmaceutical compounds. Many drugs are susceptible to chemical breakdown or physical instability in the presence of ions, leading to reduced potency and effectiveness (Bodmer et al. 1992; Tomic et al. 2016). Ionic interactions may trigger undesirable reactions, compromising the integrity of the medication. In drug release modulation, the ionic strength of the surrounding environment can influence drug release from various delivery systems, such as nanoparticles, liposomes, or micelles. High ionic strength levels have the potential to alter the conformation of polymeric carriers or disrupt electrostatic interactions within these systems. This can result in modified or premature drug release profiles, affecting the desired therapeutic outcomes (Bruno et al. 2016). Researchers and pharmaceutical scientists carefully consider and manage ionic interactions to optimize the performance and reliability of drug delivery platforms (Mahamat Nor et al. 2017). In the past few decades, researchers developed many important materials/drugs for drug delivery as follows: Hao et al. fabricated an in-situ gelling nanoparticle and an ionic-triggered deacetylated gellan gum (DGG) matrix for intranasal drug delivery. They have noticed that for in situ gel of 0.6%, DGG has superior qualities such as sole-gel transition capability, water holding capacity, and thinning shear rheology competence (Hao et al. 2016). Sastry et al. developed ionic liquids type hybrid gels from 1-dodecyl-1-methylpiperidinium acetylsalicylate and [C12mpip][AcSa] in sodium salicylate (SS) aqueous media, and from agar and chitosan with low (L) and medium (M). Sodium salicylate (SS), an additive, the micelles of an amphiphilic drug anion containing ionic liquid based on 1-dodecyl-1-methylpiperidinium acetylsalicylate may develop into sizable micellar networks with characteristic viscoelastic behavior. As a function of time, the drug anion's release from the gel state and their thin dry cast films were seen. Further, stated that the proportion of drug anion released decreased with increasing biopolymer concentration, and this was explained by the fact that at greater concentrations, the number density of network structures increases and prevents the drug anion from dissipating to the solvent system (Sastry et al. 2018).

4.6 *Magnetic and Electric Field*

To accomplish targeted and controlled drug release, magnetic and electric fields can be used in drug delivery systems. Drug-loaded magnetic nanoparticles or carriers can be directed to locations within the body. The magnetic particles can be guided toward the appropriate area by providing an external magnetic field, maximizing drug reach at the target spot while minimizing systemic exposure. Magnetic fields can be used to cause the release of pharmaceuticals from magnetic carriers in a controlled manner. Drugs may be placed onto or coated on magnetic nanoparticles or microparticles before they are exposed to an alternating magnetic field. Through magnetic hysteresis or relaxation, the field induces particle heating, causing a localized rise in temperature. With perfect control over drug release kinetics, this localized heat can cause drug release from the carriers (Mody et al. 2014; Assa et al. 2017). Electric fields can be employed to modulate drug release from carriers. Electrically responsive materials can be incorporated into drug delivery systems. When an electric field is applied, these materials change structure or permeability, thereby enabling the controlled release of the encapsulated drug. Both magnetic and electric fields hold promise in enhancing the precision, efficiency, and control of drug delivery systems. These technologies have the potential to revolutionize how drugs are administered and released, enabling tailored therapies and improved therapeutic outcomes (Alexiou et al. 2011; Jurgens et al. 2006). Supramaniam et al. utilized rice husk (an agricultural waste) for developing the Cellulose Nanocrystals (CNCs) and coprecipitated with the iron. Further, they have made alginate hydrogel beads by using magnetic cellulose nanocrystals (m-CNCs) for the drug delivery application of ibuprofen. They have observed that alginate hydrogel beads' physical and mechanical characteristics were enhanced by the inclusion of magnetic nanocellulose, boosting swelling and slowing drug release (Supramaniam et al. 2018). Varmazyar et al. numerically studied the blood flow and drug transfer behavior in arteries and examined the effect of magnetic nanoparticle distribution in the presence of a magnet. They have observed that magnetic nanoparticles (stenosis) near the magnetic field lower the amount of medicine that is accessible near the tumorous tissue and hence lower drug absorption. The medication delivery is weakened by changing flow patterns, magnetic forces, and the number of obstructions close to the target location. The cell membrane can be momentarily damaged by delivering brief electric pulses to the target tissue, improving medication absorption. Charged drug molecules or particles can migrate through a solution through electroosmosis or electrophoresis when subjected to electric fields. While charged drug molecules migrate towards the electrode with the opposite charge during electrophoresis, electroosmosis refers to the movement of solvent and drug particles caused by an applied electric field. These phenomena can be used to enable regulated drug release from electro-responsive delivery systems or to improve drug transport over biological barriers (Varmazyar et al. 2020).

5 Literature on Mechanisms Used in Drug Delivery

In the section below, we highlighted some of the author's notable contributions to the field of drug delivery systems, particularly the understanding of the molecular mechanism involved. The primary focus of these studies has been on addressing the challenge of poor solubility. Please note that unless specified otherwise, our discussion is focusing around on experimental and atomistic MM studies. We have condensed our review to concentrate solely on the mechanism aspect, with the corresponding influential parameters provided above. For more comprehensive information on the molecular mechanism and specific drug delivery techniques employed, readers are encouraged to refer to specialized literature, often referred to as "Bookworm".

The study conducted by Harland et al. delves into the analysis of a drug delivery system's swelling and dissolution behavior. This system comprises the model drugs sodium diclofenac, diprophylline, and cimetidine·HCl, along with the hydrophilic polymer polyvinyl alcohol (PVA). The analysis is conducted at various rotation speeds of 25, 50, 100, and 200 rpm. A key aspect of this research is the development of a mathematical model that predicts both the gel layer thickness and the release of the drug over time. This model provides insights into how the drug delivery system's properties evolve during the swelling and dissolution processes. The study reveals the presence of two distinct dispersed phases within the system. The first phase demarcates the transition between a glassy state and a gel-like state, while the second phase marks the boundary between a rubbery (gel-like) state and the dissolution medium (Harland et al. 1988). Dong and Hoffman pioneered the synthesis of a novel hydrogel polymer class capable of responding to both pH and temperature variations. Their research highlighted the remarkable responsiveness of poly(hydroxyethyl methacrylate-coacrylic acid) and poly(acrylamide-co-acrylic acid) hydrogels to pH changes. In the acidic gastric environment (pH 1.4), these hydrogels displayed minimal water binding. However, upon encountering the alkaline intestinal pH (7.4), the pH-sensitive polymer underwent substantial swelling, facilitating the delivery of indomethacin drugs. In-vitro investigations demonstrated that at gastric pH (1.4), the release of indomethacin was marginal within 24 h. Contrastingly, at pH (7.4), over 90% of the drug was released within just 5 h from the expanded gel.

Graham and McNeill focused on comparing the swelling behavior of poly(ethylene oxide) (PEO) cross-linked hydrogels in two different solvents: distilled water and propylene glycol. The primary objective of this research was to investigate how the choice of solvent influences the degree of swelling, particularly in response to temperature changes. The researchers found that the type of solvent played a significant role in the swelling behavior of the hydrogel. When the hydrogel was placed in distilled water, its degree of swelling was notably affected by changes in temperature. As the temperature increased, the hydrogel exhibited a dramatic increase in its swelling capacity. This suggests that the hydrogel absorbed more water and expanded to a greater extent as the temperature rose (Dong and Hoffman 1991). Bettini et al. focused on swelling and delivery experiments using the USP (United States Pharmacopeia) dissolution apparatus. The aim was to investigate controlled drug delivery from a hydrogel matrix using specific materials and to understand how the drug release kinetics can influence drug delivery outcomes. In this study, the researchers employed the drugs buflomedil pyridoxal phosphate (with a solubility of 65 gm/100 ml in water at 37 °C), along with cellulose acetate phthalate and cellulose acetate propionate for coating purposes. Additionally, HPMC polymer was used as the carrier in the controlled drug delivery system. The controlled drug delivery process from the hydrogel matrix is significantly influenced by factors such as the drug's release kinetics and the area available for drug release, which increases as the matrix swells (Bettini et al. 1994). In their experimental analysis, Colombo et al. affirmed that the drug delivery mechanism from a swellable matrix is closely related to the thickness of the gel layer. They conducted experiments using various polymers, including HPMC, methocel, polyvinyl alcohol (PVA), and carboxymethyl cellulose sodium (CMC), as carriers. The study also involved employing different drugs, such as cimetidine HCl, diclofenac sodium, and diprophylline, each with distinct solubility characteristics. The experimental analysis involves a swellable matrix comprising a combination of 65% buflomedil pyridoxal phosphate and 25% methocel. This matrix is subjected to varying rotation speeds of 50 and 250 rpm. The results indicate that drug release is higher at 250 rpm, suggesting that the rotation speed significantly influences the rate of drug release. Furthermore, the study concludes that the process of swelling is accelerated, leading to an increase in porosity as a result of faster water movement within the carriers. This increased porosity likely contributes to enhanced drug release rates (He et al. 2011; Nazir et al. 2022; Colombo et al. 1996; Conte et al. 1988). Peh and Wong prepared a drug-free film containing different proportions of sodium carboxymethyl cellulose (SCMC) or HPMC to carbopol 934P but a constant proportion of polyethylene glycol 400 as buccal drug delivery medium. Carbopol 934P ratio increases in the formation of the film and the results find out raise elasticity, softness, and bioadhesive strength but a decrease in swelling and strength of both SCMC and HPMC film. Carbopol 934P was used to modify the properties of films, and SCMC film swelling was favored in distilled water while HPMC films were in simulated saliva solution (Peh and Wong 1999).

Kiil and Dam-Johansen provided a comprehensive mathematical model that considers various factors influencing drug release from HPMC matrices. The model's success in predicting experimental results and phenomena demonstrates its potential usefulness in understanding and optimizing drug delivery systems (Kiil and Dam-Johansen 2003). Borgquist et al. described a study of a model poly(ethylene oxide) tablet containing two different drugs: slightly soluble Methyl paraben and soluble Saligenin. The study aimed to verify the accuracy of a model. They developed for predicting drug release and dissolution behavior in comparison to experimental data. The model reliability conformed from the fitting of the drug diffusion coefficient to experimental data, and they found that the values obtained from their model closely matched the measured values (Borgquist et al. 2006). Efentakis and Politis highlighted the importance of understanding drug delivery systems of different compositions. The controlled drug delivery of changed compositions meaningfully demonstrates the drug release rate and its release. Different compositions clear that increasing the core tablet's coverage area had a notable effect on the decreases of the drug's dissolution and release from the system. In other words, the cover acts as a hurdle between liquid and core surface interaction (Efentakis and Politis 2006). Nakamura et al. conducted a study to investigate the performance of a swelling polymer incorporation layer system (SPILA) containing metoprolol tartrate (ME) and morphine hydrochloride (MO) drugs. The research encompassed both in vitro and in vivo assessments to comprehensively evaluate the behavior and effectiveness of the SPILA system for drug delivery. The drug release profiles of compounds ME and MO were analyzed, revealing that in a medium with pH 1.2, the drug release occurred at a slower rate compared to a medium with pH 6.8. This information is valuable for designing and optimizing pH-responsive drug delivery systems, where the incorporation of carboxy vinyl polymer can be strategically manipulated to achieve desired drug release profiles under varying pH conditions (Nakamura et al. 2006).

Yin et al. conducted a comparison between coarse-grained molecular dynamics simulations and experiments involving polyelectrolyte gels. The simulation results revealed that under equilibrium conditions, there exists a delicate balance between electrostatic forces and the impact of counterion excluded-volume effects on the osmotic pressure within the gel-solvent system. This balance has important consequences for the overall behavior of the polyelectrolyte gel (Yin et al. 2008). In Wu's study, a molecular dynamics simulation was conducted using the GROMACS software to investigate the behavior of poly(propylene imine) (PPI) dendrimers in an aqueous medium across different pH conditions. The pH values tested ranged from high (pH 14) to neutral (pH 7) and subsequently to low (pH 5) levels. The simulation results revealed important insights into the conformational changes of the PPI dendrimer as a response to variations in pH. Specifically, Wu observed changes in parameters such as the radius of gyration (R_g), solvent-accessible surface area (SASA), and solvent-excluded volume (SEV) of the dendrimer. At high pH (pH 14), the dendrimer exhibited a certain conformation or arrangement, characterized by specific values of R_g, , SASA, and SEV. As the pH was gradually lowered to a medium range (pH 10) and then to a low level (pH 5), the parameters such as R_g, SASA, and SEV showed an increase. This increase in parameters indicated a dramatic change in the conformation of the PPI dendrimer, transitioning from a state of tension or compactness to a more relaxed or expanded position (Wu 2010). In their research, Rodriguez-Hidalgo et al. utilized coarse-grained models and employed the dissipative particle dynamics (DPD) simulation technique to investigate the drug delivery mechanism. The DPD simulations allowed them to capture the intricate changes in the internal structure of a polymeric carrier over time, specifically in response to chemical conditions. The study focused on using albendazole as a model drug and a synthetic poly(styrene–divinylbenzene) P(ST-DVD) copolymer as the carrier. The investigation revealed a specific drug delivery process that was triggered by the acidic environment. In this process, the polymeric microsphere exhibited a swelling behavior. This swelling led to the formation of pores within the carrier's structure. As a consequence of this swelling and pore formation, the drug molecules were able to be released from the carrier (Rodríguez-Hidalgo et al. 2011).

In the context of drug delivery within a heterogeneous system, Kimber's work advanced the understanding by incorporating an alternative approach called the Discrete Element Method (DEM). In the DEM approach, the drug-loaded system is represented as discrete particles, each having a specific mass. These particles are dynamic and can undergo various changes in shape, such as shrinking, expanding, or stretching. As the drug molecules interact with the polymer matrix, these particles simulate the complex interactions and movements that take place within the system (Kimber et al. 2013). Wang and colleagues presented a comprehensive three-dimensional dynamic model that addresses the processes of swelling and drug delivery within polyelectrolyte gels. This model integrates the Maxwell–Stefan equation and the Bio-heat equation, and their simulations were conducted using COMSOL Multiphysics software. The model considers the temperature aspect, where the hydrogel is subjected to temperatures ranging from approximately 37 to 40 °C. Over time, the temperature within the hydrogel gradually rises by around 40 °C, mimicking a fever-like condition. This temperature variation is significant, as it can influence the behavior of the polyelectrolyte gel (Wang et al. 2013). Gianni and co-authors conducted a study involving the utilization of irinotecan drug-loaded halloysite nanotubes. These nanotubes were coated with an anionic polymer, specifically Eudragit-S100, to facilitate controlled drug delivery via the oral route for the treatment of colon cancer. In their study, the researchers observed the release behavior of irinotecan from the Eudragit-S100 polymer-coated nanotubes under different pH conditions simulating the gastrointestinal tract. At a stomach pH of 1.2, the release of irinotecan was minimal, accounting for only 0.7% within 2 h. Similarly, at pH 4.5, the release remained low with a total of 24.8% release over a period of 4 h, distributed equally between pH 1.2 and pH 4.5. Notably, the release rate of irinotecan dramatically increased to over 90% within just one hour when the nanotubes were exposed to an intestinal pH of 7.4 (Gianni et al. 2019). Mahdavi et al. conducted a study involving molecular dynamics simulations to investigate the behavior of the anticancer drug doxorubicin (DOX) on both pristine graphene and graphene oxide surfaces. The researchers varied the surface oxygen densities of graphene oxide and examined the effects under different pH conditions in an aqueous environment. In their simulations, they observed the interactions of 24 DOX molecules with the surface of graphene oxide at two distinct pH levels: neutral pH (7) and acidic pH (5). DOX molecules were loaded onto the graphene oxide surface at a neutral pH level, which could be representative of the physiological blood pH. However, the delivery of DOX molecules was more pronounced at an acidic pH level, which could be characteristic of the environment in cancerous tumor tissues (Mahdavi et al. 2016). Rabbel et al. researched single-chain polymer nanoparticles and investigated their static properties and swelling behavior in both good and poor solvent conditions. They employed Monte Carlo simulations and a theoretical model to explore the characteristics of these nanoparticles. The research involved creating nanoparticles through simulation protocols and varying chain lengths, ranging from $N = 200$ to $N = 5000$. The Flory-Rehner theory, a well-known theoretical framework for polymer swelling behavior, was applied to analyze the data obtained from the swelling behavior of the polymer network. Specifically, the study examined the behavior of the longest simulated chain length ($N = 5000$) across a range of cross-linking densities. The findings indicated that the Flory-Rehner theory was able to accurately describe the swelling behavior of the polymer network in the case of higher cross-linking densities (Rabbel et al. 2017). In the research conducted by Hegab and colleagues, they focused on the synthesis and analysis of hydrogel films composed of a copolymer called poly(methyl methacrylate-co-acrylic acid) with different molar percentages of methyl methacrylate (MMA), specifically 10, 20, and 30 mol%. The objective of their study was to explore the potential of these hydrogel films for the controlled release of low-molecular-weight drugs. The results of their analysis indicated that as the molar percentage of MMA increased, the overall dynamic swelling percentage of the hydrogel films decreased. This suggests that higher proportions of MMA in the copolymer led to a reduction in the ability of the hydrogel films to absorb and retain solvent (Hegab et al. 2020). Beltran Villegas et al. conducted a study using coarse-grained molecular dynamics (MD) simulations to investigate the impact of altering solvophobicity on the assembly of amphiphilic block copolymers (BCPs) and inorganic nanoparticles (NPs). They explored various factors, including the composition and sequence of the BCPs, as well as the size and affinity of the inorganic nanoparticles. This research was carried out using the LAMMPS software package, a versatile tool for molecular dynamics simulations. By employing coarse-grained simulations, the researchers aimed to gain insights into the self-assembly behavior and structure of these complex systems (Beltran-Villegas and Jayaraman 2018).

In the collaborative research conducted by Deepa and colleagues, their work involved computational aspects related to the cytotoxicity and drug release behavior of cytarabine-loaded chitosan nanoparticles. They aimed to investigate the in-vitro cytotoxicity of these nanoparticles and also employed molecular dynamics simulations using the GROMACS software to gain insights into the molecular-level behavior of the system. In this study, cytarabine, a therapeutic molecule, was encapsulated within chitosan nanoparticles, and the release of the drug was investigated. The researchers focused on the Root Mean Square Deviation (RMSD) of the C_α atom of the cytarabine molecule as an indicator of its movement and diffusion within the chitosan polymer. The simulation results revealed that the binding between cytarabine and chitosan polymer influenced the diffusion behavior of the drug. At pH 6.4, the RMSD of the cytarabine molecule exhibited a trigger point around 10.2 ns (ns), indicating the initiation of drug diffusion within the chitosan polymer. The drug diffusion at pH 6.4 was relatively faster compared to pH 7.4, suggesting that the lower pH environment facilitated the release of cytarabine from the chitosan nanoparticles. Interestingly, at pH 7.4, the drug diffusion was delayed and started around 17.5 ns, indicating that the release of cytarabine was slower under more neutral pH conditions (Deepa et al. 2018). Zhang et al. focused on investigating the use of polyurethane (PU) hydrogels for the delivery of lidocaine (LID) and ofloxacin (OFX) as model drugs for skin delivery. In particular, the researchers examined the molecular interactions between the hydrogel and the drugs to understand their impact on the skin drug delivery process. The observed reduction in skin drug delivery due to hydrogen bonding interactions underscores the complex interplay between drug molecules and the carrier matrix, which can have implications for the design and optimization of drug delivery systems (Zhang et al. 2021).

6 Summary and Future Perspective

In this comprehensive review, we have extensively explored the current state of understanding regarding the molecular mechanisms underlying drug delivery, with a particular focus on the processes of swelling, diffusion, and erosion. We have provided a comprehensive overview of how various factors can influence controlled and sustained drug delivery, shedding light on the intricate interplay between molecular processes and external parameters. Through a meticulous analysis of the literature, we have highlighted the significance of key parameters in drug delivery mechanisms, emphasizing their impact on the overall efficacy and outcome of drug delivery systems. The complexities of molecular interactions, coupled with the multitude of affecting factors, have been examined to provide a comprehensive understanding of the drug delivery process. From our perspective, despite the continuous advancements in the field of drug delivery systems, the utilization of polymeric carriers is expected to persist. This projection is rooted in the profound scientific insights and extensive variability in qualitative behaviors that can be harnessed through these molecular mechanisms, encompassing a wide range of affecting factors and parameters.

Concluding this chapter, we have offered a succinct overview of the diverse applications of various molecular mechanisms in drug delivery systems. Additionally, we have highlighted the intricate influence of affecting factors on achieving controlled and sustained drug delivery. This comprehensive exploration not only contributes to the current body of knowledge but also underscores the potential for continued innovation and advancements in the field of drug delivery, as we unravel the intricacies of molecular mechanisms and their interplay with affecting parameters.

Acknowledgment This work is supported by a university research grant from MGM University, Chhatrapati Sambhajinagar, India (no. MGMU003/23-24/RO/R&D/246/24).

Declarations

Conflicts of Interest "There are no conflicts of interest to declare by the authors."

Ethical Approval The authors hereby state that the present work is in compliance with the ethical standards.

Competing Interests The authors declare no competing interests.

Availability of Data and Materials Not applicable.

References

N. Y. Abu-Thabit and A. S. H. Makhlouf, *1 - Historical Development of Drug Delivery Systems: From Conventional Macroscale to Controlled, Targeted, and Responsive Nanoscale Systems*, in *Stimuli Responsive Polymeric Nanocarriers for Drug Delivery Applications, Volume 1*, edited by A. S. H. Makhlouf and N. Y. Abu-Thabit (Woodhead Publishing, 2018), pp. 3–41.

C. Alexiou, R. Tietze, E. Schreiber, R. Jurgons, H. Richter, L. Trahms, H. Rahn, S. Odenbach, and S. Lyer, *Cancer Therapy with Drug Loaded Magnetic Nanoparticles—Magnetic Drug Targeting*, J. Magn. Magn. Mater. **323**, 1404 (2011).

C. Alvarez-Lorenzo, L. Bromberg, and A. Concheiro, *Light-Sensitive Intelligent Drug Delivery Systems†*, Photochem. Photobiol. **85**, 848 (2009).

F. Assa, H. Jafarizadeh-Malmiri, H. Ajamein, H. Vaghari, N. Anarjan, O. Ahmadi, and A. Berenjian, *Chitosan Magnetic Nanoparticles for Drug Delivery Systems*, Crit. Rev. Biotechnol. **37**, 492 (2017).

Y. H. Bae and K. Park, *Targeted Drug Delivery to Tumors: Myths, Reality and Possibility*, J. Control. Release **153**, 198 (2011).

S. Bais, R. Kumari, Y. Prashar, and N. S. Gill, *Review of Various Molecular Targets on Mast Cells and Its Relation to Obesity: A Future Perspective*, Diabetes Metab. Syndr. Clin. Res. Rev. **11**, S1001 (2017).

D. J. Beltran-Villegas and A. Jayaraman, *Assembly of Amphiphilic Block Copolymers and Nanoparticles in Solution: Coarse-Grained Molecular Simulation Study*, J. Chem. Eng. Data **63**, 2351 (2018).

H. A. E. Benson and M. S. Roberts, *Challenges and Innovations of Controlled Drug Delivery*, in *Fundamentals of Drug Delivery* (John Wiley & Sons, Ltd, 2021), pp. 1–14.

A. C. Berger and J. L. Whistler, *How to Design an Opioid Drug That Causes Reduced Tolerance and Dependence*, Ann. Neurol. **67**, 559 (2010).

R. Bettini, P. Colombo, G. Massimo, P. L. Catellani, and T. Vitali, *Swelling and Drug Release in Hydrogel Matrices: Polymer Viscosity and Matrix Porosity Effects*, Eur. J. Pharm. Sci. **2**, 213 (1994).

A. Bigham, S. A. Hassanzadeh-Tabrizi, M. Rafienia, and H. Salehi, *Ordered Mesoporous Magnesium Silicate with Uniform Nanochannels as a Drug Delivery System: The Effect of Calcination Temperature on Drug Delivery Rate*, Ceram. Int. **42**, 17185 (2016).

W. A. Birru, D. B. Warren, S. J. Headey, H. Benameur, C. J. H. Porter, C. W. Pouton, and D. K. Chalmers, *Computational Models of the Gastrointestinal Environment. 1. The Effect of Digestion on the Phase Behavior of Intestinal Fluids*, Mol. Pharm. **14**, 566 (2017).

W. A. Birru, D. B. Warren, S. Han, H. Benameur, C. J. H. Porter, C. W. Pouton, and D. K. Chalmers, *Computational Models of the Gastrointestinal Environment. 2. Phase Behavior and Drug Solubilization Capacity of a Type I Lipid-Based Drug Formulation after Digestion*, Mol. Pharm. **14**, 580 (2017).

J. S. Boateng, K. H. Matthews, H. N. E. Stevens, and G. M. Eccleston, *Wound Healing Dressings and Drug Delivery Systems: A Review*, J. Pharm. Sci. **97**, 2892 (2008).

D. Bodmer, T. Kissel, and E. Traechslin, *Factors Influencing the Release of Peptides and Proteins from Biodegradable Parenteral Depot Systems*, J. Control. Release **21**, 129 (1992).

P. Borgquist, A. Körner, L. Piculell, A. Larsson, and A. Axelsson, *A Model for the Drug Release from a Polymer Matrix Tablet—Effects of Swelling and Dissolution*, J. Control. Release **113**, 216 (2006).

G. Bruno, G. Canavese, X. Liu, C. S. Filgueira, A. Sacco, D. Demarchi, M. Ferrari, and A. Grattoni, *The Active Modulation of Drug Release by an Ionic Field Effect Transistor for an Ultra-Low Power Implantable Nanofluidic System*, Nanoscale **8**, 18718 (2016).

G. Chen, Y. Qian, H. Zhang, A. Ullah, X. He, Z. Zhou, H. Fenniri, and J. Shen, *Advances in Cancer Theranostics Using Organic-Inorganic Hybrid Nanotechnology*, Appl. Mater. Today **23**, 101003 (2021).

K. Chiotis et al., *Dual Tracer Tau PET Imaging Reveals Different Molecular Targets for 11C-THK5351 and 11C-PBB3 in the Alzheimer Brain*, Eur. J. Nucl. Med. Mol. Imaging **45**, 1605 (2018).

S. Chłopicki and R. J. Gryglewski, *Angiotensin Coverting Enzyme (ACE) and HydroxyMethylGlutaryl-CoA (HMG-CoA) Reductase Inhibitors in the Forefront of Pharmacology of Endothelium*, Pharmacol. Reports **57**, 86 (2005).

T.-C. Chou and P. Talalay, *Quantitative Analysis of Dose-Effect Relationships: The Combined Effects of Multiple Drugs or Enzyme Inhibitors*, Adv. Enzyme Regul. **22**, 27 (1984).

P. Colombo, R. Bettini, P. Santi, A. De Ascentiis, and N. A. Peppas, *Analysis of the Swelling and Release Mechanisms from Drug Delivery Systems with Emphasis on Drug Solubility and Water Transport*, J. Control. Release **39**, 231 (1996).

U. Conte, P. Colombo, A. Gazzaniga, M. E. Sangalli, and A. La Manna, *Swelling-Activated Drug Delivery Systems*, Biomaterials **9**, 489 (1988).

D. F. Costa, L. P. Mendes, and V. P. Torchilin, *The Effect of Low- and High-Penetration Light on Localized Cancer Therapy*, Adv. Drug Deliv. Rev. **138**, 105 (2019).

G. Deepa, K. C. Sivakumar, and T. P. Sajeevan, *Molecular Simulation and in Vitro Evaluation of Chitosan Nanoparticles as Drug Delivery Systems for the Controlled Release of Anticancer Drug Cytarabine against Solid Tumours*, 3 Biotech **8**, 493 (2018).
L. Dong and A. S. Hoffman, *A Novel Approach for Preparation of PH-Sensitive Hydrogels for Enteric Drug Delivery*, J. Control. Release **15**, 141 (1991).
M. Efentakis and S. Politis, *Comparative Evaluation of Various Structures in Polymer Controlled Drug Delivery Systems and the Effect of Their Morphology and Characteristics on Drug Release*, Eur. Polym. J. **42**, 1183 (2006).
X. Fu, L. Hosta-Rigau, R. Chandrawati, and J. Cui, *Multi-Stimuli-Responsive Polymer Particles, Films, and Hydrogels for Drug Delivery*, Chem **4**, 2084 (2018).
P. Gao, J. W. Skoug, P. R. Nixon, T. Robert Ju, N. L. Stemm, and K.-C. Sung, *Swelling of Hydroxypropyl Methylcellulose Matrix Tablets. 2. Mechanistic Study of the Influence of Formulation Variables on Matrix Performance and Drug Release*, J. Pharm. Sci. **85**, 732 (1996).
E. Gianni, K. Avgoustakis, M. Pšenička, M. Pospíšil, and D. Papoulis, *Halloysite Nanotubes as Carriers for Irinotecan: Synthesis and Characterization by Experimental and Molecular Simulation Methods*, J. Drug Deliv. Sci. Technol. **52**, 568 (2019).
A. Göpferich and R. Langer, *The Influence of Microstructure and Monomer Properties on the Erosion Mechanism of a Class of Polyanhydrides*, J. Polym. Sci. Part A Polym. Chem. **31**, 2445 (1993).
A. Göpferich, *Mechanisms of Polymer Degradation and Erosion*, in *The Biomaterials: Silver Jubilee Compendium*, edited by D. F. Williams (Elsevier Science, Oxford, 1996), pp. 117–128.
J. Hao, J. Zhao, S. Zhang, T. Tong, Q. Zhuang, K. Jin, W. Chen, and H. Tang, *Fabrication of an Ionic-Sensitive in Situ Gel Loaded with Resveratrol Nanosuspensions Intended for Direct Nose-to-Brain Delivery*, Colloids Surfaces B Biointerfaces **147**, 376 (2016).
R. S. Harland, A. Gazzaniga, M. E. Sangalli, P. Colombo, and N. A. Peppas, *Drug/Polymer Matrix Swelling and Dissolution*, Pharm. Res. **5**, 488 (1988).
H. He, X. Cao, and L. J. Lee, *Design of a Novel Hydrogel-Based Intelligent System for Controlled Drug Release*, J. Control. Release **95**, 391 (2004).
Q. He, Y. Gao, L. Zhang, Z. Zhang, F. Gao, X. Ji, Y. Li, and J. Shi, *A PH-Responsive Mesoporous Silica Nanoparticles-Based Multi-Drug Delivery System for Overcoming Multi-Drug Resistance*, Biomaterials **32**, 7711 (2011).
R. A. Hegab, S. Pardue, X. Shen, C. Kevil, N. A. Peppas, and M. E. Caldorera-Moore, *Effect of Network Mesh Size and Swelling to the Drug Delivery from PH Responsive Hydrogels*, J. Appl. Polym. Sci. **137**, 48767 (2020).
Y. Hou, X. Yang, R. Liu, D. Zhao, C. Guo, A. Zhu, M. Wen, Z. Liu, G. Qu, and H. Meng, *Pathological Mechanism of Photodynamic Therapy and Photothermal Therapy Based on Nanoparticles*, Int. J. Nanomedicine **15**, 6827 (2020).
E. Ilhan-Ayisigi and O. Yesil-Celiktas, *Silica-Based Organic-Inorganic Hybrid Nanoparticles and Nanoconjugates for Improved Anticancer Drug Delivery*, Eng. Life Sci. **18**, 882 (2018).
P. K. Jha, P. S. Desai, J. Li, and R. G. Larson, *PH and Salt Effects on the Associative Phase Separation of Oppositely Charged Polyelectrolytes*, Polymers (Basel). **6**, 1414 (2014).
S. C. Joshi, *Sol-Gel Behavior of Hydroxypropyl Methylcellulose (HPMC) in Ionic Media Including Drug Release*, Materials (Basel). **4**, 1861 (2011).
R. Jurgens, C. Seliger, A. Hilpert, L. Trahms, S. Odenbach, and C. Alexiou, *Drug Loaded Magnetic Nanoparticles for Cancer Therapy*, J. Phys. Condens. Matter **18**, S2893 (2006).
M. Kanamala, W. R. Wilson, M. Yang, B. D. Palmer, and Z. Wu, *Mechanisms and Biomaterials in PH-Responsive Tumour Targeted Drug Delivery: A Review*, Biomaterials **85**, 152 (2016).
R. S. Katiyar and P. K. Jha, *Phase Behavior of Aqueous Polyacrylic Acid Solutions Using Atomistic Molecular Dynamics Simulations of Model Oligomers*, Polymer (Guildf). **114**, 266 (2017).
R. S. Katiyar and P. K. Jha, *Molecular Insights into the Effects of Media–Drug and Carrier–Drug Interactions on PH-Responsive Drug Carriers*, Mol. Pharm. **15**, 2479 (2018).

S. Kiil and K. Dam-Johansen, *Controlled Drug Delivery from Swellable Hydroxypropylmethylcellulose Matrices: Model-Based Analysis of Observed Radial Front Movements*, J. Control. Release **90**, 1 (2003).
J. A. Kimber, S. G. Kazarian, and F. Štěpánek, *DEM Simulation of Drug Release from Structurally Heterogeneous Swelling Tablets*, Powder Technol. **248**, 68 (2013).
G. Kocak, C. Tuncer, and V. Bütün, *PH-Responsive Polymers*, Polym. Chem. **8**, 144 (2017).
R. S. Langer and N. A. Peppas, *Present and Future Applications of Biomaterials in Controlled Drug Delivery Systems*, Biomaterials **2**, 201 (1981).
E. Larrañeta, T. Raghu Raj Singh, and R. F. Donnelly, *1 - Overview of the Clinical Current Needs and Potential Applications for Long-Acting and Implantable Delivery Systems*, in *Long-Acting Drug Delivery Systems*, edited by E. Larrañeta, T. Raghu Raj Singh, and R. F. Donnelly (Woodhead Publishing, 2022), pp. 1–16.
M. Levina and A. R. Rajabi-Siahboomi, *The Influence of Excipients on Drug Release from Hydroxypropyl Methylcellulose Matrices*, J. Pharm. Sci. **93**, 2746 (2004).
M. Li, Z. Luo, and Y. Zhao, *Hybrid Nanoparticles as Drug Carriers for Controlled Chemotherapy of Cancer*, Chem. Rec. **16**, 1833 (2016).
J. Li, J. Zeng, X. Jia, L. Liu, T. Zhou, and P. Liu, *PH, Temperature and Reduction Multi-Responsive Polymeric Microspheres as Drug Delivery System for Anti-Tumor Drug: Effect of Middle Hollow Layer between PH and Reduction Dual-Responsive Cores and Temperature Sensitive Shells*, J. Taiwan Inst. Chem. Eng. **74**, 238 (2017).
W. Löscher and D. Schmidt, *Experimental and Clinical Evidence for Loss of Effect (Tolerance) during Prolonged Treatment with Antiepileptic Drugs*, Epilepsia **47**, 1253 (2006).
S. B. Mahamat Nor, P. M. Woi, and S. H. Ng, *Characterisation of Ionic Liquids Nanoemulsion Loaded with Piroxicam for Drug Delivery System*, J. Mol. Liq. **234**, 30 (2017).
M. Mahdavi, F. Rahmani, and S. Nouranian, *Molecular Simulation of PH-Dependent Diffusion{,} Loading{,} and Release of Doxorubicin in Graphene and Graphene Oxide Drug Delivery Systems*, J. Mater. Chem. B **4**, 7441 (2016).
D. T. Manallack, M. L. Dennis, M. R. Kelly, R. J. Prankerd, E. Yuriev, and D. K. Chalmers, *The Acid/Base Profile of the Human Metabolome and Natural Products*, Mol. Inform. **32**, 505 (2013).
P. Marizza et al., *Supercritical Impregnation of Polymer Matrices Spatially Confined in Microcontainers for Oral Drug Delivery: Effect of Temperature, Pressure and Time*, J. Supercrit. Fluids **107**, 145 (2016).
J. S. Mitcheson, Jules C Hancox, A. J. Levi, *Cultured Adult Cardiac Myocytes: Future Applications, Culture Methods, Morphological and Electrophysiological Properties*, Cardiovasc. Res. **39**, 208 (1998).
W. R. Miller, J. L. Sorensen, J. A. Selzer, and G. S. Brigham, *Disseminating Evidence-Based Practices in Substance Abuse Treatment: A Review with Suggestions*, J. Subst. Abuse Treat. **31**, 25 (2006).
V. V Mody, A. Cox, S. Shah, A. Singh, W. Bevins, and H. Parihar, *Magnetic Nanoparticle Drug Delivery Systems for Targeting Tumor*, Appl. Nanosci. **4**, 385 (2014).
Y. Mu, L. Gong, T. Peng, J. Yao, and Z. Lin, *Advances in PH-Responsive Drug Delivery Systems*, OpenNano **5**, 100031 (2021).
S. Mura, J. Nicolas, and P. Couvreur, *Stimuli-Responsive Nanocarriers for Drug Delivery*, Nat. Mater. **12**, 991 (2013).
S. Muro, *Challenges in Design and Characterization of Ligand-Targeted Drug Delivery Systems*, J. Control. Release **164**, 125 (2012).
K. Nakamura, E. Nara, and Y. Akiyama, *Development of an Oral Sustained Release Drug Delivery System Utilizing PH-Dependent Swelling of Carboxyvinyl Polymer*, J. Control. Release **111**, 309 (2006).
F. Nazir, T. A. Tabish, F. Tariq, S. Iftikhar, R. Wasim, and G. Shahnaz, *Stimuli-Sensitive Drug Delivery Systems for Site-Specific Antibiotic Release*, Drug Discov. Today **27**, 1698 (2022).
H. Omidian and K. Park, *Swelling Agents and Devices in Oral Drug Delivery*, J. Drug Deliv. Sci. Technol. **18**, 83 (2008).

P. Pan, D. Svirskis, S. W. P. Rees, D. Barker, G. I. N. Waterhouse, and Z. Wu, *Photosensitive Drug Delivery Systems for Cancer Therapy: Mechanisms and Applications*, J. Control. Release **338**, 446 (2021).
T. G. Park, *Degradation of Poly(d,l-Lactic Acid) Microspheres: Effect of Molecular Weight*, J. Control. Release **30**, 161 (1994).
K. K. Peh and C. F. Wong, *Polymeric Films as Vehicle for Buccal Delivery: Swelling, Mechanical, and Bioadhesive Properties.*, J. Pharm. Pharm. Sci. **2**, 53 (1999).
N. A. Peppas, *Hydrogels and Drug Delivery*, Curr. Opin. Colloid Interface Sci. **2**, 531 (1997).
S. Perkins, A. Evans, and A. King, *Updated List of Light-Sensitive Oral Medications*, Hosp. Pharm. **55**, 349 (2020).
O. Pillai and R. Panchagnula, *Polymers in Drug Delivery*, Curr. Opin. Chem. Biol. **5**, 447 (2001).
V. Pillay, T.-S. Tsai, Y. E. Choonara, L. C. du Toit, P. Kumar, G. Modi, D. Naidoo, L. K. Tomar, C. Tyagi, and V. M. K. Ndesendo, *A Review of Integrating Electroactive Polymers as Responsive Systems for Specialized Drug Delivery Applications*, J. Biomed. Mater. Res. Part A **102**, 2039 (2014).
H. Rabbel, P. Breier, and J.-U. Sommer, *Swelling Behavior of Single-Chain Polymer Nanoparticles: Theory and Simulation*, Macromolecules **50**, 7410 (2017).
V. V Ranade, *Drug Delivery Systems. 6. Transdermal Drug Delivery*, J. Clin. Pharmacol. **31**, 401 (1991).
M.-R. Rodríguez-Hidalgo, C. Soto-Figueroa, and L. Vicente, *Mesoscopic Simulation of the Drug Release Mechanism on the Polymeric Vehicle P(ST-DVB) in an Acid Environment*, Soft Matter **7**, 8224 (2011).
C. K. Sackett and B. Narasimhan, *Mathematical Modeling of Polymer Erosion: Consequences for Drug Delivery*, Int. J. Pharm. **418**, 104 (2011).
S. Salunke, F. O'Brien, D. Cheng Thiam Tan, D. Harris, M.-C. Math, T. Ariën, S. Klein, and C. Timpe, *Oral Drug Delivery Strategies for Development of Poorly Water Soluble Drugs in Paediatric Patient Population*, Adv. Drug Deliv. Rev. **190**, 114507 (2022).
N. V Sastry, D. K. Singh, and P. A. Trivedi, *Hybrid Hydrogel Systems of Micelles of Drug Anion Containing Ionic Liquid and Biopolymers: Rheological Behavior and Drug Release*, Colloids Surfaces A Physicochem. Eng. Asp. **555**, 668 (2018).
D. Schmaljohann, *Thermo- and PH-Responsive Polymers in Drug Delivery*, Adv. Drug Deliv. Rev. **58**, 1655 (2006).
R. Sebastian, T. Guillerm, F. Tjulkins, Y. Hu, A. J. P. Clover, A. Lyness, and C. O'Mahony, *A Comparison of Flow- and Pressure-Controlled Infusion Strategies for Microneedle-Based Transdermal Drug Delivery*, in *2022 44th Annual International Conference of the IEEE Engineering in Medicine & Biology Society (EMBC)* (2022), pp. 2573–2576.
J. Seggie, C. Canny, F. Mai, E. McCrank, and E. Waring, *Antidepressant Medication Reverses Increased Sensitivity to Light in Depression: Preliminary Report*, Prog. Neuro-Psychopharmacology & Biol. Psychiatry **13**, 537–541 (1989).
Y. Shen, X. Ma, B. Zhang, Z. Zhou, Q. Sun, E. Jin, M. Sui, J. Tang, J. Wang, and M. Fan, *Degradable Dual PH- and Temperature-Responsive Photoluminescent Dendrimers*, Chem. – A Eur. J. **17**, 5319 (2011).
J. Siepmann and F. Siepmann, *Modeling of Diffusion Controlled Drug Delivery*, J. Control. Release **161**, 351 (2012).
J. Siepmann, K. Podual, M. Sriwongjanya, N. A. Peppas, and R. Bodmeier, *A New Model Describing the Swelling and Drug Release Kinetics from Hydroxypropyl Methylcellulose Tablets*, J. Pharm. Sci. **88**, 65 (1999).
J. Siepmann, F. Lecomte, and R. Bodmeier, *Diffusion-Controlled Drug Delivery Systems: Calculation of the Required Composition to Achieve Desired Release Profiles*, J. Control. Release **60**, 379 (1999).
J. Siepmann, R. A. Siegel, and F. Siepmann, *Diffusion Controlled Drug Delivery Systems*, in *Fundamentals and Applications of Controlled Release Drug Delivery*, edited by J. Siepmann, R. A. Siegel, and M. J. Rathbone (Springer US, Boston, MA, 2012), pp. 127–152.

J. Supramaniam, R. Adnan, N. H. Mohd Kaus, and R. Bushra, *Magnetic Nanocellulose Alginate Hydrogel Beads as Potential Drug Delivery System*, Int. J. Biol. Macromol. **118**, 640 (2018).

I. Tomic, A. Vidis-Millward, M. Mueller-Zsigmondy, and J.-M. Cardot, *Setting Accelerated Dissolution Test for PLGA Microspheres Containing Peptide, Investigation of Critical Parameters Affecting Drug Release Rate and Mechanism*, Int. J. Pharm. **505**, 42 (2016).

H. Tozaki, J. Komoike, C. Tada, T. Maruyama, A. Terabe, T. Suzuki, A. Yamamoto, and S. Muranishi, *Chitosan Capsules for Colon-Specific Drug Delivery: Improvement of Insulin Absorption from the Rat Colon*, J. Pharm. Sci. **86**, 1016 (1997).

L. L. Tundisi, G. B. Mostaço, P. C. Carricondo, and D. F. S. Petri, *Hydroxypropyl Methylcellulose: Physicochemical Properties and Ocular Drug Delivery Formulations*, Eur. J. Pharm. Sci. **159**, 105736 (2021).

A. M. Vargason, A. C. Anselmo, and S. Mitragotri, *The Evolution of Commercial Drug Delivery Technologies*, Nat. Biomed. Eng. **5**, 951 (2021).

M. Varmazyar, M. Habibi, M. Amini, A. H. Pordanjani, M. Afrand, and S. M. Vahedi, *Numerical Simulation of Magnetic Nanoparticle-Based Drug Delivery in Presence of Atherosclerotic Plaques and under the Effects of Magnetic Field*, Powder Technol. **366**, 164 (2020).

E. R. Viscusi, *Improving the Therapeutic Window of Conventional Opioids: Novel Differential Signaling Modulators*, Reg. Anesth. Pain Med. **44**, 32 (2019).

C. Viseras, P. Cerezo, R. Sanchez, I. Salcedo, and C. Aguzzi, *Current Challenges in Clay Minerals for Drug Delivery*, Appl. Clay Sci. **48**, 291 (2010).

J. Wang, J. Qiu, and S. Wang, *3D Core-Shell Simulation of Hydrogel Swelling Behavior for Controlled Drug Delivery*, vol. 56222 (2013), p. V03BT03A027.

J. W. Winkelman and L. Johnston, *Augmentation and Tolerance with Long-Term Pramipexole Treatment of Restless Legs Syndrome (RLS)*, Sleep Med. **5**, 9 (2004).

M. Wittmann and P. S. Helliwell, *Phosphodiesterase 4 Inhibition in the Treatment of Psoriasis, Psoriatic Arthritis and Other Chronic Inflammatory Diseases*, Dermatol. Ther. (Heidelb). **3**, 1 (2013).

C. Wu, *PH Response of Conformation of Poly(Propylene Imine) Dendrimer in Water: A Molecular Simulation Study*, Mol. Simul. **36**, 1164 (2010).

Z. Yang, S. Sotthivirat, Y. Wu, A. Lalloo, B. Nissley, K. Manser, and H. Li, *Application of in Vitro Transmucosal Permeability, Dose Number, and Maximum Absorbable Dose for Biopharmaceutics Assessment during Early Drug Development for Intraoral Delivery*, Int. J. Pharm. **503**, 78 (2016).

D. Yang, J. S. Lee, C.-K. Choi, H.-P. Lee, S.-W. Cho, and W. Ryu, *Microchannel System for Rate-Controlled, Sequential, and PH-Responsive Drug Delivery*, Acta Biomater. **68**, 249 (2018).

D.-W. Yin, F. Horkay, J. F. Douglas, and J. J. de Pablo, *Molecular Simulation of the Swelling of Polyelectrolyte Gels by Monovalent and Divalent Counterions*, J. Chem. Phys. **129**, 154902 (2008).

T. Yuan, W. Zhan, A. Jamal, and D. Dini, *On the Microstructurally Driven Heterogeneous Response of Brain White Matter to Drug Infusion Pressure*, Biomech. Model. Mechanobiol. **21**, 1299 (2022).

S. Zhang, C. Liu, D. Yang, J. Ruan, Z. Luo, P. Quan, and L. Fang, *Mechanism Insight on Drug Skin Delivery from Polyurethane Hydrogels: Roles of Molecular Mobility and Intermolecular Interaction*, Eur. J. Pharm. Sci. **161**, 105783 (2021).

Z. Zhao, A. Ukidve, J. Kim, and S. Mitragotri, *Targeting Strategies for Tissue-Specific Drug Delivery*, Cell **181**, 151 (2020).

S. Zuleger and B. C. Lippold, *Polymer Particle Erosion Controlling Drug Release. I. Factors Influencing Drug Release and Characterization of the Release Mechanism*, Int. J. Pharm. **217**, 139 (2001).

Advances of Graphene Oxide in the Field of Microbiology

Kriti, Sujata Kumari, Mehak Ahuja, and Akarsh Verma

1 Introduction

Advancements in nanoscience and nanotechnology have paved the way for the creation of new nanomaterials, leading to the emergence of various novel drug delivery systems. Today, more and more people are devoting their time for the study of materials made of graphene. Ever since graphene was discovered, its potential in biomedical applications has garnered significant attention. In 2004, Andre Geim and Konstantin Novoselov introduced a technique to isolate graphene by peeling away individual layers from graphite's 3D structure using mechanical exfoliation. Their groundbreaking work earned them the Nobel Prize in Physics in 2010. Graphene, which first emerged in 2004, is a single layer of sp^2-hybridized carbon atoms arranged in a 2-D hexagonal pattern, captivating the scientific community. The positioning of benzene rings within graphene's structure affects its ability to form covalent or non-covalent bonds with other atoms. This bonding often results in changes to graphene's chemical or electrical characteristics (Bellier et al. 2022; Sontakke et al. 2023).

It possesses excellent physical–chemical characteristics such as high aspect ratio, large surface area, high fracture and mechanical strength, excellent thermal/electrical conductivity, fast mobility of charge carriers, biocompatibility, and optical property. Due to these characteristics, it is the perfect material for high-performance and biologically relevant nanocomposites. Graphene and its derivatives are extensively

Kriti (✉) · S. Kumari · M. Ahuja
School of Basic and Applied Sciences, K.R. Mangalam University, Gurugram, Haryana 122103, India
e-mail: kriti@krmangalam.edu.in

A. Verma
Department of Mechanical Engineering, University of Petroleum and Energy Studies, Dehradun 248007, India

Department of Mechanical Science and Bioengineering, Osaka University, Osaka 560-8531, Japan

J. Gupta and A. Verma (eds.), *Microbiology-2.0 Update for a Sustainable Future*,
https://doi.org/10.1007/978-981-99-9617-9_11

utilized in a variety of applications, including energy storage, transparent electrodes, drug delivery system, cell scaffold, biosensor, and catalysis (Zhu et al. 2010; Liu et al. 2010a; Dhar et al. 2018; Kausar 2021; Purkait and Sontakke 2023).

GO is one of the most popular graphene derivatives. GO, a highly oxidized form of chemically modified graphene is composed of a single-atom thick layer of graphene sheets consisting of different oxygen functional groups (carboxyl, hydroxyl, and epoxy) as shown in Fig. 1. It has an unoxidized benzene ring region as well as an oxidized aliphatic six-membered ring region. The proportional sizes of these two zones are determined by oxidation level and random distribution. A water-dispersible GO is produced by oxidising graphite in a solution of strong acid and an oxidizing agent. The structure of graphene is composed of planar hexagonal lattices. Four valence electrons make up each carbon atom, three of which (2s, $2p_x$, and $2p_y$) form plane sp^2 hybrid orbitals. The last orbital electron is free to wander about in the plane and forms a strong π bond. A two-dimensional carbon compound with a single atomic layer and polydisperse sheet sizes is called GO. GO has more oxygen-containing functional groups on its sheet layer than graphene, altering its structure and changing its properties. Planar sp^2 to sp^3 hybridization results in functional groups having oxygen changing the structure of graphene. As pointed out by Lerf and Klinowski, the carboxyl groups attach to the edges of the GO sheet, while hydroxyl and epoxy groups are spread across the top and bottom of the base plane, causing a slight deformation of the lattice.

Amphiphilicity, or the ability to be both hydrophobic from pristine graphite and hydrophilic from oxygen functions is one of the fascinating characteristics of GO. The interaction of hydrogen between the hydroxyl groups in GO and the solvent boundary makes GO hydrophilic and allows it to dissolve in numerous organic solvents. These characteristics are vital for bio-applications. GO possesses two unique areas due

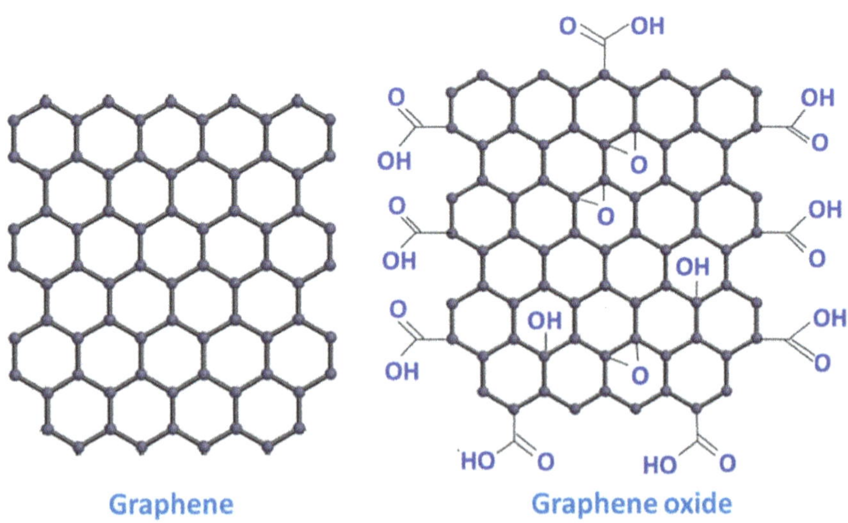

Fig. 1 Structure of graphene and graphene oxide (Kausar 2021)

to its functional groups: the aliphatic (sp^2 oxidized) zone and the aromatic (sp^2 non-oxidized) zone of graphite, which remain unchanged during GO production. The graphene structures sp^2 hybridization is partially broken by oxygen functional groups because of which the distance between the carbon layers increases and planar sp^2 transforms to sp^3 hybridization (Smith et al. 2019; Kiew et al. 2016).

The electrical or chemical reactivity of graphene depends upon the formation of covalent and non-covalent bonds with the other atoms inside the graphene structure. The formation of bonds is further influenced by the benzene ring's distribution throughout the structure. π-electrons are localised because of the partial breakdown of conjugated structure which lowers the overall electrical conductivity. However, GO displays distinct characteristics, like an attraction to aromatic rings and the ability to quench fluorescence. This is while maintaining excellent solubility in water-based solvents, owing to the remaining sp^2 domains that acquired hydrophilic groups during oxidation (Wang et al. 2011; Zhang et al. 2013).

However, the L-K structural model relies on a series of assumptions and ignores the effects of unprocessed graphene, oxidants, and oxidation methods. In their investigation of GO nano-plates, Erickson et al. found that the material also exhibited hole defects caused by excessive oxidation and lamellar peeling in addition to an oxidised zone with significant disorder and an unoxidized grapheme region. As a result, new models, such as the dynamic structure model and binary structure model have recently been proposed.

In the area of biomaterials, GO has vast area of applications, ranging from drug delivery, DNA/RNA extraction, cancer therapy, biosensing, tissue engineering, medical imaging to antiviral materials, and bacterial inhibition and so on. Graphene and its derivatives can bind to other molecules via covalent and non-covalent bonding thanks to sp^2 domains and oxygen functional groups. These characteristics make graphene nanoparticles ideal substrates for immobilizing biomolecules and increasing their environmental stability. The performance and reusability of the enzyme will be improved by immobilising it, for example, on functionalized graphene or its derivatives. Immobilized biomolecules interact with the target sample to serve as sensing components in biosensors. The outer carboxylate group gives graphene-based materials their colloidal stability and a pH-sensitive negative charge. The epoxide (–O–) and hydroxyl (–OH) groups, though uncharged, are polar and are situated on the basal plane. These groups enable weak bonds, hydrogen bonding, and various surface interactions. Moreover, the unmodified free surface electrons on graphene's basal plane are hydrophobic and allow π–π interactions, beneficial for drug loading and non-covalent functionalization. GO is a sheet-like molecule with both water-attracting and repelling properties, making it suitable for stabilizing hydrophobic substances in water solutions. However, compared to graphene or rGO, GO's mechanical, electrical, and thermal properties are inferior due to the high number of defects arising from its functional groups (Guo et al. 2011; Kim et al. 2010).

The capacity to soak up a variety of aromatic biomolecules via a π–π stacking interaction and/or electro-static interaction makes graphene-based materials ideal for building biosensors and loading drugs. They also exhibit superb electrochemical and optical properties. Additionally, GO can be easily changed with targeting ligands to enable drug delivery and targeted imaging due to the abundance of oxygen-containing groups connected to its six-atom carbon rings forming honeycomb structure.

2 Synthesis of GO

British chemist B. C. Brodie made the first attempt to create graphite oxide in 1859 while researching the reactivity of flake graphite. To identify the structure of GO, graphite was treated with a mixture of fuming nitric acid and potassium chlorate at 60 °C for four days. He performed multiple oxidative treatments (4–7) in succession, and the final oxygen, hydrogen, and carbon composition was estimated to be $C_{11}H_4O_5$ (corresponding to a C/O ratio of 2.2). The synthesized material reacted slightly with litmus paper and was termed as graphic acid as it was soluble in pure water, but flocculated in acidic media. This technique comes with multiple flaws, including a lengthy reaction time and the release of hazardous gases during the process (Fig. 2).

The Staudenmaier method represents another approach using chlorate. L. Staudenmaier refined Brodie's technique by adding several doses of potassium chlorate solution during the reaction and introducing concentrated sulfuric acid to increase the mixture's acidity. This resulted in a more oxidized GO in one reaction container, streamlining the production process. This tweaked method achieved an oxidation level similar to Brodie's multi-step oxidation approach (C: O _ 2: 1). However, this method had its drawbacks: it was lengthy, taking over a week due to the slow addition

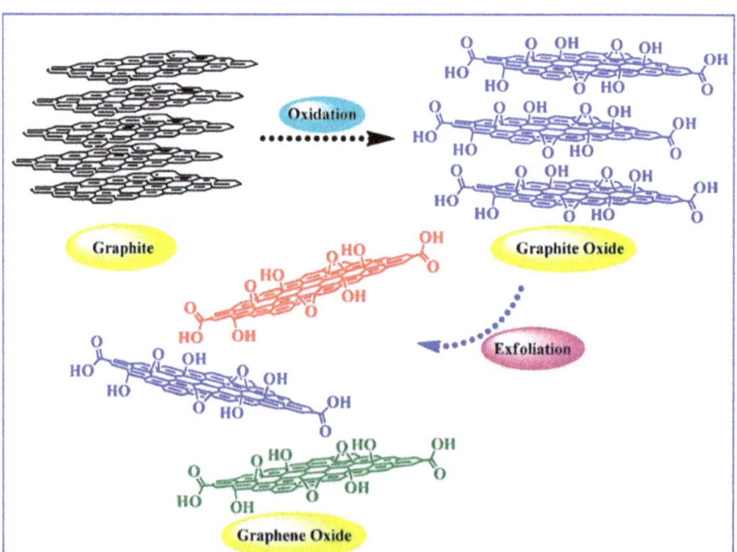

Fig. 2 Preparation of graphene oxide (Yu et al. 2020)

of potassium chlorate, and was hazardous. The generated chlorine dioxide needed to be vented using an inert gas, and there was a perpetual risk of explosion. Therefore, researchers continued to seek further refinements or entirely new oxidation procedures (Staudenmaier 1898).

In 1937, Hofmann et al. utilized concentrated sulfuric acid in conjunction with $KClO_3$ and concentrated nitric acid to oxidize graphite to produce graphite oxide. The $KClO_3$ acted as a powerful oxidizing agent that oxidizes graphite powder in acidic solutions while also acting as an in situ carrier for dioxygen, thereby serving as the reactive species. A number of studies have used the Hofmann method to synthesize GO for various applications (Hofmann and Holst 1939).

Hummers and Offeman, invented the most popular, rapid, and efficient permanganate method to synthesize GO. In this procedure, the reaction mixture contains a significant amount of potassium permanganate, sulphuric acid, and a trace of sodium nitrate. The reaction time ranges from 8 to 12 h. Because there is no explosive ClO_2, this method is much safer. At the conclusion of the procedure, weak solution of H_2O_2 is used to neutralize excess $KMnO_4$. The C/O ratios of GO produced by the Brodie's and Hummers method are 2.2 and 2.25, respectively. However, the method is not environmentally friendly due to the release of nitrogen oxide produced during the reaction. Subsequently in 2010, Tour proposed his own technique, which is discussed further below (Hummers and Offeman 1958).

Tour created a novel technique that relied more on potassium permanganate and eliminated the usage of sodium nitride. Additionally, phosphoric acid was added to the reaction vessel as a new acid, and GO with a greater degree of oxidation level was formed. Nitric acid was replaced with phosphoric acid to prevent the generation of harmful gases like NO_2, N_2O_4, ClO_2. Additionally, phosphoric acid provided more graphitic surfaces, and compared to Hummers approach, its efficiency was substantially higher. Yu et al. improved Hummers approach in a similar way by pre-oxidizing graphite to broaden the interlayer gaps. The amount of $KMnO_4$ was increased in place of $NaNO_3$, and when $NaNO_3$ was absent from the system, $KMnO_4$ eventually gave way to K_2FeO_4. $NaNO_3$ removal from the system stopped the release of harmful gases such as NO_2 and NO_4 (Yu et al. 2016a).

Electrochemical synthesis provides an alternative way to produce GO compared to traditional chemical methods, and it might pave the way for large-scale GO production. This method is more eco-friendly due to consistent electrolyte reuse and reduced equipment cleaning, making it preferable over chemical methods. Electrochemically synthesized GO, or EGO, offers higher quality due to its use of water-based electrolytes and the absence of oxidizing agents, minimizing contaminants. Plus, this method allows control over oxidation levels and defect density thanks to various experimental configurations (Yu et al. 2016b). Interestingly, for an environmentally friendly production of graphene oxide, biological processes are employed to oxidize graphite. However, oxidation isn't uniform post microbial growth. Studies have explored Acidithiobacillus ferrooxidans and Pseudomonas as potential oxidizing microbes (Zhu et al. 2014).

3 Functionalization of GO

The agglomeration, poor water solubility, and chemical reactivity during the reduction process are some restrictions in proper utilization of GO's distinctive features. However, these issues are avoided, and the fundamental features of GO are preserved through correct functionalization. In addition to preserving GO's superb properties, functional modification also adds additional functional groups to offer the material new properties. Currently, covalent and non-covalent functionalization's are the two main techniques used to modify GO.

3.1 Covalent Functionalization

The process of functionalizing graphene entails joining it with fresh groups via covalent bonds to boost and increase its functionality. This approach is typically used to change GO (GO) rather than graphene because of the existence of oxygen containing active functional groups on GO surface. This technique improves GO's exfoliation strength, hydrophilicity, and surface polarity in aqueous conditions. Additionally, the use of tiny functional molecules that act as links or bridges for covalent bonding enhances the longevity of immobilised enzymes, proteins, medicines, or tiny molecules in the overall system to enhance key qualities like biocompatibility and loading stability. Biocompatible polyethylene glycol (PEG) was the first molecule used to alter nanoscale GO to give it excellent water solubility and stability in biological fluids and serums. Numerous other substances could be used to modify peg, in addition to it, including chitosan (CS), polyethylenimine (PEI), polyacrylic acid (PAA), sulfonic groups, poly(vinyl alcohol) (PVA), DNA, and others (Sontakke et al. 2023; 徐志远, 朱绍佳, 王明伟, 李永军 2015; Shen et al. 2010).

This method includes attaching polymers or immobilising biomolecules to GO sheets by using several chemically reactive functional groups on sheet edges (carboxylic acid) and the basal plane (epoxy and hydroxyl). The hydroxyl functional groups typically entail the hydroxy-reaction of amide or isocyanate and GO to yield esters, which are then effectively modified with different groups.

Paclitaxel (PTX), a chemotherapeutic medication that cannot be dissolved in water, was examined by Xu et al. for PEGylated-GO delivery. Following PTX's initial attachment to PEG molecules, amidation was used to conjugate the molecules with GO nanosheets (Fig. 6a). The fluorescence observed in the final GO-PEG-PTX complex showed fast access to breast and lung cancer cells, demonstrating the superb biocompatibility of PEGylated-GO (徐志远, 朱绍佳, 王明伟, 李永军 2015).

To examine and study the kinetics of the anticancer medication doxorubicin (DOX) on GO and PEG-GO, Mahdavi et al. have constructed a molecular simulation as shown in Fig. 3. The research was conducted at physiological pH and body temperature conditions. Researchers looked at the connection between the PEG-GO's ability to load drugs and the size of the PEG chain. It was discovered that drug

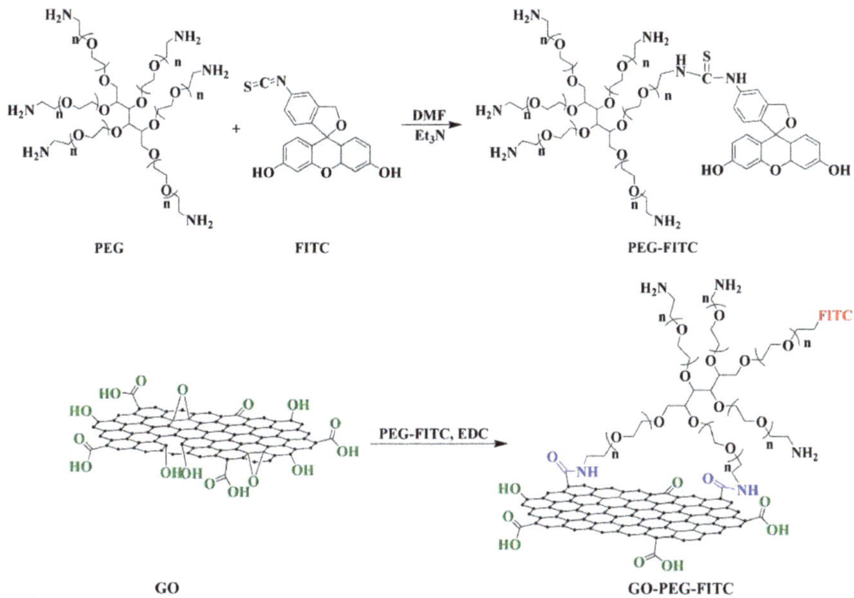

Fig. 3 Synthetic route of GO-PEG-FITC (徐志远, 朱绍佳, 王明伟, 李永军 2015)

loading increased with PEG chain length and was controlled by both covalent and non-covalent interactions (Mahdavi et al. 2020).

Different chemical reactions such as activation, amidation, or esterification can be utilized in functionalizing carboxyl groups of GO with chemical linkers OR small molecules or in polymers. Jana et al. described a versatile method for dual functionalizing carboxyl groups of GO with (Tris-NTA) Tris- [nitrilotris (acetic acid)] and biotin through covalent contacts enabling cellular transport of oligohistidine and biotintagged biomolecules (Fig. 4). This bifunctional GO was used to transport proteins and other biomolecules using oligohistidine and biotin tags (Jana et al. 2013).

To create a biocompatible GO-GOx biosensor, a neuroectodermal derivative known as the retinal pigment epithelium (RPE), was put onto the GO-GOx platform. The combination withstands remarkable biocompatibility. The GO-based glucose biosensors showed superb reproducibility, good stability, high sensitivity, broad linear range, and excellent biocompatibility to human RPE cells, showing that such biosensors have a lot of possibilities for applications in diabetic retinopathy (Liu et al. 2010a).

Biomolecules have also been used for the purpose of functionalising GO. For instance, using the common technique of diimide-activated amidation, Shen et al. utilized bovine serum albumins (BSA) to functionalize GO. A GO-BSA conjugate was produced because of this method, and there was no sign of protein denaturation. According to cyclic voltammogram studies, the combination exhibited superior water solubility and bioactivity making it appropriate for interacting with biological substrates (Shen et al. 2010).

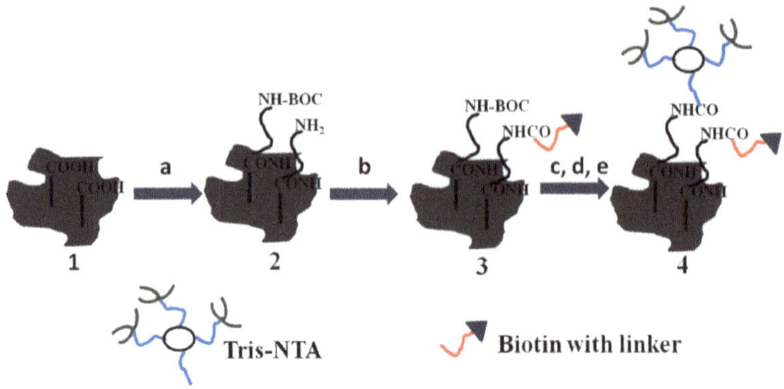

Fig. 4 Chemical reaction illustrating synthesis of dual functionalization of GO (Jana et al. 2013)

Also, Liu et al. examined the potential of modified GO as a sensitive and selective hybridized DNA biosensor. To activate GO, EDC/sulfo-NHS was used, which changes the carboxylic groups of GO in amine-reactive NHS-ester forms (Fig. 5). In a process known as hybridization, complementary ssDNA that had been tagged with gold nanoparticles was able to bind to the GO-DNA probe thanks to the functionalized carboxylic groups' affinity for amino modified ssDNA (Liu et al. 2010b).

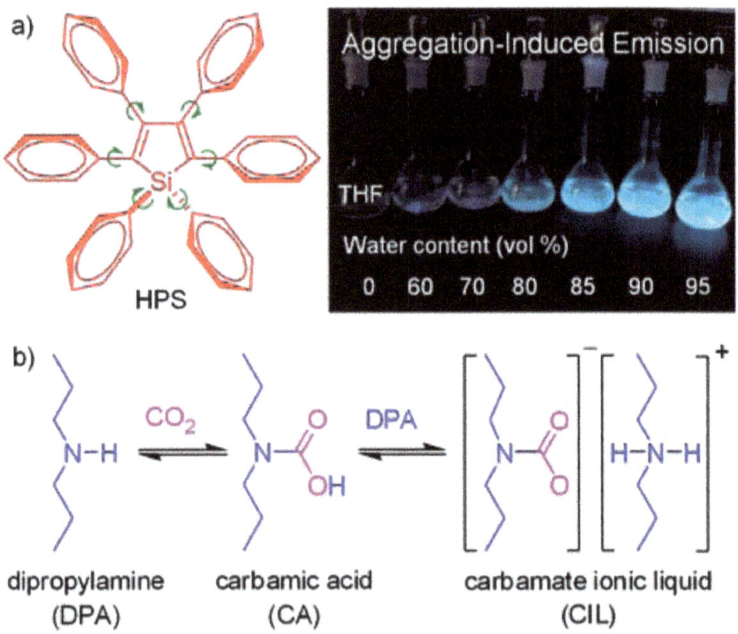

Fig. 5 **a** Nonemissive characteristics of HPS when dissolved in THF. **b** Formation of CIL (Liu et al. 2010b)

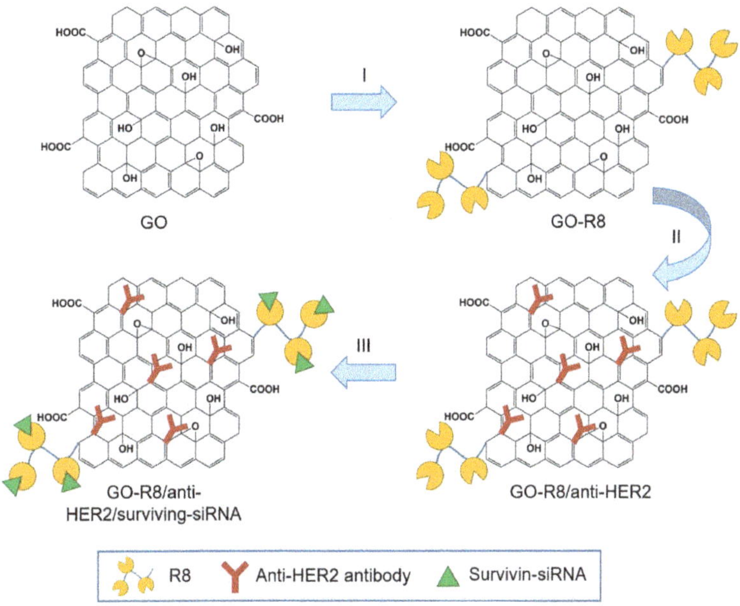

Fig. 6 Schematic illustration of covalently modified GO with anti-HER2 antibody and octaarginine (R8) (Wang et al. 2018a)

The alteration of epoxy groups includes a nucleophilic attack on the carbon of the epoxide and the use of an amine group to catalyse the ring-opening reaction. For instance, Octaarginine (R8) and antiHER2 antibody-functionalized GO were produced by Wang et al. employing covalent conjugation as a new gene delivery strategy for tumour therapy (Fig. 6). To improve the effect of siRNA distribution, R8 played the role of a cell-penetrating peptide when incorporated on the surface of GO, and anti-HER2 was tagged jointly to attach with HER2. Additionally, the gene-silencing carrier GO-R8/anti-HER2/survivin-siRNA has the potential to be effective in both in vitro and in vivo cancer therapy (Wang et al. 2018a).

As depicted in Fig. 7, for medication and gene administration, Bao et al. covalently modified GO with chitosan (CS) using a simple amidation technique. The solubility and biocompatibility of GO are increased by grafting CS onto GO sheets. In addition, iron oxide-containing inorganic nanoparticles have been coupled to the GO surface to improve the contrasts seen in T2-weighted magnetic resonance imaging (MR) scans (Bao et al. 2011).

3.2 *Non-covalent Functionalization*

Certain polymers or biomolecules with exceptionally high hydrophobicity are needed for noncovalent modifications, which typically involve coordination bonds, hydrogen

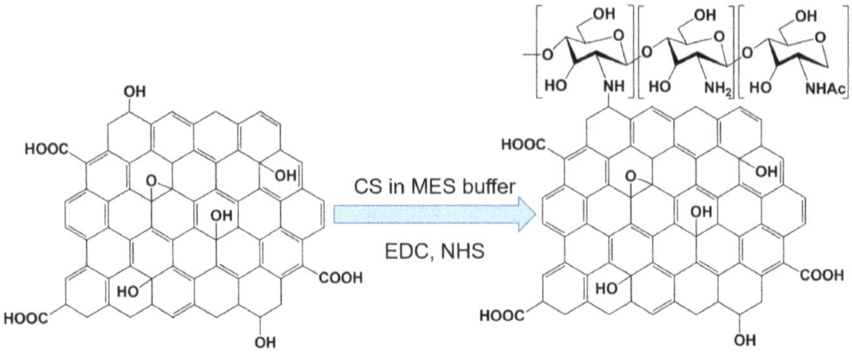

Fig. 7 Preparation of chitosan (CS)-functionalized GO (Bao et al. 2011)

bonds, Van der Waals forces, electrostatic, and π–π interactions with GO. This can be achieved by functionalizing GO non-covalently with drugs, biomolecules, and polymers, biomolecules by wrapping or absorption of such molecules on the surface of GO mostly through π–π interactions. The main benefit of this process is that it preserves the bulk structure and exceptional properties of GO while also enhancing its dispersibility and stability (Sontakke et al. 2023; Shen et al. 2010).

Wang et al. looked at functionalizing thermally reduced GO (trGO) with pyrene derivatives and ending it with the ureidopyrimidinone (UP) hydrogen bonding unit to create (trGO)-UPPY. Pyrene derivatives were added to the graphene surface, where they joined UPPY in-bonds. The Layer-by-Layer (LBL) self-assembly method was repeatedly used to create trGO-UPPY films which include both π–π and hydrogen bonds. The multilayer films that were produced showed promise as superior electron conductors to enhance electron transport (Wang et al. 2018b).

As shown in Fig. 8, Lee et al. used a tetradecene derivative that utilized a dendritic polyether branch as a modifier. He also used the synergistic combination of an aromatic cyclic fluorene skeleton alongside graphite and a polyether chain, to strip graphite, to produce high hydrophilicity, and stabilize the graphene layer. The fact that the identical indole derivative had no effect on single-walled carbon nanotube dispersibility suggests that the planar arrangement of carbon nanomaterials plays a significant role in the development of efficient π-stacking. The π–π link caused the absorption spectrum of tetraterpene derivatives to look reddened and quenched their fluorescence (Lee et al. 2011).

Several works report surface functionalization of GO-based carriers by layer-by-layer (LbL) technique. In this method, polyelectrolytes are alternately deposited on GO sheets with opposing charges through electrostatic interactions. For instance, Xie et al. (Fig. 9) prepared GO composite stabilised by oppositely charged (chitosan and dextran) polyelectrolyte for drug delivery (Xie et al. 2016).

Prasad et al. created a unique GO flake fortified with liposomal nanohybrid (GOF-Lipo). Folic acid (FA), a tumour-targeting ligand, was added to the nanohybrid. The electrostatic forces between the positively charged dipalmitoylphosphatidylcholine

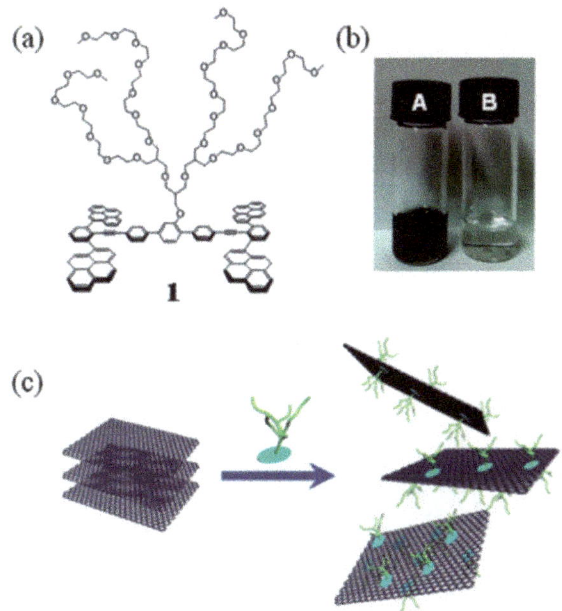

Fig. 8 Use of tetrapyrene derivative to exfoliate and stabilise graphite via π–π interaction (Lee et al. 2011)

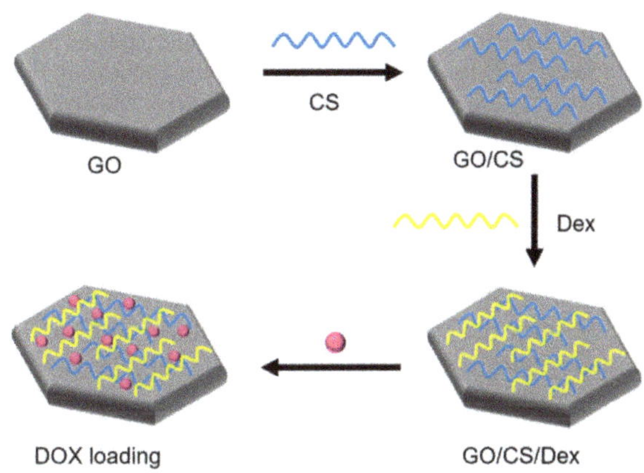

Fig. 9 Diagram illustrating the functionalization of GO with dextran (Dex) and chitosan (CS) laden with doxorubicin hydrochloride (Xie et al. 2016)

(DPPC) lipid and the negatively charged fortified GO were used to establish the functionalization. To further investigate the in-vitro cytotoxicity of the breast cancer cell lines 4 T1 and MDA-MB-231 for combined chemo-photothermal treatment, the newly obtained GOF-Lipo and GOF-Lipo-FA conjugates were mixed with the chemotherapeutic medication DOX-HCL. Superior water dispersibility, hemocompatibility, and 90% cell survival along with rapid photothermal response have all been demonstrated by the nanohybrids as described in Fig. 10 (Prasad et al. 2019).

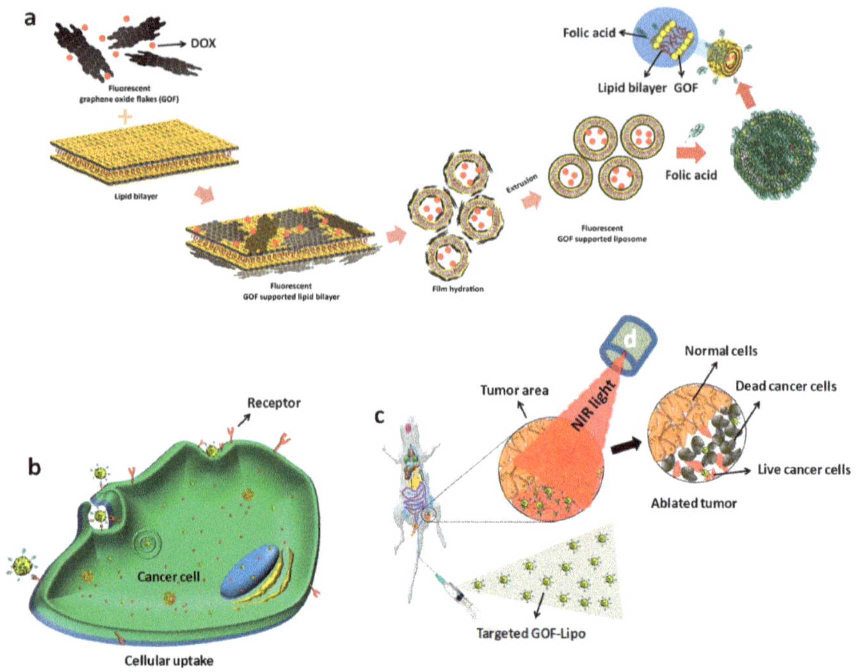

Fig. 10 Schematic illustrations of **a** the extrusion and film hydration method used to create the folic acid loaded GO flake fortified with liposomal nanohybrid (GOF-Lipo), **b** drug distribuition (red dots) and endocytosis in the interior of the cell, **c** nanohybrid in vivo biodistribuition and tumour uptake using intravenous administration and **d** localized tumour regression using NIR light (Prasad et al. 2019)

However, the primary disadvantage of this approach is the addition of new elements, such as surfactants. The possible drug delivery is less stable while interacting with the biological system both in vitro and in vivo because non-covalent contacts are weaker than covalent interactions and are susceptible to changes in the surrounding environment. Additionally, because the coated polymers have partially occupied the conjugated portions of graphene, the loading capability of non-covalently modified graphene gets reduced.

4 Applications of GO

Graphene being the thinnest material on the earth shows excellent electrical properties (Yaragalla et al. 2015; Yazyev and Louie 2010). Graphene moieties being comprised of carbon atoms possess single layer planar films with hexagonal honeycomb lattice and possess unbound free π electrons which provide high conductivity. GO-based materials have high conductivity, good flexibility and excellent sensitivity

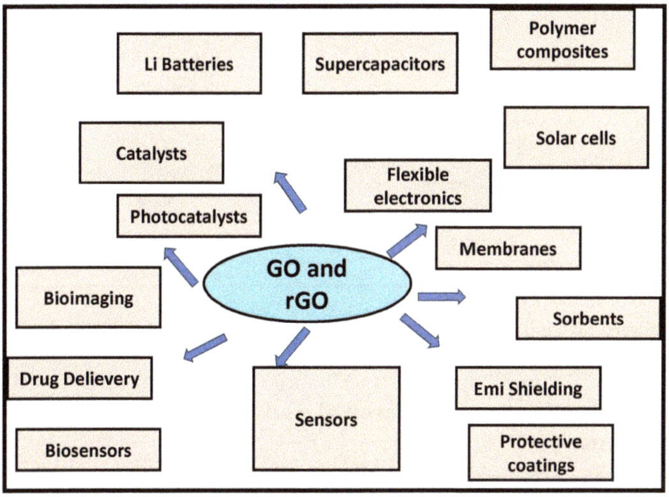

Fig. 11 Some of the prospective applications of the GO and reduced GO (Dideikin and Vul' 2019)

for biosensor applications. The hexagonal carbon arrangement in the honeycomb manner with sp^2 hybridized carbon atoms and pure pz electrons allow it to hold biomoleculesproviding powerful bactericidal effects against pathogenic microorganisms (Valishina et al. 2018). The biosensors based on GO possess large surface area and have excellent electrical conductivity which increases the rate of electron transfer providing easy pathway for the accurate and selective detection of biomolecules (Kuila et al. 2011). These GO-based sensors are generally used for the detection of biomolecules like proteins, DNA and enzymes. The biomolecular interactions of these graphene-based derivatives play an important role in detecting sensitivity and specificity towards a particular active site and it depends on the immobilization of bioreceptors (Liu et al. 2022). Further GO-based materials also have bio-medical applications as they undergo hydrophobic interactions with other insoluble drugs which doesn't impact their potential activity (Liu et al. 2008a). Their tendency for non-covalent linking is observed due to availability of abundant free π electrons, large specific surface area, non-toxicity, biocompatibility and instant response to environmental stimuli (Zhang et al. 2010). Some of the applications of these materials are given below in Fig. 11.

4.1 Antimicrobial Activity

The antimicrobial activity of GO is observed because of presence of polar oxygen groups, and it depends on factors like composition of the species, nature of targeted microorganism, surface modification and the characteristic environment in which the interaction between the cell and graphene moieties takes place.

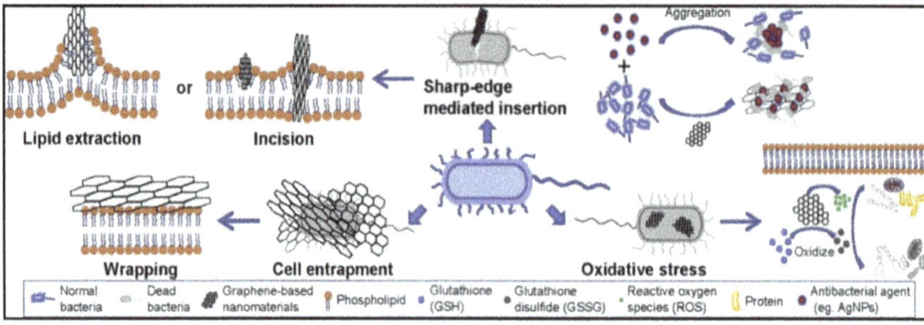

Fig. 12 GO-based materials—their antimicrobial mechanism (Xia et al. 2019)

4.1.1 Mechanism for the Antimicrobial Activity of the GO-Based Materials

The mechanism of interaction to study the antibacterial activity of graphene-based materials is generally oxidative stress and extraction of phospholipids from bacteria. This depends on the insertion mode of mechanism for the action; where the two-dimensional GO sheets with their sharp edges destroy bacterial cells and led to cell death due to leakage of intracellular material (Hernandez et al. 2013). The pictorial representation for the mechanism showing antimicrobial activity of GO-based materials (Xia et al. 2019), as depicted in Fig. 12. It has also been reported by Li et al. that the GO sheets have the tendency of penetration in the bacterial cell through the lipid bilayer bypiercing the cell-wall of bacteria through its sharp edges. The antibacterial activity also depends on the incorporating GO sheets size as it allows different orientation. If the size lies in the nanometres range it adopts parallel position according to the lipid bilayer while if the size is in micrometres range, then there is perpendicular orientation to the cell wall (Xia et al. 2019). Graphene moieties show the tendency of sinking between the lipid tails which is observed because of interaction between the lipid's hydrocarbon tail and graphene flat lipophilic tails.

4.1.2 Factors Affecting the Antimicrobial Activity of GO

(a) The Number of corresponding GO layers:

The antibacterial activity of GO-based materials is affected with the number of layers, due to the presence of three layers in the multilayer arrangement they experience more energy barrier in penetrating in comparison to the single layer. Lesser extent of interaction between the subsequent units due to accumulation of graphene sheets results in better antibacterial activity (Yi and Gao 2015; Marković et al. 2018; Wang et al. 2013a).

(b) Size of GO sheets:

Smaller size GO sheets show better entrapment to the bacterial cells as they are more cytotoxic, have higher cellular internalization and show suitable antimicrobial activity in surface coating.

4.1.3 Graphene-Oxide Based Materials and Their Antimicrobial Applications

GO, along with its associated polymer composites, and graphene reinforced polymer composites undergo potential antimicrobial applications (Fig. 13), some of which are discussed below:

(a) GO based antimicrobial hydrogels:

Multifunctional GO-based hydrogel with an agarose polysaccharide was reported as an efficient stabilizer and cross-linking agent which prevents the bacterial growth (Wang et al. 2013b). Benzalkonium bromide doping on GO improved its antimicrobial properties, moreover silver/GO nanocomposites also improved antimicrobial activity (Wang et al. 2015).

(b) In packaging:

Flexible type of packaging which is one of the emerging areas for food, technology and science, the thermomechanical and barrier properties of the polymer are enhanced

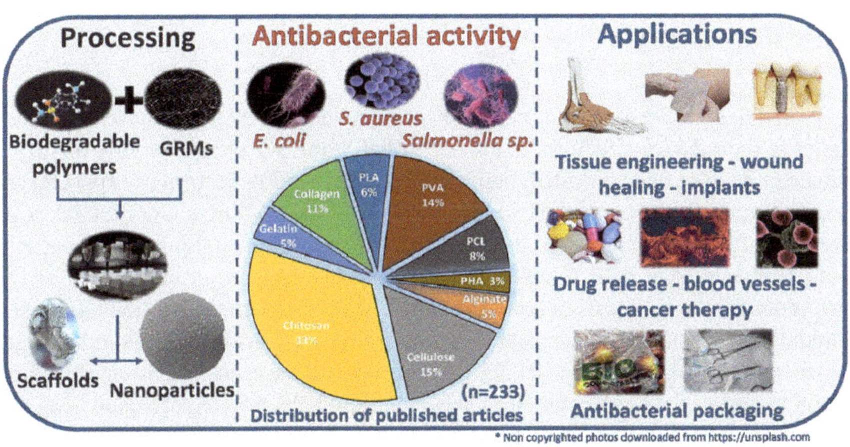

Fig. 13 Antimicrobial applications of GO based materials

when GO is added. Various GO-based composites like GO/polyvinyl alcohol (PVA), LLDPE/GR have worked efficiently in bio-based packaging. Solution casted PLA film with GO also worked as antimicrobial film for the clove oil packaging (Bera and Maji 2017). It has also been reported by Wang et al. that the multilayer film of MTAC/rGO/EVOH with rGO and ethylene covinyl alcohol show excellent antimicrobial and mechanical properties (Wang et al. 2018c).

(c) Bandages and wound dressing:

GO based materials significantly work for wound management, they have the tendency to maintain moisture across the wound, accelerate the closure of wound and stimulate the wound healing process by minimizing the rate of formation of SCAR (Tabish 2018). Graphene based materials with silver and methylene bis(acrylamide) cross linked GO moieties enhanced the rate of wound healing process (Shams et al. 2017).

(d) Antimicrobial films and coatings:

These GO based materials can be employed in antimicrobial films as they work on two mechanisms either they trigger the approach of bacteria and light through the process of electron transfer and the second process which is physical destruction. The antimicrobial coating of these CVD graphene being merged with silver nanowires worked as an effective disinfectant; this film effectively showed antimicrobial activity against E.coli on applying to ordinary film of polyethylene terephthalate (Zhao et al. 2016).

4.2 Solar Cells

Low-cost flexible organic solar cells constitute reduced GO as the transparent electrodes in their device structure. In the recent literature GO and chemically modified reduced GO have been reported, being the components for electron transport layers in organic perovskite based solar cells (Assadi et al. 2018). These GO-based derivatives have noticeably improved the quantum efficiencyand durability of the organic and perovskite solar cells (Han et al. 2015). Kakavelakis et al. reported effective procedure for the synthesis of nano structured TiO_2 in which the applied GO sheets worked as a support and dispersant for the organic solar cells and photo catalysis systems (Kakavelakis et al. 2017). These materials have also been employed as useful materials for counter electrodes for the liquid electrolyte based and low-cost dye-sensitized solar cells (Williams et al. 2008).

4.3 *Biosensors*

Biosensors are the devices which possess sensitivity for the biological substances and for the detection convert their electrical signals into concentration. These biosensors play important role in the early detection of some diseases and the components present in modern medicine. As the GO materials possess strong adsorption capacity, good ability for chemical functionalization and excellent conductivity they experience higher sensitivity towards detection of technologies for biological activity (Justino et al. 2017). GO works as efficient sensor for the detection of nitrogen dioxide, hydrogen, and in humidity sensing (Yu et al. 2014; Rathi and Pal 2017). Gold nanoparticles/bicyclodextrin functionalized reduced GO nanocomposites were reported as ultrasensitive sensors for the environmental monitoring of bisphenols (Yun 2016; Sinha et al. 2018). A nanocomposite ($Ni(OH)_2$)-PEDOT-rGO based electrochemical glucose sensor was fabricated which worked with high responsivity and selectivity (Sheng et al. 2018). Enzyme reactions also play an important role in selectivity and specificity for biological and chemical sensors which would be very difficult to detect in complex mixtures. GO based materials have different sensing mechanisms which are based on optical and electrochemical signalling which play important role in disease development. This is important for accurate detection of biomolecules for therapy and the disease diagnosis (Liu et al. 2012). GO being a promising electrode material for the biomolecule's detection and because of high sensitivity, quick response, low cost and easy operation the electrochemical approach is adopted for the detection of biomolecules (Liu et al. 2005). The application of GO based materials in different biosensors has been discussed in detail in the further section.

4.3.1 For Small Biomolecules and H_2O_2 Detection

Hydrogen peroxide being an essential component for mediating biological and clinical processes, therefore a need has risen for the electrochemical H_2O_2 sensor for which GO possess latent capacity (Liao et al. 2009). Zhou et al. reported H_2O_2 on a GO modified electrode being a highly sensitive electrochemical sensor which showed increased electron transfer rate in comparison to bare electrodes and graphite (Zhou et al. 2009). Shao et al. also observed that in the case of N-doped graphene the electro catalytic activity toward the reduction of H_2O_2 is improved in comparison to the graphene because of reduced over-potentials and the existence of nitrogen and oxygen-containing functional groups and the presence of structural defects (Shao et al. 2010). GO also works as an excellent electrode material for the diagnosis and therapy of diabetes and as an oxidase-based biosensors, likewise for the determination of glucose, which is very important. Kang et al. also fabricated a glucose oxidase (GOD)-graphene chitosan nanocomposite modified electrode by thermally splitting graphene for which the immobilized enzyme exhibited reversible two-proton/two-electron transfer reaction. The obtained biosensor not only retained the GOD bioactivity, along with this also exhibited a broad range with the detection limit of 0.02 mM and higher sensitivity in comparison to other nanostructured supports (Kang et al. 2009). A graphene-based glucose biosensor was reported by Shan et al., which was synthesized after immobilization of GOD (negatively charged) onto the modified electrode surface which works through the electrostatic interaction with the polyvinyl pyrrolidone protected graphene (Shan et al. 2009). The nitrogen doped graphene resulted in excellent electrocatalytic activity for the reduction of H_2O_2 as it was reported by Wang et al. which showed high sensitivity and selectivity for glucose biosensing (Wang et al. 2010). This remarkable performance observed by these derivatives for glucose detection make them promising materials for biosensors.

4.3.2 For Nicotinamide Adenine Dinucleotide (NAD+) and Dopamine Detection

Nicotinamide adenine dinucleotide (NAD+) and its reduced form (NADH) are the cofactor of many dehydrogenases and possess considerable interest for the development of amperometric biosensors, biofuel cells, and bioelectronic devices (Wang and Lin 2008). The chemically reduced GO (CRGO) modified glass carbon electrode (GCE) possess the potential capacity of reducing the oxidation peak potential of NADH in comparison to GCE and graphite/GCE as the presence of defective sites of high density on CRGO, which provided efficient electron transfer in case of biological species (Tang et al. 2009). Liu et al. further reported improved performance of a GO-based electrode for the oxidation of NADH, this is observed as a consequence of increased dispersity of GO through the noncovalent functionalization of the graphene with the methylene green (Liu et al. 2009). Pumera et al. studied the molecular dynamics of NAD+/NADH behaviour on the adsorption through the high-resolution X-ray photoelectron spectroscopy studies. It was observed that NAD+ molecules present on the edges of graphene undergo better interaction with oxygen-containing groups such as carboxylic groups, while GO edges substituted only with hydrogen undergo passivation (Pumera et al. 2009). Hence, the oxygen-containing groups play a positive role for the enhanced activity of graphene. Further GO-based materials play important role for the Dopamine (DA) detection, it works as a monoamine neurotransmitter and a hormone that is widely distributed in the central nervous system of mammals. Very slight changes in the DA concentration affects the human's health hence its rapid and sensitive detection is very important. Due to its electrochemical activity, it undergoes easy detection but ascorbic acid (AA) due to its similar oxidation potential disturbs the dopamine detection. Wang et al. also fabricated a modified GO electrode, which selectively detected dopamine and the enhanced activity was observed because of presence of phenyl structure which allows efficient π–π stacking interaction with the GO (Wang et al. 2009). Some of the other sensors like the porphyrin functionalized and carbon nanotubes GO-based biosensors for dopamine detection have been reported with excellent activity which show edge plane sites/defects in the vertical GO nanoflakes.

4.3.3 GO Based Optical Biosensors

As GO fluoresces over a wide range of wavelength from near-infrared to ultraviolet region it shows interesting optical properties due to which it can effectively lower the fluorescence tendency of other fluorescent dyes (Liu et al. 2008b). Due to these characteristics these are the suitable materials for fabricating fluorescence resonance energy transfer (FRET) sensors. Vast literatures have been reported for FRET sensors based on graphene for monitoring of the target ssDNA. This ssDNA gets efficiently adsorbed onto the surface of GO as a consequence of hydrophobic and π–π interactions among the corresponding aromatic regions of the graphene and the nucleobases (Tang et al. 2010). As the addition of a complementary sequence of DNA takes place, the formed DNA gets expelled out from the GO surface resulting in weaker interaction with graphene.

4.4 Biomedical Applications

In the 21st century, Malignant tumour adversely affected people's health, treatments like operative treatment, chemotherapy and radiotherapy were adopted but chemotherapy was the most widely adopted technique as they used most suitable chemical drugs and the drug delivery systems (DDSs) with little side effects. The drugs being consumed for cancer therapy bear poor solubility in water, low specificity and side effects which finally gives low therapeutic effect (Thomas et al. 2015; Mainardes and Silva 2004). These GO based materials when employed undergo very small side effects, high therapeutic effect, and low toxicity and work as efficient nano-carriers for the drug delivery systems (Fig. 14) overcoming some of the limitations of chemotherapy (Akhavan et al. 2012, 2014). Traditional cancer therapy like the radiation therapy and surgery effect the normal cells and healthy tissues while it can be avoided with phototherapy techniques selecting PDT and PTT as the drugs, which cures cancer with less adverse effects and reduced the systemic toxicity. GO-based materials have high photo-thermal conversion efficiency and excellent optical absorption in the NIR region, low cost and high specific area is widely used materials in phototherapy (Zhang et al. 2016). Their pharmaceutical applications are dependent on the extent of hydrophobic interactions, π–π stacking and electrostatic interactions between the graphene-based nanomaterials and drugs which are responsible for the loading of drug/gene. A smart drug carrier having optimal dosage and time needs to be developed which is effective to both internal stimuli like body temperature, pH and specific chemical reactions and external factors like light, field, and ultrasound, also the acidic and basic environment affects the drug delivery mechanism differently (Dreaden et al. 2011).

Mechanism: The primary method through which drug loading occurs on GO nanoplatforms is through straightforward physisorption, facilitated by either π–π stacking or hydrophobic interactions. Following exposure to an acidic environment, the integrity of drug molecules and their interactions with the graphene surface

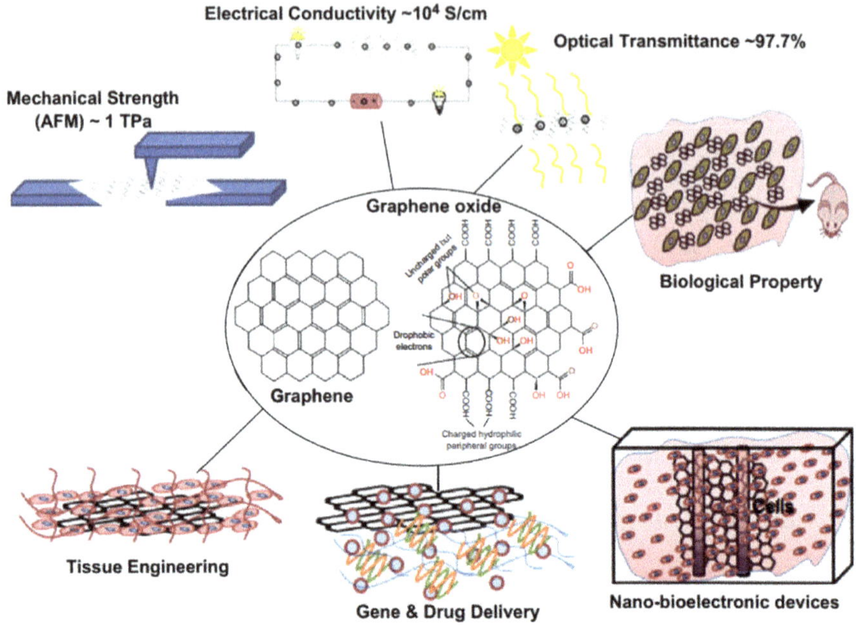

Fig. 14 A diagram illustrating the medicinal and non-medicinal applications of GBNs (Dasari Shareena et al. 2018)

are compromised due to the process of protonation, which affects hydrophobic medicines. Consequently, the phenomenon of drug release was shown to take place at a lower pH level, which might be attributed to the increased hydrophilicity and solubility of the drug (Hoseini-Ghahfarokhi et al. 2020). Various strategies, such as employing simple physisorption through π–π stacking, have been employed to load numerous hydrophobic medicines, including doxorubicin and docetaxel, onto antibodies to achieve targeted cytotoxicity against cancer cells.

4.4.1 Drug Delivery

Over the course of the previous decade, there has been a significant amount of research conducted on drug delivery systems utilizing nanomaterials. The primary objective of these investigations has been to enhance the effectiveness of cancer treatment while simultaneously minimizing the adverse consequences associated with toxicity. The GO exhibits a significantly increased surface area of 2600 m^2g^{-1} rendering it a promising candidate for investigation in the field of drug delivery (Kuila et al. 2012). In the existing body of literature, two primary alterations have been documented for drug administration utilizing GO. These modifications include chemical alteration through electrostatic contact and attachment to the aromatic molecule by π–π stacking interaction (Yang et al. 2013a, b). Another advantage of drug delivery

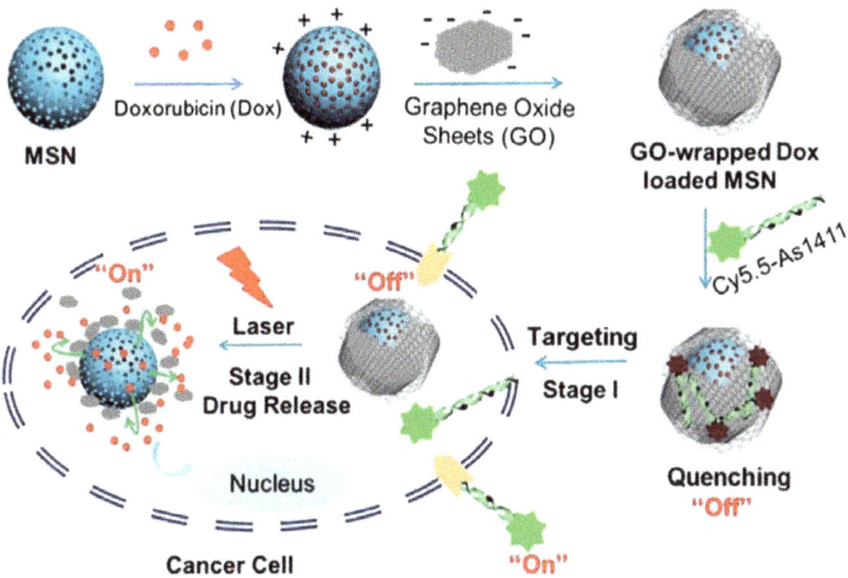

Fig. 15 GO-wrapped Dox-loaded MSN-NH2 bound and the corresponding NIR light-monitored intracellular drug release (Wu et al. 2014)

using GO is the ability to regulate the rate of drug release, hence enabling sustainable drug release (Moore et al. 2014). The researchers found that this system could release DOX from PEGylated GO at specific places in real time when triggered from the outside by glutathione (GSH). Another group of researchers found that doxorubicin (DOX)-loaded GO releases more drug at a pH of 5.3 (Wu et al. 2014), as depicted in Fig. 15.

This can be attributed to the diminished interaction between DOX and the drug carrier. The utilization of GO loaded with DOX demonstrates an augmented level of cellular toxicity and exhibits potential efficacy in inhibiting tumour growth, resulting in a substantial percentage of cell death ranging from 66 to 91%.

In order to attain a mutually reinforcing anticancer outcome, a newly developed nanoplatform consisting of graphene oxide (GO) was modified through receptor-mediated targeting and juxtaposition of activator agents. Specifically, tumor necrosis factor-related apoptosis-inducing ligand (TRAIL) was combined with DOX (Jiang et al. 2015). As depicted in Fig. 16, nanoplatforms exhibited a tendency to accumulate near the tumor location, where they proceeded to enclose the cell surface through the process of TRAIL binding to the respective receptors. The TRAIL molecule initiates caspase-dependent apoptosis following the enzymatic cleavage of the peptide linker by Furin and subsequent activation of the apoptosis trigger factor. Finally, the DOX-containing GOs are internalized into the cytoplasm, and drug release begins due to the low pH of endosomes.

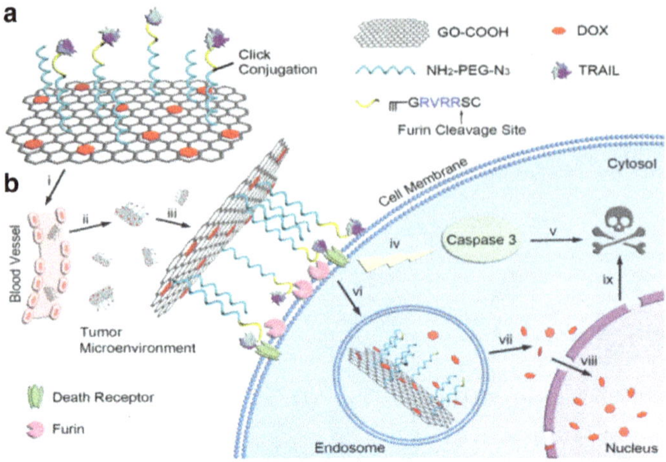

Fig. 16 Schematic representation of the TRAIL/DOX-fGO nano platform. The primary constituents of TRAIL/DOX-fGO and the sequential process of its activity, starting from its injection in the vessel to the subsequent release of the drug in the cell nucleus (Jiang et al. 2015)

Chemotherapy medicines, namely paclitaxel and methotrexate, were successfully loaded onto GO using π–π stacking and amide bonding. These drug-laden GO formulations have shown remarkable efficacy against lung cancer and breast cancer, leading to a substantial suppression of tumour development ranging from 66 to 90%. When tested on human acute lymphoblastic leukemia cell lines (CEM) and Michigan Cancer Foundation 7 cell lines (MCF-7), the biocompatibility of functionalized GO, which was made by linking NSAID ibuprofen to chitosan-functionalized GO through amide linkages, was found to be significantly higher (20%) than that of GO sheets (Rana et al. 2011). The utilization of GO, coupled with a second-generation photosensitizer known as chlorine 6 (Ce6), led to an increased concentration of GO within tumour cells. Consequently, this enhanced accumulation resulted in a better efficacy of photodynamic therapy when subjected to irradiation. Singh et al. provided a comprehensive compilation of publications pertaining to graphene-based nanomaterials (GBNs) and their use in medication delivery systems (Singh 2016).

It is predicted that by the year 2023, the number of newly diagnosed cancer cases (Sathishkumar et al. 2022) in India will reach 1,461,427. In contrast to normal tissues, tumour tissues typically exhibit distinct microstructural characteristics, a unique microenvironment, and specific physicochemical qualities. These features may include inconsistent temperature gradients, weak acidity, and overexpression of proteins and enzymes (Burger and Kipps 2006; Bissell and LaBarge 2005; Kessenbrock et al. 2010). The modification of tumour intracellular microenvironment, including the pH conditions within endosomes and lysosomes, is considered during the development of anticancer drugs that are designed to be released specifically at the intended spot. Numerous studies have made significant contributions to the development and implementation of GO based chemotherapeutics in the field of cancer treatment. The promising potential of GO in cancer therapeutics has motivated numerous researchers to explore the integration of multifunctionalities for the treatment of cancer. Furthermore, the utilization of photodynamic therapy (PDT) and photothermal therapy (PTT) in conjunction with gene ontology has been deliberated in the context of anticancer treatment. Shim et al. (2016) presented a compilation of several anticancer therapies that have been administered utilizing graphene nanosheets by either physical absorption or chemical conjugation methods. The compounds that were incorporated in the study encompassed doxorubicin, camptothecin, paclitaxel, 1,3-bis(2-chloroethyl)-1-nitrosourea, fluorouracil, methotrexate, lucanthone, b-lapachone, and ellagic acid. Numerous studies have made significant contributions to the development and administration of GO based chemotherapeutics in the context of cancer treatment. The promising potential of GO in cancer therapeutics has motivated numerous researchers to explore the integration of multifunctionalities for the purpose of treating cancer. Several classes of anticancer medicines are frequently used in conjunction with GO, and Shim et al. presented a thorough enumeration of these classes. Included are a wide range of chemotherapeutic medicines such as anthracycline antibiotics, quinolone alkaloids, taxanes, platinum complexes, nitrosourea compounds, pyrimidine analogs, polyphenolic compounds, quinone compounds, and many more (Moore et al. 2014). Zhang et al. devised a straightforward approach to produce a three-dimensional nanoscale carrier, known as GON-Cy-ALG-PEG, which exhibits biocompatibility and responsiveness to reduction99. This nanocarrier is employed for the efficient delivery of the anticancer medication DOX, with a notable capacity for high drug loading and controlled release upon triggering. The potential for enhanced efficacy in combined chemo- and photothermal therapy surpasses that of conventional therapy. The utilization of a multifunctional nanocomposite has the potential to facilitate the precise treatment and timely detection of various types of cancers. Chemotherapy and radiation therapy are prominent therapeutic modalities utilized in contemporary medicine for the management of several forms of invasive malignancies. Nevertheless, a significant drawback of chemotherapy and radiotherapy lies in their inherent lack of selectivity towards cancer cells, resulting in the destruction and potential harm to healthy tissues and organs (Adhikari et al. 2012; Zhang et al. 2011).

(a) Photothermal therapy (PTT):

In recent years, there has been a growing interest in phototherapies that utilize the distinctive optical and chemical characteristics of graphene. Graphene materials, especially GO, are much better than noble metal nanoparticles and carbon nanotubes in many ways. These include better light absorption in the near-infrared (NIR) range, better photo-thermal conversion, a larger specific area, and lower cost (Paszko et al. 2011; Shi et al. 2014). In recent years, there has been significant research conducted on the robust intrinsic optical absorbance of GO-hybrid materials, with a particular focus on their potential applications in in vivo cancer phototherapy (Zhang et al. 2016; Rahmanian et al. 2014) (Fig. 17).

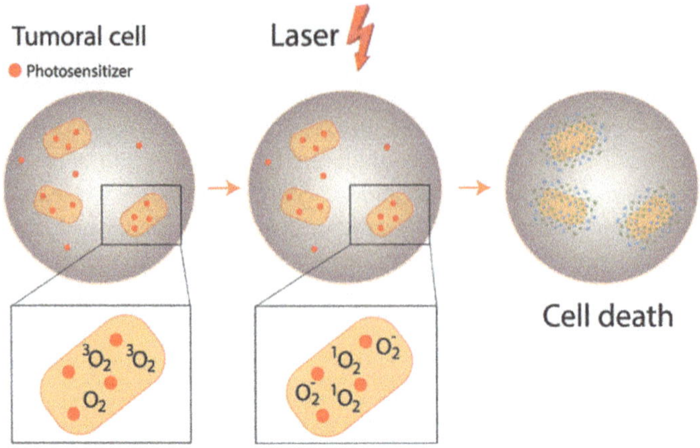

Fig. 17 Mechanism of photothermal therapy (Pinto and Pocard 2018)

Zhang et al. have successfully synthesized a nanomaterial, namely a dual-drug-loaded, doxorubicin (DOX)-loaded PEGylated nano GO (NGO–PEG (polyethylene glycol)–DOX), that possesses the capability to transfer both heat and the drug to the tumorigenic zone (Rana et al. 2011). This innovative system enables the integration of chemotherapy and photo-thermal treatment within a single platform. The results obtained in vivo provided evidence that the proposed technique exhibited superiority over the individual treatments of chemotherapy or photo-thermal therapy. In their study, Yang et al. developed a nanocomposite probe consisting of iron oxide nanoparticles (CRGO-IONP) with the purpose of integrating tumour bioimaging with photothermal therapy (PTT). This study demonstrated the successful ablation of solid tumours using an 808-nm NIR laser at a low power density of 0.5 Wcm^2, facilitated by the utilization of RGO–IONP–PEG and guided by magnetic resonance imaging (MRI). In their study, Hu et al. successfully fabricated a nanocomposite consisting of quantum-dot-tagged carbon-reduced GO (QD–CRGO). This innovative material exhibits the dual functionality of enabling bright fluorescence bioimaging of cells and tumours, as well as facilitating PTT (Wang et al. 2014a). The composite material successfully controlled the distance between the QDs and the RGO, which decreased the toxicity of the QDs and prevented the suppression of fluorescence.

The composite material's inclusion of a folic acid moiety allowed it to selectively target MCF-7 cells. Following exposure to radiation at a wavelength of 808 nm, the cells were effectively eliminated due to the thermal energy produced by the quantum dot-reduced GO (QD-RGO) material. The elevated temperature led to a significant reduction in the brightnessof the QDs, thereby serving as a method for in situ heat and temperature detection, as well as an indication of the advancement of the PTT.

The efficacy of chemo PTT was demonstrated by the testing of Au nanoribbon (AuNR)–PEG–GO nanocomposites in both in vitro and in vivo settings. A nanocomposite, denoted as AuNR–PEG–GO, was produced by mixing gold nanoribbon (AuNR) with GO. This nanocomposite was shown to be appropriate for PTT due to the significant near-infrared (NIR) absorption properties exhibited by both AuNR and GO. The incorporation of AuNR–PEG–GO composite materials should prove advantageous in facilitating the attachment of functional groups that can selectively target cancer cells.

The surfaces of AuNRs have the potential to serve as a viable platform for the attachment of proteins and other molecules, enabling the targeted binding to specific cancer cells by incorporating suitable cross-linkers (Dembereldorj et al. 2014). In their study, Wang et al. devised a novel approach including the utilization of chitosan (CS)-modified graphene nanogel for the purpose of achieving non-invasive controlled drug release. The researchers developed a drug delivery platform that is activated by near-infrared (NIR) light. This platform is based on chemically reduced GO (CRGO) that has been modified with chitosan (CS), and it is inserted into a nanogel that is sensitive to changes in temperature (thermo-sensitive). The polymer poly(N-isopropyl acrylamide) (PNIPAM) exhibited a reversible, abrupt change in its phase in aqueous solution, transitioning from a hydrophilic state to a hydrophobic one, in response to variations in temperature. This observation provided evidence for the thermo-sensitivity of PNIPAM hydrogel. The carbon-graphene nanocomposite (CGN) demonstrated a thermal effect generated by near-infrared (NIR) radiation, like that observed in crystalline grain-oriented electrical steel (CRGO). Additionally, CGN exhibited reversible thermo-responsive properties within the temperature range of 37–42 °C. Furthermore, CGN shown a high loading capacity for the drug doxorubicin (48 wt.%). The DOX-loaded CGN (DOX–CGN) exhibited a higher rate of DOX release at 42 °C compared to 37 °C. Fluorescence images confirmed the presence of DOX-CGN expression in the cytoplasm and nucleus of cancer cells when incubated at temperatures of 37 and 42 °C, respectively. A prompt and recurring release of DOX from the DOX-CGN was detected upon exposure to NIR light at a wavelength of 808 nm. In addition, it was shown that the cancer cells treated with DOX-CGN and exposed to NIR light exhibited a notably higher level of cytotoxicity compared to those without irradiation. This enhanced cytotoxicity can be attributed to the rise in temperature induced by NIR light, which in turn promotes the release of DOX within the cell nucleus. The findings of this study indicate that the application of CGN for medication release upon exposure to NIR light has great promise (Wang et al. 2013c).

Jinet et al. synthesized microcapsules of polylactic acid (PLA) modified with GO (GO–PLA), which were subsequently loaded with gold nanoparticles (AuNPs). Under the direction of ultrasound (US) and computed tomography (CT) bimodal imaging, these microcapsules were used for PTT. Following the administration of micro-capsules, US/CT imaging can provide precise measurements and spatial information regarding the tumor, while offering real-time guidance and monitoring (Jin et al. 2013). Subsequently, the diagnostic imaging data can be utilized to perform NIR laser-induced PTT without causing harm to healthy tissues. The approach exhibited considerable potential as a viable strategy for the treatment of tumours.

4.4.2 Tissue Engineering

Tissue engineering is a nascent field within the realm of life sciences that focuses on the creation of biological replacements with the aim of modifying tissue function to restore and preserve its inherent features. The biological alternatives, sometimes referred to as scaffolds, are composed of a biodegradable substance (Song et al. 2012). The conventional approach to transplantation is constrained in its ability to address tissue damage resulting from trauma, infection, tumour growth, and deformities.

The unique properties exhibited by GO, such as their remarkable elasticity, flexibility, and adaptability to both flat and irregular surfaces, make them ideal for the reinforcement of structural materials that are essential to tissue engineering. This reinforcement has the potential to enhance several aspects such as adhesion, differentiation, and cellular functionality, as supported by previous studies (Wang et al. 2011, 2014b; Qiu et al. 2012). GO can be readily changed within the class of graphene-based nanomaterials due to the presence of functional groups, such as hydroxyl, epoxy, and carboxyl, on its surface. Furthermore, GO, RGO, and other composites containing graphene exhibit a high degree of chemical modifiability due to the presence of functional groups on their surfaces. These functional groups enable effective interactions with a wide range of biological molecules, including DNA, proteins, peptides, and enzymes. In contrast, the utilization of rGO and other GO composites in tissue engineering is attributed to their inherent adaptability in fabrication processes. Biomaterials such as GO could elicit targeted biological responses, guide the development of cells, and regulate connections between cells (Li et al. 2017). The application of GO has been documented in various fields of tissue engineering, including cardiac, neural, bone, cartilage, skeletal muscle, and skin/adipose tissue engineering. According to the reports, it has been suggested that GO may possess osteogenic and neurological capabilities (Nayak et al. 2011; Park et al. 2011). The antibacterial activity of GBNs contributes to its efficacy in the field of tissue engineering, as it helps to mitigate the illnesses caused by microorganisms, hence promoting human health (Pinto and Pocard 2018). The efficacy of GO in promoting stem cell differentiation was successfully proven by researchers. The study conducted by Park et al. provided evidence for the utilization of GBNs as a substrate for stem cell culture, specifically in promoting the cardiomyogenic differentiation of mesenchymal cells. In a separate study, Shin and colleagues successfully fabricated three-dimensional multilayer tissue constructions and demonstrated robust spontaneous beating and frequency dependence when subjected to a low external electric field (Shin et al. 2016). The GO nanocomposite films create a good setting for cell growth and the production of extracellular matrix by mesenchymal stem cells (mMSCs). This helps mMSCs change into osteoblasts, which is good for bone regeneration. Golafshan et al. looked at what scaffolds do to PC12 cell cultures in their work. The results of their study showed that these structures helped PC12 cells (Golafshan et al. 2017) stick together, spread out, and make more copies of themselves. In the area of regenerative medicine, it is very important that GBNs can keep cells alive for long periods of time after they have differentiated (Pinto and Pocard 2018). In a separate study, Zhou and colleagues conducted an evaluation of graft-based therapies utilizing stem cells for the purpose of treating bone disorders (Zhou et al. 2009). The researchers have verified that the chondrogenic capabilities of cartilage cells remain intact when they are seeded on the hybrid scaffold composed of graphene-based nanomaterials (GBNs). Furthermore, they have determined that this scaffold is a good substrate for supporting the growth and development of these cells. The issue of low survival rates of mesenchymal cellswhen transplanted in myocardial tissue was addressed by Park et al. through the utilization of GO and fibronectin-reduced GO-mesenchymal stem cell (RGO-MSC) hybrids, resulting in enhanced restoration of cardiac function (Dembereldorj et al. 2014; Park et al. 2009).

5 Conclusions

In brief, GO demonstrates exceptional characteristics that render it well-suited for a wide range of biomedical applications. These include its notable colloidal stability, favourable biocompatibility, and inherent antibacterial capabilities. GO exhibits promising potential as a nanocarrier due to its ability to undergo conjugation, embedding, or loading with various substances such as medicines, proteins, metals, and biomolecules. This chapter aims to examine the characteristics of GO and explore the prospective approaches for its applications after employing both covalent and noncovalent bonding mechanisms. Through the process of functionalization, GO has been deliberately modified to possess certain properties and serve as effective nanocarriers for therapeutic biomolecules. These biomolecules include anticancer medications and genes, which may be transported and delivered to targeted sites within the body. Additionally, GO can be further modified to enable applications in phototherapy, bioimaging, biosensing, tissue engineering, and antibacterial treatments. Therefore, it is evident that nanocomposite nanocarriers based on GO hold significant promise in the realm of biomedical applications. Nevertheless, it is imperative to conduct additional investigations to thoroughly understand their impact in vivo, thereby propelling the advancement of this discipline.

Declarations

Conflicts of Interest "There are no conflicts of interest to declare by the authors."

Ethical Approval The authors hereby state that the present work is in compliance with the ethical standards.

Competing Interests The authors declare no competing interests.

Availability of data and materials Not applicable.

References

Adhikari, B., Biswas, A. and Banerjee, A., 2012. GO-based hydrogels to make metal nanoparticle-containing reduced GO-based functional hybrid hydrogels. *ACS applied materials & interfaces*, *4*(10), pp. 5472–5482.

Akhavan, O., Ghaderi, E., Aghayee, S., Fereydooni, Y. and Talebi, A., 2012. The use of a glucose-reduced GO suspension for photothermal cancer therapy. *Journal of Materials Chemistry*, *22*(27), pp. 13773–13781.

Akhavan, O., Meidanchi, A., Ghaderi, E. and Khoei, S., 2014. Zinc ferrite spinel-graphene in magneto-photothermal therapy of cancer. *Journal of Materials Chemistry B*, *2*(21), pp. 3306–3314.

Assadi, M.K., Bakhoda, S., Saidur, R. and Hanaei, H., 2018. Recent progress in perovskite solar cells. *Renewable and Sustainable Energy Reviews*, *81*, pp. 2812–2822.

Bao, H., Pan, Y., Ping, Y., Sahoo, N.G., Wu, T., Li, L., Li, J. and Gan, L.H., 2011. Chitosan-functionalized GO as a nanocarrier for drug and gene delivery. *Small*, *7*(11), pp. 1569–1578.

Bellier, N., Baipaywad, P., Ryu, N., Lee, J.Y. and Park, H., 2022. Recent biomedical advancements in GO-and reduced GO-based nanocomposite nanocarriers. *Biomaterials research*, *26*(1), p. 65.

Bera, M. and Maji, P.K., 2017. Graphene-based polymer nanocomposites: materials for future revolution. *MOJ Polymer Science*, *1*(3), p. 00013.

Bissell, M.J. and LaBarge, M.A., 2005. Context, tissue plasticity, and cancer: are tumor stem cells also regulated by the microenvironment?. *Cancer cell*, *7*(1), pp. 17–23.

Brodie, B.C., 1859. XIII. On the atomic weight of graphite. *Philosophical transactions of the Royal Society of London*, (149), pp. 249–259

Burger, J.A. and Kipps, T.J., 2006. CXCR4: a key receptor in the crosstalk between tumor cells and their microenvironment. *Blood*, *107*(5), pp. 1761–1767.

Dasari Shareena, T.P., McShan, D., Dasmahapatra, A.K. and Tchounwou, P.B., 2018. A review on graphene-based nanomaterials in biomedical applications and risks in environment and health. *Nano-micro letters*, *10*, pp. 1–34.

Dembereldorj, U., Choi, S.Y., Ganbold, E.O., Song, N.W., Kim, D., Choo, J., Lee, S.Y., Kim, S. and Joo, S.W., 2014. Gold nanorod-assembled PEGylated graphene-oxide nanocomposites for photothermal cancer therapy. *Photochemistry and photobiology*, *90*(3), pp. 659–666.

Dhar, P., Gaur, S.S., Kumar, A. and Katiyar, V., 2018. Cellulose nanocrystal templated graphene nanoscrolls for high performance supercapacitors and hydrogen storage: An experimental and molecular simulation study. *Scientific Reports*, *8*(1), p. 3886.

Dideikin, A.T. and Vul', A.Y., 2019. GO and derivatives: the place in graphene family. *Frontiers in Physics*, *6*, p. 149.

Dreaden, E.C., Mackey, M.A., Huang, X., Kang, B. and El-Sayed, M.A., 2011. Beating cancer in multiple ways using nanogold. *Chemical Society Reviews*, *40*(7), pp. 3391–3404.

Golafshan, N., Gharibi, H., Kharaziha, M. and Fathi, M., 2017. A facile one-step strategy for development of a double network fibrous scaffold for nerve tissue engineering. *Biofabrication*, *9*(2), p. 025008.

Guo, F., Kim, F., Han, T.H., Shenoy, V.B., Huang, J. and Hurt, R.H., 2011. Hydration-responsive folding and unfolding in GO liquid crystal phases. *Acs Nano*, *5*(10), pp. 8019–8025.

Han, G.S., Song, Y.H., Jin, Y.U., Lee, J.W., Park, N.G., Kang, B.K., Lee, J.K., Cho, I.S., Yoon, D.H. and Jung, H.S., 2015. Reduced GO/mesoporous TiO2 nanocomposite based perovskite solar cells. *ACS applied materials & interfaces*, *7*(42), pp. 23521–23526.

Hernandez, V., Crépin, T., Palencia, A., Cusack, S., Akama, T., Baker, S.J., Bu, W., Feng, L., Freund, Y.R., Liu, L. and Meewan, M., 2013. Discovery of a novel class of boron-based antibacterials with activity against gram-negative bacteria. *Antimicrobial agents and chemotherapy*, *57*(3), pp. 1394–1403.

Hofmann, U. and Holst, R., 1939. Über die Säurenatur und die Methylierung von Graphitoxyd. *Berichte der deutschen chemischen Gesellschaft (A and B Series)*, *72*(4), pp. 754–771.

Hoseini-Ghahfarokhi M, Mirkiani S, Mozaffari N, AbdolahiSadatlu MA, Ghasemi A, Abbaspour S, Akbarian M, Farjadian F, Karimi M. Applications of Graphene and GO in Smart Drug/Gene Delivery: Is the World Still Flat?. Int J Nanomedicine. 2020;15:9469-9496 https://doi.org/https://doi.org/10.2147/IJN.S265876

Hummers Jr, W.S. and Offeman, R.E., 1958. Preparation of graphitic oxide. *Journal of the american chemical society*, *80*(6), pp. 1339–1339.

Jana, B., Mondal, G., Biswas, A., Chakraborty, I., Saha, A., Kurkute, P. and Ghosh, S., 2013. Dual Functionalized GO Serves as a Carrier for Delivering Oligohistidine-and Biotin-Tagged Biomolecules into Cells. *Macromolecular Bioscience*, *13*(11), pp. 1478–1484.

Jiang, T., Sun, W., Zhu, Q., Burns, N.A., Khan, S.A., Mo, R. and Gu, Z., 2015. Furin-mediated sequential delivery of anticancer cytokine and small-molecule drug shuttled by graphene. *Advanced materials*, *27*(6), pp. 1021–1028.

Jin, Y., Wang, J., Ke, H., Wang, S. and Dai, Z., 2013. GO modified PLA microcapsules containing gold nanoparticles for ultrasonic/CT bimodal imaging guided photothermal tumor therapy. *Biomaterials*, *34*(20), pp. 4794–4802.

Justino, C.I., Gomes, A.R., Freitas, A.C., Duarte, A.C. and Rocha-Santos, T.A., 2017. Graphene based sensors and biosensors. *TrAC Trends in Analytical Chemistry*, *91*, pp. 53–66.

Kakavelakis, G., Maksudov, T., Konios, D., Paradisanos, I., Kioseoglou, G., Stratakis, E. and Kymakis, E., 2017. Efficient and highly air stable planar inverted perovskite solar cells with reduced GO doped PCBM electron transporting layer. *Advanced Energy Materials*, *7*(7), p. 1602120.

Kang, X., Wang, J., Wu, H., Aksay, I.A., Liu, J. and Lin, Y., 2009. Glucose oxidase–graphene–chitosan modified electrode for direct electrochemistry and glucose sensing. *Biosensors and Bioelectronics*, *25*(4), pp. 901–905.

Kausar, A., 2021. Conjugated polymer/graphene oxide nanocomposites—State-of-the-art. *Journal of Composites Science*, *5*(11), p. 292.

Kessenbrock, K., Plaks, V. and Werb, Z., 2010. Matrix metalloproteinases: regulators of the tumor microenvironment. *Cell*, *141*(1), pp. 52–67.

Kiew, S.F., Kiew, L.V., Lee, H.B., Imae, T. and Chung, L.Y., 2016. Assessing biocompatibility of GO-based nanocarriers: A review. *Journal of Controlled Release*, *226*, pp. 217–228.

Kim, F., Cote, L.J. and Huang, J., 2010. GO: surface activity and two-dimensional assembly. *Advanced Materials*, *22*(17), pp. 1954–1958.

Kuila, T., Bose, S., Khanra, P., Mishra, A.K., Kim, N.H. and Lee, J.H., 2011. Recent advances in graphene-based biosensors. *Biosensors and bioelectronics*, *26*(12), pp. 4637–4648.

Kuila, T., Bose, S., Mishra, A.K., Khanra, P., Kim, N.H. and Lee, J.H., 2012. Chemical functionalization of graphene and its applications. *Progress in Materials Science*, *57*(7), pp. 1061–1105.

Lee, D.W., Kim, T. and Lee, M., 2011. An amphiphilic pyrene sheet for selective functionalization of graphene. *Chemical Communications*, *47*(29), pp. 8259–8261.

Li, D., Liu, T., Yu, X., Wu, D. and Su, Z., 2017. Fabrication of graphene–biomacromolecule hybrid materials for tissue engineering application. *Polymer Chemistry*, *8*(30), pp. 4309–4321.

Lin, W.J., Liao, C.S., Jhang, J.H. and Tsai, Y.C., 2009. Graphene modified basal and edge plane pyrolytic graphite electrodes for electrocatalytic oxidation of hydrogen peroxide and β-nicotinamide adenine dinucleotide. *Electrochemistry Communications*, *11*(11), pp. 2153–2156.

Liu, G., Riechers, S.L., Mellen, M.C. and Lin, Y., 2005. Sensitive electrochemical detection of enzymatically generated thiocholine at carbon nanotube modified glassy carbon electrode. *Electrochemistry Communications*, *7*(11), pp. 1163–1169.

Liu, Z., Robinson, J.T., Sun, X. and Dai, H., 2008a. PEGylated nanoGO for delivery of water-insoluble cancer drugs. *Journal of the American Chemical Society*, *130*(33), pp. 10876–10877.

Liu, Z., Liu, Q., Huang, Y., Ma, Y., Yin, S., Zhang, X., Sun, W. and Chen, Y., 2008b. Organic photovoltaic devices based on a novel acceptor material: graphene. *Advanced materials*, *20*(20), pp. 3924–3930.

Liu, H., Gao, J., Xue, M., Zhu, N., Zhang, M. and Cao, T., 2009. Processing of graphene for electrochemical application: noncovalently functionalize graphene sheets with water-soluble electroactive methylene green. *Langmuir*, *25*(20), pp. 12006–12010.

Liu, Y., Yu, D., Zeng, C., Miao, Z. and Dai, L., 2010a. Biocompatible graphene oxide-based glucose biosensors. *Langmuir*, *26*(9), pp. 6158–6160.

Liu, F., Choi, J.Y. and Seo, T.S., 2010b. GO arrays for detecting specific DNA hybridization by fluorescence resonance energy transfer. *Biosensors and Bioelectronics*, *25*(10), pp. 2361–2365.

Liu, Y., Dong, X. and Chen, P., 2012. Biological and chemical sensors based on graphene materials. *Chemical Society Reviews*, *41*(6), pp. 2283–2307.

Liu, J., Bao, S. and Wang, X., 2022. Applications of graphene-based materials in sensors: A review. *Micromachines*, *13*(2), p. 184.

Mahdavi, M., Fattahi, A., Tajkhorshid, E. and Nouranian, S., 2020. Molecular insights into the loading and dynamics of doxorubicin on PEGylated GO nanocarriers. *ACS applied bio materials*, *3*(3), pp. 1354–1363.

Mainardes, R.M. and Silva, L.P., 2004. Drug delivery systems: past, present, and future. *Current drug targets*, *5*(5), pp. 449–455.

Marković, Z.M., Jovanović, S.P., Mašković, P.Z., Danko, M., Mičušík, M., Pavlović, V.B., Milivojević, D.D., Kleinová, A., Špitalský, Z. and Marković, B.M.T., 2018. Photo-induced antibacterial activity of four graphene based nanomaterials on a wide range of bacteria. *RSC advances*, *8*(55), pp. 31337–31347.

Moore, T.L., Podilakrishna, R., Rao, A. and Alexis, F., 2014. Systemic administration of polymer-coated nano-graphene to deliver drugs to glioblastoma. *Particle & Particle Systems Characterization*, *31*(8), pp. 886–894.

Nayak, T.R., Andersen, H., Makam, V.S., Khaw, C., Bae, S., Xu, X., Ee, P.L.R., Ahn, J.H., Hong, B.H., Pastorin, G. and Ozyilmaz, B., 2011. Graphene for controlled and accelerated osteogenic differentiation of human mesenchymal stem cells. *ACS nano*, *5*(6), pp. 4670–4678.

Park, S., An, J., Jung, I., Piner, R.D., An, S.J., Li, X., Velamakanni, A. and Ruoff, R.S., 2009. Colloidal suspensions of highly reduced GO in a wide variety of organic solvents. *Nano letters*, *9*(4), pp. 1593–1597.

Park, S.Y., Park, J., Sim, S.H., Sung, M.G., Kim, K.S., Hong, B.H. and Hong, S., 2011. Enhanced differentiation of human neural stem cells into neurons on graphene. *Advanced materials*, *23*(36), pp. H263–H267.

Paszko, E., Ehrhardt, C., Senge, M.O., Kelleher, D.P. and Reynolds, J.V., 2011. Nanodrug applications in photodynamic therapy. *Photodiagnosis and photodynamic therapy*, *8*(1), pp. 14–29.

Pinto, A. and Pocard, M., 2018. Photodynamic therapy and photothermal therapy for the treatment of peritoneal metastasis: a systematic review. *Pleura and peritoneum*, *3*(4), p. 20180124.

Prasad, R., Yadav, A.S., Gorain, M., Chauhan, D.S., Kundu, G.C., Srivastava, R. and Selvaraj, K., 2019. GO supported liposomes as red emissive theranostics for phototriggered tissue visualization and tumor regression. *ACS Applied Bio Materials*, *2*(8), pp. 3312–3320.

Pumera, M., Scipioni, R., Iwai, H., Ohno, T., Miyahara, Y. and Boero, M., 2009. A mechanism of adsorption of β-nicotinamide adenine dinucleotide on graphene sheets: Experiment and theory. *Chemistry–A European Journal*, *15*(41), pp. 10851–10856.

Purkait, M.K. and Sontakke, A.D., 2023. Carbon-Based Nanocarriers for Drug Delivery. CRC Press.

Qiu, L., Liu, J.Z., Chang, S.L.Y., Wu, Y., Li, D. Biomimetic superelastic graphene-based cellular monoliths. Nat. Commun. 3, 1241 (2012). https://doi.org/10.1038/ncomms2251

Rahmanian, N., Hamishehkar, H., Dolatabadi, J.E.N. and Arsalani, N., 2014. Nano GO: a novel carrier for oral delivery of flavonoids. *Colloids and Surfaces B: Biointerfaces*, *123*, pp. 331–338.

Rana, V.K., Choi, M.C., Kong, J.Y., Kim, G.Y., Kim, M.J., Kim, S.H., Mishra, S., Singh, R.P. and Ha, C.S., 2011. Synthesis and drug-delivery behavior of chitosan-functionalized GO hybrid nanosheets. *Macromolecular Materials and Engineering*, *296*(2), pp. 131–140.

Rathi, K. and Pal, K., 2017. Impact of doping on GO: Fast response–recovery humidity sensor. *ACS omega*, *2*(3), pp. 842–851.

Sathishkumar, K., Chaturvedi, M., Das, P., Stephen, S. and Mathur, P., 2022. Cancer incidence estimates for 2022 & projection for 2025: Result from national cancer Registry Programme, India. *The Indian journal of medical research, 156*(4–5), p. 598.

Shams, E., Yeganeh, H., Naderi-Manesh, H., Gharibi, R. and Mohammad Hassan, Z., 2017. Polyurethane/siloxane membranes containing GO nanoplatelets as antimicrobial wound dressings: in vitro and in vivo evaluations. *Journal of Materials Science: Materials in Medicine, 28*, pp. 1–14.

Shan, C., Yang, H., Song, J., Han, D., Ivaska, A. and Niu, L., 2009. Direct electrochemistry of glucose oxidase and biosensing for glucose based on graphene. *Analytical chemistry, 81*(6), pp. 2378–2382.

Shao, Y., Zhang, S., Engelhard, M.H., Li, G., Shao, G., Wang, Y., Liu, J., Aksay, I.A. and Lin, Y., 2010. Nitrogen-doped graphene and its electrochemical applications. *Journal of Materials Chemistry, 20*(35), pp. 7491–7496.

Shen, J., Shi, M., Yan, B., Ma, H., Li, N., Hu, Y. and Ye, M., 2010. Covalent attaching protein to GO via diimide-activated amidation. *Colloids and Surfaces B: Biointerfaces, 81*(2), pp. 434–438.

Sheng, L., Li, Z., Meng, A. and Xu, Q., 2018. Ultrafast responsive and highly sensitive enzyme-free glucose sensor based on a novel Ni (OH) 2@ PEDOT-rGOnanocomposite. *Sensors and Actuators B: Chemical, 254*, pp. 1206–1215.

Shi, S., Chen, F., Ehlerding, E.B. and Cai, W., 2014. Surface engineering of graphene-based nanomaterials for biomedical applications. *Bioconjugate chemistry, 25*(9), pp. 1609–1619.

Shim, G., Kim, M.G., Park, J.Y. and Oh, Y.K., 2016. Graphene-based nanosheets for delivery of chemotherapeutics and biological drugs. *Advanced drug delivery reviews, 105*, pp. 205–227.

Shin, S.R., Li, Y.C., Jang, H.L., Khoshakhlagh, P., Akbari, M., Nasajpour, A., Zhang, Y.S., Tamayol, A. and Khademhosseini, A., 2016. Graphene-based materials for tissue engineering. *Advanced drug delivery reviews, 105*, pp. 255–274.

Singh, Z., 2016. Applications and toxicity of graphene family nanomaterials and their composites. *Nanotechnology, science and applications*, pp. 15–28.

Sinha, A., Wu, L., Lu, X., Chen, J. and Jain, R., 2018. Advances in sensing and biosensing of bisphenols: A review. *Analyticachimicaacta, 998*, pp. 1–27.

Smith, A.T., LaChance, A.M., Zeng, S., Liu, B. and Sun, L., 2019. Synthesis, properties, and applications of GO/reduced GO and their nanocomposites. *Nano Materials Science, 1*(1), pp. 31–47.

Song, P., Zhang, X., Sun, M., Cui, X. and Lin, Y., 2012. GO modified TiO 2 nanotube arrays: enhanced visible light photoelectrochemical properties. *Nanoscale, 4*(5), pp. 1800–1804.

Sontakke, A.D., Tiwari, S. and Purkait, M.K., 2023. A comprehensive review on GO-based nanocarriers: Synthesis, functionalization and biomedical applications. *FlatChem*, p. 100484.

Staudenmaier, L., 1898. Process for the preparation of graphitic acid. *Reports of the German Chemical Society , 31* (2), pp. 1481–1487.

Tabish, T.A., 2018. Graphene-based materials: The missing piece in nanomedicine?. *Biochemical and biophysical research communications, 504*(4), pp. 686–689.

Tang, L., Wang, Y., Li, Y., Feng, H., Lu, J. and Li, J., 2009. Preparation, structure, and electrochemical properties of reduced graphene sheet films. *Advanced Functional Materials, 19*(17), pp. 2782–2789.

Tang, Z., Wu, H., Cort, J.R., Buchko, G.W., Zhang, Y., Shao, Y., Aksay, I.A., Liu, J. and Lin, Y., 2010. Constraint of DNA on functionalized graphene improves its biostability and specificity. *small, 6*(11), pp. 1205–1209.

Thomas, R.G., Moon, M.J., Lee, H., Sasikala, A.R.K., Kim, C.S., Park, I.K. and Jeong, Y.Y., 2015. Hyaluronic acid conjugated superparamagnetic iron oxide nanoparticle for cancer diagnosis and hyperthermia therapy. *Carbohydrate polymers, 131*, pp. 439–446.

Valishina, A., Burganova, R., Petrova, A., Nedopekin, O. and Tayurskii, D., 2018, November. Ab initio study of electronic properties of graphene/MoS2 heterostructure under biaxial deformations. In *AIP Conference Proceedings* (Vol. 2041, No. 1). AIP Publishing.

Wang, J. and Lin, Y., 2008. Functionalized carbon nanotubes and nanofibers for biosensing applications. *TrAC Trends in Analytical Chemistry*, *27*(7), pp. 619–626.
Wang, Y., Li, Y., Tang, L., Lu, J. and Li, J., 2009. Application of graphene-modified electrode for selective detection of dopamine. *Electrochemistry communications*, *11*(4), pp. 889–892.
Wang, Y., Shao, Y., Matson, D.W., Li, J. and Lin, Y., 2010. Nitrogen-doped graphene and its application in electrochemical biosensing. *ACS nano*, *4*(4), pp. 1790–1798.
Wang, Y., Li, Z., Wang, J., Li, J. and Lin, Y., 2011. Graphene and GO: biofunctionalization and applications in biotechnology. *Trends in biotechnology*, *29*(5), pp. 205–212.
Wang, J., Wei, Y., Shi, X. and Gao, H., 2013a. Cellular entry of graphene nanosheets: the role of thickness, oxidation and surface adsorption. *Rsc Advances*, *3*(36), pp. 15776–15782.
Wang, Y., Zhang, P., Liu, C.F. and Huang, C.Z., 2013b. A facile and green method to fabricate graphene-based multifunctional hydrogels for miniature-scale water purification. *RSC advances*, *3*(24), pp. 9240–9246.
Wang, C., Mallela, J., Garapati, U.S., Ravi, S., Chinnasamy, V., Girard, Y., Howell, M. and Mohapatra, S., 2013c. A chitosan-modified graphene nanogel for noninvasive controlled drug release. *Nanomedicine: Nanotechnology, Biology and Medicine*, *9*(7), pp. 903–911.
Wang, X., Sun, X., Lao, J., He, H., Cheng, T., Wang, M., Wang, S. and Huang, F., 2014a. Multifunctional graphene quantum dots for simultaneous targeted cellular imaging and drug delivery. *Colloids and Surfaces B: Biointerfaces*, *122*, pp. 638–644.
Wang, L., Lu, C., Zhang, B., Zhao, B., Wu, F. and Guan, S., 2014b. Fabrication and characterization of flexible silk fibroin films reinforced with GO for biomedical applications. *RSC advances*, *4*(76), pp. 40312–40320.
Wang, X., Liu, Z., Ye, X., Hu, K., Zhong, H., Yuan, X., Xiong, H. and Guo, Z., 2015. A facile one-pot method to two kinds of GO-based hydrogels with broad-spectrum antimicrobial properties. *Chemical Engineering Journal*, *260*, pp. 331–337.
Wang, X., Sun, Q., Cui, C., Li, J. and Wang, Y., 2018a. Anti-HER2 functionalized GO as survivin-siRNA delivery carrier inhibits breast carcinoma growth in vitro and in vivo. *Drug design, development and therapy*, pp. 2841–2855.
Wang, S., Yang, L., Wang, Q., Fan, Y., Shang, J., Qiu, S., Li, J., Zhang, W. and Wu, X., 2018b. Supramolecular self-assembly of layer-by-layer graphene film driven by the synergism of π–π and hydrogen bonding interaction. *Journal of Photochemistry and Photobiology A: Chemistry*, *355*, pp. 249–255.
Wang, H., Chen, M., Jin, C., Niu, B., Jiang, S., Li, X. and Jiang, S., 2018c. Antibacterial [2-(methacryloyloxy) ethyl] trimethylammonium chloride functionalized reduced GO/poly (ethylene-co-vinyl alcohol) multilayer barrier film for food packaging. *Journal of agricultural and food chemistry*, *66*(3), pp. 732–739.
Williams, G., Seger, B. and Kamat, P.V., 2008. TiO_2-graphene nanocomposites. UV-assisted photocatalytic reduction of GO. *ACS nano*, *2*(7), pp. 1487–1491.
Wu, H., Lu, W., Shao, J.J., Zhang, C., Wu, M.B., Li, B.H. and Yang, Q.H., 2014. Ph-dependent size, surface chemistry and electrochemical properties of GO. *Carbon*, *67*, p. 795.
Xia, M.Y., Xie, Y., Yu, C.H., Chen, G.Y., Li, Y.H., Zhang, T. and Peng, Q., 2019. Graphene-based nanomaterials: the promising active agents for antibiotics-independent antibacterial applications. *Journal of Controlled Release*, *307*, pp. 16–31.
Xie, M., Lei, H., Zhang, Y., Xu, Y., Shen, S., Ge, Y., Li, H. and Xie, J., 2016. Non-covalent modification of GO nanocomposites with chitosan/dextran and its application in drug delivery. *RSC advances*, *6*(11), pp. 9328–9337.
Yang, K., Feng, L., Shi, X. and Liu, Z., 2013a. Nano-graphene in biomedicine: theranostic applications. *Chemical Society Reviews*, *42*(2), pp. 530–547.
Yang, K., Gong, H., Shi, X., Wan, J., Zhang, Y. and Liu, Z., 2013b. In vivo biodistribution and toxicology of functionalized nano-GO in mice after oral and intraperitoneal administration. *Biomaterials*, *34*(11), pp. 2787–2795.

Yaragalla, S., Anilkumar, G., TV, V., Kalarikkal, N. and Thomas, S., 2015. Preparation of epoxy graphene and its structural and optical properties. *Advanced Materials Letters*, *6*(10), pp. 848–852.

Yazyev, O.V. and Louie, S.G., 2010. Electronic transport in polycrystalline graphene. *Nature materials*, *9*(10), pp. 806–809.

Yi, X. and Gao, H., 2015. Cell interaction with graphenemicrosheets: near-orthogonal cutting versus parallel attachment. *Nanoscale*, *7*(12), pp. 5457–5467.

Yu, H.W., Kim, H.K., Kim, T., Bae, K.M., Seo, S.M., Kim, J.M., Kang, T.J. and Kim, Y.H., 2014. Self-powered humidity sensor based on GO composite film intercalated by poly (sodium 4-styrenesulfonate). *ACS applied materials & interfaces*, *6*(11), pp. 8320–8326.

Yu, H., Zhang, B., Bulin, C., Li, R. and Xing, R., 2016a. High-efficient synthesis of graphene oxide based on improved hummers method. *Scientific reports*, *6*(1), p. 36143.

Yu, P., Tian, Z., Lowe, S.E., Song, J., Ma, Z., Wang, X., Han, Z.J., Bao, Q., Simon, G.P., Li, D. and Zhong, Y.L., 2016b. Mechanically-assisted electrochemical production of graphene oxide. *Chemistry of Materials*, *28*(22), pp. 8429–8438.

Yu, W., Sisi, L., Haiyan, Y. and Jie, L., 2020. Progress in the functional modification of graphene/graphene oxide: A review. *RSC advances*, *10*(26), pp. 15328–15345.

Yun, Y., 2016. Electrochemical sensor for ultrasensitive determination of bisphenol a based on gold nanoparticles/β-cyclodextrin functionalized reduced GO nanocomposite. *Int J ElectrochemSci*, *11*(4), pp. 2778–2789.

Zhang, W., Guo, Z., Huang, D., Liu, Z., Guo, X. and Zhong, H., 2011. Synergistic effect of chemo-photothermal therapy using PEGylated GO. *Biomaterials*, *32*(33), pp. 8555–8561.

Zhang, Y., Wu, C., Guo, S. and Zhang, J., 2013. Interactions of graphene and GO with proteins and peptides. *Nanotechnology Reviews*, *2*(1), pp. 27–45.

Zhang, B., Wang, Y. and Zhai, G., 2016. Biomedical applications of the graphene-based materials. *Materials Science and Engineering: C*, *61*, pp. 953–964.

Zhang, L., Xia, J., Zhao, Q., Liu, L. and Zhang, Z., 2010a. Functional GO as a nanocarrier for controlled loading and targeted delivery of mixed anticancer drugs. *small*, *6*(4), pp. 537–544.

Zhao, C., Deng, B., Chen, G., Lei, B., Hua, H., Peng, H. and Yan, Z., 2016. Large-area chemical vapor deposition-grown monolayer graphene-wrapped silver nanowires for broad-spectrum and robust antimicrobial coating. *Nano Research*, *9*, pp. 963–973.

Zhou, M., Zhai, Y. and Dong, S., 2009. Electrochemical sensing and biosensing platform based on chemically reduced GO. *Analytical chemistry*, *81*(14), pp. 5603–5613.

Zhu, Y., Murali, S., Cai, W., Li, X., Suk, J.W., Potts, J.R. and Ruoff, R.S., 2010. Graphene and GO: synthesis, properties, and applications. *Advanced materials*, *22*(35), pp. 3906–3924.

Zhu, C., Liu, L., Fan, M., Liu, L., Dai, B., Yang, J. and Sun, D., 2014. Microbial oxidation of graphite by Acidithiobacillus ferrooxidans CFMI-1. *RSC advances*, *4*(98), pp. 55044–55047.

徐志远, 朱绍佳, 王明伟, 李永军, 史萍 and 黄晓宇, 2015. Delivery of Paclitaxel Using PEGylated GO as a Nanocarrier.

Tribological Study of Nano Copper-Hydroxyapatite Polyoxymethylene (nCuHAp POM) for Biomedical Applications

Anik Banerjee, Tanveer Ahamed Rohit, Rajeswar Bandyopadhyay, and Piyush Uniyal

1 Introduction

Hydroxyapatite (HAp) is one of the most widely used bioceramics for teeth and bone fillers. Because of the similarity with the composition of bones, HAp is considered biocompatible and proved to be osteoconductive, facilitating bone cell growth (Nayak 2010). The porous nature of HAp further improves its biocompatibility. These pores facilitate the osteo cell growth and bone remodeling. Hydroxyapatite can be used as a porous block, hybrid composites, and powders to fill bone voids and defects. Hydroxyapatite can be produced from various sources. Natural sources rich in calcium are widely used for its preparation such as eggshells, coral, and seashells. Eggshells are rich in calcium and readily available and chosen as the source (Rivera et al. 1999). Various methods can produce hydroxyapatite, including sol–gel, hydrothermal, wet precipitation, and dry method (Vecchio et al. 2007). Among these, the best method is

Anik Banerjee, Tanveer Ahamed Rohit, Rajeswar Bandyopadhyay these authors contributed equally to this work and share first authorship.

A. Banerjee (✉)
Department of Aerospace Engineering and Applied Mechanics, Indian Institute of Engineering Science and Technology, Shibpur, Howrah, West Bengal, India
e-mail: anikbanerjee.phd@gmail.com

T. A. Rohit
IFB Industries Limited, Verna Industrial Estate, Plot-N 76, Phase 4, Verna, Goa 403722, India

R. Bandyopadhyay
B32, Infosys, Electronic City Phase 1, Bengaluru, Karnataka 560100, India

P. Uniyal (✉)
Center for Biomedical Engineering, Indian Institute of Technology Ropar, Rupnagar, Punjab 140001, India
e-mail: piyu.uniyal@gmail.com

J. Gupta and A. Verma (eds.), *Microbiology-2.0 Update for a Sustainable Future*,
https://doi.org/10.1007/978-981-99-9617-9_12

the wet precipitation method, as water is the by-product, the scopes of contamination are minimal (Granizo et al. 2018), and the hydroxyapatite yield is relatively high. The additional advantage includes that this process is quite simple and not harmful to the environment (Ansari et al. 2011).

One of the most crucial substances in the human body is the trace elements responsible for immunity, metabolism, endocrine function, and human growth (Alvira et al. 2010). Trace elements found in bones and teeth, apart from giving information regarding the eating habits and geographic conditions of humans (Guzmán-Venegas et al. 2018), also give an image of health conditions, especially oral health (Dolphin et al. 2013). It has been found that metal trace elements like Iron (Fe), Copper (Cu), Strontium (Sr), and Zinc (Zn) present can prevent many dental diseases and has an anti-bacterial property. Thus, making these trace elements very vital (Báez et al. 2004). The low levels of these trace elements lead to osteoporosis, which causes porosity and weakness in bones. At multiple points, a shallow serum of copper, manganese, and zinc level showed speedy healing (Rath et al. 2000). In the present study, the trace element copper has been chemically incorporated in HAp and the composite was prepared by reinforcing the copper hydroxyapaptite (CuHAp). This study further ball milled the sample to produce nano CuHAp (nCuHAp).

Polyoxymethylene (POM) exhibits superior mechanical properties, high yield stress, stiffness and resistance to fatigue. Previous studies have described novel composites of POM and its extensive use in biomedical applications due to its biocompatibility (Banerjee et al. 2022). Mercuri et al. 2015 have successfully conducted a total temporomandibular joint (TMJ) replacement with polyoxymethylene TMJ implant. This study has also validated successful bone remodeling and osseointegration of the implant-bone construct. Apart from POM, ultra-high-molecular-weight polyethylene (UHMWPE) is another widely used biopolymer in TMJ implants. Shen et al. 2008, have compared both the material for joint replacement and claimed POM as more suitable for joint replacement due to its low wear rate. In this study, the base matrix for the proposed composite is considered to be POM. The composite pin was prepared by injection molding. nCuHAp was taken in five different percentages—1, 2, 3, 4 and 5% and reinforced in the POM polymer matrix. Five pins of each percentage were fabricated to conduct the tribology test in the pin-on-disc tribometer. Investigations have held bio-compatible materials suitable to adjust with the physiological environment and function daily activities similar to bone. Co-Cr-Mo, Ti-4V-6Al, and AISI 316L are some of these materials classified to perform the function. Multiple researchers have conducted tribological studies with the corresponding materials.

Multiple tests are performed to validate each parameter in this study and results are compared with previous similar findings. Scanning Electron Microscope and XRD Imaging of nCuHAp are conducted to obtain a suitable size and property of nCuHAp. To confirm the mechanical property of the fabricated pin, this study conducted a hardness test. To confirm the wear and study the pattern, roughness test was conducted for the pin surface in contact with the disc. Porosity test was conducted to study the nature of the cracks and its relationship with the wear. This study, based on previous studies, conducted a lymphocyte cell proliferation test to investigate the toxicity. The test was conducted on all the six composite pins with five different HAp percentages considered in the study including a pure POM sample.

In this study, nCuHAp was prepared from eggshells through chemical synthesis. The novelty of this study is developing a new biomaterial composite for its uses in the total replacement of articulating joints like total TMJ replacement. The study, to the best of its knowledge, confirms the validity of such a composite with POM, hydroxyapatite, and trace element-Cu. This study is not a comparative study but instead proposes a new composite which will be beneficial for orthopaedic research. The authors also confirm the validation of individual testing carried out to present the composite as a suitable material for TMJ implant. Each test has been followed as per the standard and the authors provide details on its success.

2 Materials and Methods

2.1 *Synthesis of nCuHAp*

The synthesis of nCuHAp was done as per the described in the literature (Lu et al. 2019). It involved wet precipitation using eggshells. Thus, the presence of copper apart from calcium also had a significant effect. Eggshells consist of calcium carbonate and must be thermally reduced to calcium oxide, further used in nCuHAp synthesis. First, the eggshells were washed to remove dirt and impurities. The eggshells were mechanically crushed to thermally treat the compound inside a muffle furnace to produce the calcium precursor. It was isothermally heated at a maximum of 900 °C for 6 h in a muffle furnace (Rivera et al. 1999).

Calcium hydroxide was then prepared by mixing calcium oxide and water in a ratio of 1:25. Copper sulfate was then added to calcium hydroxide by maintaining a stoichiometric ratio of (Cu + Ca)/P of 1.67. The doping of copper ions was 5% to Calcium ions and was then stirred thoroughly to have a homogenous mixture. A uniform temperature of 60 °C was maintained while adding diluted phosphoric acid dropwise to this solution until a pH of 7 was achieved. Then, sodium hydroxide was added after removing the heat until a pH of 11 was achieved. This was kept undisturbed for 24 h for the precipitate to settle down. The precipitate was then decanted and washed with distilled water several times until the pH was brought down to 7. After that, the precipitate was obtained using filter paper and kept for drying in a petri dish. It was then sintered inside the muffle furnace at 900 °C and cooled to obtain the nCuHAp. It was ball milled for 3 min to obtain fine nCuHAp powder. The composition was confirmed by X-Ray Diffractometer test of the sample (Fig. 1a).

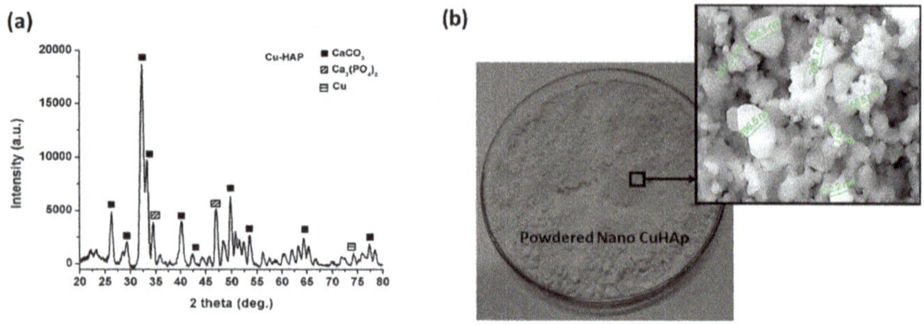

Fig. 1 **a** Plot of X-ray diffractometer (Bruker, D8 Advance, Billerica, Massachusetts, USA) imaging of nano copper hydroxyapatite (nCuHAp) obtained after synthesis, **b** nano particle of CuHAp under Scanning electron microscope (FEI Quanta FEG 200: Eindhoven, Netherlands; Detector type: PIXcel 1D-Medipix3 Detector)

2.2 *Fabrication of nCuHAp Composite Sample Pin*

The POM pin is fabricated with nCuHAp as per the weight percentage ratios. DSC (Differential Scanning Calorimetry) is carried out for the polymer in an aluminum crucible to ensure the material and measure the melting point. Figure 2d and e demonstrate a few cylindrically HAP pins with a height of 35 and 10 mm diameter. Injection molding is done according to the standards (500 bar pressure and 170 °C temperature. Six pins were fabricated for each material concentration (1, 2, 3, 4 and 5%) including six samples of pure POM.

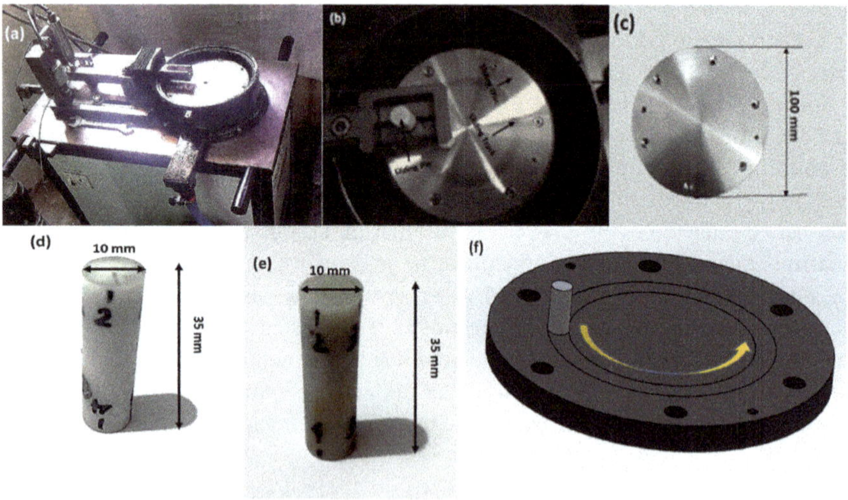

Fig. 2 **a** Pin-on-disc experimental setup, **b** closed view of the pin sliding on the disc, **c** 316L stainless steel metal plate, **d** 0% nCuHAp POM(C) or Pure POM(C) Pin, **e** 2% nCuHAp POM Pin, **f** CAD model of the entire setup

2.3 Tribological Test of nCuHAp Composite Pin

The tribological testing was carried out in Pin on Disc Tribometer (Ducom, Model TR20) according to ASTM G-99 standards. A friction contact testing was conducted between the disc and composite pins. The time of the experiment for each loading condition is 60 min. The Load and Speed were taken as fixed variables with an average load of 80N and 156 RPM speed. The plane of the pin is rotated about its axis in four different alignments keeping the same plane in contact with the disc Fig. 2f. The alignment was fixed at four different angles on the plane—90°, 180°, 270° and 360°. Each sample testing was done for 1 h 15 min for each orientation. The orientation was changed to 90° every 15 min and the wear debris remained intact during the change. Six samples of each HAp concentrated composite were fabricated. The average wear rate and COF of the six samples are calculated by the tribometer software (V 1.0.0, WINDUCOM, Bengaluru, India). Individual specimens were polished before every testing and the complete setup was cleaned with acetone after 1-h 15 min. All the tribological tests were carried out in dry conditions. An AISI 316L Stainless steel disc of dimensions 100 mm in diameter and 7 mm thickness was used in the study due to its wide variety of biomedical applications. The experimental instruments were described in Fig. 2a, b.

2.4 Measuring the Hardness, Microporosity, and Roughness of the Composite Pin

The individual pin's shore hardness was found in shore-D-hardness Durometer (Model: DIN 53 505). The disc's characterization was carried out using Optical Emission Spectrometer (EQ ID-OIAL/MMN/CH-01). The micro-porosity for each sample was observed under Metallurgical Microscope-OIAL/MMN/MET/01 at 100X as per ASTM E3-11, ASTM E407-7 (RA15) & ASM HANDBOOK VOL. 9-04. The arithmetic means deviation of the roughness profile and maximum height was taken in a non-contact 3D surface profilometer (Taylor Hobson, Leicester, UK).

2.5 Toxicity Test of nCuHAp Composite Pin

Cell Proliferation capability of human lymphocytes was tested in vitro in the presence of pure POM and the composites of nCuHAp-POM (1–5%) and nano nCuHAp (1–5%) in a composition. The composition of 10% FBS in RPMI solution consists of RPMI-1640—440 ml, FBS—50 ml and Penicillin/Streptomycin—10 ml. FBS is the nutrient-providing medium in cell culture techniques, RPMI solution contains the human lymphocyte cells for culture, and streptomycin (or penicillin) is also added to prevent microbial growth in the test samples that may hinder the culture process.

All the test samples were first UV sanitized to kill any bacteria or microorganisms present in them. After that, the 10% FBS in the RPMI solution was prepared. Each test sample was kept under an equal amount of the solution containing an equal number of cells. All the test samples were kept in the media for 48 h at an ambient temperature of 25.7 °C. After the due time, the samples were observed under an optical microscope at 10X zoom, and the cell count in the respective test samples was obtained from the light-microscopy image. The Optical Density (at 570 nm—absorbance) acts as a measure of biofilm formation.

3 Result and Discussion

3.1 Characterization of nCuHAp and POM

The composition of nCuHAp powder was characterized using the Scanning electron microscope (SEM) (FEI Quanta FEG 200: Eindhoven, Netherlands; Detector type: PIXcel 1D-Medipix3 Detector). The constituent phases were also examined. The sintered copper hydroxyapatite powder was examined in an X-Ray Diffractometer (XRD) (Bruker, D8 Advance, Billerica, Massachusetts, USA) in Fig. 1a. The XRD profile was found to be $Cu\text{-}Ca_{10}(PO_4)_6(OH)_2$, the chemical naming for nCuHAp and matched the JCPDS 72-1243 file number. The polymer POM was examined in Differential Scanning Calorimeter (DSC 214 Polyma) to confirm the melting point and material characteristics. This result has shown the melting point of the polymer at 161.8 °C. The POM was processed in the injection molding to make a cylindrical composite pin. The nanoparticle size of the copper hydroxyapatite after ball-milling was confirmed with a scanning electron microscope image, as shown in Fig. 1b.

3.2 Study the COF and Wear Rate of the Sample

The results for the tribological experiments were tabulated for each experiment. Figure 3a depicts a bar graph drawn between the coefficient of friction (COF) and respective nCuHAp percentage in the composite. The lowest COF was 0.18 at 1%, followed by 0.2 at 2% nCuHAp. There is a rise at 3% and 4%, 0.22 and 0.23 respectively and a sudden dip of 0.20 at 5%. The COF of pure POM is 0.2. Figure 3b was a plot between specific wear rates for each percentage of nCuHAp in the composite. The present work showed a gradual decrease in wear rate from 0 to 2%, and constant for 3% and 4% as depicted in Fig. 3b, followed by a sharp rise at 5%. The pure POM accounted for 3.3 m^2/N and 2.8 m^2/N for 1% nCuHAp in POM. 2% counted the least with 1.8 m^2/N while 3% and 4% measured 2.4 m^2/N and 2.5 m^2/N. On the other hand, 5% nCuHAp counted 3.1 m^2/N.

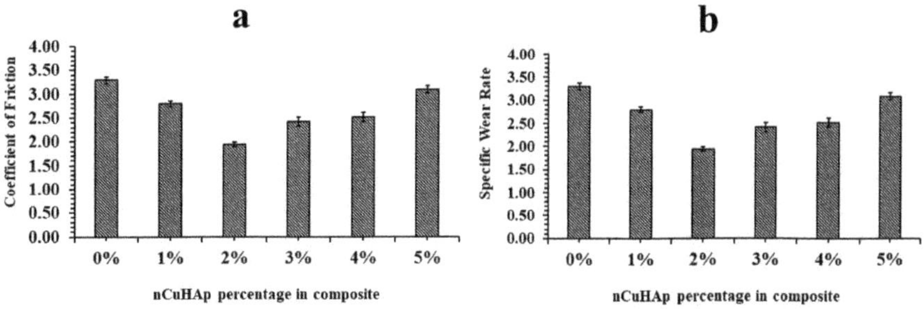

Fig. 3 **a** Plot of coefficient of friction (COF) with five different nCuHAp percentage in POM composite and pure POM, and **b** plot of specific wear rate with five different nCuHAp percentage in composite and pure POM

3.3 Study the Roughness, Microporosity, and Hardness of Composite

The topographical roughness variation of the composite pin surfaces in contact with the 316L stainless steel plate. Arithmetic mean deviation of the roughness profile is the average roughness (Ra), and the maximum height of roughness profile (Rz) is the difference between the peak and valley of the roughness curve along a given direction on the surface plane. According to Table 1, the Ra increased from 0 to 2% and then decreased to 5%, at which both Ra and Rz are maximum. This shows that 2% and 4% have relatively better surface characteristics than 5%. The Ra value for pure POM is 0.439 μm. The value for 2%, 4% and 5% is 0.621 μm, 0.527 μm and 1.569 μm respectively. The height at 5% was maximum at 10.627 μm, and 0%, 2% and 4% counting 3.254 μm, 3.27 μm, and 10.627 μm. The profile for the respective composite is depicted in Fig. 4.

Table 2 describes the Shore-D-hardness for each percentage of nCuHAp in the POM matrix. The hardness was comparable and didn't show any specific characteristics. Although 1% counted the least count of 76, it is almost constant from 2 to 5%. Figure 5 describes the micro-voids generally occurring in between the polymer fibers. The size of the micro-void in pure POM to 5% nCuHAp-POM was observed to have a subsequent growth in the size of porosity between 2 and 20 μm. Although the trend is non-uniform, 1%, 2%, 3%, and 4% accounted for 12 μm, 2 μm, 4 μm, and 12 μm respectively.

Table 1 Non-contact surface roughness (Ra and Rz) for nCuHAp in POM composite pins

Criteria	0%	1%	2%	5%
Ra: Arithmetic mean deviation of the roughness profile (μm)	1.679	0.746	0.615	1.328
Rz: Maximum height of roughness profile (μm)	6.881	4.217	3.788	6.841

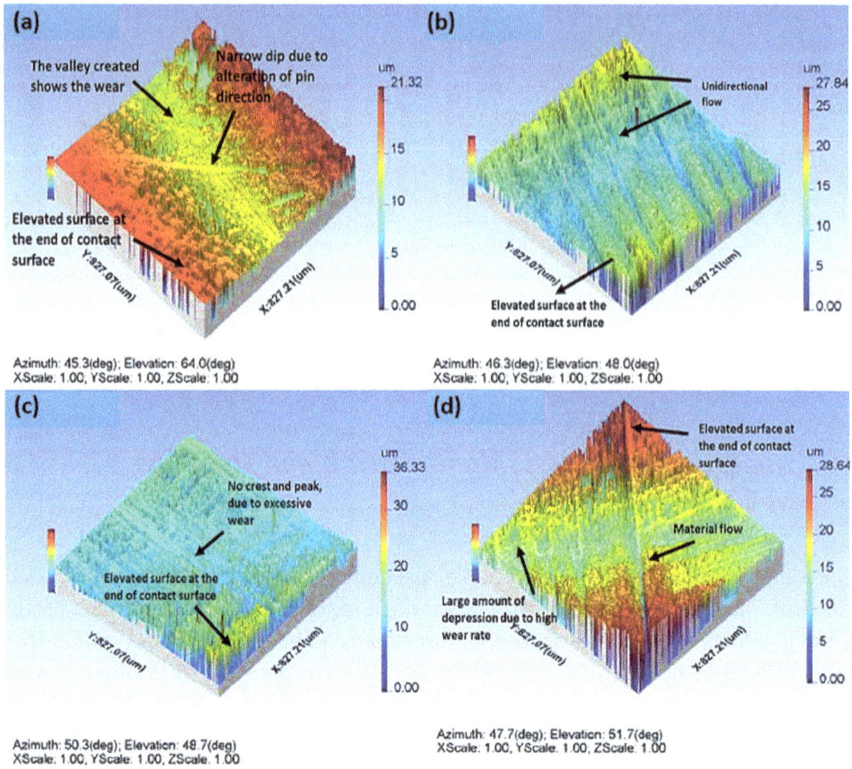

Fig. 4 3D surface roughness images for nCuHAp POM composite pins after the wear experiment. The 3D profiling in 3D surface profilometer (Taylor Hobson, Leicester, UK) of the surface interface of the pin with 316L steel is shown—**a** 0% nCuHAp, **b** 1% nCuHAp, **c** 2% nCuHAp, and **d** 5% nCuHAp

Table 2 Table of shore-D-hardness for composite pins of 0% to 5%

S. no	Sample no (%)	Shore-D-hardness
1	0	78
2	1	76
3	2	79
4	3	79
5	4	80
6	5	80

3.4 Toxicity Test of nCuHAp

In the toxicity test, new cell growth was first observed around the 1% nCuHAp with an assembled cluster. However, dead cells were noticed accumulating in the sample in

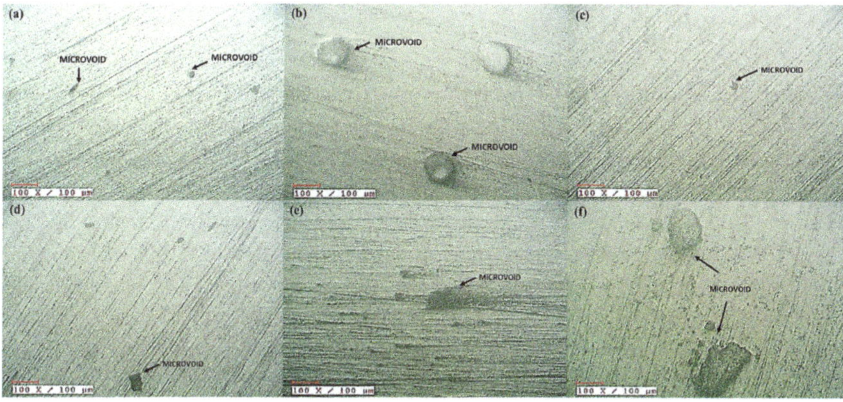

Fig. 5 Optical microscope images Metallurgical Microscope (OIAL/MMN/MET/01) at 100× indicating the presence of voids in **a** POM without nCuHAp, **b** 1% nCuHAp, **c** 2% nCuHAp, **d** 3% nCuHAp, **e** 4% nCuHAp, **f** 5% nCuHAp

a few regions. Conversely, 2% and 4% nCuHAp showed controlled assembled clusters of cells proliferating around the sample. 3% nCuHAp has increased anomalous density in dead cells around the sample (Fig. 6).

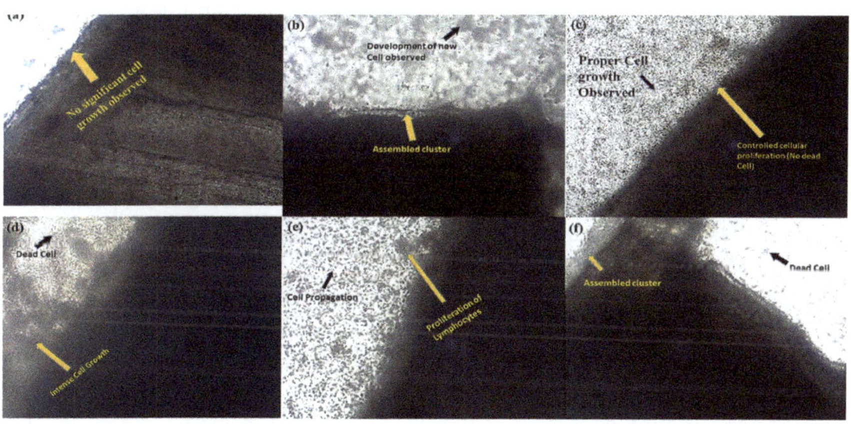

Fig. 6 Optical microscope images at 10× zoom indicating the growth of cells in: **a** POM without nCuHAp, **b** 1% nCuHAp, **c** 2% nCuHAp, **d** 3% nCuHAp, **e** 4% nCuHAp, and **f** 5% nCuHAp

4 Discussion

The fabricated composite nCuHAp has antibacterial properties and may be used as biomedical TMJ implants, is reinforced with the polyoxymethylene matrix (Buciumeanu et al. 2018). The results from XRD and SEM confirmed the synthesis of nCuHAp (Fig. 1a, b). The manufacturing technique requires producing cylindrical pin from POM with a melting point of 161.8 °C. This was confirmed from DSC test of the POM samples. This study has opted to fabricate 5 different samples with varied weight ratio of nCuHAp reinforced in POM. The injection molding technique was found suitable for these operations. The tribology test was conducted with a similar setting of a joint articulation such as temporomandibular joint. Individual nCuHAp-POM polymer composite pin was placed normally, with the base in contact with the 316L steel plate (Fig. 1). The temporomandibular joint comprising of two articulating surfaces—mandible and articular fossa (Banerjee et al. 2021, 2022). In TMJ, the condylar fracture is reconstructed with several TMJ implants like Christensen TMJ implant (Banerjee et al. 2023). However, the articulating surface of fossa is made up of materials with properties like low coefficient of friction, low wear rate, and bioactive. This helps the implant to easily osseointegrate with the bone and increase the implant's life. In this study, the proposed composite material has similar properties.

In the present study, the interacting bodies are stainless steel 316L and polymer; 316L is low carbon steel. The 316L steel used in the study is less brittle and softer than austenite steel. This makes stainless steel 316L accurate in the machining operation, and hence a vastly polished surface can be machined; Moreover, the presence of 16.30% chromium in the 316L alloy lowers the chances of fatigue. This ultimately makes the wear rate very low as POM is used as a pin (Fig. 3b). In the interaction of polymer and metal surface, metal has higher young's modulus, and polymer layer tends to transfer on the steel surface (Vecchio et al. 2007). Therefore, to conclude, the nCuHAp-POM composite and 316L steel transfer mechanism take place in 3 different ways. Firstly, the sliding causes the polymer particles to transfer to the 316L disc. Secondly, due to the continuous sliding, the polymer transferred gets oriented in the direction of the slide, tangential to the locus of the pin. During continuous sliding, the preceding layers of the composite get transferred, accumulated, and create grooves in the sliding direction. This describes the instability of pure POM and the high wear rate and friction coefficient. Upon the addition of nCuHAp, the mechanical properties of POM are enhanced, and the hexagonal structure becomes more stable because of an effective nucleating site during polymerization processes (Ripamonti et al. 2009). With the addition of nCuHAp, the Young's Modulus of the pin increases to decrease the chances of failure and has a low wear rate. The decreasing trend in the Coefficient of Friction (Fig. 3a) at 5% nCuHAp is due to the crystallinity and transfer mechanism between the polymer and the disc. POM is a polymer with a continuous structure, and therefore a uniform layer is transferred on the disc (Granizo et al. 2018; Yadav et al. 2021). The interaction between metal and polymer leads to minor adhesive wear. Because adhesive forces occur more profoundly between two similar surfaces, the composite layers get transferred on the plate, and subsequently, the adhesive wear increases slowly (Dolphin et al. 2013). The nCuHAp added to POM acts as a nucleating agent in the crystallization process. The nucleating agent regulates the mechanical properties of a composite. It has been reported in previous studies that nCuHAp inhibits the composite wear by inhibiting the subsurface crack growth in the composite pin material and therefore reducing the wear rate from 1 to 2% nCuHAp in POM, and 3 and 4% almost constant. On the other hand, the sudden increase in the specific wear at 5% can be confirmed by the phenomenon of heterogeneous nucleation (Fig. 3b). In heterogeneous nucleation, the crystallization process lost its homogeneity due to the excess nCuHAp concentrations, resulting in various micro defects in the crystal (Rath et al. 2000). These defects increase the residual stress and vibration in the cracks. Ultimately, the chattering effect causes surface failure at the micro-level and results in comparatively higher wear (Yadav et al. 2019).

POM is a continuous structure and transfers a smooth layer. 0% nCuHAp POM undergoes continuous plastic deformation when the average load is applied. The yield point of the polymer tends to decrease with increasing plastic deformation. This occurs due to intrinsic strain softening and at considerable strain due to progressive re-hardening. Because of this, the surface in immediate contact gets removed, and the micro-voids size of pure POM is much less. The nucleating agent nCuHAp introduces crystallinity in the polymer and increases the brittleness, and material transfer occurs. But deformation within the subsequent layer in the polymer continues to occur. This dislocation piles up to a finite distance and finally leads to micro-voids. This explains subsequent growth in the size of microporosity between 2 to 20 μm, in 2 to 5%. There was an increase in the hardness as in Fig. 5 upon the addition of nano nCuHAp in POM Hardness is the resistance to any surface failure or deformation in any materials. Thus, average hardness measured from the deformation of the testing surface on the penetration of the indenter played a vital role in controlling the wear rate. The crystallization process is due to heterogeneous nucleation on mixing nCuHAp in POM. At the same time, the composite was cooled down. This resulted in the generation of residual stresses within layers. Cracks began to grow in the lateral plane, generating microvoids (Fig. 5).

This also indicates the cell growth to understand the bioactivity of the biocomposite. The black portion resembles the composite, while the cell activity can be noticed around the sample. Significant cell growth is observed in each sample. Pure POM observed no significant cell growth, indicating that the reinforcement enhances the cell growth in the samples. New cell growth was first observed around the 1% nCuHAp with an assembled cluster. However, dead cells were noticed accumulating the sample in a few regions. Conversely, 2 and 4% nCuHAp showed controlled assembled clusters of cells proliferating around the sample. 3% nCuHAp has increased anomalous density in dead cells around the sample (Kumar et al. 2013; Uniyal et al. 2021).

At a particular wavelength, the number of cells is depicted by the Optical density. The increase in nCuHAp percentage results in causing assembled clusters and rapid cell growth. This also leads to cell death and failed cell growth. The degradation rate constant determines the biodegradability of the sample. A specified formula calculates the biodegradation kinetics, the first-order degradation rate constant (min^{-1}) (Eq. 1).

$$K_{deg} = \frac{1}{t}\left(\frac{w_i - w_f}{w_i - w_t}\right) \tag{1}$$

where w_i is the initial scaffold weight fraction, w_f is the final scaffold weight fraction, and w_t is the scaffold weight fraction at a time 't.' A proportionate rate is required for implant material so that the wear rate is optimum. The interconnecting pores formation and steric hindrance of (–OH) of Hydroxyapatite cause the rate to remain similar. Due to heterogeneous distribution and residual stress development, the wear rate increased and the degradation rate constant. Therefore, with the pattern in Fig. 7, 2% nCuHAp-POM reinforced composite can be held as the selected sample as it showed the optimized results for cell proliferation. Hence, it can be used as a scaffold. It relates the optical density at 570 nm to the cell concentrations in each sample of nCuHAp (Uniyal et al. 2022a, b). The optical density of more than 3% was due to the polymer matrix sample in the study region. However, a significant percentage of the light is not transmitting across the region of cell growth which linearly correlates to the concentrations.

This study has several limitations to be considered in future studies. Due to the length of the research, this study is not a comparative study. The material selection

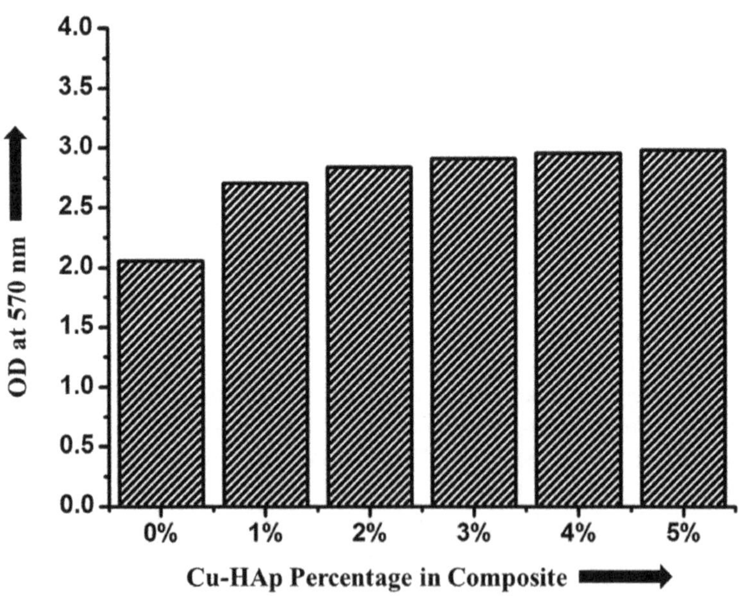

Fig. 7 Statistical difference in optical density of cells obtained in POM composite with varied percentages of nCuHAp

for the matrix is based on previous studies. Therefore, a study could be considered by altering the matrix material and adding nCuHAp as reinforcement. Secondly, this study did not mention any clinical study. Implant and scaffolds could be fabricated with this composite and in-vitro results could be compared to validate its use in medical science. Due to the limitation of this study to restrict it to the tribology, an in-vitro validation for the biocompatibility is not considered.

5 Conclusions

This study found various aspects of the mechanical property of nCu-HAp composite reinforced in POM and cell proliferation results in these samples to confirm its use in the human body. Decrease in the coefficient of friction from 0 to 1% nCuHAp, increasing up to 4% and then a sudden decrease in 5%. It was lowest at 1% nCuHAp. The specific wear rate gradually decreased to 4% nCuHAp, with a sudden increase at 5%. It was observed lowest at 2%. The evidence of the trace elements and composite signify the nucleation of nCuHAp particles in the POM pin, and its concentration in the pin plays a significant role in controlling the tribological property of the composite. Cell proliferation increases with the increase in the percentage doping of nCuHAp. The best result was shown at 2%. However, the optimized result for 2% is similar to 4%. According to this study, this can be a biomedical application alternative such as implants, and 2% nCuHAp reinforced in POM may be ideal. The study presents a future scope of development with similar composite material and encourages parametric qualitative study.

Acknowledgements The authors acknowledge the support of Dr. Subrajit Bhaumik for his support and supervision during the completion of this project. The author would also like to acknowledge the support of Dr. Kautav Sarkar, Department of Biotechnology at SRM Institute of Science and Technology, Kattankulathur, in conducting the cell-proliferation test.

Declarations

Conflicts of Interest

"There are no conflicts of interest to declare by the authors."

Ethical Approval

The authors hereby state that the present work is in compliance with the ethical standards.

Competing Interests

The authors declare no competing interests.

Availability of data and materials

Not applicable.

References

Alvira, F.C., Rozzi, F.R. and Bilmes, G.M., 2010. Laser-induced breakdown spectroscopy microanalysis of trace elements in Homo sapiens teeth. *Applied Spectroscopy*, *64*(3), pp. 313–319.

Ansari, M., Naghib, S.M., Moztarzadeh, F. and Salati, A., 2011. Synthesis and characterisation of hydroxyapatite-calcium hydroxide for dental composites. *Ceramics-Silikáty*, *55*(2), pp. 123–126.

Báez, A., Belmont, R., García, R. and Hernández, J.C., 2004. Cadmium and lead levels in deciduous teeth of children living in México City. *Revista internacional de contaminación ambiental*, *20*(3), p. 0.

Banerjee, A., Choudhury, S., Rana, M., Chakraborty, A., Gupta, A. and Chowdhury, A.R., 2023. Biomechanical analysis of mandibular bone-implant construct with three implant screw design: A finite element study. *Journal of Engineering and Science in Medical Diagnostics and Therapy*, *6*(3).

Banerjee, A., Rana, M., Chakraborty, A., Singh, A.P. and Roy Chowdhury, A., 2022. Influence of implant parameters on biomechanical stability of TMJ replacement: A finite element analysis. *The International Journal of Artificial Organs*, *45*(8), pp. 715–721.

Dolphin, A.E., Naftel, S.J., Nelson, A.J., Martin, R.R. and White, C.D., 2013. Bromine in teeth and bone as an indicator of marine diet. *Journal of archaeological Science*, *40*(4), pp. 1778–1786.

Granizo, R.M., Muñoz, D.C.C. and Reyes, E.V., 2018. Rearthroscopy of the temporomandibular joint: A retrospective study of 600 arthroscopies. *Journal of Cranio-Maxillofacial Surgery*, *46*(9), pp. 1555–1560.

Guzmán-Venegas, R.A., Palma, F.H. and de la Rosa, F.J.B., 2018. Spectral components in electromyograms from four regions of the human masseter, in natural dentate and edentulous subjects with removable prostheses and implants. *Archives of Oral Biology*, *90*, pp. 130–137.

Lu, Y., Dong, W., Ding, J., Wang, W. and Wang, A., 2019. Hydroxyapatite nanomaterials: synthesis, properties, and functional applications. In *Nanomaterials from clay minerals* (pp. 485–536). Elsevier.

Mercuri, L.G., Mathew, M.T., Kerwell, S., Lundberg, H. and Sukotjo, C., 2015. Temporomandibular joint replacement device research wear and corrosion technology transfer from orthopedics. *Journal of Bio-and Tribo-Corrosion*, *1*, pp. 1–3.

Nayak, A.K., 2010. Hydroxyapatite synthesis methodologies: an overview. *International Journal of ChemTech Research*, *2*(2), pp. 903–907.

Rath NC, Huff GR, Huff WE, Balog JM. Factors regulating bone maturity and strength in poultry. Poultry science. 2000 Jul 1;79(7):1024–32.

Ripamonti, U., Crooks, J., Khoali, L. and Roden, L., 2009. The induction of bone formation by coral-derived calcium carbonate/hydroxyapatite constructs. *Biomaterials*, *30*(7), pp. 1428–1439.

Rivera, E.M., Araiza, M., Brostow, W., Castano, V.M., Dıaz-Estrada, J.R., Hernández, R. and Rodrıguez, J.R., 1999. Synthesis of hydroxyapatite from eggshells. *Materials Letters*, *41*(3), pp. 128–134.

Shen, J.W., Wu, T., Wang, Q. and Pan, H.H., 2008. Molecular simulation of protein adsorption and desorption on hydroxyapatite surfaces. Biomaterials, 29(5), pp. 513–532.

Uniyal, P., Sharma, A. and Kumar, N., 2022a. Investigation on the sensitivity of indentation devices for detection of fatigue loading induced damage in bovine cortical bone. *Journal of Biomechanics*, *143*, p. 111274.

Uniyal, P., Sihota, P. and Kumar, N., 2022b. Effect of organic matrix alteration on strain rate dependent mechanical behaviour of cortical bone. *Journal of the Mechanical Behavior of Biomedical Materials*, *125*, p. 104910.

Uniyal, P., Sihota, P., Tikoo, K. and Kumar, N., 2021. Anatomical variation in intracortical canal network microarchitecture and its influence on bone fracture risk. *Journal of the Mechanical Behavior of Biomedical Materials*, *123*, p. 104770.

Vecchio, K.S., Zhang, X., Massie, J.B., Wang, M. and Kim, C.W., 2007. Conversion of bulk seashells to biocompatible hydroxyapatite for bone implants. *Acta biomaterialia*, *3*(6), pp. 910–918.

Yadav, J.P., Yadav, R.N., Uniyal, P., Chen, H., Wang, C., Sun, C.C., Kumar, N., Bansal, A.K. and Jain, S., 2019. Molecular interpretation of mechanical behavior in four basic crystal packing of Isoniazid with homologous cocrystal formers. *Crystal Growth & Design, 20*(2), pp. 832–844.
Yadav, R.N., Sihota, P., Uniyal, P., Neradi, D., Bose, J.C., Dhiman, V., Karn, S., Sharma, S., Aggarwal, S., Goni, V.G. and Kumar, S., 2021. Prediction of mechanical properties of trabecular bone in patients with type 2 diabetes using damage based finite element method. *Journal of Biomechanics, 123*, p. 110495.

Sustainable Innovations and Production Strategies of White Rot Fungi-Derived Laccase

Junmoni Das, Marjum Badak, and R. K. Singh

1 Introduction

Laccases (Polyphenol oxidase, EC 1.10.3.2) are copper-containing enzymes belonging to the oxidoreductase group (Thurston 1994), which have been extensively studied since the late nineteenth century. First discovered in 1883 in the latex of the Japanese lacquer tree *Rhus vernicifera* (Yoshida 1883), laccase was later demonstrated to be a fungal enzyme by Bertrand and Laborde in 1896. This versatile enzyme acts as an electron acceptor and can oxidize a wide range of substrates, including monophenols, diphenols, monolignols, polyphenols, and hydroxyindols (Perna et al. 2020). During its reaction cycle, laccase oxidizes a phenol to form a highly reactive phenoxy radical (Munk et al. 2015), which can further react non-enzymatically with other phenols, leading to the formation of polymerized products. Laccases are recognized as green catalysts with notable bioactive properties thus, plays a vital role in advancing cleaner and sustainable industrial processes. In addition, the broad scope of substrates that laccases can act upon leads to a diverse array of biological functions, emphasizing their extensive potential in various biotechnological and environmental applications. Furthermore, researchers have also focussed on the laccase-catalyzed synthesis of metal nanoparticles (Chaurasia et al. 2022). The aforementioned shed light on the wide spectrum of functions and distributions associated with laccases, offering promising avenues for applications in biotechnology and environmental management.

Scientists have recently regained interest in laccase with them getting identified in various eukaryotes and prokaryotes, including higher plants, bacteria, insects, fungi, and mammals (Janusz et al. 2020). In this regard, laccase producing Wood

J. Das (✉) · M. Badak · R. K. Singh (✉)
Department of Botany, Rajiv Gandhi University, Rono Hills, Doimukh 791112, India
e-mail: junmoni.das@rgu.ac.in

R. K. Singh
e-mail: rajiv.singh@rgu.ac.in

J. Gupta and A. Verma (eds.), *Microbiology-2.0 Update for a Sustainable Future*,
https://doi.org/10.1007/978-981-99-9617-9_13

Rot Basidiomycetes (WRB), known for their ability to degrade different components of wood, including lignin, cellulose, and hemicellulose, are categorized as: Brown Rot Fungi (BRF) and White Rot Fungi (WRF) (Rodríguez-Couto 2017). Here, while BRF primarily degrade cellulose and hemicellulose components, WRF are predominantly lignin degraders, though they also partially breakdown cellulose and hemicellulose (Cha et al. 2018). They perform various functions, including lignin degradation, morphogenesis, and pathogenesis (Eggert et al. 1996) and find utility in wood pulp delignification and bioleaching, substituting environmentally harmful chlorine-based methods. The paper, textile, and food industries also benefit from its extensive applications. Distinct functions executed by laccase point to the existence of diverse isozymes, assuming varied roles in fungal physiology. These isozymes are distributed across cellular compartments and can be secreted into the surrounding medium.

1.1 Occurance

The majority of laccase is found extracellular however; evidence of intracellular production was identified for *Neurospora crassa* (Froehner and Eriksson 1974). They suggested that the intracellular laccase functioned as a precursor for extracellular laccase since there were no differences between them other than their occurrence. Blaich and Esser (1975) also confirmed the production of both extracellular and intracellular laccases with isozymes, showing similar patterns of activity staining after isoelectric focusing from most WRF species. *Trametes versicolor* cultivated on glucose, wheat straw, and beech leaves exhibited laccase production both extracellularly and intracellularly (Schlosser et al. 1997). In addition, *Neurospora crassa, Rigidoporus lignosus,* and one laccase isoenzyme in *Pleurotus ostreatus* likely localize intracellularly or on cell walls (Palmieri et al. 2000). In *Irpex lacteus*, laccase activity is predominantly associated with cell walls (Svobodová 2005). This localization influences laccase's physiological role and substrate accessibility. Additionally, cell wall and spore-associated laccases may contribute to melanin and other protective compounds (Galhaup and Haltrich 2001).

1.2 Characteristics, Mode of Action and Functional/Structural Diversity

Fungal laccase, exhibit varying catalytic activity and stability, influenced by factors like source, temperature, pH, and culture medium. Operating well within acidic pH (3–6), with an activity peak at pH 3, they display a broad temperature range (20–55 °C) with an optimal temperature of 55 °C. Nonetheless, thermally stable variants (60–70 °C) have also been identified (Nyanhongo et al. 2002; Saraiva et al. 2012).

Also, fungal growth phase, inducers, and production conditions contribute to the presence of isoforms where, isoforms within a strain can differ in pI values and molecular weights while culture age and substrate dictate the proportion produced. Genes encoding these isoforms exhibit differential regulation, being constitutively expressed or inducible throughout the cell's lifespan. Consequently, the intricate interplay of these variables governs the characteristics and functional diversity of laccases in fungal systems (Castanera et al. 2012; Piscitelli et al. 2011).

Typically, mature laccase protein functions as a holoenzyme, adopting monomeric, dimeric, or tetrameric active forms, each containing four copper atoms per monomer (Kunammeni et al. 2008). Molecular weights span 50–100 kDa containing 520–550 amino acids, excluding the signaling peptide sequence. Most laccases are glycoproteins (glycosylation levels ~10–30%), influencing secretion, proteolytic stability, copper retention, and thermal stability (Morozova et al. 2007; Thurston 1994). Glycosylation also impacts laccase isoform pI values (3–7). These enzymes vary in kinetic parameters as shown in Table 1.

Fungal laccases comprise of protein subunits forming dimeric or tetrameric arrangements, with each monomer containing four copper atoms. These copper atoms are organized into three distinct centers within the enzyme (refer to Fig. 1): Type-1 (blue copper center), Type-2 (normal copper), and Type-3 (coupled binuclear copper centers) (Ferraroni et al. 2007). Type-1 is characterized by its vivid blue color and is coordinated by specific amino acids, including histidine, cysteine, and methionine (Mot and Silaghi-Dumitrescu 2012). In contrast, Type-2 and Type-3 centers possess unique coordination and electronic properties. Additionally, Type-1 displays absorption at 600 nm, leading to the characteristic blue colour while Type-2 lacks visible absorption but shows paramagnetic traits in EPR studies. Type-3 exhibits absorbance at 330 nm and participates in molecular oxygen reduction, releasing water (Guimarães et al. 2017). Furthermore, Type-2 copper coordinates with two histidines, while Type-3 coordinates with six histidines (Claus 2004).

They also exhibit high redox potential (0.4–0.8) volts (Messerschmidt 1997) consequently, considered a better contestant in biotechnology and bioremediation applications. From a chemical perspective, they play a pivotal role in synthetic processes, and their mechanisms are often simplified as described below. These mechanisms rely on of the copper cluster's redox activity in the active site (Fig. 2). In substrates with suitable redox potential, laccases can directly oxidize them to form corresponding organic radicals (Fig. 2a). However, if direct oxidation is hindered due to steric hindrance or incompatible redox potentials, "chemical mediators" comes into play. These mediators, resembling natural redox intermediates such as the cytochrome chain, are employed successfully subsequently, the resulting radicals from mediator-driven oxidations can engage with bulky or high redox-potential substrate targets (Fig. 2b). This versatile redox approach (laccase-mediator-system (LMS)) finds extensive applications in organic synthesis, e.g., in converting activated hydroxyl groups to carbonyls or acids, as well as in various biotechnological and manufacturing processes (Hilgers et al. 2018; Jin et al. 2016). Furthermore, LMS finds use in regenerating cofactors during coupled, multienzymatic redox conversions (Fig. 2c).

Table 1 Physical and biochemical properties of some fungal laccase

Fungus	MW (kDa)	Km(μM)	pH	Temp (°C)	Reference
ABTS					
Trametes gallica Lac 1	60	12	2.2	70	Dong and Zhang (2004)
Ganoderma lucidum	57	3000	5	40–60	Himanshu et al. (2023)
Trametes sp.	65	66.9 ± 3.2	4.0–4.5	40	Li et al. (2016)
T. versicolor	45 & 83	95.75 ± 11.53	3	40	Backes et al. (2023)
T. trogii	56	69	4	55	Yan et al. (2014)
Trametes hirsuta	66	0.02 ± 0.005	5	30	Navada and Kulal (2021)
Trametes sp. LAC-01	59	30.28	2.6	40	Ling et al. (2015)
T. polyzona	66	8.66	5.5–6.5	40–50	Ezike et al. (2020)
Guaiacol					
Cerrena maxima	57–67	160–300		50	Shleev et al. (2004)
G. lucidum	57	21,360	5	40–60	Himanshu et al. (2023)
Trametes sp.	65	108.9 ± 3.4	4.0–4.5	40	Li et al. (2016)
G. leucocontextum	65	1658	3.0–4.0	70	Umar and Ahmed (2022)
T. hirsuta	66	0.4 ± 0.06	5	30	Navada and Kulal (2021)
T. versicolor, Lcc1	60	15	3–3.5	22–24	Martínez-Morales et al. (2015)
T. versicolor, Lcc2	100	15			
G. lucidum	38.3	94	5	55	Manavalan et al. (2013)
2,6-dimethoxyphenol (DMP)					
T. gallica Lac II	60	410	3	70	Dong and Zhang (2004)
Trametes sp.	65	38.1 ± 1.7	4.0–4.5	40	Li et al. (2016)
Phoma herbarum	298	216	8	50	Debnath et al. (2021)
T. versicolor, Lcc1	60	10	3–3.5	22–24	Martínez-Morales et al. (2015)
T. versicolor, Lcc2	100	47			
Syringaldazine					
Daedalea quercina	69	131	7	55	Baldrian (2004)
T. versicolor, Lcc1	60	17	3–3.5	22–24	Martínez-Morales et al. (2015)
T. versicolor, Lcc2	100	39			

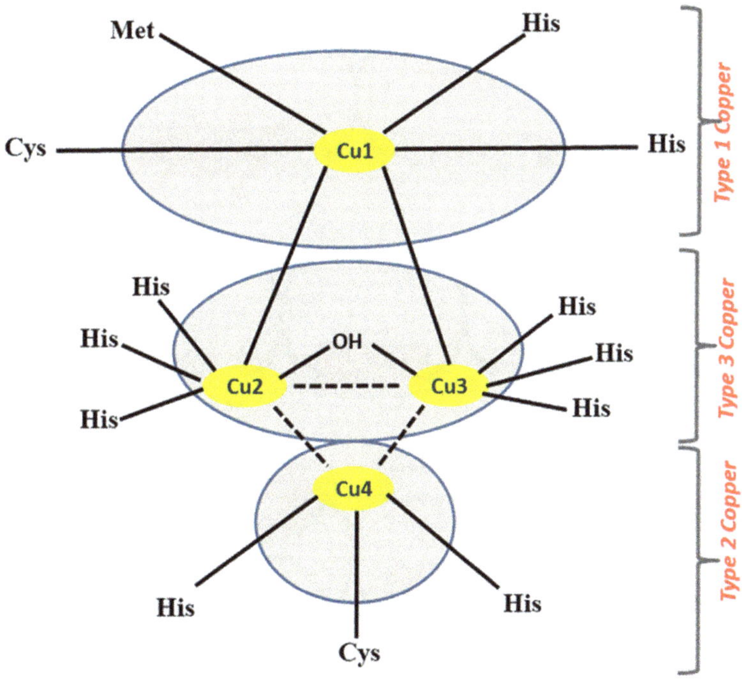

Fig. 1 Schematic representation of active site in laccases

Fungal laccases are broadly classified into three types: blue, yellow, and white. Blue (yellow) laccases do (don't) absorb at 600 nm (Mot et al. 2020), and white laccases absorb at 400 nm (Agrawal et al. 2019). Blue laccases (Jin et al. 2016) require a mediator to oxidize non-phenolic substrates, while yellow laccases do not (Ike et al. 2015). Yellow laccases have a unique loop that may or may not contain a tyrosine residue (Radveikiene et al. 2021) and are catalytically and environmentally advantageous than blue laccases since they don't require a mediator.

2 Strategies for Laccase Production

Laccase are primarily extracellular glycoproteings although a significant portion of researches have reported intracellular ones (Kunamneni et al. 2007). They are typically produced all through the secondary metabolism of diverse fungi, and are influenced by different features, such as fungal species, cultivation type (submerged or solid state fermentation), aeration and culture time, composition of growth medium, and supplementation of growth medium with carbon, nitrogen, and inducer agent concentration and surfactants. Fungal growth in Solid-State Fermentation (SSF) is different from fungal growth in Submerged Fermentation (SmF) because of different

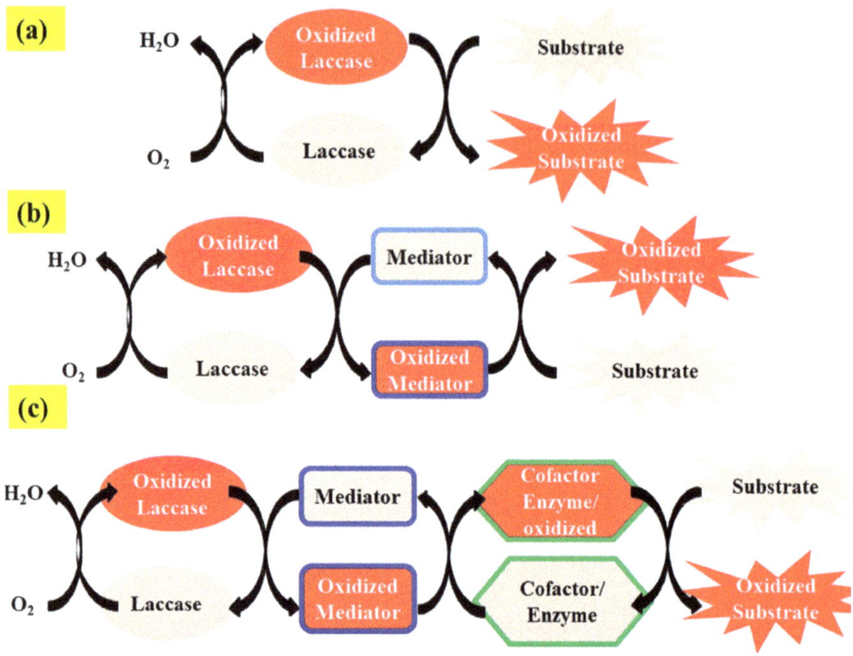

Fig. 2 Redox cycles in laccase catalysis for substrate oxidation: **a** direct oxidation; **b** chemical mediator facilitated oxidation; **c** regeneration of cofactor in multienzymatic coupled redox biotransformations

surface phenomena like moisture content and chemical composition of the supporting substrate (Wang et al. 2019). Moreover, SSF reproduces the natural living conditions of the WRF since, it is a statis process and thus does not consume mechanical energy and provides a better oxygen circulation (Mitchell et al. 2011). It is a cheap, simple, eco-friendly, and cost-effective technique, where various lignocellulosic agricultural wastes are used as substrate because they provide suitable environment for the natural growth of the organism. Researchers have used diverse agro-residues like wheat straw, white bran rubber wood dust, Peanut shells Grape seeds, Sunflower seed hulls, Pine wood chips and orange peels (1:1) *Eichhornia crassipes* biomass: sawdust (1.8:1), Corncob waste, Pineapple peel, Tea residues, rice straw, rice bran, empty fruit bunch, sugarcane bagasse, etc. as growth media due to the presence of vitamins, inducers, soluble carbohydrates, vitamins, and nitrogen for laccase production by various fungi (Wang et al. 2019; Singhania et al. 2009).

Fermentation conditions have been suitably optimized by either i.e., Classical method or Statistical method for high laccase production. Here, the classical method, also called one factor–at–time (OFAT) involves changing one independent factor while keeping the others constant. However, OFAT is time-consuming and incapable of detecting the true optimum due to its inability to detect interactions among the factors. On contrary, statistical method like Response Surface Methodology (RSM)

(Box and Wilson 1951) helps in screening of the main variables in a multivariable system thus, offering an effective strategy for medium optimization. It establishes a relationship among continuously varying factors and predicts the resultant response to gain optimum production points. Also, statistical method based experimental designs have many advantages, including better results with less process variability, less development time and overall costs compared to the classical method. RSM has been successfully applied in the optimization of media conditions for laccase production from different microorganism. The extensive industrial applications need large amount of laccase. Therefore, a good strategy to increase laccase production would be the optimization of the fermentation medium together with application of different inducers. In order to stimulate optimized production, Carbon and nitrogen sources, different metal ions, phenolic compounds, aromatic compounds and physical factors (pH, temperature and moisture ratio etc.) have been suitably manipulated by researchers for various fungi including.

2.1 *Effect of pH, Temperature, Moisture Ratio*

The pH of the culture medium is critical for the growth and production of laccase by different WRF and the optimum pH ranges 3.0–6.0 (Naidu et al. 2020). Schneider et al. (2018) reported that a higher pH of the medium results in poor mycelial growth thereby decreasing enzyme production. Change in pH of the medium might alter the three-dimensional structure of laccase, thus leading to a decrease in enzyme activity (Chhaya and Gupte 2013).

In enzyme production, optimal temperature ensures proper growth and enzyme secretion. WRF have tremendous ability to produce laccase enzyme under sub-optimal temperatures of 15 to 35 °C (Yasmeen et al. 2013). Other works also reported suitable temperature for WRF laccase production ranging between 25 and 30 °C (Ergun and Urek 2017; Patel et al. 2019).

During SSF, moisture content is significant for the growth of microorganisms where, optimum moisture ratio needs to be maintained since, lower moisture ratio tends to reduce nutrient diffusion, enzyme stability and substrate swelling (Chhaya and Gupte 2013). It also decreases the decomposition rate of organic matter thereby, affecting enzyme production (Raimbault 1998). Contrastingly, increase in substrate moisture content makes mobile nutrients easily available to the fungus; though it also reduces the air content of the substrate which may negatively affect microbial growth (Battan et al. 2007; Sharma et al. 2015).

2.2 *Effect of Carbon and Nitrogen Sources*

Carbon (C) sources play an important role in laccase production since they can promote mycelial growth and induce transcription of the laccase gene (Kaluskar et al. 1999). Fructose, glucose, galactose, mannitol, maltose, lactose, sucrose, cellulose and soluble starch, have been used as carbon source by various researchers to increase laccase production. Among them, glucose has been reported to be a good for enhancing ligninolytic enzyme production and promoting growth of organism to degrade lignin (Cilerdzic et al. 2019).

The nitrogen (N) source and C/N ratio of the medium are important factors that affect fungal growth and laccase production. The optimal nitrogen source and C/N ratio will vary depending on the fungal species (Jaramillo et al. 2017). In general, organic nitrogen sources are more effective than inorganic ones (Piscitelli et al. 2011) as they are more easily assimilated by fungi and can be utilized to synthesize proteins quickly. In addition, low C/N will favor laccase production and the production of several, while a high C/N will inhibit the production compounds (Moiseenko et al. 2018) because low C/N will force the fungus to use its carbon resources for growth and reproduction, rather than for laccase production. The optimal C/N ratio by a particular fungal species can be determined by experimentation. Furthermore, the ratio of the medium can be adjusted by adding different amounts of carbon and nitrogen sources. Many researchers have reported that inorganic nitrogen sources {$NaNO_3$, $(NH_4)_2SO_4$, NH_4Cl, NH_4NO_3 and $C_4H_{12}N_2O_6$} lead to low laccase production despite of the availability of sufficient biomass, whereas organic nitrogen sources (urea, yeast extract, peptone, beef extract) render high laccase production together with good fungal growth (Piscitelli et al. 2011). Organic nitrogen sources such as yeast extract and peptone have also been reported to be more suitable for various fungi (Karp et al. 2015). In case of *Trametes versicolor* and *Coriolopsis polyzona*, Elisashvili and Kachlishvili (2008) reported significant improvement in laccase production upon addition of NH_4NO_3, $(NH_4)_2SO_4$, KNO_3 and peptone as supplementary nitrogen sources. Other researchers have found $(NH_4)_2SO_4$ as best nitrogen source for laccase production by various fungal stain (Meehnian et al. 2017). Organic nitrogen source like Casein was found good for laccase production by *Pleurotus sajor-caju* (Bettin et al. 2009). Supplementation of the medium with an additional nitrogen source like L-asparagine has also been found to increased laccase production (Janusz et al. 2007).

2.3 Effect of Incuders and Surfactant

Copper stands out as the prime inducer of laccase activity in a wide array of fungal species. However, other metal ions like Mn^{+2}, Zn^{+2}, Fe^{+2}, Mg^{+2} and Ca^{+2} have also shown the capacity to modulate laccase expression due to their distinct affinities for fungal systems (Akpinar and Urek 2017). The expression of laccase enzyme is linked to the hyphal branching of fungi. The presence of heavy metals and xenobiotics can also induce laccase synthesis, leading to the creation of new isoforms and expanding enzyme functionality. While certain fungi like *Trametes villosa* and *Pleurotus sajor-caju* exhibit constitutively expressed laccase genes, others require intricate inducer cocktails rather than single inducers. Laccase exhibits increased gene expression upon exposure to Copper ions, primarily operating at the transcriptional level (Zhu et al. 2016). Copper's inducing effect has been observed in various fungi including *Trametes versicolor*, *Pycnoporus sanguineus*, *Coriolus versicolor*, *Ceriporiopsis subvermispora*, *Pleurotus* species, *Panus osteratus*, and *Ganoderma* species (Wang et al. 2019). These studies collectively emphasize the distinctiveness of optimal copper concentration for augmented laccase production among WRF species. However, elevated copper concentrations can also exert toxicity on fungi, causing growth inhibition, enzyme inactivation, metabolic perturbations, and the generation of oxygen-reactive species (Krumova et al. 2012). Interestingly, Fe^{+2} has also been found to significantly elevate laccase production in *Pleurotus eryngii* and *Trametes velutina*, yet excessive Fe^{+2} levels hinder enzyme activity (Stajić and Vukojević 2011). Similarly, Manganese emerges as a potent laccase inducer in several WRF, exemplifying the organism-dependent nature of metal ion requirements (Imran et al. 2016). Overall, the most efficient laccase inducer depends on the fungal species and the desired level of laccase production. Thus, it is important to carefully consider all of the factors involved before selecting an inducer. Also, genes are always active, regardless of the presence or absence of inducers. Thus, a complex cocktail of inducers may be required in some cases, to achieve the desired level of laccase production. Finally, inducer use can entail toxicity and supplementary costs, presenting inherent challenges to the production process. Hence, it is observed that the multifaceted landscape of laccase induction and production in fungi arises from the interplay between metal ions, gene expression, fungal physiology, and inducer intricacies.

Laccase expression has also been increased on addition of phenolic and aromatic compounds to the cultivation medium, such as guaiacol, pyrogallol, catechol, resorcinol, orcinol and veratryl alcohol in SSF. Aromatic inducers have been reported to induce secondary metabolism which in turn results in increased laccase concentration (Tavares et al. 2005). Addition of Veratryl alcohol to cultivation media of many WRF has been reported to increase laccase production (Arora and Gill 2010; Thakkar et al. 2020). On the other hand, Pyrogallol and Ferulic acid have been reported to be the most effective inducers for laccase production by *Ganoderma lucidum* (Elisashvili et al. 2010; Sharma et al. 2014) and *Pleurotus sajor-caju* (Zucca et al. 2011). Consequently, several hypotheses have been formulated concerning alcohol's role in laccase production: (i) enhancement in permeability of membrane and protein secretion, (ii) growth in phenolic monomers as a consequence of melanin formation inhibition, and (iii) activation of oxidative stress, perhaps indirectly responsible for laccase induction (Lee et al. 1999).

The introduction of xylidine as an inducer yielded remarkable enhancements in laccase production across various fungi, including *Trametes villosa*, *T. versicolor*, *Pleurotus sajor-caju* (Da Cunha et al. 2003), *Cariolopsis polyzona* (Jaouani et al. 2006), and *Cariolopsis caperata* RCR2011 (Nandal et al. 2013). Also, utilization of lignin and tannic acid as an inducer led to heightened laccase gene expression and increased enzyme production (Gianfreda et al. 1999).

Various surfactants, including Tween 20, Tween 40, Tween 80 and Triton X-100, at their optimal concentrations, enhance laccase production (Elsayed et al. 2012; Teodoro et al. 2018). Surfactants improve cell membrane permeability, affect protein secretion, and reduce unproductive enzyme adsorption via hydrophobic interactions with solid substrates (Sharma et al. 2005).

3 Sustainable Innovations for Fungal Laccase Applications

Various strategies aim to reduce production costs, enhance their economic viability for industrial applications despite diverse origins, sizes, and traits of fungal laccases. In this regard, laccase immobilization is an effective approach, offering stability during storage, biocatalyst recovery, and support for sustainable processes (Masjoudi et al. 2021). In addition, immobilization permits reuse and stability, mimicking enzymes' natural occurrence in cell membranes, stabilizing structure and activity and enhances enzyme performance under the optimal conditions of various industrial processes (Wen et al. 2019) Additionally, it simplifies product separation, enabling continuous processes and and preventing losses preventing protein or activity loss in subsequent steps (Polizzi et al. 2007). Nevertheless, effective immobilization demands method choice, supporting material, and initial laccase activity consideration (Yavaşer and Karagözler 2021).

The procedure of immobilization is categorized into physical and chemical interactions (Fernández et al. 2013). The different modes of immobilization are displayed in Fig. 3. Prominent physical interactions like entrapment, encapsulation and adsorption maintain enzyme activity with weak forces like hydrogen bonds, ionic, and hydrophobic interactions (Zhou et al. 2021). Chemical interactions (covalent binding and crosslinking) create covalent bonds between enzyme and support via functional groups ($-NH_2$, –SH, –OH) (Daronch et al. 2020). This strong attachment, especially in changing conditions, makes chemical immobilization preferable for industrial use like wastewater treatment (Wahab et al. 2020). Additionally, it reduces enzyme leakage and enhances reusability (Zdarta et al. 2018).

4 Application of WRF-Derived Laccases

Laccase has gained attention for its potential applications in various fields, including biotechnology, environmental remediation, food industry, etc. Their versatility and ability to catalyze the oxidation of various substrates make them valuable tools in addressing various challenges across different sectors. Some of their applications are discussed below.

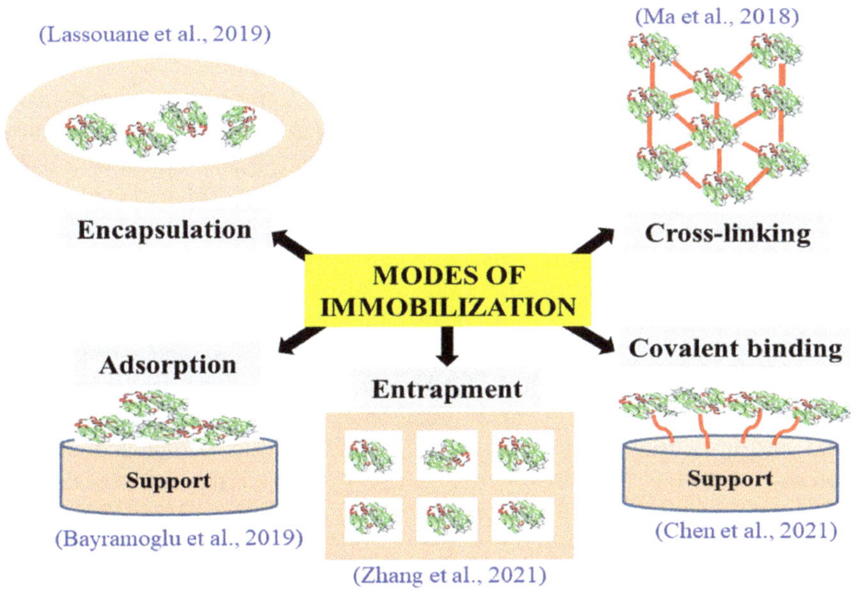

Fig. 3 Pictorial representation of different immobilization modes

4.1 Paper and Pulp Industry

The WRF-derived laccases (WRF-L) have potentials for degradisation of wood's lignocellulosic materials in paper and pulp industry. Here, the most successful usage of enzymes in called kraft pulp bleaching. However, the processes of lignin oxidation and isolation are generally carried out using Chlorine and other oxidising agents. Interestingly, bleaching process from enzymatic treatment with no Chlorine can result in lighter pulp with low lignin content (Bajpai et al. 2006). Also, addition of laccase to lignocellulosic fibres enhances the chemical and physical properties of kraft pulp fibre materials (Kumar and Chandra 2020). In addition, WRF-L directly attacks lignin's phenolic components and consequently, there is approximately no loss of cellulose and large-scale saving of chemicals is possible as compared to xylanase (Singh et al. 2015). The combination of laccase derived from *Trametes villosa* (20 U/g) and plant phenols viz., p-hydroxycinnamic compounds sinapic acid, ferulic acid, coniferyl aldehyde and sinapyl aldehyde as redox-mediators was evaluated at pH-4, 50 °C, for 4 h to investigate their effect on sisal pulp for potential biobleaching or functionalization of its fibres and compared with results from synthetic 1-hydroxybenzotriazole (HBT) (Aracri et al. 2009). They observed less efficiency of natural mediators than HBT in facilitating pulp bleaching however, increase fibre strength due to biografting was realized. A comprehensive review on utilization of different laccases including (WRF-L) in paper and pulp industry was presented by (Singh and Arya 2019).

4.2 Textile Industry

Textile industry waste is a major contamination source, contributing to two-thirds of dye production. During clothing manufacturing, water and chemical use generates substantial pollution (Mojsov 2014). Dye used in clothing is tough to decolorize due to complex synthetic nature and presence of toxic substances (Mahuri et al. 2023). In this regard, utilization of laccase can lead to economic utilization of water and energy, shorter processing time, enhanced product quality, and possibility of process integration in the textile production (Mojsov 2014). The potential and efficiency of *Trametes hirsuta* PW17-41 in decolourisation of mixed textile dyes i.e., Navy EC-R, Ruby S3B, and Super Black G from real industrial wastewater samples was investigated according to American Dye Manufacturer's Institute standard methodology (Thampraphaphon et al. 2022). They observed the optimal decolourisation of 95.39% from using palm sugar and ammonium nitrate as carbon and nitrogen sources. Nonetheless, Manganese peroxidase (4942 U/L) was reported to be the critical enzyme for decolourisation process as compared to WRF-L (2389 U/L) at intial dye concentration of 21,200 ADMI. A system for decolorization of azo dye (Direct Red 5B) was reported using *Ganoderma lucidum* EN2 and alkali lignin (Sun et al. 2019). They observed laccase to be a critical facor in the decolurisation process. Additionally, inclusion of lignin improved the process via augmentation in laccase activity from degradation of lignin into small molecular aromatic compounds and phenol and guaiacol acting as mediator.

4.3 Biosensors

A general overview on laccase structure and function, especially emphasizing on its suitable electrochemistry features such as, bioreceptors for electrochemical devices, was reported (Castrovilli et al. 2019). The authors defined how crosscutting technology like nanotechnology, immobilization approaches, and material science can result in augmenting the analytical performance of laccase-based biosensors. In this context, an assessment was conducted on a laccase-driven paper biosensor using *Trametes versicolor* (UC-3) laccase absorption combined with 3-Methyl-2-benzothiazolinone hydrazone hydrochloride hydrate (MBTH) solution on Whatman No. 1 paper. The resultant pink hues indicated phenol oxidation products. Optimal enzyme concentration was 2 mg/mL, and MBTH concentration was 24 mM for phenolic content detection. The biosensor achieved a minimum detection limit of 0.1 μM for phenolic content. This biosensor successfully detected phenolic content in water, wastewater, textile dyeing effluents, and textile dyes (Malleswari and Perumal 2016). In addition, in an electrochemical paper-based biosensors, replacing a carbon screen-printed electrode with graphene-cellulose microfibers (Palanisamy et al. 2017) saw how the laccase from *Trametes versicolor* and immobilized on graphene-functionalized cellulose formed a well-defined redox pair of Cu^{I}/Cu^{II}. They observed that in comparison to unmodified electrodes, where any evident redox pair for laccase occurred, the superior electrochemical behavior of the designed sensor offered a 7-fold higher catalytic activity with a reduced oxidation potential towards catechol. Additionally, this led to a more effective sensor architecture with improved analytical performance via good sensitivity (0.932 $\mu M \mu A^1/cm^2$), low detection limit (0.085 μM), and quick response time (2 s) in the linear range between 0.2 and 209.7 M of catechol concentration.

4.4 *Bioremediation of Soil and Water*

The SSF process with *Trametes versicolor* for producing high activity laccase was studied by (Zeng et al. 2017) and reported a mixed substrate of wheat bran and corn straw to be effective for the desired result. In addition, they demonstrated the potential utilization of SSF process by *T. versicolor* for simultaneous laccase production and Bisphenol A degradation. They also found the degradation process to be more effective than *in-vitro* degradation with laccase in terms of pH, metal ions, mediators, and laccase production. The byproduct of olive mill waste in olive oil extraction contained phenolic components toxic to aquatic organisms. In this context, (Elisashvili et al. 2018) demonstrated removal of phenolics from olive mill waste by WRB (*Cerrena unicolor* BCC 301, *C. unicolor* BCC 302, *C. unicolor* BCC 303, *Fomes fomentarius* BCC 38, *Ganoderma lucidum* BCC 447, *Pleurotus ostreatus* BCC 2175, *Pycnoporus coccineus* BCC 310, *Trametes trogii* BCC 146, and *Trametes versicolor* BCC 13) derived laccase through SmF. They observed that utilizing substrate-specific WRB, especially strains of *C. unicolor*, for separate treatment of olive pomace (OP) and olive mill wastewater (OMW) holds great potential. This approach offers a viable option for simultaneously generating lignin-modifying enzymes and hydrolases, while effectively diminishing total phenolic content. The the findings suggested by-product fermentation via substrate-specific WRB as an economic process for production of cellulase, detoxified materials, xylanase, and lignin-modifying enzymes, crucial in biotechnological and agronomic applications.

4.5 *Food Industry*

Due to their energy-efficient and environmentally degradable nature, laccase-centered biocatalysts align effectively with the advancement of food sectors where their utilization is progressively expanding to create economical and healthful food products (Brijwani et al. 2010). Due to the elevated redox potential (~+800 mV) in comparison to laccases from plants or bacteria, fungal laccases have a prominent role in various biotechnological applications linked to food processing like enhancing the cost-effectiveness and ecological friendliness. For example, during bread production, it's common to incorporate additives for dough enhancement, leading to better taste, texture, size, and longevity of the bread dough. Motivated work by (Selinheimo et al. 2006) reported *Trametus hirsute*-derived laccase induces oxidation within the dough components, thereby enhancing the resilience of gluten formations in baked goods. The inclusion of laccase in dough formulation leads to enhancements in crumb structure, volume, stability, strength, and softness of the baked item, while also reducing stickiness. Laccase reduces dough extensibility in both flour and gluten dough, simultaneously bolstering the dough's resistance until reaching its peak. Also, (Minussi et al. 2007) demonstrated the potential of *Trametes versicolor*-derived laccase for phenolic removal of must, necessary for stabilisation of wine. The results demonstrated that treating white must with laccase causes higher reduction in total phenol as compared to total antioxidant potential allowing softer and ecologically correct treatments and consequently, reduce processing cost and avoid deterioration over long storage times.

4.6 Biomedical Industry

Due to their specificity and natural origin, the potential applications of laccases are sparking dedicated research in biomedical applications. An inhibitory activity towards HIV-1 reverse transcriptase and tumor cells by laccase derived from *Pleurotus cornucopiae* was reported by (Wong et al. 2010). They conveyed the inhibition of proliferation of murine leukemia cell line L1210) and human hepatoma cell line (HepG2) along with reduction in activity of HIV-1 reverse transcriptase with a half-maximal inhibitory concentration (IC_{50}) of 22 μm. Additionally, there was absence of mitogenic activity toward mouse splenocytes and hemagglutinating/hemolytic activity toward rabbit erythrocytes. In another work, (El-Fakharany et al. 2010) examines the inhibition capacity of *Pleurotus ostreatus* against entry of Hepatitis C virus to peripheral blood cells and hepatoma cells. Subsequently, they observed that laccase exhibited the ability to hinder virus replication at doses of 1.25 and 1.5 mg/ml during the initial 4-day treatment and at doses of 0.75, 1.0, 1.25, and 1.5 mg/ml during the subsequent 4-day treatment.

4.7 Dye Degradation

Globally, there's significant focus on addressing wastewater with dye content due to the potential for dyes to induce cancer, genetic mutations, chromosomal damage, birth defects, and respiratory harm (Sharma et al. 2015). In this context, WRF's are particularly proficient at breaking down synthetic dyes, as the dye structure resembles wood lignin (Diwaniyan et al. 2010). The decolourization hence, reduced toxicity of malachite green by laccase produsing *Ganoderma sp.* rckk-02 was reported (Sharma et al. 2015). They speculated the results to be beneficial for textile effuents and fish farms. In another work, the effect of *Trametes versicolor*-derived laccase immobilization on magnetic nanoparticles (Fe_3O_4 and $CuFe_2O_4$) and its effect on the biocatalytic degradation of organic dyes were studied by (Alsaiari et al. 2021). The immobilization conditions were optimized to obtain the maximum activity of immobilized laccase enzymes. The results exhibited superior activity of immobilized enzymes in comparision to free laccase for a wide range of pH, temperature, and incubation period at 4 °C. In addition, $CuFe_2O_4$ supported laccase showed enhanced activity as compared to Fe_3O_4 and free laccase. They additionally examined and optimized the nanobiocatalyst for degradation of anionic dye Direct Red 23 (DR23), where High-Performance Liquid Chromatography (HPLC) was used to confirm the degradation. The output indicated good reusability and excellent stability of immobilized laccase for DR23 degradation with decolorization efficiency of 70% up to six successive cycles.

5 Conclusion

This chapter explored the multifaceted world of laccases especially, WRF-Laccase, emphasizing their diverse applications and potential for sustainable advancements. These enzymes exhibit exceptional catalytic ability, oxidizing a broad spectrum of aromatic compounds. Their significance extends to diverse applicability in textiles, pulp and paper, biomedical, food industries along with bioremediation and biosensor applications. In this regard, understanding their structural and functional diversity, production optimization and scaling, and immobilization is paramount for augmented performance in applications. Overall, WRF-Laccase hold immense potential for sustainable, eco-friendly innovations, promising a greener future.

Declarations

Conflicts of Interest

"There are no conflicts of interest to declare by the authors."

Ethical Approval

The authors hereby state that the present work is in compliance with the ethical standards.

Competing Interests

The authors declare no competing interests.

Availability of Data and Materials

Not applicable.

References

Agrawal, K., Bhardwaj, N., Kumar, B., Chaturvedi, V. and Verma, P., 2019. Process optimization, purification and characterization of alkaline stable white laccase from *Myrothecium verrucaria* ITCC-8447 and its application in delignification of agroresidues. *International Journal of Biological Macromolecules,* 125, pp. 1042–1055.

Akpinar, M. and Urek, R. 2017. Induction of fungal laccase production under solid-state bioprocessing of new agroindustrial waste and its application on dye decolorization. *3 Biotechnology*, 7, 98.

Alsaiari, N.S., Amari, A., Katubi, K.M., Alzahrani, F.M., Harharah, H.N., Rebah, F.B. and Tahoon, M.A. 2021. The biocatalytic degradation of organic dyes using laccase immobilized magnetic nanoparticles. *Applied Sciences*, 11(17), 8216.

Aracri, E., Colom, J.F. and Vidal, T., 2009. Application of laccase-natural mediator systems to sisal pulp: An effective approach to biobleaching or functionalizing pulp fibres? *Bioresource Technology,* 100(23), pp. 5911–5916

Arora, D.S. and Gill, P.K., 2010. Effects of various media and supplements on laccase production by some white rot fungi. *Bioresource Technology*, 73(3), pp. 283-285.

Backes, E., Kato, C. G., de Oliveira Junior, V. A., Uber, T. M., dos Santos, L. F. O., Corrêa, R. C. G. and Peralta, R. M. 2023. Overproduction of laccase by *Trametes versicolor* and *Pycnoporus sanguineus* in Farnesol-Pineapple Waste Solid Fermentation. *Fermentation*, *9*(2), 188.

Bajpai, P., Anand, A. and Bajpai, P.K., 2006. Bleaching with lignin-oxidizing enzymes. *Biotechnology Annual Review,* 12, pp. 349–378.

Baldrian, P., 2004. Purification and characterization of laccase from the white-rot fungus *Daedalea quercina* and decolorization of synthetic dyes by the enzyme. *Applied Microbiology and Biotechnology,* 63, pp. 560–563

Battan, B., Sharma, J., Dhiman, S.S. and Kuhad, R.C. 2007. Enhanced production of cellulase-free thermostable xylanase by *Bacillus pumilus* ASH and its potential application in the paper industry. *Enzyme and Microbial Technology*, 41, pp. 733-739.

Bettin, F., Montanari, Q., Calloni, R., Gaio, T.A., Silveira, M.M. and Dillon, A.J.P. 2009. Production of laccases in submerged process by *Pleurotus sajor-caju* PS-2001 in relation to carbon and organic nitrogen sources, antifoams and Tween 80. *Journal of Industrial Microbiology and Biotechnology*, 36(1), pp. 1–9.

Blaich, R. and Esser, K. 1975. Function of enzymes in wood-destroying fungi. 2. Multiple forms of laccase in white rot fungi. *Archives of Microbiology*, 103, pp. 271–277.

Box, G.E.P. and Wilson, K.B. 1951. On the experimental attainment of optimum conditions. *Journal of Royal Statistical Society*, 13, pp. 1–45.

Brijwani, K., Rigdon, A. and Vadlani, P.V. 2010. Fungal laccases: production, function, and applications in food processing. *Enzyme Research*, pp. 1–10.

Castanera, R., Pérez, G., Omarini, A., Alfaro, M., Pisabarro, A. G., Faraco, V., Amore, A. and Ramírez, L. 2012. "Transcriptional and enzymatic profiling of *Pleurotus ostreatus* laccase genes in submerged and solid-state fermentation cultures." *Applied and Environmental Microbiology*, 78(11), pp. 4037–4045.

Castrovilli, M.C., Bolognesi, P., Chiarinelli, J., Avaldi, L., Calandra, P., Antonacci, A. and Scognamiglio, V., 2019. The convergence of forefront technologies in the design of laccase-based biosensors – An update. *Trends in Analytical Chemistry*, 119, 115615

Cha, J-H., Yoon, J.J. and Cha, C.J. 2018. Functional characterization of a thermostable endoglucanase belonging to glycoside hydrolase family 45 from *Fomitopsis palustris. Applied Microbiology and Biotechnology*, 102, pp. 6515–6523.

Chaurasia P.K., Bharati, S.L. and Yadava, S. 2022. Nano-reduction of gold and silver ions: A perspective on the fate of microbial laccases as potential biocatalysts in the synthesis of metals (gold and silver) nanoparticles. *Current Research in Microbial Sciences*, 3, 100098.

Chen, Xiaochu, et al. 2020. Performance and kinetic of pesticide residues removal by *Microporous* starch immobilized laccase in a combined adsorption and biotransformation process. *Environmental Technology & Innovation*, 101235.

Chhaya, U. and Gupte, A. 2013. Effect of different cultivation conditions and inducers on the production of laccase by the litter-dwelling fungal isolates *Fusarium incarnatum* LD-3 under solid substrate fermentation. *Annals of Microbiology*, 63, pp. 215–223.

Cilerdzic, J., Galic, M., Ivanovic, Z., Brceski, I., Vukojevic, J. and Stajic, M. 2019. Stimulation of Wood Degradation by *Daedaleopsis confragosa* and *D. tricolor*. *Applied Biochemistry* and *Biotechnology*. 187, pp. 1371–1383.

Claus, H. 2004. Laccases: structure, reactions, distribution. *Micron*, 35, pp. 93–96.

Da Cunha, M.A.A., Barbosa, A.M., Giese, E.C. and Dekker, R.F.H. 2003. The effect of carbohydrate carbon sources on the production of constitutive and inducible laccases by *Botryosphaeria* sp. *Journal of Basic Microbiology*, 43(5), pp. 385–392.

Daronch, N.A., Kelbert, M., Pereira, C.S., et al. (2020). Elucidating the choice for a precise matrix for laccase immobilization: a review. *Chemical Engineering Journal*, 397, 125506.

Debnath, R., Das, S., Mukhopadhyay, A. and Saha, T. 2021. Enrichment of laccase production by *Phoma herbarum* isolate KU4 under solid-state fermentation by optimizing RSM coefficients using genetic algorithm. *Letters in Applied Microbiology*, 73(4), pp. 515–528.

Diwaniyan, S., Kharb, D., Raghukumar, C. and Kuhad, R.C. 2010. Decolorization of synthetic dyes and textile effluents by basidiomycetous fungi. *Water, Air and Soil Pollution*, 210, pp. 409–419.

Dong, J.L. and Zhang, Y.Z. 2004. Purification and characterization of two laccase isoenzymes from a ligninolytic fungus *Trametes gallica*. *Preparative Biochemistry and Biotechnology*, 34(2), pp. 179–194.

Eggert, C., Temp, U. and Eriksson, K.L. 1996. The ligninolytic system of the white rot fungus *Pycnoporus cinnabarinus*: Purification and characterization of the laccase. *Applied and Environmental Microbiology*, 62, pp. 1151–1158.

El-Fakharany, E.M., Haroun, B.M., Ng, T.B. and Redwan, E.-R.M. 2010. Oyster mushroom laccase inhibits hepatitis C virus entry into peripheral blood cells and hepatoma cells. *Protein and Peptide Letters*, 17, pp. 1031–1039.

Elisashvili, V. and Kachlishvili, E. 2008. Physiological regulation of laccase and manganese peroxidase production by white-rot Basidiomycetes. *Journal of Biotechnology*, 144, pp. 37–42.

Elisashvili, V., Kachlishvili, E., Asatiani, M., Darlington, R. and Kucharzyk, K. 2018. Physiological Peculiarities of Lignin-Modifying Enzyme Production by the White-Rot Basidiomycete *Coriolopsis Gallica* Strain BCC 142. *Microorganisms*, 5, 73.

Elisashvili, V., Kachlishvili, E., Khardziani, T. and Agathos, S.N. 2010. Effect of aromatic compounds on the production of laccase and manganese peroxidase by white-rot basidiomycetes. *Journal of Industrial Microbiology and Biotechnology*, 37, pp. 1091–1096.

Elsayed, M.A., Hassan, M.M., Elshafei, A.M., Haroun, B.M. and Othman, A.M. 2012. Optimization of cultural and nutritional parameters for the production of laccase by *Pleurotus ostreatus* ARC280. *British Biotechnology Journal*, 2, pp. 115–132.

Ergun, O.S. and Urek, R.O. 2017. Production of ligninolytic enzymes by solid state fermentation using *Pleurotus ostreatus*. *Annals of Agricultural Sciences*, 15(2), pp. 273–277.

Ezike, T. C., Ezugwu, A. L., Udeh, J. O., Eze, S. O. O. and Chilaka, F. C. 2020. Purification and characterisation of new laccase from *Trametes polyzona* WRF03. *Biotechnology reports, 28*, e00566.

Fernández-Fernández, M., Sanromán, M.Á. and Moldes, D. 2013. Recent developments and applications of immobilized laccase. *Biotechnology Advances, 31*(8), pp. 1808–1825.

Ferraroni, M., Myasoedova, N.M., Schmatchenko, V., Leontievsky, A.A., Golovleva, L.A., Scozzafava, A. and Briganti, F. 2007. Crystal structure of a blue laccase from Lentinus tigrinus: evidences for intermediates in the molecular oxygen reductive splitting by multicopper oxidases. *BMC structural biology, 7*, 60.

Froehner, S.C. and Eriksson, K.E.L. 1974. Purification and properties of Neurospora crassa laccase. *Journal of Bacteriology*, 120(2), pp. 458–465.

Galhaup, C. and Haltrich, D. 2001. Enhanced formation of laccase activity by the white-rot fungus Trametes pubescens in the presence of copper. *Applied Microbiology and Biotechnology*, 56, pp. 225–232.

Gianfreda, L., Xu, F. and Bollag, J.M. 1999. Laccases: a useful group of oxidoreductive enzymes. *Bioremediation Journal*, 3(1), pp. 1–25.

Guimarães, I., Fukushima, R., Santamaria, M., Mazza, R. and Paulo, H. 2017. Comparison between some methods for quantifying the lignin content in forage. *Journal of Agricultural and Food Chemistry*, 52, pp. 3713–3720.

Hilgers, R., Vincken, J.P., Gruppen, H. and Kabel, M.A. 2018. Laccase/mediator systems: their reactivity toward phenolic lignin structures. *ACS Sustainable Chemistry & Engineering*, 6, pp. 2037–2046.

Himanshu, Chamoli, S., Singh, A., Kapoor, R.K., Singh, S., Singh, R.K. and Saini, J.K. 2023. Purification and characterization of laccase from *Ganoderma lucidum* and its application in decolorization of malachite green dye. *Bioresource Technology Reports*, 21, 101368.

Ike, P.T., Moreira, A.C., de Almeida, F.G., Ferreira, D., Birolli, W.G., Porto, A.L. and Souza, D.H. 2015. Functional characterization of a yellow laccase from *Leucoagaricus gongylophorus*. *Springerplus*, 4, 654.

Imran, S.G., Aziz, G.M. and Razuqi, N.S. 2016. Optimization of ligninolytic enzymes production from *Aspergillus terreus* SG-777 by solid state fermentation. *Iraqi Journal of Veterinary Medicine*, 40, pp. 116–124.

Janusz, G., Pawlik, A., Świderska-Burek, U., Polak, J., Sulej, J., Jarosz-Wilkołazka, A. and Paszczyński, A., 2020. Laccase properties, physiological functions, and evolution. *International Journal of Molecular Sciences*, *21*(3), 966.

Janusz, G., Rogalski, J. and Szczodrak, J. 2007. Increased production of laccase by *Cerrena unicolor* in submerged liquid cultures. *World Journal of Microbiology and Biotechnology*, 23, pp. 1459–1464.

Jaouani, A., Tabka, M.G. and Penninckx, M.J. 2006. Lignin modifying enzymes of *Coriolopsis polyzona* and their role in olive oil mill waste waters decolourisation. *Chemosphere*, 62(9), pp. 1421–1430.

Jaramillo, A.C., Cobas, M., Hormaza, A. and Sanroman, M.A. 2017. Degradation of adsorbed azo dye by solid-state fermentation: Improvement of culture conditions, a kinetic study, and rotating drum bioreactor performance. *Water Air Soil Pollution*, 228, 205.

Jin, X., Yu, X., Zhu, G., Zheng, Z., Feng, F. and Zhang, Z. 2016. Conditions optimizing and application of laccase-mediator system (LMS) for the laccase-catalyzed pesticide degradation. *Scientific Reports*, 6, 35787.

Kaluskar, V.M., Kapadnis, B.P., Jaspers, C. and Penninckx, M.J. 1999. Production of laccase by immobilized cells of *Agaricus* sp.: Induction effect of xylan and lignin derivatives. *Applied Biochemistry and Biotechnology*, 76(2), pp. 161–170.

Karp, S.G., Faraco, V., Amore, A., Letti, L.A.J., Soccol, V.T. and Soccol, C.R. 2015. Statistical optimization of laccase production and delignification of sugarcane bagasse by Pleurotus ostreatus in solid-state fermentation. *BioMed Research International, 12*, pp. 16–19.

Krumova, E.T., Stoitsova, S.R., Paunova-Krasteva, T.S., Pashova, S.B. and Maria, B.A. 2012. Copper stress and filamentous fungus *Humicola lutea* 103 – ultrastructural changes and activities of key metabolic enzymes. *Canadian Journal of Microbiology*, 58(11), pp. 1335–1343.

Kumar, A. and Chandra, R. 2020. Ligninolytic enzymes and its mechanisms for degradation of lignocellulosic waste in the environment. *Heliyon*, 6(2).

Kunamneni, A., Ghazi, I., Camarero, S., Ballesteros, A., Plou, F.J. and Alcalde, M. 2007. Decolorization of synthetic dyes by laccase immobilized on epoxy-activated carriers. *Process Biochemistry*, 43, pp. 169–178.

Lassouane, F., Aït-Amar, H., Amrani, S. and Rodriguez-Couto, S. 2019. A promising laccase immobilization approach for Bisphenol A removal from aqueous solutions. *Bioresource Technology*, 271, pp. 360–367.

Lee, I.Y., Jung, K.H., Lee, C.H. and Park, Y.H. 1999. Enhanced production of laccase in *Trametes vesicolor* by the addition of ethanol. *Biotechnology Letters*, 21(10), pp. 965–968.
Li, X., Xu, Q.-M., Cheng, J. and Yuan, Y.-J. 2016. Improving the bioremoval of sulfamethoxazole and alleviating cytotoxicity of its biotransformation by laccase producing system under coculture of *Pycnoporus sanguineus* and *Alcaligenes faecalis*. *Bioresource Technology*, 220, pp. 333–340.
Ling, Z., Wang, S., Zhu, M., Ning, Y.-J., Wang, S.-N., Li, B. and Zhao, X. 2015. An extracellular laccase with potent dye decolorizing ability from white rot fungus *Trametes* sp. LAC-01. *International Journal of Biological Macromolecules*.
Ma, W., Cao, B.-K., Xu, S.-N., Chen, D.-M. and Chen, Y. 2017. Activated carbon/polyvinyl formal sponge as a carrier for laccase immobilization for dyes decolourization. In *Proceedings of the 2nd Annual 2016 International Workshop on Materials Science and Engineering (IWMSE 2016)* (pp. 339–346).
Mahuri, Monalisa, et al. 2023. A review of microbial laccase production and activity toward different biotechnological applications. *Systems Microbiology and Biomanufacturing*, pp. 1–19.
Malleswari R.B. and Perumal K. 2016. A novel and simple laccase based paper biosensor for the detection of selected pollutants in textile dyeing effluents, water and waste water. *Octa Journal of Biosciences*, 4(2), pp. 71–74.
Manavalan, T., Manavalan, A., Thangavelu, K.P. and Heese, K. 2013. Characterization of optimized production, purification and application of laccase from *Ganoderma lucidum. Biochemical Engineering Journal*, 70, pp. 106–114.
Martínez-Morales, F., Bertrand, B., Pasión Nava, A.A, Tinoco, R., Acosta-Urdapilleta, L. and Trejo-Herna´ndez MR. 2015. Production, purification and biochemical characterization of two laccase isoforms produced by *Trametes versicolor* grown on oak sawdust. *Biotechnology Letters*, 37, pp. 391–396
Masjoudi, M., Golgoli, M., Ghobadi Nejad, Z., Sadeghzadeh, S. and Borghei, S.M. 2021. Pharmaceuticals removal by immobilized laccase on polyvinylidene fluoride nanocomposite with multi-walled carbon nanotubes. *Chemosphere*, 263, 128043.
Meehnian, H., Jana, A.K. and Jana, M.M. 2017. Pretreatment of cotton stalks by synergistic interaction of *Daedalea flavida* and *Phlebia radiata* in co-culture for improvement in delignification and saccharification. *International Biodeterioration and Biodegradation*, 117, pp. 68–77.
Messerschmidt, A. 1997. Multi-copper oxidases. *World Scientific*, Singapore.
Minussi, RC., Rossi, M., Bologna, L., Rotilio, D., Pastore, GM. and Durán, N. 2007. Phenols removal in musts: strategy for wine stabilization by laccase. *Journal of Molecular Catalysis B*, 45, pp. 102–107.
Mitchell, D.A., De Lima, L.F. and Krieger, N. 2011. Bioreactors for solid-state fermentation. In *Comprehensive Biotechnology* (Ed. Moo-Yong M), (pp. 347–360).
Moiseenko, K.V., Vasina, D.V., Farukshina, K.T., Savinova, O.S., Glazunova, O.A., Fedorova, T.V. and Tyazhelova, T.V. 2018. Orchestration of the expression of the laccase multigene family in white-rot basidiomycete *Trametes hirsuta* 072: Evidences of transcription level subfunctionalization. *Fungal Biology*, 122(5), pp. 353–362.
Mojsov, K. 2014. Biotechnological applications of laccases in the textile industry. *Savremene tehnologije*, 3(1), pp. 76–79.
Morozova, O.V., Shumakovich G.P., Gorbacheva M.A., Shleev S.V. and Yaropolov A.I. 2007. Blue laccases. *Biochemistry (Moscow)*, 72, pp. 1136–1150.
Mot, A. and Silaghi-Dumitrescu, R. 2012. Laccases: complex architectures for one-electron oxidations. *Biochemistry*, 77, pp. 1395–1407.
Mot, A.C., Coman, C., Hădade, N.D., Damian, G., Radu Silaghi-Dumitrescu and Heering, H.A. 2020. Yellow laccase from Sclerotinia sclerotiorum is a blue laccase that enhances its substrate affinity by forming a reversible tyrosyl-product adduct. *PLOS ONE*, 15(1), e0225530.
Munk, L., Sitarz, A.K., Kalyani, D.C., Mikkelsen, J.D. and Meyer, A.S. 2015. Can laccases catalyze bond cleavage in cignin? *Biotechnology Advances*, 33, pp. 13–24.

Naidu, Y., Siddiqui, Y. and Idris, A.S. 2020. Comprehensive studies on optimization of ligno-hemicellulolytic enzymes by indigenous white rot hymenomycetes under solid-state cultivation using agro-industrial wastes. *Journal of Environmental Management*, *259*, 110056.

Nandal, P., Ravella, S.R. and Kuhad, R.C. 2013. Laccase production by *Coriolopsis caperata* RCK2011: Optimization under solid state fermentation by Taguchi DOE methodology. *Scientific Reports, 3,* 1386.

Navada, K. K. and Kulal A. 2021. Kinetic characterization of purified laccase from Trametes hirsuta: A study on laccase catalyzed biotransformation of 1,4-dioxane. *Biotechnology Letters.* 43, pp. 613–626.

Nyanhongo, G.S., Gomes, J., Gübitz, G.M., Zvauya, R., Read, J. and Steiner, W. 2002. Decolorization of textile dyes by laccases from a newly isolated strain of *Trametes modesta. Water Research*, 36, pp. 1449–1456.

Palanisamy, S., Ramaraj, S.K., Chen, S.M., Yang, T., Yi-Fan, P., Chen, T.W., Velusamy, V. and Selvam, S. 2017. A novel laccase biosensor based on laccase immobilized graphene-cellulose microfiber composite modified screen-printed carbon electrode for sensitive determination of catechol. *Scientific Reports, 7,* 41214.

Palmieri, G., Giardina, P., Bianco, C., Fontanella, B. and Sannia, G. 2000. Copper Induction of laccase isoenzymes in the ligninolytic fungus *Pleurotus ostreatus. Applied and Environmental Microbiology*, 66(3), pp. 920–924.

Patel, H.K., Sharma, P.J. and Kalaria, R.K. 2019. Bioprospecting and molecular characterization of laccase-producing bacteria from industrial contaminated sites. *Global Journal of Environmental Science and Management*, 5, pp. 203–212.

Perna, V., Meyer, A. S., Holck, J., Eltis, L. D., Eijsink, V. G. and Agger, J.W. 2020. Laccase-catalyzed oxidation of lignin induces production of H2O2. *ACS Sustainable Chemistry & Engineering*, 8, pp. 831–841.

Piscitelli, A., Giardina, P., Lettera, V., Pezzella, C., Sannia, G. and Faraco, V. 2011. Induction and transcriptional regulation of laccases in fungi. *Current Genomics*, 12, pp. 104–112.

Polizzi, K.M., Bommarius, A.S., Broering, J.M. and Chaparro-Riggers, J.F. 2007. Stability of biocatalysts. *Current Opinion in Chemical Biology*, 11(2), pp. 220–225.

Radveikiene, I., Vidziunaite, R., Meškiene, R., Meškys, R. and Casaitė, V. 2021. Characterization of a yellow laccase from *Botrytis cinerea* 241. *Journal of Fungi*, 7, 143.

Raimbault, M. 1998. General and microbiological aspects of solid substrate fermentation. *Electronic Journal of Biotechnology*, 1(3), pp. 26–27.

Rodríguez-Couto, S. 2017. Industrial and environmental applications of white-rot fungi. *Mycosphere*, 8, pp. 456–466.

Saraiva, J.A., Tavares, A.P. and Xavier, A.M. 2012. Effect of the inducers veratryl alcohol, xylidine, and ligninosulphonates on activity and thermal stability and inactivation kinetics of laccase from *Trametes versicolor*. *Applied Biochemical and Biotechnology*, 167, pp. 685–693.

Schlosser. D., Grey R. and Fritsche W. 1997. Patterns of ligninolytic enzymes in T. versicolor. Distribution of extra- and intracellular enzyme activities during cultivation on glucose, wheat straw, and beech wood. *Applied Biochemical and Biotechnology*, 47, pp. 412–418.

Schneider, W.D.H., Fontana, R.C., Mendonça, S., Siqueira, F.G., Dillon, A.J.P. and Camassola, M. 2018. High-level production of laccases and peroxidases from the newly isolated white-rot basidiomycete *Marasmiellus palmivorus* VE111 in a stirred-tank bioreactor in response to different carbon and nitrogen sources. *Process Biochemistry*, 69, pp. 1–11.

Selinheimo, E., Kruus, K., Buchert, J., Hopia, A. and Autio, K. 2006. Effects of laccase, xylanase, and their combination on the rheological properties of wheat doughs. *Journal of Cereal Science*, 43(2), pp. 152–159.

Sharma, A., Shrivastava, B. and Kuhad, R.C. 2015. Reduced toxicity of malachite green decolorized by laccase produced from *Ganoderma* sp. rckk-02 under solid-state fermentation. *3 Biotechnology*, 5, pp. 621–631.

Sharma, A., Thakur, V.V., Shrivastava, A., Jain, R.K., Mathur, R.M., Gupta, R. and Kuhad, R.C. 2014. Xylanase and laccase-based enzymatic kraft pulp bleaching reduces adsorbable organic halogen (AOX) in bleach effluents: a pilot scale study. *Bioresource Technology*, 169, pp. 96–102.

Sharma, K.K., Kapoor, M. and Kuhad, R.C. 2005. In vivo enzymatic digestion, in vitro xylanase digestion, metabolic analogues, surfactants and polyethylene glycol ameliorate laccase production from *Ganoderma sp.* kk-02. *Letters in Applied Microbiology, 41*, pp. 24–31.

Sharma, R.K. and Arora, D.S. 2010. Antioxidant properties and nutritional value of wheat straw bioprocessed by *Phanerchaete chrysosporium* and *Daedalea flavida of General and Applied Microbiology*, 56, pp. 519–523

Shleev, S.V., Morozova, O., Nikitina, O., Gorshina, E.S., Rusinova, T., Serezhenkov, V.A., Burbaev, D.S., Gazaryan, I. G. and Yaropolov, A.I. 2004. Comparison of physico-chemical characteristics of four laccases from different basidiomycetes. *Biochimie*, 86, pp. 693–703.

Singh G. and Arya, S.K. 2019. Utility of laccase in pulp and paper industry: A progressive step towards green technology. *International Journal of Biological Macromolecules*, 134, pp. 1070–1084.

Singh, G., Kaur, K., Puri, S. and Sharma, P. 2015. Critical factors affecting laccase-mediated biobleaching of pulp in paper industry. *Applied Biochemical and Biotechnology*, 99, pp. 155–164.

Singhania, R.R., Patel, A.K., Soccol, C.R. and Pandey, A. 2009. Recent advances in solid-state fermentation. *Biochemical Engineering Journal*, 44, pp. 13–18.

Stajić, M. and Vukojević, J. (2011). Interaction of trace elements and ligninolytic enzymes in *Pleurotus eryngii. Biological Trace Element Research*, 143, pp. 1202–1208.

Sun, D., Sun, S.-C., Wang, B., Sun, S.-F., Shi, Q., Zheng, L., Sun, R.-C. 2019. Effect of various pretreatments on improving cellulose enzymatic digestibility of tobacco stalk and the structural features of co-produced hemicelluloses. *Bioresource Technology*, 122471.

Svobodová, K. 2005. The implications of ligninolytic enzymes in the decolorization of synthetic dyes by the white-rot fungus *Irpex lacteus*. PhD Thesis, Charles University, Prague, Czech Republic.

Tavares, A.P. M., Coelho, M.A.Z., Coutinho, J.A.P. and Xavier, A.M.R.B. 2005. Laccase improvement in submerged cultivation, induced production and kinetic modeling. *Journal of Chemical Technology and Biotechnology*, 80(6), pp. 669–676.

Teodoro, T.S., Oliveira, F.D., Poffo, C., Braga, L.P., Arbigaus, A., Rampinelli, J.R., Wisbeck, E. and Bonatti-Chaves, M., Furlan, S.A. 2018. The influence of Tween 80 on laccase production by *Pleurotus sajor-caju* and the efficiency of crude enzyme broth in the removal of bisphenol-A. *Arquivos do Instituto Biológico*, 85, pp. 1–10.

Thakkar, A.T., Pandya, D.C. and Bhatt, S.A. 2020. Optimization of laccase enzyme production by *Amesia atrobrunnea* A2: A first report. *Biosciences Biotechnology Research Asia*, 17, pp. 65–72.

Thampraphaphon, B., Phosri, C., Pisutpaisal, N., Thamvithayakorn, P., Chotelersak, K., Sarp, S. and Suwannasai, N. 2022. High potential decolourisation of textile dyes from wastewater by manganese peroxidase production of newly immobilised Trametes hirsuta PW17-41 and FTIR analysis. *Microorganisms, 10*(5), 992.

Thurston, C.F. 1994. *The structure and function of fungal laccases. Microbiology*, 1, pp. 19–26.

Tišma, M., Žnidaršič-Plazl, P., Šelo, G., Tolj, I., Šperanda, M., Bucić-Kojić, A. and Planinić, M. 2021. *Trametes versicolor* in lignocellulose-based bioeconomy: State of the art, challenges, and opportunities. *Bioresource Technology*, 330, 124997.

Umar, A. and Ahmed, S. 2022. Optimization, purification, and characterization of laccase from *Ganoderma leucocontextum* along with its phylogenetic relationship. *Scientific Reports*, 12(1).

Wahab, R.A., Elias, N., Abdullah, F. and Ghoshal, S.K. 2020. On the taught new tricks of enzymes immobilization: An all-inclusive overview. *Reactive & Functional Polymers.*, 152, 104613.

Wang, F., Xu, L., Zhao, L., et al. 2019. Fungal laccase production from lignocellulosic agricultural wastes by solid-state fermentation: A review. *Microorganisms*. 7

Wen, X., Zeng, Z., Du, C., et al. 2019. Immobilized laccase on bentonite-derived mesoporous materials for removal of tetracycline. *Chemosphere*, 222, pp. 865–871.

Wong, J.H., Ng, T.B., Jiang, Y., Liu, F., Sze, S.C.W. and Zhang, K.Y. 2010. Purification and characterization of a laccase with inhibitory activity toward HIV-1 reverse transcriptase and tumor cells from an edible mushroom (*Pleurotus cornucopiae*). *Protein Peptide Lett*, 17, 1.

Yan, J.P., Chen, D.D., Yang, E., Niu, J.Z., Chen, Y.H. and Chagan, I. 2014. Purification and characterization of a thermo-tolerant laccase isoform in *Trametes trogii* strain and its potential in dye decolorization. *International Biodeterioration and Biodegradation*, 93, pp. 186–194.

Yasmeen, Q., Asgher, M., Sheikh, M. A. and Nawaz, H. 2013. Optimization of ligninolytic enzymes production through response surface methodology. *Bioresource*, 8, pp. 944–968.

Yavaşer, R. and Karagözler, A.A. 2021. Laccase immobilized polyacrylamide-alginate cryogel: A candidate for treatment of effluents. *Process Biochemistry.*, 101, pp. 137–146.

Yoshida, H. 1883. Chemistry of lacquer (*urushi*). *Journal of the Chemical Society*, 43, pp. 472–486.

Zdarta, J., Meyer, A.S., Jesionowski, T. and Pinelo, M. 2018. A general overview of support materials for enzyme immobilization: Characteristics, properties, practical utility. *Catalysts*, 8, 92.

Zeng, S., Zhao, J. and Xia, L. 2017. Simultaneous production of laccase and degradation of bisphenol A with *Trametes versicolor* cultivated on agricultural wastes. *Bioprocess and Biosystems Engineering*, 40, pp. 1237–1245.

Zhang, Y., Hu, P., Muhammad, Y., Tang, Y., Shao, S., Gao, Z., et al. 2021. High-density immobilization of laccase on hollow nano-sphere NH2-MIL88(Fe) host with interfacial defects to improve enzyme activity and stability for remazol brilliant blue R decolorization. *Chemical Engineering Journal* 405, 127003.

Zhou, W., Zhang, W. and Cai, Y. 2021. Laccase immobilization for water purification: A comprehensive review. *Chemical Engineering Journal.*, 403, 126272.

Zhu, C, Bao, G. and Huang, S. 2016. Optimization of laccase production in the white-rot fungus *Pleurotus ostreatus* (ACCC 52857) induced through yeast extract and copper. *Biotechnology & Biotechnological Equipment*, 30, pp. 270–276.

Zucca, P., Rescigno, A., Olianas, A., Maccioni, S., Sollai, F.A. and Sanjust, E. 2011. Induction, purification, and characterization of a laccase isozyme from *Pleurotus sajor-caju* and the potential in decolorization of textile dyes. *Journal of Molecular Catalysis B: Enzymatic, 68*, pp. 216–222.

MICROBIAL GENOMICS-the Changing Technological Landscape of Microbiology via NGS

Tanvi Bajaj, Sayantan Bhattacharya, and Juhi Gupta

1 Introduction

Since the ancient ages, the curiosity to know about the different microbiomes that surround us has played a major role in scientific development, either to live disease-free or as a prevention to avoid contact with extremely contagious organisms or to adapt better food crop techniques and preserve food for the long term. This led to the discovery of animalcules by Antoine van Leeuwenhoek by using magnifying lenses, and furthermore, Louis Pasteur in 1864 proved biogenesis, and Robert Koch proposed Koch's postulates, establishing the causative relationship between a particular organism and a specific disease. The term microbe was coined in the late nineteenth century.

Microorganisms are ubiquitous organisms that play vital roles in both the human body and the environment. While they are often associated with disease and decay, they also provide many beneficial services, such as food production, waste decomposition, and antibiotic production (Sloan et al., 2006). Microorganisms play an indispensable role in our ecosystem by, transforming animal and plant remnants into more elementary molecules that is utilized through recycling processes.

Our Understanding of the biology of the human microbiome was limited by the technologies available for studying it and traditional methods; culturing microorganisms is slow and labour-intensive and often restricted to a single organism present

These authors "Tanvi Bajaj, Sayantan Bhattacharya" have equally contributed to the current work.

T. Bajaj
Strand Lifesciences Pvt. Ltd., Bengaluru, India

S. Bhattacharya
NIPGR, New Delhi, India

J. Gupta (✉)
Amity Institute of Environmental Sciences, Amity University, Noida 201303, India
e-mail: jgupta2@amity.edu

J. Gupta and A. Verma (eds.), *Microbiology-2.0 Update for a Sustainable Future*,
https://doi.org/10.1007/978-981-99-9617-9_14

in a sample. The human microbiome represents an intricate ecological system of paramount importance in shaping human well-being and illness which consists of trillions of viruses, bacteria, fungi, and other microorganisms which engage in intricate interactions with one another and with their human host, exerting influence over a diverse array of physiological functions, spanning from digestive and metabolic processes to immune responses and even brain development. Furthermore, it's worth noting that the techniques employed in previous studies exhibited bias toward specific microorganism types and were unable to detect those that resisted cultivation.

Next-generation sequencing is an innovative technology with the potential to supplant numerous conventional microbiological processes. NGS has also paved the way for fresh research into the function of the microbiome in human well-being and illness. By sequencing the DNA of microbial communities in different parts of the body, researchers can explore how changes in the microbiome are associated with various health outcomes, such as obesity, diabetes, and inflammatory bowel disease. NGS has also been used to study the microbiomes of other animals providing insights into the role of these microbes in animal health and production. In addition to its applications in human and animal health, NGS has also been used to study the microbiomes of various environments, such as soil, water, and air. This has provided insights into the ecological processes that govern microbial communities in these environments as well as their potential applications in bioremediation and other environmental processes.

Clinical metagenomics next-generation sequencing (mNGS) possesses the ability to identify most pathogens in a sample providing higher sensitivity for Identification of novel infectious agents. This efficient method has enhanced pathogen identification for diagnostic and surveillance objectives. This technology amalgamates the benefits of distinct sequencing chemistry, diverse sequencing matrices, and advanced bioinformatics techniques. Such a combination enables the high-throughput parallel sequencing of DNA or RNA sequences of varying lengths, or even entire genomes, within a remarkably short timeframe. While Sanger sequencing represents a pioneering sequencing approach, this article will also outline the most frequently used sequencing terminology in Table 1 (Buermans & den Dunnen, 2014).

2 The Evolution of Sequencing

Between 1975 and 2005, the Sanger method, also known as "first-generation" technology, was the benchmark for DNA sequencing as it Produced DNA sequences that were both long (500–1000 bp) and of high quality (Sanger et al., 1977). The next-generation sequencing (NGS) revolution began in 2005, with the release of pyrosequencing technology by 454 Life Sciences (Margulies et al., 2005), revolutionized DNA sequencing by enabling the rapid and efficient generation of millions of short reads in a single run without cloning. This technology has been expanded upon and refined in the years since, with a variety of NGS platforms now available that can generate reads of different lengths, ranging from 50 to 100,000 base pairs.

Table 1 Basic sequencing terms

Oligonucleotide	A short DNA or RNA sequence
Fragment	A concise nucleic acid fragment that emerges as a result of longer stretches being fragmented and subsequently subjected to sequencing
Fragmentation	Genetic material can be fragmented into desired sizes using either mechanical (sonication) or enzymatic methods
Sanger sequencing	A sequencing technique with limited throughput, which depends on the specific inclusion of chain-terminating dideoxy nucleotides by DNA polymerase while conducting in vitro DNA replication
Semiconductor sequencing	This Ion-Torrent Sequencing platform detects hydrogen ions released during nucleotide addition
Metagenomics	A cutting-edge genomic approach that involves the comprehensive analysis of genetic material collected from diverse environmental samples
Amplicon sequencing	A molecular biology technique that focuses on selectively amplifying and subsequently sequencing specific DNA regions of interest, typically generated through PCR (polymerase chain reaction)
Omics	The analysis and study of numerous biomolecules such as genes, RNA, methylated DNA, histones, and proteins is referred to as genomics. Transcriptomics, epigenomics, and proteomics
Sequencing by synthesis	Sequencing by synthesis (SBS) technology concurrently examines millions of clusters on the flow cell surface by employing four nucleotides, each labelled with a distinct fluorescent marker
Pyrosequencing	Sequencing involves the detection of nucleotide incorporation through enzymatic reactions, followed by the emission of light by the substrate
De Novo assembly genome	De Novo assembly refers to the procedure through which numerous brief DNA sequence fragments, like those produced by next-generation sequencers, are reconstructed into a representation of the original genomic sequence, without relying on any pre-existing reference sequences
CNV	Copy-number variations (CNVs) are structural modifications in the DNA resulting from the addition or removal of an extended DNA segment
Single-read	A technique for analysing a fragment in which the reading is conducted exclusively from a single end during the sequencing process
Read	Data generated from the analysis of a single sequence
Read accuracy	Highlights the presence of errors in percentages following the initial analysis
Read length	The quantity of bases read per fragment or the utmost fragment length that can be sequenced concurrently (expressed in base pairs)
Run	A single operational cycle of the sequencer, encompassing the entire process from startup to the acquisition of raw data
Adapter	Short, sequence-specific oligonucleotides in a library that are ligated to both the 5′ and 3′ ends of each fragment. These adapters are complementary to the concise sequence on the surface of Illumina flow cells
Lane	A designated pathway on a flow cell chip (Illumina), which can be a single, dual, or octet channel

(continued)

Table 1 (continued)

Oligonucleotide	A short DNA or RNA sequence
Long-reads	Sequencing strategy employed for samples prepared using the mate-pair-end method, resulting in longer sequence reads
Error rate	The percentage of error per sequencing read
Flow cell	A consumable component in NGS instruments, often a glass slide or solid surface
Index (barcode)	Distinct individual DNA sequences ligated to each sample to facilitate distinction and classification during data analysis
Coverage depth	The average number of reads encompassing a specific nucleotide in the reconstructed sequence
Insert	The DNA fragments of a specified size (typically 200–500 bp or larger) generated from fragmented sample DNA and ligated between two adapter sequences
Cluster generation	The localized amplification process of single DNA molecules into numerous clonal clusters across the flow cell, facilitated by unique oligonucleotides (P5 and P7)
Cluster	A clonal aggregation of template DNA bound to the flow cel's surface, originating from a single template DNA strand, and undergoing clonal amplification through bridge amplification until the cluster reaches million copies. Each cluster on the flow cell generates a single sequencing read
Contig	A contiguous consensus sequence obtained from the assembly of multiple concise, overlapping DNA fragments
Unique dual index	Pairs of indexes in which each i5 index and each i7 index are employed only once, enabling the identification and filtering of indexed hop reads, thus enhancing confidence in multiplexed samples
Alignment	The sequencing reads are mapped with reference genomes
High-quality SNP	A single nucleotide polymorphism that meets specific criteria, including sequence coverage, sequence quality, population and allelic frequency, ensuring its reliability
Paired-end reading	A sequencing method where DNA fragment is sequenced from both ends in a single sequencing run

* Modified table from Buermans & den Dunnen, 2014 and Illumina technological notes

Next-generation sequencing (NGS) technologies, such as short-read sequencing and massively parallel sequencing, generate billions of short, random reads of DNA in a single run (van Dijk et al., 2014). Despite the fact that the systems' biochemistry and arrays differ, the procedures are similar. Short-read sequencing platforms vary widely with respect to their design, sequencing chemistry, throughput (read length and number of sequences), accuracy, and cost (Buermans and den Dunnen, 2014). The Illumina technology, employed extensively within the NGS sector, depends on the detection of reversibly blocked terminator nucleotides through fluorescence and the subsequent synthesis of the complementary strand as part of the sequencing process (Vincent et al., 2017). The NovaSeq 6000 system is an addition to Illumina's flow cell technology which uses two-channel chemistry or a dual flow system

Table 2 Popular sequencing technologies

Traditional sequencing
• **Sanger sequencing**:
– Benchmark for sequencing concise and specific segments up to 1000 base pairs of DNA accuracately with ease, but sequencing entire genomes on a regular basis is a challenging task that requires a significant investment of time and resources
• **Shotgun sequencing**:
– Initial whole genome sequencing involved fragmentation of lengthy strands of DNA into many shorter segments to proceed with Sanger sequencing, It is even used in NGS methods
Next-generation sequencing technologies
• **Pyrosequencing (Roche 454)**:
– The first high-throughput sequencer to use sequencing by synthesis by detection of the pyrophosphate release upon addition of a complementary nucleotide to read the template sequence
– It has low throughput (700 and 100 bp) and subsequently high cost per base and lower accuracy in homopolymer regions, hence it was discontinued in December 2016
• **SOLiD sequencing (Life Technologies)**:
– Sequencing by oligonucleotide ligation and detection (SOLiD) generates 80–100 gbp per run using a ligation-based method, but it had a high cost per run, less data output, and lower read length
• **Ion semiconductor sequencing (Life Technologies Ion Torrent)**:
– Applies sequencing-by-synthesis method, detects pH change upon hydrogen ion release during synthesis of complementary DNA
– It is popular due to effective time and costs of sequencing, but it has an additional emulsion PCR library amplification prior to sequencing which is slow and complicated, though automation can be performed using the separate Ion Chef system
– It has high error rates (especially for homopolymers) and extremely AT-or GC-rich regions have poor coverage issues
• **Sequencing by synthesis by bridge amplification (Illumina)**:
– Applies sequencing-by-synthesis process with reversible chain termination by detection of fluorescent labels released upon incorporation of nucleotides to read the sequence
– It is the most popular and widely used DNA sequencing method on the market since it sequences a large amount of DNA quickly and accurately, at a relatively low cost
– Its limitations include difficult to assemble complete genomes or detect complex genetic variants., take longer to sequence DNA, it can dephase sequences, which can make it difficult to study gene interactions, requires multiple platforms to achieve different throughput levels, which can be expensive and complex
• **Gene reader (Qiagen) Intelligent BioSystems cyclic reversible termination platform**:
– First to automate the entire process from sample preparation to analysis. It is also highly scalable and adaptable, allowing labs to process different batch sizes and load multiple flow cells at the same time
– This platform is specifically designed for clinical applications, such as cancer biology. It was introduced at a later stage with third-generation sequencing technologies
• **Single-molecule real-time sequencing (Pacific Biosciences)**:

(continued)

Table 2 (continued)

Traditional sequencing
– Novel sequencing method uses natural DNA synthesis to produce long reads (up to 40 kb) with fluorescent labels attached to the phosphate group
– The errors are higher and random but can be corrected by overlapping reads to produce a high-accuracy consensus sequence
– Assembles gen omes and complete bacterial genomes more accurately than traditional methods, without using PCR to close gaps. However, it is expensive to set up and slow, so it is not widely used
• **Nanopore sequencing (Oxford Nanopore)**:
– Nanopore sequencing is a leading benchtop sequencing technology under development. It works by detecting changes in an electric current as DNA passes through a tiny protein channel
– It has rapid and smart sample prep in under 10 min, small size and low cost, but has ability to generate long reads
– Two portable and affordable benchtop DNA sequencers: the MinION is the size of a USB stick, while the GridION is rack-mountable

delivering highest throughput across multiple samples. A consistent fragment length should be maintained while operating on the NovaSeq platform. The NovaSeq X series is the latest instalment from the Illumina company. It is equipped with Illumina's state of the art XLEAP-SBS chemistry, which is by far, Illumina's most advanced, delivering high rate of accuracy, robustness and ultra-high throughput real time sequencing which is also easy to use and provides precise genomic information. It has improved genome coverage and better sensitivity for indels and SNP detection. Due to their high throughput with accuracy and comparatively low cost. Illumina systems are the most frequently utilised in both clinical and research sites. To achieve optimal cost efficiency in real-time testing, it is important to select a sequencing platform that is compatible with the laboratory's sample throughput. Thermo Fisher's IonTorrent platform is also based on sequencing by synthesis, but detection relies on solid-state pH meters that monitor hydrogen ions generated during DNA polymerization (semiconductor technology) (Schadt et al., 2010). The IonTorrent PGM and S5 instruments exhibit similarities to the Illumina MiniSeq and MiSeq, respectively, while the Ion Proton aligns with the Illumina NextSeq in terms of their respective capabilities and functions. When contrasted with the Illumina Platform, the IonTorrent library readiness technique can be automated using the IonChef Sample preparation system, which, when done manually, can increase the hours, but on the other hand, the sequencing run is considerably shorter, irrespective of the generation, single-end reads are the norm, with read lengths typically extending 100 base pairs beyond the maximum read length achievable with Illumina sequencing (Besser et al., 2018). The semiconductor sequencing method employed by the IonTorrent platform tends to yield elevated error rates in base calls, particularly in extended homopolymer regions. Short-read methodologies are frequently employed for creating synthetic bacterial genomes for the purpose of variant calling in diagnostic and infectious disease surveillance applications. Nevertheless, low-throughput technology can be advantageous in specialized applications such as customized amplicon sequencing, 16S sequencing for the detection of antimicrobial resistance markers, metagenomics sequencing, and RNA sequencing for gene expression analysis. Short reads produced by second-generation sequencing systems often lead to genome assemblies that are only partially contiguous. To attain complete reference genomes, longer reads become indispensable. Longer reads are especially advantageous when sequencing complex genomic regions that contain repetitive elements and phages. Third-generation sequencing systems have been developed to fill the gap by targeting individual DNA molecules directly, eliminating the necessity for PCR amplification. Pacific Biosciences' PacBio RSII platform applies single-molecule real-time sequencing technology (van Dijk et al., 2014). PacBio sequencing is a long-read sequencing technology that uses single-molecule real-time (SMRT) sequencing to sequence DNA fragments of up to 20 kb or greater. In SMRT sequencing, single-stranded DNA molecules are deposited into tiny crevices where a single DNA polymerase molecule is immobilized. Nucleotides labeled with unique fluorescent dyes are synthesized, and the fluorescence is detected as the nucleotides are incorporated into the DNA strand. This allows for the real-time sequencing of DNA molecules. While PacBio sequencing offers the advantage of generating long reads, it has a lower throughput and higher error rate than short-read sequencing technologies. As a result, PacBio sequencing is often used in sequencing core facilities to generate high-quality completed genomes. Pacific Biosciences' Sequel sequencing platform is a high-throughput and more affordable version of the RSII platform. Sequel achieves a lower error rate by generating more sub-reads per sequence. However, the length of the reads decreases as the coverage increases.

Oxford Nanopore Technologies Ltd. has developed fourth-generation DNA sequencing technology that offers label-free, ultralong reads, high throughput, and low material requirements. Nanopore sequencing is a single-molecule sequencing technology that can sequence DNA or RNA without the need for PCR amplification or chemical labeling of the sample. This makes it a promising technology for field diagnostics applications, as it offers the potential for relatively low-cost, portable, and rapid sequencing with real-time results. Nanopore sequencing provides low-cost genotyping, testing mobility, and swift sample preparation with the facility to display insights in real time, making it a compelling choice for field diagnostics applications. Electrophoresis is used in nanopore sequencing to transport a sample through a 9–10 m diameter aperture. An electric current is identified in the system upon the application of a steady electric field, and an electrolytic solution is constantly present in a nanopore system. Nanopore sequencing is a method of sequencing DNA or RNA by measuring the changes in electrical current as each nucleotide passes through a nanopore. The size of the nanopore and the composition of the DNA or RNA molecule affect the magnitude of the current change. This allows for the identification of individual nucleotides, which can then be assembled into a sequence. The MinION device, invented by Oxford Nanopore Technologies, is the smallest long-read sequencing technology available, measuring the size of a big thumb drive. The MinION demonstrates a read-length distribution similar to PacBio, but it exhibits a higher error rate, ranging from 12 to 38%. In contrast to PacBio, enhancing coverage does not ameliorate the error rate, as the MinION is inherently restricted to two sequencing passes. The present iteration of the MinION faces challenges when sequencing GC-rich regions and necessitates intricate assembly and error correction algorithms to generate high-quality assemblies. Furthermore, its current throughput is relatively modest and exhibits some instability.

Given the wide range of NGS platforms available and the diverse workflows required for different applications, choosing the optimal platform for a specific workflow can be challenging (Bharagava, 2019). Table 2 provides a summary of the categories of commercially available NGS platforms (Dubey et al., 2022).

3 Basic NGS Workflow

3.1 Microbial Whole Genome Sequencing

A vast majority of Next Generation Sequencing techniques follow multiple similar operations in the sequencing of a particular gene or genome. Pathogen identification, comparative genomics, and analysis of antibiotic resistance genes are all possible with the help of microbial whole-genome sequencing. For microbial analysis, the workflow involves: (i) Nucleic acid isolation; (ii) NGS Library construction which is fragmentation of the nucleic acids to the desired range of nucleotide base pairs using mechanical or enzymatic shearing, addition of adaptors, barcodes/indexes and amplification; (iii) template preparation, either by bridge amplification or emulsion PCR; and (iv) automated sequencing, (v) Data analysis and storage (Didelo et al., 2012; Gilchrist et al., 2015).

3.2 Identification & Preparation of Microbial Sample

For microbial analysis of any disease, firstly, the causative pathogen is identified and based on the type of the condition, the sample collection is executed, for example, if the patient suffers respiratory distress, the clinical sample should be a swab. Clinical samples can be of different types-swab, sputum, stool, urine or tissue which harbours the gene of interest, the target DNA of the pathogen that is responsible for disease progression/ infection. The sample, once collected, needs to be maintained at a particular temperature until it safely reaches the microbiology laboratory.

For detection of novel genes in microbes thriving in nature and to unravel their genomic characteristics, researchers also perform microbial analysis on environmental samples with the help of NGS. To perform metagenomic analysis in a microbial community, samples are extracted from soil, air and water and made ready for sequencing. Since DNA needs to be extracted from the microbial community as a whole, for example, from a biofilm settled on a rock, sometimes it becomes difficult to extract high molecular weight DNA from limited environmental samples as the samples will vary in terms of cell wall rigidity, with each species of bacteria having varying layers of peptidoglycan. Not only bacteria, but NGS methodologies are often employed to decipher complex microbial genomes of both culturable and unculturable microorganisms in their natural niche (Bag et al., 2016). The type of sample will depend on the purpose of study and application; e.g.-In plant–microbe interactions, samples need to be collected from the soil or bark-wood. For microbiome analysis and ecological studies, community DNA is extracted from biofilms. Air quality assessment requires sampling of air for the detection of harmful microbial genes in the community. To study water quality or water borne infections, sampling of sewage water is also executed to study the vast repertoire of microorganisms and understand the molecular basis of their virulence genes through Next Generation Sequencing.

3.3 *Nucleic Acid Isolation*

The process of DNA extraction is varied and depends on the sample type and the genetic material that is required for the sequencing. Certain steps are either added or removed for a specific extraction process. The composition and strength of the lysis buffer used can be specific for a fungal, viral, bacterial sample. For a bacterial sample, it is grown in a culture medium and harvested by centrifugation before extraction whereas a viral sample can be any human body fluid, such as blood or swab dipped in Viral Transport Media, which needs to be processed prior to DNA extraction for the elimination of unwanted debris and host cells (Fig. 1).

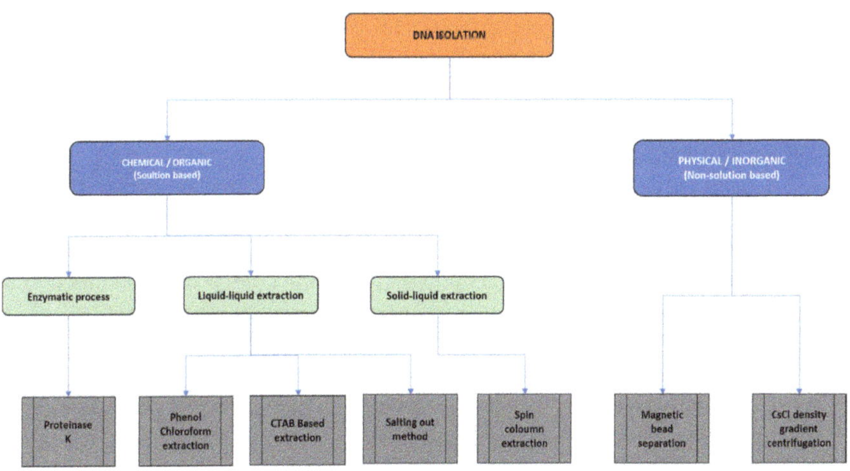

Fig. 1 Classification of DNA isolation methods

Viral DNA extraction also requires additional steps compared to bacterial extraction, such as the infection stage, viral particle lysis stage and nucleic acid release, followed by selective amplification of viral DNA using PCR. Viral DNA yield is generally low as compared to bacterial DNA which has higher biomass in a culture. However, it is pure and free of most other impurities which are present in case of bacterial isolation, that may interfere with downstream applications. The yield of fungal DNA is also generally low owing to the fact that it has a rigid and a complex cell wall that provides structural support to the cell. Fungal DNA extraction requires extra steps such as freeze drying with liquid nitrogen to break open the mycelium, which hinders the smooth conduction of the process. The overall yield of the process is low, but the DNA quality is good and pure (Kumar et al., 2018). After the microbial sample has been collected, we need to extract the genetic material from the selected microbial colonies by using either manual methods or commercial kits of DNA isolation and other methodologies that are governed by the type of NGS workflow being followed. PureLink™ Genomic DNA Mini Kit, Qiagen DNeasy Blood and Tissue Kit, Bio-Rad Genomic DNA Isolation Kit and Illumina TruSeq DNA PCR-Free Library Preparation Kit are some of the most commonly used DNA isolation kits. DNA extraction can be of various types-depending on the sample that we have at hand.

1. PCI: The most popular or widely used method is the organic method of extraction, which is performed by using Phenol Chloroform Isoamyl alcohol. It works on the principle of liquid–liquid extraction where, phenol, being the organic solvent, dissolves lipids and proteins and pushes them towards the organic phase after phase separation using Chloroform. DNA can be isolated from almost every known biological material (mostly plant & bacterial cells) using this method and the yield rate generally remains above 90%. The method can be time-consuming and hazardous due to the use of toxic chemicals.
2. Proteinase K extraction: Developed by Ebeling et al. in 1974, this method of extraction uses serine protease-proteinase K which directs an enzymatic catalytic reaction that denaturates both nuclear and cellular proteins. It is a rapid method of extraction, can be used for any sample type, safe to use and yields are fairly high. The only con of this method is its high cost.
3. Salting out method: This is a common method used in extraction of genomic DNA from peripheral blood in clinical cell lysates. It involves precipitation of proteins and other cellular contaminants by using a high concentration of salts due to their relative hydrophobicity. As DNA doesn't precipitate (Sun et al., 2010), it generally results in low yield, although it is simpler and more cost effective. It does not work effectively with plant samples (Gautam et al., 2022).
4. Silica based adsorption method: Started by *McCormick* in 1989, this method works on the principle of solid liquid extraction. It uses silica (stationary phase) based spin columns and a supplied lysis buffer, where the DNA gets adsorbed or immobilized over the silica membrane after lysis of the cell contents. It is a fast and efficient method that produces high-quality DNA, but can be expensive. Yield

is good but not better than the above-mentioned techniques, and any biological sample (viral, bacterial, plant, etc.) can be used.

5. Magnetic bead DNA extraction: Originally developed by Hawkins T in 1998, this method uses the principle of solid phase extraction to extract DNA with the help of specialised beads. The beads, coated with specific polymers, attach to the DNA present in the sample when it is incubated in the presence of a magnetic field. The DNA bound with the bead remains at the bottom, while the supernatant is discarded. The process is faster, any biological sample can be used in this technique and yield is better than spin column-based extraction. However, it requires advanced machinery for its set-up.
6. CTAB based DNA extraction: Plant samples are mainly used for this extraction process. This method uses a simple precipitation technique by virtue of which the DNA gets precipitated by CTAB, a non-ionic detergent (Cetyltrimethylammonium bromide) followed by chloroform addition and alcohol precipitation (Sharma et al., 2023).

To check whether the Genetic material (DNA or RNA) is of good quality and can be processed further for sequencing, a spectrophotometric analysis and gel electrophoresis should be carried out to determine the purity of the extracted DNA or RNA. DNA quantitation is performed by Picogreen fluorescence chemistry (Thermo Fisher Scientific, Waltham, Massachusetts) (Yohe et al., 2017). The purity ratio, or A260/A280 value, should be 1.8–2 for DNA and close to 2 for RNA. The bands on the gel should be sharp and intact, without any contamination. In the event of contamination, the DNA or RNA should be further purified. Using oligo-dT primers and reverse transcriptase, RNA is converted to cDNA. A favourable outcome hinges on choosing the right strategy and appropriate method selection, thereby minimizing the occurrence of biases and false negatives.

3.4 Library Construction for NGS

Library construction is the process of generating DNA fragments that can be sequenced. This is typically done by amplifying DNA fragments and adding barcodes and adapters to the ends. The most common method is to sheer the nucleic acids into smaller pieces and then add adapters at the ends of the sequences. Two approaches that are followed during the fragmentation of DNA are Enzymatic Fragmentation and Mechanical shearing. The former uses restriction enzymes, which cut or nicks the DNA at specific positions within the genome, while the latter uses Focused Acoustic shearing or Hydrodynamic Shearing. In the case of mechanical shearing, the DNA sample is acted upon by ultrasonic sound energy, which shears the DNA into fragments ranging from 150 to 5000 bp. This method is capital-intensive, cumbersome and time consuming as it requires advanced machinery. The latter approach using restriction enzymes requires less capital, is easy to use and parallelly used for multiple samples at the same time. However, this method is prone to contamination in the DNA sample (Lynne, 2017). The adaptor ligated fragments include molecular bar codes (to assign a unique identity for pooling of multiple samples together), universal PCR primers, complementary DNA regions that can hybridize sequences onto a rigid surface for the initial process of sequencing. When the sequencing-ready DNA fragments are flanked by adaptors on both ends, these fragments are referred to as libraries. The size of the DNA fragment within the upstream and downstream adaptors is known as the insert size. The length of the DNA sequence between the upstream and downstream adapters is known as the size of the construct (Yohe et al., 2017). Cloned libraries should be prepared and quantified and the accuracy of the library preparation largely determines the efficiency of the NGS workflow. To decrease the impact of sequencing errors or degradation of the samples, nucleic acids are made sequencer compatible. The first step for DNA based Sequencing is to quantify DNA. The amount of DNA that is used for formation of a library is typically 25 to 300 ng. DNA fragmentation or shearing which breaks up genomic DNA into small fragments upon treatment by restriction enzymes or sonication. Adapter molecules specifically bind to both ends of each fragment. Indexing of the samples should be done after adapter ligation which involves introduction of unique sequences, called barcodes to each library, which are amplified by PCR, and a cleanup is performed to remove any unwanted materials in the final library. Each library can be differentiated from other libraries on the basis of the presence of these barcodes. and thus, sequencing libraries are formed A library quality check is performed using the Agilent Tapestation or any other Advanced Analytical Fragment Analyzer.

3.5 QC Parameters for the NGS Library

Quality control or QC is a technique for assigning quality parameters to the prepared NGS library that will be sequenced upon qualifying the standards. QC provides information about the quality of the library profile in terms of library size, distribution, concentration and contamination, which needs to be determined through qPCR, Fluorometrics and Microfluidics. An unsuitable library for sequencing is one in which the amount of DNA, i.e., concentration is lower or higher than a certain optimum level which either leads to under clustering or over clustering during sequencing, leading to technical problems while loading the sample in the flow cell. In a similar way, a library fit for sequencing contains short read fragments, particularly in the range of 200 to 1000 bp. Shorter fragments are less acted upon by polymerases and are difficult to assemble. Library concentration can be estimated by qPCR by selective amplification of the adapter ligated fragments which can provide insights about the fragment size and total amount of DNA that will be required for Sequencing. Microfluidics are mainly used to check the fragment sizes and the concentration of the DNA by using fluorescence generated in an electrophoretic assembly chemistry, thereby evaluating the quality of the samples thoroughly. A quality failed library is indicative of contaminants present in the sample-unwanted biases, adapter or primer contamination, and primer dimers—which generate errors in sequencing and hamper downstream analysis.

3.6 Nucleic Acid Sequencing

Sequencing for microbial samples is mostly performed using either of these platforms: Illumina sequencers (San Diego, California)-Novaseq, MiSeq, NextSeq or the Ion Torrent sequencing method, which includes the IonPGM, IonProton and IonS5. (Thermo Fisher Scientific) The first step in sequencing DNA on both Illumina and Ion Torrent platforms is to attach each DNA fragment to a solid surface and amplify it many times over (Yohe et al., 2017). The most commonly used NGS techniques for sequencing clinical samples containing microbial colonies are short read platforms (such as Illumina and Ion Torrent) and long read platforms. (Single molecule real time sequencing (Pacific Biosciences) and nanopore (Oxford Nanopore) sequencing) (Levy et al., 2016) Generally, the majority of genetic diversity is captured by short reads, but long-read sequencing allows the detection of complex variants like inversions, deletions, or translocations. First, the sample or nucleic acid is sheared into numerous fragments through sonication, nebulization, or restriction digestion (Escalona et al., 2016, Motro et al., 2017). In this particular step, the adapter ligated DNA fragments (cloned libraries) are loaded into a special chamber, which is called the flow cell. A flow cell is typically a glass slide with lanes, each lane containing sequences that are complementary to one of the adapter sequences in the fragmented DNA, which ensures proper attachment of the target DNA with the flow cell. Each single stranded DNA fragment in the flow cell is clonally amplified in a process called Bridge amplification, which involves amplifying the clusters of DNA fragments. (Cluster generation) Due to distinctive sequences found in the synthetic adapter molecules, the templates are primed after fragment amplification, resulting in a free 3′ OH for enzymatic extension on the templates together with on-instrument data detection (McCombie et al., 2019). dNTPs (sometimes chemically modified) attach to a region complementary to the template strand or the target DNA to be sequenced, downstream to the sequencing primer on the new strand. This technique is also called Sequencing by Synthesis, Illumina sequencers use this technique of sequencing aided by fluorescence detection. Deoxynucleotide Triphosphates (A,T,G,C), fluorescently labelled, are loaded and a reversible terminator is also involved to prevent the inclusion of the subsequent base. Reversible terminators (ddATP, ddGTP, ddCTP, and ddTTP) have a cleavable fluorescent dye which is protected at the 3′-OH group. During DNA synthesis, modified reversible terminators are integrated into the expanding DNA strands and produce fluorescence that is detected by a charge-coupled device camera (Zhong et al., 2021). This is explained by the fact that a blocker or a terminator limits the addition of nucleotides to one per sequencing cycle. Laser light emits fluorescence signals that is detected and according to the type of fluorescence emitted, we get to know about the type of nucleotide that has been incorporated (Mardis ER, 2013; Goodwin et al., 2016).

The fluorescent labels and blockers are removed, and the next sequencing cycle begins. During each cycle, one base pair is identified from each DNA cluster. The short reads (non-bound nucleotides) are removed after sequencing of the sense DNA strand by a washing step, following which the process is reiterated for the antisense strand. This method is called paired-end sequencing.

3.7 Sequence Data Analysis

NGS technologies employ a wide array of interplay between sequencing chemistry or technique, hardware and optical sensors. After bridge amplification, when the amplified clones have been subjected to sequencing by synthesis. Quality Check of the sequenced data is executed to determine the arrangement of the DNA nucleotides for identification of the genomic sequence of the pathogen. This feature makes use of computer softwares (Phred Q scores) to convert signals (say, light) into actual sequence data with a particular score for identity and sequence coverage (Ewing et al., 1998). The likelihood of incorrect substitutions or indels (insertions/deletions) at a specific place within a read is determined using error-rate models (Johnson et al., 2014; Illumina Technical note, 2011; Escalona et al., 2016).

Every sequencing platform features unique error rate models which sometimes vary within the same method employed (Escalona et al., 2016; Motro et al., 2017). In today's generation of biological sciences and in the current world of sequencing technologies, we have multiple bioinformatics tools (like Multi Align) to analyse and interpret the NGS data. These tools provide sequence alignment between the already known gene sequences and target DNA to be sequenced, variant calling, data visualization or interpretation.

3.8 Data Release, Output and Report

Different file formats may be utilised for storing the output of NGS reads. SFF (or Standard flowgram format) data may be collected from Roche life sciences platform while FASTA or FASTQ files are required for Ion Torrent, Illumina or Oxford Nanopore, depending on the particular NGS technology emulated. Other potential output alignment files can be of the format MAF (multiple alignment format) or SAM and BAM formats.

The stored data for the whole workflow should be released as a final report and such documentation should be archived, keeping in mind the confidentiality of the report (Escalona et al., 2016; Motro et al., 2017).

3.9 QA/QC

The NGS/WGS data should pass the QC metrics parameters for reaching a particular standard. For example, to predict the antibiotic resistance genes of some bacterial species, the quality and quantity of raw reads must be evaluated to ensure that the bacterial genome is adequately covered (e.g., >30-fold coverage), and the quality of the de novo assembly must also be evaluated along with the detection of contamination from other DNA samples which may originate from improper handling of bacterial isolates or from handling & running of samples in the sequencer (Ellington et al., 2017). Some of the most followed and important QC parameters include number of reads, average read length, depth of coverage & total DNA sequence, total number of contigs, etc.

4 Applications of NGS

The accessibility of a microbial genetic toolkit at a wide range will help supplement existing perception of the microbiota and aids in real-time surveillance of the processes for determining the exact health status of the host. In addition, developing infectious illnesses like antimicrobial resistance (AMR) ensue for improved and safer health-care solutions and with the advent of microbial genomics, one can anticipate that bacteria will be a major player in therapeutics and the development of innovative diagnostic procedures in the next years. One promising area of microbial research is the development of new probiotics that can improve human health and metabolic function in ways that are specific to the microbiome. Another important microbiological event that could occur in the next few years is the development of alien microorganisms or the adaptation of terrestrial bacteria to extraterrestrial environments. The rapid pace of urbanization, overpopulation and global trade is pressuring the environment and human health and there is an urgent need to slow or reverse these trends as well as there is also a need to implement sustainable measures to limit the recurrence of diseases and pandemics (Kumar et al., 2021). Next Generation Sequencing techniques can be harnessed in multiple ways to meet the challenges of disease diagnostics in the fields of microbiology. Firstly, it serves to discover and detect novel pathogens or a community of unknown microbes at a specific area or in a microbial sample using Metagenomic approaches, thus providing a comprehensive view of the microbial community in a sample. The amount of data generated using Next Generation Sequencing can be utilized to infer the taxonomy, virulence pattern and antimicrobial resistance phenotypes of the pathogen (Bitar et al., 2023). It has been also observed that rapid interpretation of the resistance profile of an infectious agent, facilitated by Oxford Nanopore's MinION Real time sequencing system assists in faster diagnostics procedure (Avershina et al., 2022). Also, NGS techniques can be widely used to investigate an outbreak of any invasive microbial species that has the potential to cause an epidemic or a pandemic and is routinely used to perform mutation analysis of genomes. Sequencing through NGS monitors the emergence and spread of outbreak strains in healthcare settings, allowing for early detection and intervention. NGS can be used to identify viral, bacterial, fungal, and parasitic pathogens in microbial samples. The use of NGS techniques have modified with time and has been readily replacing traditional techniques like Sanger and DNA Microarray analysis. Some of the utility applications of NGS include Chromatin immunoprecipitation sequencing (ChIP-Seq) and Transcriptome sequencing (RNA-seq). The former is mainly used for the detection of DNA binding sites for Transcription factors or other related proteins and prediction of methylation sites, thereby assisting as a diagnostic tool for studying diseases like cancer and its progression while the latter is used to study gene expression and other complex biological processes (Fig. 2).

Another major application existing in the fields of microbial Next Generation Sequencing & microbiology is the study of ancient DNA of entire microbiomes affixed to solid surfaces, such as rocks, deep sea boulders, glaciers which has been shaping the Earth's microbial diversity since centuries. Efforts have been made towards sampling of these ecologically diverse microbiota so that these rare genomes can be sequenced and further information about their evolution can be obtained. For Example, ice samples were collected by using drill machines and aging of the ice glacier was done through computer softwares. The combined technology, together

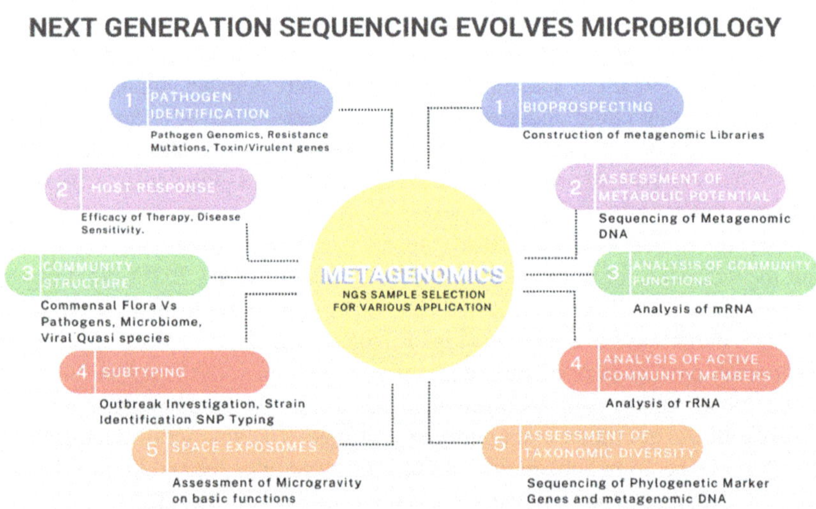

Fig. 2 Next generation sequencing evolves microbiology, the various applications (left side) of NGS that will evolve Microbiology studies and solve many clinical as well as environmental curiosities with the right technical use (Right side)

with sequencing of microbial species in ice throws light on the evolutionary history and ecological diversity of ancient microbial populations residing since the past 2000 years (Garcia-Lopez et al., 2021).

4.1 Prominent NGS Techniques Used in Microbiology

Novel strategies to disclose the whole evolutionary ecosystem of microorganisms and their interactions that impact everyday life. There are currently just a few ways being used to characterise these bacteria. We begin by using various methods, including fluorescence-activated cell sorting (FACS), serial dilution, micromanipulation, optofluidics, and laser-capture microdissection of tissue samples, to isolate single cells or colonies from environmental samples. To work with the DNA samples, cell lysis is followed after separation using procedures like alkaline lysis (Clingenpeel et al., 2014). Whole genome amplification (WGA) using methods such as isothermal multiple displacement amplification (MDA) or multiple annealing and looping-based amplification cycles (MALBAC) can be used to amplify the DNA from a single cell, generating enough DNA for subsequent analyses. This allows for the discovery of microbial dark matter, which is the unculturable and understudied fraction of the microbiome, in a wide range of environmental samples (Zong et al., 2012).

Targeted NGS panels make use of panels that are predesigned in such a way that it can detect known sets of pathogens from microbial samples, the information for the same being available in published literature. Targeted NGS gene panels can achieve higher range, responsiveness and accuracy for the sequenced genes resulting in a higher diagnostic yield than exome sequencing (ES) or genome sequencing (GS) for many applications (Zhong et al., 2021). These panels are highly sensitive and specific, have high throughput and a faster processing rate of generating read data. Targeted panels have sequences which are specific or complementary to the genome sequences of the target organism-can be bacterial, viral or even eukaryotes. Also, the panels can be calibrated according to the type of sample that is used for sequencing. (For example, CSF, nasal swabs, sputums, etc.) (Mitsura et al., 2021).

Massive metagenomics sequencing is being used to obtain the genomic information of microorganisms that are difficult or impossible to grow in the laboratory and the sequencing technique which has the potential to detect all infectious agents—bacteria, viruses, fungi and parasites—in a sample is the metagenomic NGS (mNGS) (Chiu et al., 2019). This method requires procurement of both clinical & environmental samples directly from the patients or from places like sewage, sea bed, rocks, surfaces, etc. After massive sequencing, the DNA of all known species in the sample, which includes the host cells, can be amplified using PCR. This reveals the previously unknown diversity of microbial dark matter, both in terms of its phylogenetic relationships and its functional capabilities. This allows researchers to identify microorganisms or microbial processes that were previously undetectable, and that may be specific to a particular population or habitat. Given the challenges of recreating natural conditions in the laboratory, scientists are developing new methods to cultivate bacteria in their natural environment. Metagenomic approaches define the total nucleic acid content (both DNA and RNA) in a sample, which helps us characterize the transcriptome in such samples and develop a method to analyse the entire microbiome as well (Kumar et al., 2021; Mitsura et al., 2021). mNGS has the ability to detect pathogenic sequences that constitute a very meagre fraction of all sequenced reads. In addition, it is useful when results from targeted sequencing are not satisfactory, as from a single sample, this method can detect a wide variety of microbial agents. Another sequencing technique follows genome wide sequencing of microbial samples, which is a broad sequencing procedure as it generates information about the entire genome of the pathogen, including extrachromosomal plasmids. (or cosmids) Mainly carried out in a laboratory for detecting the presence of antimicrobial or antibiotic resistance genes in clinical isolates, this method is useful for preparing first line drugs for therapeutic purposes. Inadequate information about genetic diversity further complicates the precise prediction of resistance cassettes by whole genome sequencing and results in diminished susceptibility to a particular antimicrobial agent (Ellington et al., 2017).

5 Future of Next Generation Sequencing in Microbiology

Microbiology is the study of the framework and operations of microbiological communities, the interplay of components and behaviours of internal communities, and the communications of microbes with their surroundings or hosts (Alexander, 1962; Niel, 1966). The microbiome is an ecosystem of all microbial species in a certain environment, including their microbial genome data and identities. Microorganisms participate in a variety of life activities and have an impact on their surroundings as well as other organisms. Human health, agricultural development, livestock cultivation, sustainability, industrial chemical manufacturing, and food production all rely on microorganisms. Microbes were first spotted and studied with microscopes in the nineteenth century. With the implementation of high-throughput sequencing, collection of massive volumes of data about microbial resources has been made possible. Consequently, machine-learning techniques are also been readily used in research pertaining to microbiology. Obtaining relevant OTU in microbial research based on gathered samples is a critical stage in the examination of microbial data (Yu et al., 2015, 2016; Shi et al., 2016; Su et al., 2018; Fan et al., 2019). Despite significant progress, many challenges remain in microbiological research. To overcome these challenges and advance the field, collaboration between researchers from different disciplines, such as biology, informatics, and medicine, is essential.

Thanks to advancements in microbial sequencing, the microbiome has been increasingly prominent in a variety of studies in recent years. The gut microbiota inside humans is proposed to be a possible moderator of health and illness; hence, the future of microbiological sciences will be rich with pharmaceutical breakthroughs based on microbiome research (Kho and Lal, 2018). As the global market for therapies continues to grow, the potential for personalized medicine based on individual gut microbiota is also expanding (Ihekweazu and Versalovic, 2018). Humans are predetermined to seek space research, yet extended space voyages can expose astronauts to a range of illnesses. Stress factors such as microgravity and interstellar cosmic radiation compose 'the space exposome' (Crucian et al., 2018). Space exposome, in conjunction with ship sterility, food in space, and environmental pressures, may produce alterations in the gut microbiota, compromising immunity and placing astronauts at risk of sickness. Pathogen virulence increases in microgravity, perhaps making extraterrestrial flight unfeasible (Saei and Barzegari, 2012). Space medicine researchers are paying increasing attention to the development of effective life support systems for long-duration space missions (Kumar et al., 2021). Conventional methods such as microscopy and cell culture are costly and require significant time, thus, machine-learning is an upcoming futuristic technology combined with the powers of artificial intelligence that can be utilised with the right resources in microbial investigations to speed up the process and uncover a lot of untapped knowledge in microbiology.

6 Limitations

Although NGS has become the most effective diagnostic tool, revolutionizing the field of microbiology, it has its limitations. One of the greatest challenges of today's sequencing technology is the inability/difficulty to sequence "inaccessible" regions of the genome, specifically the regions harbouring GC rich repeats & repetitive elements. Currently, it is becoming increasingly difficult to distinguish homologous regions of the genome. Homologous regions bear a remarkable sequence similarity with the actual gene of interest, which differs from the target construct sequence by only a few base pairs. They mainly result from short read platforms and are called pseudogenes. Improper differentiation of homologous regions leads to pitfalls in disease diagnostics like increased probability of false positives & false negatives due to poor mapping quality or mis-mapping. GC rich regions are involved with unwanted background noises which appear as many smaller, undefined peaks under the peaks of interest. Another major con of NGS technology lies in the fact that the decreased sensitivity of several advanced platforms fetch short sequencing reads which leads to gaps in sequences and often sequence alignment failures arise with respect to repetitive areas of the genome (Zhong et al., 2021). To add to the complexities, it also accounts for misinterpretation of mutations (like methylation of cytosines in the DNA or (SNVs)) and other genomic information (Yohe et al., 2017).

In a molecular diagnosis of certain microbial diseases, NGS has been criticised for its inability to interpret results due to impact of background bacterial contamination, leading to improper sequence analysis (Duan et al., 2021). In microbial diagnostics, antibiotic resistant genes are sequenced to find out the potential number of MDR strains that has been rampaging the healthcare facility for quite some time. For detection of MDR/AMR isolates using WGS, it requires culturing of the pathogen. This extra step affects the rate of diagnosis of antimicrobial susceptibility. Current methodologies lack a proper workflow of generating antimicrobial susceptibility data from NGS data which leads to delayed report generation in health care centres (Ellington et al., 2017).

Popular techniques to know functional analysis of gene and protein expression includes RNA-seq which serves to study the entire transcriptome of the concerned species. Even this classical approach has its own drawbacks as the sequenced transcriptome is not exactly the correct measure of the rate of protein synthesis or transcriptional activity as the platform sequences "steady-state" RNA (Van Dijk et al., 2014). In metagenomic sequencing, the starting material or the sample that is prepared pre-sequencing analysis decides the sensitivity of the following approach as this process is heavily dependent on the level of background impurities contained within the sample which generates a lot of noise in the data. The presence of host DNA and commensal bacteria in samples can interfere with mNGS results by reducing the proportion of microbial reads, which can lead to decreased sensitivity (Chiu et al., 2019). Some other common relevant limitations in mNGS techniques include a prolonged optimization of the entire process control unit with standardizations executed at every individual step-right from sample preparation to data analysis and storage. Along with this, mNGS is still not a very cost-effective procedure and it involves a lot of complications (Mitsura et al., 2021).

On the basis of the type of platform used for sequencing, 454 by Roche Life-sciences had problems of random INDEL addition and the cost of reagents used for 454 sequencing per megabase is reasonably high. In-case of Illumina and SOLiD platforms, the nucleic acid detection is slower, occurring one at a time fetching short-read length due to dephasing of optical signals (Shokralla et al., 2012).

In the field of metagenomics, community DNA extraction can sometimes hinder the smooth process of sequencing multiple genomes as it requires an exhaustive sampling method involving countless species-each species having their own physical structure which holds their genetic material. The prepared sample will be heterogeneous in nature which may affect further downstream processes in sequencing. Since an entire ecosystem of microbes exist in a particular sample, targeted sequencing for a specific genome becomes difficult due to poor sequencing of under-represented genomes. Existing methods for extracting genetic material from environmental samples are not yet robust and efficient enough to be used at scale. Also, the cost of maintenance & sequencing of such diverse samples also is on the rise. Another example of insufficient sample for sequencing arises during sequencing of glacial microbiomes, where sample has to be drilled out, filter sterilised and converted to melt-water prior to DNA isolation. For microbial analysis of water, sampling & sequencing of sewage water/fertile water (on land) is needed which increases the probability of harbouring zillions of impurities within the sample-which again fails to incorporate knowledge about harmful antimicrobial resistance genes and virulence factors of those microbes which tends to harm crop plants and human health.

7 Conclusion

In conclusion, NGS has several applications in microbiology, including Identification, antibiotic resistance profiling, epidemic investigation, and metagenomic analysis are all part of the process. By providing a strong tool for researching microbial populations at high resolution, Researchers may employ NGS to create accurate and useful insights on infectious illnesses by carefully evaluating these criteria. NGS has transformed microbiology by enabling rapid, affordable, and precise sequencing of microbial genomes. NGS technologies have enabled researchers to study microbial communities in unprecedented detail, revealing the diversity and complexity of microbial ecosystems. However, despite its many advantages, NGS has several limitations that must be considered when interpreting results. These include sequencing errors, bias and artifacts, assembly and annotation challenges, coverage depth, sample preparation and sequencing quality, data storage and analysis, and cost. By carefully considering these factors, researchers can use NGS to generate accurate and meaningful insights into microbial diversity and function.

Conflicts of Interest Declaration The authors have no competing interests, received no funding, are solely responsible for the findings, and uses trade names for identification only.

References

Alexander, M., 1962. Introduction of soil microbiology. *Soil Science*, *93*(1), 74. https://doi.org/10.1097/00010694-196201000-00034

Avershina, E., Frye, S.A., Ali, J., Taxt, A.M. and Ahmad, R., 2022. Ultrafast and cost-effective pathogen identification and resistance gene detection in a clinical setting using Nanopore Flongle sequencing. *Frontiers in Microbiology*, *13*, p.822402.

Bag, S., Saha, B., Mehta, O., Anbumani, D., Kumar, N., Dayal, M., Pant, A., Kumar, P., Saxena, S., Allin, K.H. and Hansen, T., 2016. An improved method for high quality metagenomics DNA extraction from human and environmental samples. *Scientific reports*, *6*(1), p.26775.

Besser, J., Carleton, H.A., Gerner-Smidt, P., Lindsey, R.L. & Trees, E. 2018. Next-generation sequencing technologies and their application to the study and control of bacterial infections. Clinical Microbiology and Infection, 24(4): 335–341. https://doi.org/10.1016/j.cmi.2017.10.013.

Bharagava, R.N., Purchase, D., Saxena, G. and Mulla, S.I., 2019. Applications of metagenomics in microbial bioremediation of pollutants: from genomics to environmental cleanup. In *Microbial diversity in the genomic era* (pp. 459–477). Academic Press.

Bitar, I., Papagiannitsis, C.C. and Antonelli, A. eds., 2023. *The use of next-generation sequencing in fighting antimicrobial resistant pathogens* (Vol. 16648714). Frontiers Media SA.

Buermans, H.P.J. & den Dunnen, J.T. 2014. Next generation sequencing technology: Advances and applications. *Biochimica et Biophysica Acta (BBA)-Molecular Basis of Disease*, 1842(10): 1932–1941. https://doi.org/10.1016/j.bbadis.2014.06.015.

Chiu, C.Y. and Miller, S.A., 2019. Clinical metagenomics. *Nature Reviews Genetics*, *20*(6), pp.341-355.

Clingenpeel, S., Clum, A., Schwientek, P., Rinke, C. and Woyke, T. 2015. Reconstructing each cell's genome within complex microbial communities—dream or reality?. *Frontiers in microbiology*, *5*, 771. https://doi.org/10.3389/fmicb.2014.00771

Crucian, B.E., Choukèr, A., Simpson, R.J., Mehta, S., Marshall, G., Smith, S M., ... and Sams, C. 2018. Immune system dysregulation during spaceflight: potential countermeasures for deep space exploration missions. *Frontiers in immunology, 9,* 1437. https://doi.org/10.3389/fimmu.2018.01437

Didelot, X., Bowden, R., Wilson, D.J., Peto, T.E. and Crook, D.W., 2012. Transforming clinical microbiology with bacterial genome sequencing. *Nature Reviews Genetics, 13*(9), pp.601-612.

Duan, H., Li, X., Mei, A., Li, P., Liu, Y., Li, X., Li, W., Wang, C. and Xie, S., 2021. The diagnostic value of metagenomic next- generation sequencing in infectious diseases. *BMC infectious diseases, 21*(1), pp.1-13.

Dubey, A., Malla, M.A. & Kumar, A. 2022. Role of Next-Generation Sequencing (NGS) in Understanding the Microbial Diversity. *Molecular Genetics and Genomics Tools in Biodiversity Conservation*: 307–328. https://doi.org/10.1007/978-981-16-6005-4_16

Ellington, M.J., Ekelund, O., Aarestrup, F.M., Canton, R., Doumith, M., Giske, C., Grundman, H., Hasman, H., Holden, M.T.G., Hopkins, K.L. and Iredell, J., 2017. The role of whole genome sequencing in antimicrobial susceptibility testing of bacteria: report from the EUCAST Subcommittee. *Clinical microbiology and infection, 23*(1), pp.2-22.

Escalona, M., Rocha, S. and Posada, D., 2016. A comparison of tools for the simulation of genomic next-generation sequencing data. *Nature Reviews Genetics, 17*(8), pp.459-469.

Ewing, B. and Green, P., 1998. Base-calling of automated sequencer traces using phred. II. Error probabilities. *Genome research, 8*(3), pp.186–194.

Ewing, B., Hillier, L., Wendl, M.C. and Green, P., 1998. Base-calling of automated sequencer traces usingPhred. I. Accuracy assessment. *Genome research, 8*(3), pp.175–185.

Fan, C., Lei, X., Guo, L. and Zhang, A. 2019. Predicting the associations between microbes and diseases by integrating multiple data sources and path-based HeteSim scores. *Neurocomputing, 323,* 76-85. https://doi.org/10.1016/j.neucom.2018.09.054

Garcia-Lopez, E., Moreno, A., Bartolomé, M., Leunda, M., Sancho, C. and Cid, C., 2021. Glacial ice age shapes microbiome composition in a receding southern European glacier. *Frontiers in Microbiology, 12,* p.714537.

Gautam, A., 2022. Isolation of DNA from Blood Samples by Salting Method. In *DNA and RNA Isolation Techniques for Non-Experts* (pp. 89-93). Cham: Springer International Publishing.

Gilchrist, C.A., Turner, S.D., Riley, M.F., Petri Jr, W.A. and Hewlett, E.L., 2015. Whole-genome sequencing in outbreak analysis. *Clinical microbiology reviews, 28*(3), pp.541-563.

Goodwin, S., McPherson, J.D. and McCombie, W.R., 2016. Coming of age: ten years of next-generation sequencing technologies. *Nature Reviews Genetics, 17*(6), pp.333–351.

Gupta, N., 2019. DNA extraction and polymerase chain reaction. Journal of cytology, 36(2), pp.116–117.

Ihekweazu, F.D. and Versalovic, J. 2018. Development of the pediatric gut microbiome: impact on health and disease. *The American journal of the medical sciences, 356*(5), 413-423. https://doi.org/10.1016/j.amjms.2018.08.005

Illumina. Technical note: Sequencing. Quality scores for next-generation sequencing: assessing sequencing accuracy using Phred quality scoring. Illumina http://www.illumina.com/documents/products/technotes/technote_Q-Scores.pdf (2011).

Johnson, S., Trost, B., Long, J.R., Pittet, V. and Kusalik, A., 2014, September. A better sequence-read simulator program for metagenomics. In *BMC bioinformatics* (Vol. 15, No. 9, pp. 1–10). BioMed Central.

Kho, Z.Y. and Lal, S.K. (2018). The human gut microbiome—A potential controller of wellness and disease. *Frontiers in microbiology, 9,* 1835. https://doi.org/10.3389/fmicb.2018.01835

Kumar, R., Sood, U., Kaur, J., Anand, S., Gupta, V., Patil, K.S. & Lal, R. 2021. The rising dominance of microbiology: what to expect in the next 15 years? *Microbial Biotechnology*, 15(1): 110–128. https://doi.org/10.1111/1751-7915.13953.

Kumar, M. and Mugunthan, M., 2018. Evaluation of three DNA extraction methods from fungal cultures. *medical journal armed forces india, 74*(4), pp.333–336.

Ledergerber, C. and Dessimoz, C., 2011. Base-calling for next-generation sequencing platforms. *Briefings in bioinformatics*, *12*(5), pp.489-497.

Levy, S.E. and Myers, R.M., 2016. Advancements in next-generation sequencing. *Annual review of genomics and human genetics*, *17*, pp.95-115.

Lynne Apone, Ph.D., Eileen Dimalanta, Ph.D., and Fiona Stewart, Ph.D., New England Biolabs, Inc. Improving Enzymatic DNA Fragmentation for Next Generation Sequencing Library Construction.

Mardis, E.R., 2013. Next-generation sequencing platforms. *Annual review of analytical chemistry*, *6*, pp.287-303.

Margulies, M., Egholm, M., Altman, W.E., Attiya, S., Bader, J.S., Bemben, L.A., Berka, J., Braverman, M.S., Chen, Y.-J., Chen, Z., Dewell, S.B., Du, L., Fierro, J.M., Gomes, X.V., Godwin, B.C., He, W., Helgesen, S., Ho, C.H., Irzyk, G.P., Jando, S.C., Alenquer, M.L.I., Jarvie, T.P., Jirage, K.B., Kim, J.-B., Knight, J.R., Lanza, J.R., Leamon, J.H., Lefkowitz, S.M., Lei, M., Li, J., Lohman, K.L., Lu, H., Makhijani, V.B., McDade, K.E., McKenna, M.P., Myers, E.W., Nickerson, E., Nobile, J.R., Plant, R., Puc, B.P., Ronan, M.T., Roth, G.T., Sarkis, G.J., Simons, J.F., Simpson, J.W., Srinivasan, M., Tartaro, K.R., Tomasz, A., Vogt, K.A., Volkmer, G.A., Wang, S.H., Wang, Y., Weiner, M.P., Yu, P., Begley, R.F. & Rothberg, J.M. 2005. Genome sequencing in microfabricated high-density picolitre reactors. *Nature*, 437(7057): 376–380. https://doi.org/10.1038/nature03959.

McCombie, W.R., McPherson, J.D. and Mardis, E.R., 2019. Next-generation sequencing technologies. *Cold Spring Harbor perspectives in medicine*, *9*(11). (p)

Mitsura, V.M., 2021. Next-generation sequencing (NGS) methods and their application in clinical microbiology, infectology and epidemiology. *Health and Ecology Issues*, *18*(4), pp.26-32.

Motro, Y. and Moran-Gilad, J., 2017. Next-generation sequencing applications in clinical bacteriology. *Biomolecular detection and quantification*, *14*, pp.1–6. (o)

Ravi, R.K., Walton, K. and Khosroheidari, M., 2018. MiSeq: a next generation sequencing platform for genomic analysis. *Disease gene identification: methods and protocols*, pp.223–232.

Saei, A.A. and Barzegari, A. 2012. The microbiome: the forgotten organ of the astronaut's body–probiotics beyond terrestrial limits. *Future microbiology*, *7*(9), 1037-1046. https://doi.org/10.2217/fmb.12.82

Sah S, Chen L, Houghton J, et al. Functional DNA quantification guides accurate next-generation sequencing mutation detection in formalin-fixed, paraffin-embedded tumor biopsies. Genome Med. 2013;5(8):77.

Sanger, F., Nicklen, S. & Coulson, A.R. 1977. DNA sequencing with chain-terminating inhibitors. *Proceedings of the National Academy of Sciences*, 74(12): 5463–5467. https://doi.org/10.1073/pnas.74.12.5463.

Schadt, E.E., Turner, S. & Kasarskis, A. 2010. A window into third-generation sequencing. *Human Molecular Genetics*, 19(R2): R227–R240. https://doi.org/10.1093/hmg/ddq416.

Sharma, R., Sharad, S., Minhas, G., Sharma, D.R., Bhatia, K. and Sharma, N.K., 2023. DNA, RNA isolation, primer designing, sequence submission, and phylogenetic analysis. In *Basic Biotechniques for Bioprocess and Bioentrepreneurship* (pp. 197–206). Academic Press.

Shokralla, S., Spall, J.L., Gibson, J.F. and Hajibabaei, M., 2012. Next-generation sequencing technologies for environmental DNA research. *Molecular ecology*, *21*(8), pp.1794-1805.

Sloan, W.T., Lunn, M., Woodcock, S., Head, I.M., Nee, S. & Curtis, T.P. 2006. Quantifying the roles of immigration and chance in shaping prokaryote community structure. *Environmental Microbiology*, 8(4): 732–740. https://doi.org/10.1111/j.1462-2920.2005.00956.x.

Sun, W., 2010. Nucleic extraction and amplification. In *Molecular diagnostics* (pp. 35–47). Academic Press.

van Dijk, E.L., Auger, H., Jaszczyszyn, Y. & Thermes, C. 2014. Ten years of next-generation sequencing technology. *Trends in Genetics*, 30(9): 418–426. https://doi.org/10.1016/j.tig.2014.07.001.

van Niel, C.B. 1966. Microbiology and molecular biology. *The Quarterly review of biology*, *41*(2), 105–112. https://doi.org/10.1086/404937

Vincent, A.T., Derome, N., Boyle, B., Culley, A.I. & Charette, S.J. 2017. Next-generation sequencing (NGS) in the microbiological world: How to make the most of your money. *Journal of Microbiological Methods*, 138: 60–71. https://doi.org/10.1016/j.mimet.2016.02.016.

Wilson, M.R., Sample, H.A., Zorn, K.C., Arevalo, S., Yu, G., Neuhaus, J., Federman, S., Stryke, D., Briggs, B., Langelier, C. and Berger, A., 2019. Clinical metagenomic sequencing for diagnosis of meningitis and encephalitis. *New England Journal of Medicine, 380*(24), pp.2327-2340.

Yang, L., Haidar, G., Zia, H., Nettles, R., Qin, S., Wang, X., Shah, F., Rapport, S.F., Charalampous, T., Methé, B. and Fitch, A., 2019. Metagenomic identification of severe pneumonia pathogens in mechanically-ventilated patients: a feasibility and clinical validity study. *Respiratory research, 20*, pp.1–12.

Yohe, S. and Thyagarajan, B., 2017. Review of clinical next-generation sequencing. *Archives of pathology & laboratory medicine, 141*(11), pp.1544-1557.

Yu, L., Huang, J., Ma, Z., Zhang, J., Zou, Y. and Gao, L. 2015. Inferring drug-disease associations based on known protein complexes. *BMC medical genomics, 8*(2), 1-13. https://doi.org/10.1186/1755-8794-8-S2-S2

Yu, L., Ma, X., Zhang, L., Zhang, J. and Gao, L. 2016. Prediction of new drug indications based on clinical data and network modularity. *Scientific reports, 6*(1), 32530. https://doi.org/10.1038/srep32530

Shi, J.Y., Li, J.X. and Lu, H.M. 2016. Predicting existing targets for new drugs base on strategies for missing interactions. *BMC bioinformatics, 17*, 601-609. https://doi.org/10.1186/s12859-016-1118-2

Su, R., Wu, H., Xu, B., Liu, X. and Wei, L. 2018. Developing a multi-dose computational model for drug-induced hepatotoxicity prediction based on toxicogenomics data. *IEEE/ACM Transactions on computational biology and bioinformatics, 16*(4), 1231-1239. https://doi.org/10.1109/TCBB.8857 https://doi.org/10.1109/TCBB.2018.2858756

Zhong, Y., Xu, F., Wu, J., Schubert, J. and Li, M.M., 2021. Application of next generation sequencing in laboratory medicine. *Annals of Laboratory Medicine, 41*(1), pp.25-43.

Zhu, J., Ofek, G., Yang, Y., Zhang, B., Louder, M.K., Lu, G., McKee, K., Pancera, M., Skinner, J., Zhang, Z. and Parks, R., 2013. Mining the antibodyome for HIV-1–neutralizing antibodies with next-generation sequencing and phylogenetic pairing of heavy/light chains. *Proceedings of the National Academy of Sciences, 110*(16), pp.6470-6475.

Zong, C., Lu, S., Chapman, A.R., Xie, X.S., 2012. Genome-wide detection of single-nucleotide and copynumber variations of a single human cell. *Science 338*, pp. 1622–1626.

Recent Advances in Bioethanol: Current Scenario, Sources and Production Techniques

Amit Kumar and Akarsh Verma

1 Introduction

The remarkable growth in industrialization, modernization of lifestyle, and increase in vehicular population has led to a substantial upsurge in the demand for energy. At present, over 80% of the worldwide energy demand is fulfilled by fossil-based fuels, with petroleum constituting 33% of the overall energy provision. Figure 1 shows the amount of fossil fuel consumed around the world (Ritchie et al. 2022). To begin with, the strong reliance on fossil fuels such as coal, oil, and natural gas greatly adds to greenhouse gas emissions and exacerbates climate change. The depletion of finite fossil fuel reserves also raises concerns about long-term sustainability. Furthermore, the unequal distribution of energy access remains, with millions of people in developing countries lacking access to dependable and modern energy services, stifling their socioeconomic advancement. Furthermore, the current energy infrastructure requires significant modifications and investments to meet rising demand while maintaining energy security and stability in the face of geopolitical uncertainty (Chauhan and Pratap Singh 2023). Addressing these challenges requires a worldwide transition toward cleaner and renewable energy sources, as well as encouraging energy efficiency and equitable energy access, in order to establish a sustainable and resilient energy future for future generations. The escalating cognizance of the predicament of global warming has also acted as a catalyst in augmenting the curiosity in the

A. Kumar (✉)
Department of Mechanical, Material and Aerospace Engineering, Indian Institute of Technology Dharwad, Dharwad 580011, India
e-mail: me23dp001@iitdh.ac.in

A. Verma
Department of Mechanical Engineering, University of Petroleum and Energy Studies, Dehradun 248007, India

Department of Mechanical Science and Bioengineering, Osaka University, Osaka 560-8531, Japan

J. Gupta and A. Verma (eds.), *Microbiology-2.0 Update for a Sustainable Future*,
https://doi.org/10.1007/978-981-99-9617-9_15

evolution of methodologies to curtail the discharge of greenhouse gases into the atmosphere. Mitigating the utilization of fossil fuels would undoubtedly lead to a substantial decrease in the amount of CO_2 emissions, thereby ameliorating the pollution levels as well. Figure 2 shows the graph of the change in the amount of global greenhouse gas emissions over time. As a measure of greenhouse gases, carbon dioxide equivalents (CO_2e) are used (Ritchie et al. 2020). In order to mitigate such emissions, it is imperative that our endeavours be directed towards the advancement of technologies that encompass the following aspects: Reduction of energy consumption, Enhancement of energy conversion or usage efficiency, Transition to fuels with lower carbon content, Augmentation of natural CO_2 sinks, and Facilitation of CO_2 capture and storage (Sawarkar et al. 2022).

Biofuels can play a significant role in reducing the emissions. Biofuels such as biogas, biodiesel, and bioethanol are made from organic resources called biomass. Biomass energy is one of the most promising possibilities. Despite its widespread availability and low cost in some areas, biomass energy has intrinsic limitations that limit its widespread use: less energy conversion of less energy, difficulty in logistics like storage and transportation, and the negative impacts of direct burning of biomass. Biomass energy, frequently referred to as conventional energy, is regularly employed in rural localities where alternative energy sources are inaccessible or financially unfeasible. Different biochemical and thermochemical processes, such as bio methanation, fermentation, gasification and pyrolysis, can be used to turn biomass into biofuels. The quick bioenergy cycle makes it possible to produce biofuel from

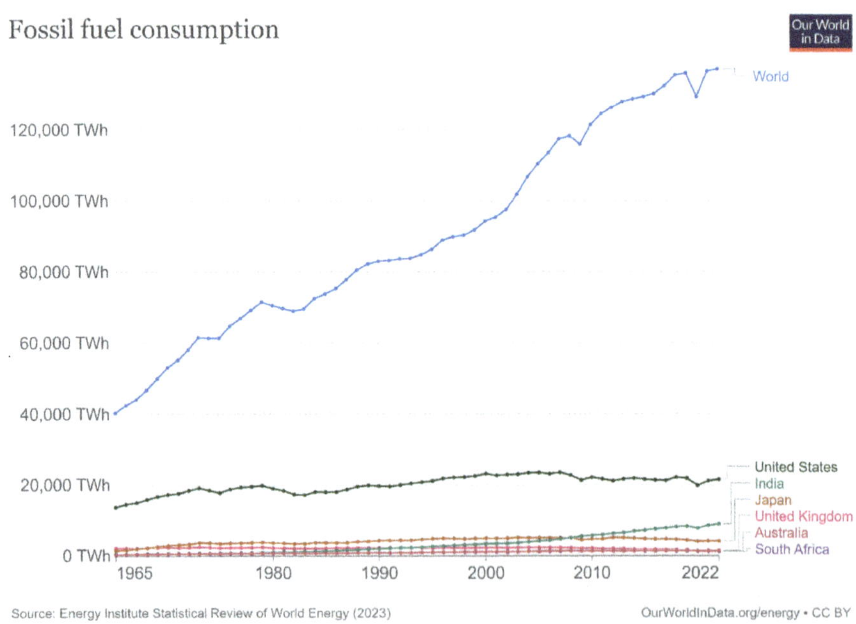

Fig. 1 Fossil fuel consumption around the world (Ritchie et al. 2022)

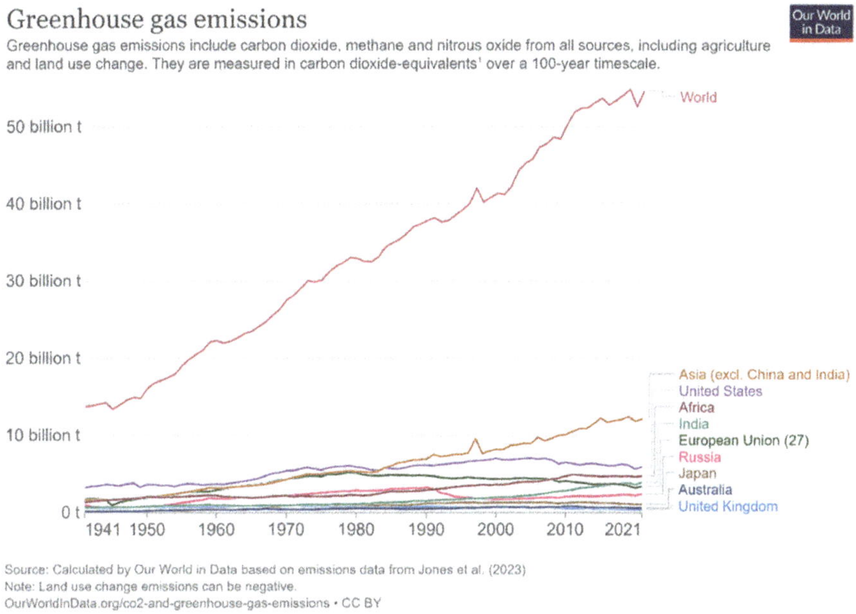

Fig. 2 Evolution of global greenhouse gas emissions (Ritchie et al. 2020)

biomass with minimal environmental effect, which is a sustainable solution to fulfil the rising need for fuel (Kurnia et al. 2016).

Bioethanol is one of the biofuels that may be used as a substitute for gasoline in automobiles. Bioethanol, usually known simply as ethanol, is a sustainable, bio-based alcohol fuel created by the fermentation of diverse biological resources. Its physical and chemical properties are shown and compared with the gasoline in Table 1. From the table it can be It is typically obtained from agricultural products such as sugarcane, corn, wheat, and other plant-based feedstocks. The manufacture of bioethanol is a biochemical process in which the sugars present in the raw materials are transformed into ethanol by the activity of yeast or other microorganisms. Although it is very simple to convert simple sugars or starch-based biomass to ethanol via fermentation, these first-generation feedstocks are no longer favoured due to the vast quantity of land required for agriculture, which creates unwanted competition with the food chain. One of the most important applications of bioethanol is as a replacement for traditional fossil fuels such as gasoline. When used as a transportation fuel, it can be combined with gasoline in various ratios or utilized as a high-percentage mixture, such as E85 (85% ethanol and 15% gasoline) (Chauhan et al. 2023).

In context of rising worldwide energy needs and environmental issues, this review article emphasizes the critical relevance of bioethanol as a sustainable and renewable energy source, its sources and production. Bioethanol is a significant participant in the conversion to a low-carbon and resilient energy environment because of its capacity

Table 1 Various properties comparison of gasoline and bioethanol

Properties	Gasoline	Bioethanol
Chemical formula	C_4–C_{12}	C_2H_5OH
Molecular weight	100–105	46.07
Density kg/m^3	779	792
Vapour pressure at 37.8 °C	55–103	15.84
Octane number	90–100	
Cetane number	<15	8
Calorific value KJ/Kg	43,729	29,773
Flash Point °C	−43	12.8
Boiling point °C	30–225	78.3
Auto ignition temperature °C	257.2	422.8

to considerably cut greenhouse gas emissions, potential to provide security in energy conservation, and adaptability in a variety of applications.

2 Current Scenario of Bio-ethanol in Developed and Developing Countries

Bioethanol, although not a recent energy source, has a long history of utilization in Europe and the United States during the early 1900s. However, it was neglected due to its elevated production expenses in comparison to gasoline. Due to the crisis of oil in the 1970s, bioethanol manufacturing was then resumed. Table 2 shows the production of bio-ethanol of different region (Huschitt et al. 2023). Figure 3 depicts the rise of biofuel demand. The disparity in demand growth between established and emerging economies is obvious. Brazil and India are seeing significant growth in ethanol consumption. The United States has attained the status of the foremost producer of bioethanol within the global arena. In 2019, ethanol output in the United States reached 59.5 billion litres. A similar amount was projected for 2020, but the pandemic has had a significant impact on the production of bioethanol. In spite of another record corn harvest, a decrease of 12% in production, resulting in 52 billion litres in 2020, which represents the lowest recorded output since 2014. This reduction is attributed to a 10% decline in gasoline consumption.

In the year 2020, it is anticipated that the utilization of ethanol shall constitute roughly 10% of the fluctuating gasoline demands. It is predicted that the United States shall witness an escalation in ethanol production to roughly 55–58 billion litres by 2021–22, in response to the recovery of gasoline demand. In the 2023–25 period, it is estimated that the production shall stabilize at around 55 billion litres, owing to the improvement in passenger car economy which shall subsequently reduce the demand for gasoline, thereby lowering the necessity for ethanol blending. The reduced output was a reaction to the sudden plunge in gasoline demand, negative

Table 2 Production of bio-ethanol worldwide (in million gallons)

Countries	2017	2018	2019	2020	2021	% of world production
United States	15,936	16,091	15,778	13,941	15,016	55
Brazil	6,730	8,060	8,860	8,100	7,320	27
European Union	1,250	1,300	1,350	1,280	1,350	5
China	850	810	1,010	930	870	3
India	230	430	460	540	850	3
Canada	460	460	497	429	434	2
Thailand	380	390	430	390	360	1
Argentina	290	290	290	210	270	1
Rest of World	644	709	655	650	820	3
Total	26,770	28,540	29,330	26,470	27,290	

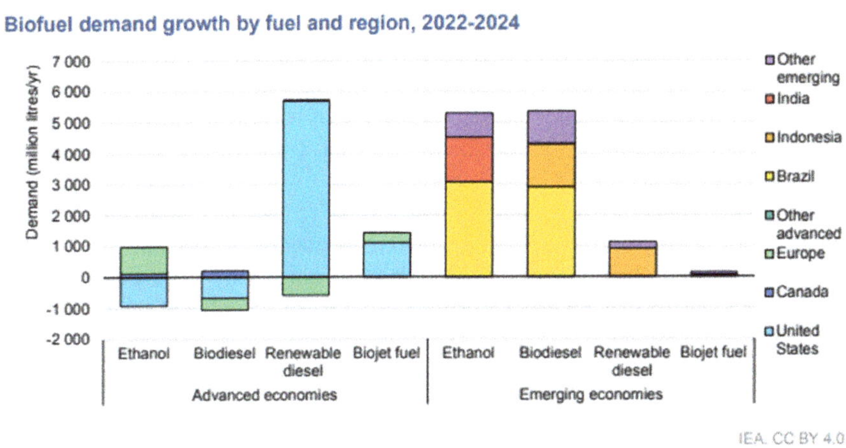

Fig. 3 Biofuel demand growth by fuel and region (IEA (2023), *Renewable Energy Market Update-June 2023, IEA, Paris* https://www.iea.org/reports/renewable-energy-market-update-june-2023, *License: CC BY 4.0* 2023)

operating margins, and restricted storage capabilities. However, the output has lately witnessed an increase, and in September, it was merely 9% below the 2019 level. Between February and April, output observed a decrease of almost 40% as several facilities either halted or reduced their output in response to the rapid decline in gasoline demand, negative operating margins, and limited storage capacity (Newes et al. 2022).

Brazil was investigated for the prospect of producing ethanol to replace 5% of world gasoline consumption by 2025. In addition to additional land (specifically, accessible pastures), rising sugarcane output would be necessary as well. According to the current article, average production in Brazil is significantly below what it should be (Cerqueira Leite et al. 2009). In 2019, Brazil was able to manufacture a significant amount of ethanol, around 36 billion litres to be exact. However, it is anticipated that this year's production will decrease by 16.5% to 30 billion litres. The country has been confronted with two major challenges in 2020. Firstly, there has been an 8% reduction in gasoline demand, resulting in a decrease in ethanol consumption that is mixed under the 27% requirement. Secondly, the low oil prices have resulted in a decrease in gasoline costs, negatively impacting the competitiveness of unblended ethanol at the pump. In addition, Brazilian ethanol output is expected to decrease this year due to mills prioritizing sugar production over ethanol during the 2020–21 sugar cane harvest season (Salles-Filho et al. 2017).

Indonesia has also announced two projects to expedite the implementation of biofuels by the year 2022. To initiate, the organization intends to elevate its biodiesel blending obligation to 35% by 2023, a significant increase from the current 30% set for 2022. Secondly, the entity has declared a strategy labelled as Ethanol for Energy Security, which aims to broaden the utilization of ethanol. However, specific details on the implementation of this strategy are scarce. The National Energy Policy of Indonesia is built on the principles of equity, sustainability, and environmental friendliness, as well as the construction of national energy security. Beginning in early 2015, the Government of Indonesia adopted Ministerial Energy and Mineral Resources Regulation 12/2015 mandating the use of biofuel as a substitute or mix on convention fuel, non-convention fuel, industrial, commercial, and power plant sectors (Sugiarto et al. 2018).

The European Union aims to elevate its target of renewable energy utilization in the transport sector from 14 to 29% by 2030. Nonetheless, a conclusive consensus has not been reached, and it will necessitate a reasonable time for the member countries to synchronize their internal strategies. The prices of ethanol, biodiesel, and renewable diesel persist to be more expensive than those of gasoline and diesel, thereby rendering the customers with the benefit of biofuel consumption only if it is subsidized. Furthermore, in 2022, the European Union imported more than 40% of its biodiesel (or the feedstocks used to manufacture it) (Purkus et al. 2019). Concerns regarding the difficulties of validating the sustainability credentials of these imports have resurfaced, casting more doubt on raising blending regulations. In 2017, all of the imported bioethanol was intended for biofuel purposes. The feedstock utilized for this imported bioethanol was primarily generated in the European Union. Of the bioethanol generated, 46% was derived from wheat, 87% of which originated from the European Union. Additionally, 33% was derived from maize, 60% of which had its origins in the European Union. Furthermore, 9% was derived from sugar beet, with 96% originating from the European Union. Moreover, 11% was derived from sugar cane, with all of its origins being traced back to Latin America. Lastly, 1% was derived from rye and triticale, with almost all of it originating from the European Union (Cadillo-Benalcazar et al. 2021).

With an annual growth rate of 7.1%, China's bioethanol output has climbed from 3137 million litres in 2001 to 5000 million litres in 2005 and 10,000 million litres in 2018. China's rapid expansion in bioethanol production is being pushed by measures to reduce stale maize/wheat inventories, as well as crude oil, maize, and bioethanol pricing. Maize is one of China's most significant crops and the primary source of bioethanol, accounting for 39.1% of grain output, 42.2% of cereal production, and 71.7% of bioethanol production. Domestic maize is used to make bioethanol (Han et al. 2022). However, maize is an important feed source in China, and the country's fast economic expansion has resulted in increased meat and feed consumption. Domestic maize production in China in 2019 was 260.8 mmt, with net imports of 4.8 mmt, for a self-sufficiency rate of 98.2%. China is currently pushing non-grain feedstock bioethanol, such as cassava, sweet sorghum, and sweet potato. Corn and wheat-based bioethanol continue to account for the majority of China's bioethanol production capacity, but future increase is prohibited. Also, the production of cellulosic bioethanol on a large commercial scale is currently hindered by technological challenges, including the deconstruction of lignocellulose waste and high production costs, which limits its current status to experimental phase (Ge and Lei 2017).

India currently ranks as the fifth largest energy consumer globally, following the United States, China, Russia, and Japan, with its energy consumption contributing to 4% of the world's total energy consumption. The country's energy demand has been steadily increasing with an annual growth rate of 4.8%, which places it on a trajectory to become the third largest energy user in the world by the year 2030. Economic growth in India is intricately intertwined with industrial energy consumption, which is the largest consumer of energy in the country, followed by the transportation sector that relies on petroleum products to meet approximately 40% of its energy requirements. India has also pushed the usage of non-food feedstocks for ethanol production like rice straw, wheat straw, food waste etc. which will help in reducing the crop burning (Chauhan and Pratap Singh 2023). Because of the consistent and plentiful supply of biomass resources in India, it offers potential as a future energy source. Once improved biofuel conversion technology is in place, it will be able to use forest and agricultural leftovers to create enough ethanol to supply the whole country's gasoline demands. Given the importance of biofuels, Indian ministries are now participating in the promotion, development, and policymaking of indigenous biomass for optimal development and utilization. India puts out a yearly plan to expand the domestic production of ethanol in accordance with the updated National Policy on Biofuels (2018) and its EBP (Ethanol Blended Petrol) Programme, which have an objective to accomplish a blending of 80:20 of petrol and ethanol i.e., E20 by the year 2025–26. In June 2022, India accomplished blending of 10% ethanol in petrol i.e., E10. Also, E20-tuned engines will be rolled out by April 2025 (Ministry of Petroleum and Natural Gas 2018).

3 Sources of Bioethanol

The accessibility of bioethanol feedstocks varies greatly from season to season and by geographic location. Feedstock price fluctuations can have a significant impact on bioethanol production costs. Another aspect to take into account is the utilization of food sources as the principal feedstocks for first generation biofuels, which could potentially lead to competition. The production of biomass feedstock for first generation biofuels only exploits a small percentage of the world's arable land, approximately 2%, and this may result in a surge in the prices of food and animal feeds. Nevertheless, the extent to which biofuels have a direct or indirect impact on the escalation of food prices remains ambiguous. First generation bioethanol necessitates the cultivation of crops such as maize, sugar beets, and others capable of meeting bioethanol demand without interfering with the production of food. Cereal grains are now the utmost plentiful crops utilized for ethanol production of first generation (Dash and Lingfa 2018). The infrastructure for cultivating, gathering, and refining maize is firmly ingrained, and the transformation of corn starch and syrups into ethanol is a relatively facile process. Corn kernels furnish approximately 65–76% w/w starch, while other beneficial sources include wheat (66–82%), barley (55–74%), sorghum (68–80%), oat (45–69%), and rice (74–85%). Sugarcane is the second most commonly utilized source of material in bioethanol manufacturing. Fuels which are generated biologically for example propanol, butanol and ethanol, are obtained by the process of fermentation of sugars, starches, or cellulose by the activity of enzymes and microorganisms (Bušić et al. 2018). First-generation biofuel methods are important, but they are unable to produce sufficient biofuel without jeopardizing the supply of food for the large population and biodiversity. Lignocellulosic crops serve as a primary source for the production of second-generation biofuels. The aforementioned method entails the separation of a plant's lignin and cellulose, subsequently allowing the cellulose to undergo fermentation into alcohol. These biofuels possess the potential to be derived from a diverse range of biomass, which is characterized as any organic carbon source. As share of the carbon cycle, this may be quickly replenished (Evans 2007). Based on the generation of feedstock, sources of bioethanol are shown in Fig. 4.

Due to the considerable constraints of first-generation biofuel production, second-generation bioethanol technologies have been developed. Second-generation bioethanol represents a promising solution to the challenges posed by traditional

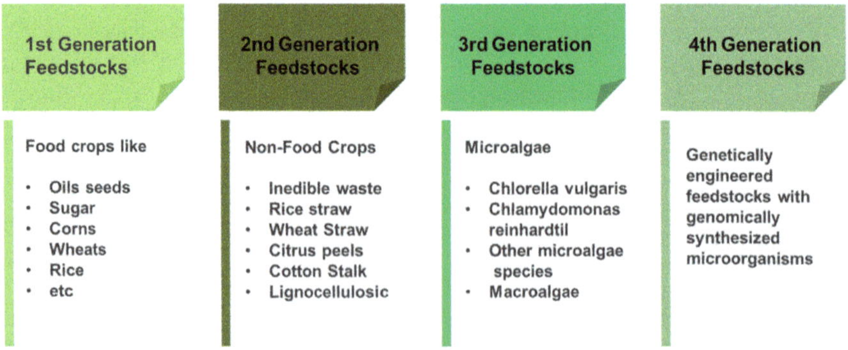

Fig. 4 Generation of bioethanol feedstocks

biofuels. By leveraging advanced techniques, this next iteration of bioethanol production stands to offer both sustainable and economic benefits while simultaneously improving environmental outcomes. Specifically, the goal of second-generation bioethanol techniques is to enable the generation of greater quantities of ethanol from biomass sources in a sustainable manner. Biomass, to be precise, denotes the non-edible segments of modern-day harvests that persist after extraction of the edible produce, for instance, stalks, foliage, and shells. Furthermore, certain crops, which are not primarily earmarked for consumption, such as switch grass, wheat straw, oats, among others, that exhibit limited grain production, can be harnessed for their potential as biomass (Ministry of Petroleum & Natural Gas 2018). The manufacture of second-generation bioethanol was done from lignocellulosic biomass which was obtained from corncobs and paper waste. To produce bioethanol, they employed SSF (simultaneous saccharification and fermentation) of cellulose via bacterial processes (Chandrasiri et al. 2022). Kut et al. investigated alternative solutions for the sustainable and cost-effective synthesis of bioethanol from apple pomace, a by-product of the apple processing industry. They believe that using Pichia stipitis as a microorganism for bioethanol synthesis from apple pomace hydrolyzate might assist to lessen the environmental effect of apple pomace disposal while simultaneously offering a sustainable energy source (Kut et al. 2022).

The reintroduction of third-generation biofuels into the industry has been observed since their initial development in the 1950s. Algae have been making a strong argument since they can radically transform the energy sector. Algae have several benefits because their growth simply require carbon dioxide, water, nutrients and sunlight. Certain strains of algae have been found to exhibit growth when subjected to wastewater, thereby reducing the costs incurred by eliminating the need for a freshwater medium. This innovative approach not only enhances the economic feasibility but also promotes sustainable practices (Hebbale and Ramachandra 2022). Furthermore, it has been demonstrated that renewable biofuels exhibit a notably advantageous energy output relative to conventional fuels, resulting in a significantly greater energy gain in proportion to energy input, yet concurrently maintaining a significantly reduced carbon footprint (Dahman et al. 2019). Laboratory studies have indicated that algae have the potential to yield up to 30 times more energy per acre than conventional land crops like soybeans. Despite this, there has been no commercial production of algal fuel thus far, indicating that there are still significant challenges to be addressed in this field. For example, in order to make algal fuel, it should be mixed consistently, which influence the growth of biomass if it is done through agitation. Several corporations and government organizations are sponsoring attempts to minimize capital and cost of operations and commercialize the production of algal fuel. Algae has also been shown to have substantially higher lipid levels than typical biofuel feedstock and to develop much quicker within the time period incubation of 1–10 days, resulting in providing the 200-fold gain over the next highest-yielding oil crop (Aoanan 2015). This means that algae have a far better chance of becoming a key role in world energy use. Because microalgae develop quickly and do not require any particular defined season for growth, they can be collected several times a year. Because of this microalgae is able to produce large amount of biofuel approximately around 1200 gallons of biofuel per acre per year (Borowitzka 2013; Surendhiran and Sirajunnisa 2019).

Fourth generation bioethanol are produced from genetically engineered feedstocks. The genetic modification of lignocellulosic feedstocks is a compelling technique that may enhance carbohydrate accessibility in the course of conversion procedures. Secondary cell walls, due to their intricate chemical composition and structure, establish a formidable physical barrier that is difficult to overcome. To minimize biomass recalcitrance, many approaches for genetically modifying lignocellulosic feedstocks and improving conversion efficiencies have been proposed. Furthermore, this technique enables for the transmission of tuned qualities to succeeding generations, by making the process further efficient over time (Surendhiran and Sirajunnisa 2019). Irrespective, the genetic alterations induced in these organisms can damage plant cell growth and/or its developmental processes, in doing so output of biomass is reduced. Plants have the potential for genetic modification to encourage the production of polymers in the cell wall, specifically cellulose, decrease the crystallinity of cellulose, and alter the composition of the lignin polymer. Midst these approaches, modification of lignin has been specifically designed to reduce biomass recalcitrance. By modifying the phenolic metabolism, the lignin structure has been modified. Lignin polymer is composed of p-hydroxphenyl (H), guaiacyl (G), and syringyl (S) units derived from the previously stated p-coumaryl, coniferyl, and sinapyl alcohols. The proportion of these units changes according to tissue and plant type. For example, gymnosperm lignin's typically have a larger proportion of G units, but angiosperm lignin is mostly made of both G and S units (Duque et al. 2021; Umezawa 2018).

The genetic modification of feedstocks, such as plant cells, holds significant promise for enhancing the effectiveness of lignocellulose conversion procedures in the context of a bio-based economy. However, further investigation is necessary to gain a comprehensive understanding of the potential of these approaches and to counteract the adverse effects of such alterations (Vohra et al. 2014).

4 Production Methods of Bioethanol

The bioethanol production method consists of a number of sophisticated processes, each of which is critical in converting diverse feedstocks into the valuable fuel source that is bioethanol. These steps, including pre-treatment, hydrolysis, fermentation, and ethanol recovery, all work together to produce an ecologically benign and sustainable energy source. Figure 5 shows the process of bio-ethanol production.

4.1 Pretreatment

The production of bioethanol from lignocellulose biomass involves a costly phase called pretreatment, which can amount to approximately $0.30 per gallon of ethanol

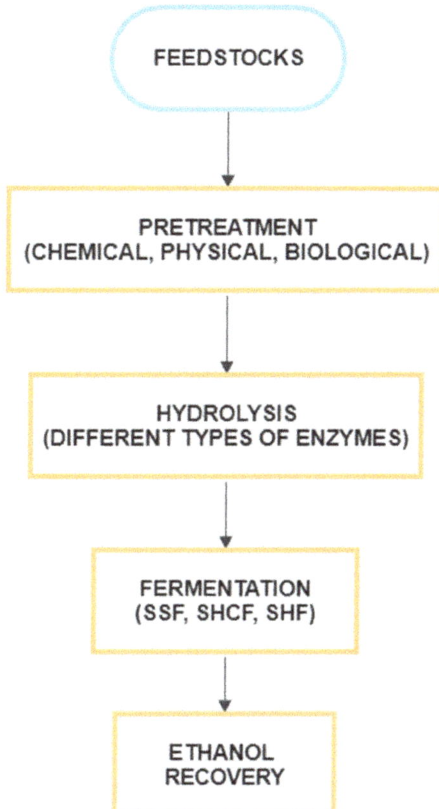

Fig. 5 Production process of bioethanol from feedstocks

produced. To enhance the reactivity of cellulose and the potential production of fermentable sugars, various pretreatment procedures are available. These methods are classified into traditional and advanced methods. Traditional pretreatment methods are categorized into three divisions based on the techniques used, which are chemical, physical, physicochemical, and biological. On the other hand, advanced pretreatment techniques are either acid-based fractionation or ionic liquid-based fractionation (ILF). The selection of the appropriate pretreatment technique is crucial to optimize the production of bioethanol from lignocellulose biomass while minimizing the overall cost (Yang and Pan 2012). Chemical pretreatment methods are the most efficient and so often utilized among conventional pretreatment procedures. Pretreatment advantages encompass the prevention of pentose sugar deterioration, the guarantee of bioethanol production process viability via the utilization of reactors of moderate size and the reduction of heat and power necessities, in addition to the minimization of inhibitor formation that ultimately leads to the reduced yield of hydrolysis and fermentation of sugar to ethanol (Edeh 2016).

The pretreatment technique is primarily classified into three processes: (1) physical pretreatment- this comprises the reduction of the size of the lignocellulosic biomass and crystallinity by procedures such as milling, grinding, irradiation, and extrusion. The consequence is increased surface area and pore size of the biomass, allowing for an improvement in enzymatic hydrolysis efficiency. To improve the effectiveness of lignocellulose deconstruction, physical pretreatment may be combined with chemical pretreatment. (2) Chemical pretreatment: acid, alkali, oxidative delignification, and organic acid (organosolvation) techniques are examples. They are very selective for certain feedstock types and are used to deconstruct and remove lignin and/or hemicellulose from polymer matrices. Chemical pretreatments are unquestionably effective, but they need severe working conditions that may have an unfavourable influence on downstream processing, and by-products may necessitate particular disposal processes. (3) Biological pretreatment: This entails using microorganisms to break down lignocellulosic biomass in preparation for subsequent enzymatic hydrolysis. White-rot, brown-rot, and soft-rot fungi, as well as bacteria, are among the organisms (Bensah and Mensah 2013).

4.2 Hydrolysis

Following the processing of lignocellulosic biomass, polymeric carbohydrate (cellulose and hemicellulose) is hydrolysed to yield sugar monomers. This stage is necessary because the enzymes required in the following stage (fermentation) can only digest sugar monomers. Acid or enzymes can be used to catalyse the process. The most prevalent approach is acid-catalysed hydrolysis, which uses either concentrated or dilute acid. The downside of this technology is the high production cost as a result of the difficulties in acid recovery, disposal, concentration control, and recycling. Another issue with concentrated acid-catalysed hydrolysis is its tendency to destroy sugar monomers owing to the acidic environment (Oliveira et al. 2020). High temperature and low acid concentration are required for dilute acid-catalysed hydrolysis. Dilute acid is the most often used acid. The drawback with this technique of hydrolysis is that it produces inhibitors as compared to acid-catalysed hydrolysis. Lignocellulose is acid hydrolysed in two steps. In the first step, hemicellulose is hydrolysed using dilute acid, and cellulose is degraded using concentrated acid in the second stage.

Under mild operating conditions of 45–50 °C and pH 4.8–5.0, enzymes hydrolyse polymeric carbohydrate to sugar monomers. Enzymatic hydrolysis is the process of using enzymes to extract fermentable sugars from biomass. This approach is efficient and produces excellent sugar recovery with no inhibitor development or corrosion risk. The effectiveness of enzyme-catalysed hydrolysis is influenced by variables such as pH, enzyme loading, duration, temperature, and the presence of a catalyst. Endo-1,4-glucanases, cellobiohydrolases, and β-glucosidases constitute three distinct categories of cellulase enzymes, each capable of driving the hydrolytic process. This enzymatic breakdown is particularly sought-after in industries such as paper, textile, and food processing, driving significant demand. However, the cost associated with these enzymes poses a considerable challenge, given the substantial quantities needed. This cost factor significantly impacts the overall manufacturing expenses. To address this, cost-effective strategies have emerged, prominently involving the utilization of microorganisms renowned for their cellulolytic enzyme secretion. This approach has gained momentum in contemporary applications. Notable examples of these microorganisms encompass Clostridium, cellulomonas, Erwinia, Thermonospora, Bacteriodes, Bacillus, Ruminococcus, Acetovibrio, and Streptomyces, each contributing to this enzymatic process. Additionally, fungi like Trichoderma, Penicillium, Fusarium, Phanerochaete, Humicola, and Schizophillum sp. have also emerged as valuable contributors. Notably, among these microbial candidates, the Trichoderma species has risen to prominence as a preeminent enzymatic workhorse. In essence, the integration of these diverse microorganisms has paved the way for more economically viable and sustainable cellulase enzyme production, thereby addressing the challenge posed by exorbitant enzyme costs.

4.3 Fermentation

This is a biological process that includes employing microorganisms such as yeast, fungus, or bacteria to convert the monomeric units of sugars derived from the hydrolysis stage into ethanol, acids, and gases. Yeast, particularly Saccharomyces cerevisiae, is the most often employed microbe due to its high ethanol output and tolerance limitations (Yang and Pan 2012). Saccharomyces cerevisiae transforms glucose, mannose, or fructose acquired from cellulose hydrolysis to ethanol, but xylan derived from hemicellulose hydrolysis can be converted to xylose. Separate hydrolysis and fermentation (SHF), simultaneous saccharification and fermentation (SSF), simultaneous saccharification and cofermentation (SSCF), non-isothermal simultaneous saccharification and fermentation, simultaneous saccharification, filtration, and fermentation, and consolidated bioprocessing (CBP) are the technologies used for the fermentation of monomeric units of sugar to ethanol. The first three are the most regularly utilized (Kumar et al. 2022).

In SHF enzymatic hydrolysis is conducted separately from the fermentation stage. Because hydrolysis and fermentation take place in separate containers, each process may be carried out at optimal conditions. More precisely, it allows enzymes to perform at peak levels in order to create more substrates for yeast fermentation (Günan Yücel and Aksu 2015). However, one of the disadvantages of SHF is the build-up of hydrolysis products. Because glucose and cellobiose impede cellulase activity, the rate of hydrolysis decreases gradually (Tavva et al. 2016). SSF is a significant bioethanol production process approach in which enzyme hydrolysis and fermentation take place in the same vessel. Unlike SHF, the end-product inhibition from cellobiose and glucose in hydrolysis is gradually digested by the yeast during the fermentation process. As a result, as compared to SHF, SSF requires less enzyme and produces more bioethanol. Additionally, the greater bioethanol content in SSF synthesis lowers external contamination. The bioethanol yield of SSF was increased by phosphoric acid-acetone pretreatment, which also decreased inhibitory chemicals in the hydrolysis and fermentation with a high solids content (>15% dry matter). The downside of SSF is that the optimal temperature required for effective cellulase and microorganism performance during hydrolysis and fermentation varies. The high temperature required by cellulase for hydrolysis may limit the microorganisms utilized for fermentation, such as yeast.

SSCF entails performing hydrolysis and saccharification in the same unit, as well as co-fermentation of pentose sugars. Because regular Saccharomyces cerevisiae cannot ferment pentose sugar, genetically engineered Saccharomyces cerevisiae strains that can ferment xylose are typically employed. SSCF, like SSF, has cheaper costs, better ethanol production, and shorter processing time. Furthermore, because most bacteria eat xylose, SSCF serves to decrease the inhibition induced by sugars during the enzymatic hydrolytic process and enhances the xylose to glucose concentration ratio. SHCF is a method that combines the benefits of SHF and SSCF (Bondesson and Galbe 2016). SHCF's hydrolysis and fermentation operations are carried out in separate containers to ensure that each phase is carried out under ideal circumstances. Furthermore, because the microorganisms in SHCF successfully use both pentoses and hexoses in the co-fermentation process, the bioethanol output is larger than in SHF. There have been few publications on SHCF operations to date, although the specifics may be derived by referring to SHF and SSCF protocols (Harun et al. 2009).

Other fermentation methods which can be utilised for bioethanol production include batch, fed-batch, continuous, and solid-state fermentation.

4.4 Ethanol Recovery

Typically, monomeric sugar fermentation is followed by ethanol recovery from the fermentation broth. Typically, the water content of the broth is decreased to around 0.5% by volume, allowing for the generation of anhydrous ethanol with a minimum concentration of 99.5% by volume. This procedure is limited by the azeotropic nature of ethanol–water solution and may be carried out using the distillation concept (i.e., utilizing the difference in boiling points of the solution components). The azeotropic solution's difficulty is solved by utilizing a separating agent that changes the relative volatility of the main component (Nitsche et al. n.d.). Adsorption distillation, azeotropic distillation, diffusion distillation, extractive distillation, vacuum distillation, membrane distillation, and chemical dehydration are some of the procedures used to recover pure ethanol from fermentation broth. Azeotropic distillation, liquid–liquid extraction, and extractive distillation are examples of traditional procedures. For large-scale operations, extractive distillation is the most often utilized method. Other approaches are gaining traction for future usage, particularly because they use less energy. Pervaporation and salt distillation are two examples (Nagy et al. 2015). Closing this, the authors of this chapter have considerable experience in the field of bio-composites (Arpitha et al. 2022; Bharath et al. 2020; Chaturvedi et al. 2023;

Dogra et al. 2021; Dora et al. 2023; Jain et al. 2019, 2023; Marichelvam et al. 2021; Nagaraju et al. 2023; Prabhakaran et al. 2022; Sati et al. 2023; Sharma et al. 2023; TG et al. 2023; Verma et al. 2017, 2018, 2019; Verma et al. 2023).

5 Conclusion

The worldwide pursuit of greener energy alternatives has pushed significant breakthroughs in bioethanol technology, driving it to the forefront of the renewable energy sector. As previously said, the manufacturing of bioethanol is a multifaceted process that involves the conversion of several feedstocks into a sustainable fuel source. The use of numerous feedstocks, such as sugarcane, corn, lignocellulosic materials, and even algae, demonstrates the versatility of this technology and its potential to harness a diverse range of resources for energy generation. The essential significance of feedstock selection, emphasizing the need of long-term environmental advantages and sustainable sourcing to avoid rivalry with food production. As the world's population grows, the convergence of food security and energy demands becomes more prominent, necessitating research into innovative feedstocks and biorefinery technologies to optimize resource utilization. Furthermore, the use of cutting-edge technologies such as genetic engineering and metabolic engineering has the potential to improve feedstock yield, fermentation efficiency, and overall bioethanol production rates.

Looking ahead, the future of bioethanol production is promising, albeit with numerous problems that will require creative solutions. The push for second and third-generation bioethanol production methods that utilise non-food feedstocks and improved conversion techniques provides a way to address concerns about land use competition and environmental impact. Furthermore, the introduction of integrated biorefineries, in which bioethanol production is combined with the generation of value-added co-products, represents a comprehensive approach to resource usage and economic feasibility. Policymakers' continuous support, together with shifting regulatory frameworks, will be critical in shaping the bioethanol industry's trajectory, supporting investment and innovation in this sector.

The exploration of various feedstock sources, coupled with advancements in production techniques, paves the way for a sustainable future. With ongoing research and innovation, bioethanol production is poised to play an increasingly pivotal role in the global energy matrix, fostering a greener and more sustainable world for generations to come.

Acknowledgements Author "Akarsh Verma" is grateful to the monetary support provided by the University of Petroleum and Energy Studies (UPES)-SEED Grant program.

Declarations

Conflicts of Interest "There are no conflicts of interest to declare by the authors."

Ethical Approval The authors hereby state that the present work is in compliance with the ethical standards.

Competing Interests The authors declare no competing interests.

Availability of Data and Materials Not applicable.

References

Aoanan, J., 2015. Microalgae Biofuels vs. Other Crop-Based Biofuels [online]. *ISB high school interns*. Available from: https://baliga.systemsbiology.net/see-interns/hs2015/projects-2/chlamydomonas-reinhardtii/microalgae-biofuels-vs-other-crop-based-biofuels/.

Arpitha, G.R., Verma, A., MR, S., Gorbatyuk, S., Khan, A., Sobahi, T.R., Asiri, A.M. and Siengchin, S., 2022. Bio-composite film from corn starch based vetiver cellulose. *Journal of Natural Fibers*, *19*(16), pp. 14634–14644.

Bensah, E.C. and Mensah, M., 2013. Chemical Pretreatment Methods for the Production of Cellulosic Ethanol: Technologies and Innovations. *International Journal of Chemical Engineering*, 2013, 1–21.

Bharath, K.N., Madhu, P., Gowda, T.G., Verma, A., Sanjay, M.R. and Siengchin, S., 2020. A novel approach for development of printed circuit board from biofiber based composites. *Polymer Composites*, *41*(11), pp. 4550–4558.

Bondesson, P.M. and Galbe, M., 2016. Biotechnology for Biofuels Process design of SSCF for ethanol production from steam - pretreated , acetic - acid - impregnated wheat straw. *Biotechnology for Biofuels*, 1–12.

Borowitzka, M.A., 2013. Energy from Microalgae: A Short History BT - Algae for Biofuels and Energy. *In*: M.A. Borowitzka and N.R. Moheimani, eds. Dordrecht: Springer Netherlands, 1–15.

Bušić, A., Mardetko, N., Kundas, S., Morzak, G., Belskaya, H., Šantek, M.I., Komes, D., Novak, S., and Šantek, B., 2018. Bioethanol production from renewable raw materials and its separation and purification: A review. *Food Technology and Biotechnology*, 56 (3), 289–311.

Cadillo-Benalcazar, J.J., Bukkens, S.G.F., Ripa, M., and Giampietro, M., 2021. Why does the European Union produce biofuels? Examining consistency and plausibility in prevailing narratives with quantitative storytelling. *Energy Research and Social Science*, 71 (September 2020), 101810.

Cerqueira Leite, R.C. de, Verde Leal, M.R.L., Barbosa Cortez, L.A., Griffin, W.M., and Gaya Scandiffio, M.I., 2009. Can Brazil replace 5% of the 2025 gasoline world demand with ethanol? *Energy*, 34 (5), 655–661.

Chandrasiri, Y.S., Weerasinghe, W.M.L.I., Madusanka, D.A.T., and Manage, P.M., 2022. Waste-Based Second-Generation Bioethanol: A Solution for Future Energy Crisis. *International Journal of Renewable Energy Development*, 11 (1), 275–285.

Chaturvedi, S., Kataria, A., Chaudhary, V., Verma, A., Jain, N., Sanjay, M.R. and Siengchin, S., 2023. Bionanocomposites reinforced with cellulose fibers and agro-industrial wastes. In *Cellulose Fibre Reinforced Composites* (pp. 317–342). Woodhead Publishing. https://doi.org/10.1016/B978-0-323-90125-3.00017-3

Chauhan, K., Kumar, A., Goswami, K., Negi, L., Chauhan, A., Madan, K., and Jain, S., 2023. Lignin extraction from lignocellulosic biomass (sugarcane bagasse) and its potential application as a feedstock for fuel production. *Materials Today: Proceedings*, 78 (xxxx), 688–694.

Chauhan, K. and Pratap Singh, V., 2023. Prospect of biomass to bioenergy in India: An overview. *Materials Today: Proceedings*, (xxxx).

Dahman, Y., Syed, K., Begum, S., Roy, P., and Mohtasebi, B., 2019. *Biofuels: Their characteristics and analysis*. Biomass, Biopolymer-Based Materials, and Bioenergy: Construction, Biomedical, and other Industrial Applications. Elsevier Ltd.

Dash, S.K. and Lingfa, P., 2018. *Advanced Manufacturing and Materials Science*. Springer International Publishing.

Dogra, V., Kishore, C., Verma, A., Rana, A.K. and Gaur, A., 2021. Fabrication and experimental testing of hybrid composite material having biodegradable bagasse fiber in a modified epoxy resin: evaluation of mechanical and morphological behavior. *Applied Science and Engineering Progress*, *14*(4), pp. 661–667.

Dora TL, Singh SK, Mishra RR, Verma A, 2023. Role of Crystal Orientation, Temperature, and Strain Rate on the Mechanical Characterization of Nickel: An Atomistic-scale investigation. Journal of Micromanufacturing. https://doi.org/10.1177/25165984231195519.

Duque, A., Álvarez, C., Doménech, P., Manzanares, P., and Moreno, A.D., 2021. Advanced bioethanol production: From novel raw materials to integrated biorefineries. *Processes*, 9 (2), 1–30.

Edeh, I., 2016. Bioethanol Production: An Overview. *Intech*, 11 (tourism), 13.

Evans, G., 2007. International biofuels strategy project. *Liquid transport biofuels–technology status report …*, (November).

Ge, J. and Lei, Y., 2017. Policy options for non-grain bioethanol in China: Insights from an economy-energy-environment CGE model. *Energy Policy*, 105 (February 2017), 502–511.

Günan Yücel, H. and Aksu, Z., 2015. Ethanol fermentation characteristics of Pichia stipitis yeast from sugar beet pulp hydrolysate: Use of new detoxification methods. *Fuel*, 158, 793–799.

Han, X., Chen, Y., and Wang, X., 2022. Impacts of China's bioethanol policy on the global maize market: a partial equilibrium analysis to 2030. *Food Security*, 14 (1), 147–163.

Harun, R., Liu, B., and Danquah, M.K., 2009. Analysis of Process Configurations for Bioethanol Production from Microalgal Biomass.

Hebbale, D. and Ramachandra, T. V, 2022. Chapter 15 - Third-generation bioethanol: status, scope, and challenges. *In*: S.B.T.-H. of B. Sahay, ed. Academic Press, 295–312.

Huschitt, E., Oestmann, J., and Schwarck, R., 2023. 2023 Ethanol Industry Outlook.

IEA (2023), Renewable Energy Market Update - June 2023, IEA, Paris https://www.iea.org/reports/renewable-energy-market-update-june-2023, *License: CC BY 4.0*, 2023.

Jain, N., Verma, A. and Singh, V.K., 2019. Dynamic mechanical analysis and creep-recovery behaviour of polyvinyl alcohol based cross-linked biocomposite reinforced with basalt fiber. *Materials Research Express*, *6*(10), p. 105373.

Jain, N., Sharma, P., Verma, A. and Gupta, J., 2023. Enhancement of thermo-mechanical, creep-recovery, and anti-microbial properties in PVA-based biodegradable films through cross-linking with oxalic acid: implications for packaging application. *Biomass Conversion and Biorefinery*, pp.1–10. https://doi.org/10.1007/s13399-023-04748-y.

Kumar, R.S., Singh, P., and Ghosh, S., 2022. Sequential fermentation for enhanced volumetric productivity of bioethanol from mixed sugars. *Fuel*, 308, 121984.

Kurnia, J.C., Jangam, S. V., Akhtar, S., Sasmito, A.P., and Mujumdar, A.S., 2016. Advances in biofuel production from oil palm and palm oil processing wastes: A review. *Biofuel Research Journal*, 3 (1), 332–346.

Kut, A., Demiray, E., Ertuğrul Karatay, S., and Dönmez, G., 2022. Second generation bioethanol production from hemicellulolytic hydrolyzate of apple pomace by Pichia stipitis. *Energy Sources, Part A: Recovery, Utilization, and Environmental Effects*, 44 (2), 5574–5585.

Marichelvam, M.K., Manimaran, P., Verma, A., Sanjay, M.R., Siengchin, S., Kandakodeeswaran, K. and Geetha, M., 2021. A novel palm sheath and sugarcane bagasse fiber based hybrid composites for automotive applications: An experimental approach. *Polymer Composites*, *42*(1), pp. 512–521.

Ministry of Petroleum & Natural Gas, 2018. National Biofuel Policy. https://mopng.gov.in/files/uploads/NATIONAL_POLICY_ON_BIOFUELS-2018.pdf, 2018 (D), 1–23.

Nagaraju, S.B., Sathyanarayana, K., Somashekara, M.K., Pradeep, D.G., Puttegowda, M. and Verma, A., 2023. Artificial neural networks for predicting mechanical properties of Al2219-B4C-Gr composites with multireinforcements. *Proceedings of the Institution of Mechanical Engineers, Part C: Journal of Mechanical Engineering Science.* https://doi.org/10.1177/09544062231196038.

Nagy, E., Mizsey, P., Hancsók, J.Å., Boldyryev, S., and Varbanov, P., 2015. Chemical Engineering and Processing : Process Intensi fi cation Analysis of energy saving by combination of distillation and pervaporation for biofuel production, 98, 86–94.

Newes, E., Clark, C.M., Vimmerstedt, L., Peterson, S., Burkholder, D., Korotney, D., and Inman, D., 2022. Ethanol production in the United States: The roles of policy, price, and demand. *Energy Policy*, 161 (September 2021), 112713.

Nitsche, M., Gbadamosi, R., Column, P., and Guide, D., n.d. Extractive and Azeotropic Distillation, 153–164.

Oliveira, T.C.G., Hanlon, K.E., Interlandi, M.A., Torres-Mayanga, P.C., Silvello, M.A.C., Lachos-Perez, D., Timko, M.T., Rostagno, M.A., Goldbeck, R., and Forster-Carneiro, T., 2020. Subcritical water hydrolysis pretreatment of sugarcane bagasse to produce second generation ethanol. *The Journal of Supercritical Fluids*, 164, 104916.

Prabhakaran, S., Sharma, S., Verma, A., Rangappa, S.M. and Siengchin, S., 2022. Mechanical, thermal, and acoustical studies on natural alternative material for partition walls: A novel experimental investigation. *Polymer Composites*, *43*(7), pp. 4711–4720.

Purkus, A., Gawel, E., and Thrän, D., 2019. The role of a renewable energy target for the transport sector beyond 2020: Lessons learned from eu biofuel policy. *The European Dimension of Germany's Energy Transition: Opportunities and Conflicts*, 527–542.

Ritchie, H., Roser, M., and Rosado, P., 2020. CO_2 and Greenhouse Gas Emissions [online]. *OurWorldInData.org*. Available from: https://ourworldindata.org/co2-and-greenhouse-gas-emissions.

Ritchie, H., Roser, M., and Rosado, P., 2022. Energy [online]. *OurWorldInData.org*. Available from: https://ourworldindata.org/energy.

Salles-Filho, S.L.M., Castro, P.F.D. de, Bin, A., Edquist, C., Ferro, A.F.P., and Corder, S., 2017. Perspectives for the Brazilian bioethanol sector: The innovation driver. *Energy Policy*, 108 (May), 70–77.

Sati, P., Verma, A., Zindal, A., Chauhan, S. and Singh, V.K., 2023. PVA biopolymer-acidic functionalized graphene hybrid nano composite for vibration isolation application: An experimental approach with variable reflux and vacuum timings. *Chemical Physics Impact*, *6*, p. 100212.

Sawarkar, A.N., Kirti, N., Tagade, A., and Tekade, S.P., 2022. Bioethanol from various types of banana waste: A review. *Bioresource Technology Reports*, 18.

Sharma, S., Verma, A., Rangappa, S.M., Siengchin, S. and Ogata, S., 2023. Recent progressive developments in conductive-fillers based polymer nanocomposites (CFPNC's) and conducting polymeric nanocomposites (CPNC's) for multifaceted sensing applications. *Journal of Materials Research and Technology*, 26, 5921–5974. https://doi.org/10.1016/j.jmrt.2023.08.300

Sugiarto, B., Setyo Wibowo, C., Zikra, A., Budi, A., and Mulya, T., 2018. Characteristic of gasoline fuels in Indonesia blend with varying percentages of bioethanol. *E3S Web of Conferences*, 67, 1–4.

Surendhiran, D. and Sirajunnisa, A.R., 2019. Chapter 18 - Role of Genetic Engineering in Bioethanol Production From Algae. *In*: R.C. Ray and S.B.T.-B.P. from F.C. Ramachandran, eds. Academic Press, 361–381.

TG, Y.G., Ballupete Nagaraju, S., Puttegowda, M., Verma, A., Rangappa, S.M. and Siengchin, S., 2023. Biopolymer-Based Composites: An Eco-Friendly Alternative from Agricultural Waste Biomass. *Journal of Composites Science*, *7*(6), p. 242. https://doi.org/10.3390/jcs7060242

Tavva, S.S.M.D., Deshpande, A., Durbha, S.R., Palakollu, V.A.R., Goparaju, A.U., Yechuri, V.R., Bandaru, V.R., and Muktinutalapati, V.S.R., 2016. Bioethanol production through separate hydrolysis and fermentation of Parthenium hysterophorus biomass. *Renewable Energy*, 86, 1317–1323.

Umezawa, T., 2018. Lignin modification in planta for valorization. *Phytochemistry Reviews*, (January 2017).

Verma, A., Gaur, A. and Singh, V.K., 2017. Mechanical properties and microstructure of starch and sisal fiber biocomposite modified with epoxy resin. *Materials Performance and Characterization*, *6*(1), pp. 500–520.

Verma, A., Joshi, K., Gaur, A. and Singh, V.K., 2018. Starch-jute fiber hybrid biocomposite modified with an epoxy resin coating: fabrication and experimental characterization. *Journal of the Mechanical Behavior of Materials*, *27*(5–6), p. 20182006. https://doi.org/10.1515/jmbm-2018-2006

Verma, A., Negi, P. and Singh, V.K., 2019. Experimental analysis on carbon residuum transformed epoxy resin: Chicken feather fiber hybrid composite. *Polymer Composites*, *40*(7), pp. 2690–2699.

Verma, A. and Ogata, S., 2023. Magnesium based alloys for reinforcing biopolymer composites and coatings: a critical overview on biomedical materials. *Advanced Industrial and Engineering Polymer Research*. https://doi.org/10.1016/j.aiepr.2023.01.002

Vohra, M., Manwar, J., Manmode, R., Padgilwar, S., and Patil, S., 2014. Bioethanol production: Feedstock and current technologies. *Journal of Environmental Chemical Engineering*, 2 (1), 573–584.

Yang, Q. and Pan, X., 2012. Bioresource Technology Pretreatment of Agave americana stalk for enzymatic saccharification. *Bioresource Technology*, 126, 336–340.

Biofuel: Generations, Current Status, Sustainability Impact, Feedstock, and Production Technologies

Kunal Chauhan, Kriti, and Akarsh Verma

1 Introduction

Global population has increased more than three times since the middle of the twentieth century. The world's growing population, accelerating migration, and rising urbanization come with much-increasing fossil fuel demand. Figure 1 shows the daily oil consumption (Barrels) of various countries. The United States has the highest consumption, followed by China and India, and the lowest is of Niue. Developed and developing nation requires oil for the proper functioning of the country for various aspects such as transportation and electricity. Global oil consumption is expected to reach 105.7 million barrels per day in 2028, up 5.9 million barrels per day from 2022 levels (Chandel and Sukumaran 2017).

Rising energy needs will be a problem for the globe. The most feasible solution to this problem will be producing biofuel from readily available sources such as municipal solid and liquid waste, biomass waste, etc. (Chauhan and Pratap Singh 2023). Biofuels are gaining popularity as potential solutions to petroleum-derived fuels for transportation to solve the energy security, cost of energy, and environmental problems arising from liquid fossil fuels (Velvizhi et al. 2023).

K. Chauhan (✉)
Department of Mechanical and Industrial Engineering, Indian Institute of Technology Roorkee, Roorkee 247667, India
e-mail: kunal_c1@me.iitr.ac.in

Kriti
School of Basic and Applied Sciences, K.R. Mangalam University, Gurugram, Haryana 122103, India

A. Verma
Department of Mechanical Engineering, University of Petroleum and Energy Studies, Dehradun 248007, India

Department of Mechanical Science and Bioengineering, Osaka University, Osaka 560-8531, Japan

J. Gupta and A. Verma (eds.), *Microbiology-2.0 Update for a Sustainable Future*,
https://doi.org/10.1007/978-981-99-9617-9_16

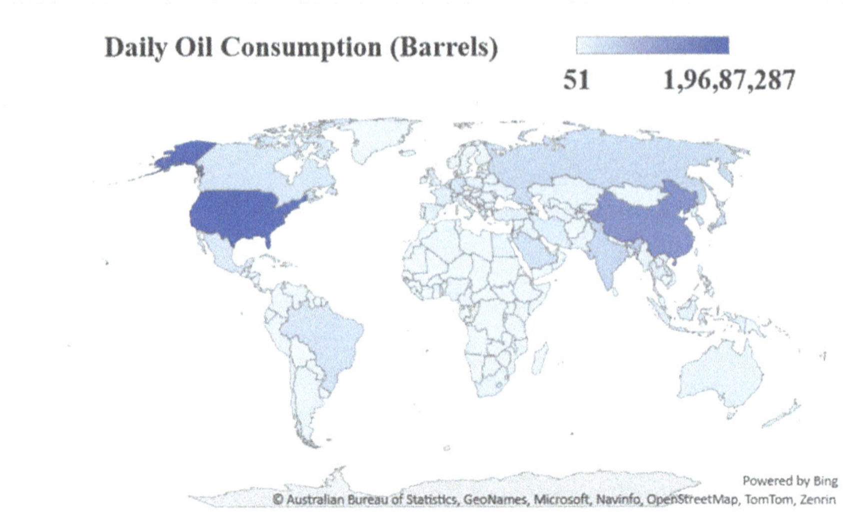

Fig. 1 Daily oil consumption (Oil Consumption by Country 2023) (Dark colour indicate high consumption and light colour indicates low consumption)

Biofuel is a renewable energy source generated from microbial, plant, or animal waste. The most popular kinds of biofuels are liquid and gaseous because they are the easiest to carry, deliver, and burn cleanly. Biofuels can also be solid, liquid, or gaseous (Chen 2022). Some examples of biofuels are:

- Ethanol: In the US and Brazil, sugarcane and maize are frequently used to make ethanol (Roberts and Smagala 2022). The majority of the carbohydrates used to make ethanol come from sugar or starch crops like maize, and sugarcane (Prasad and Ingle 2023).
- Biodiesel: Biodiesel is generated from liquid animal fats and vegetable oils. The most popular biofuel in Europe is biodiesel, which is produced by transesterifying oils or fats (Roberts and Smagala 2022).
- Green Diesel: Green diesel is produced using plants, including algae. Commercial production of algae biofuels has not yet started, although 2013 the US started commercial production of cellulosic biofuel (*Biofuel*, no date),.
- Biogas: Methane produced by the digestion of organic matter, such as animal dung, is known as biogas. Biogas is a sustainable fuel made from biomass, which includes plant material, animal waste, and microbes (Brito Cruz, Souza and Barbosa Cortez 2014; Roberts and Smagala 2022).

The generalized life cycle of biofuels is depicted in Fig. 2. Their Generation and utilization are becoming more important due to their capacity to replace fossil fuels and fulfil global energy needs while limiting the role of fuels to global warming. In addition to sustainability, contamination of the environment lessening, and reduction in GHG emissions, biofuels are preferred to fossil fuels due to a variety of other

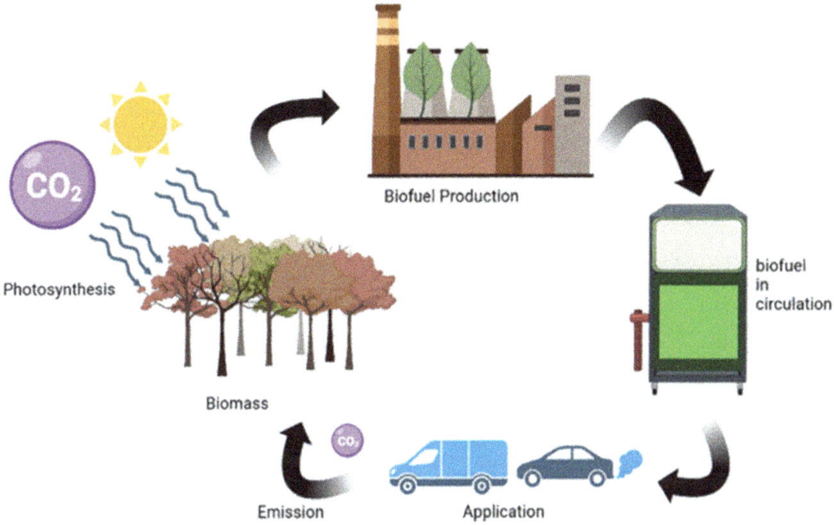

Fig. 2 The generalized life cycle of biofuels (Mukrimaa et al. 2023)

advantages, such as their environmentally friendly nature, future availability security, and cost-effectiveness; all of these benefits are particularly noteworthy in light of frequently rising costs of petroleum and international fluctuations in substantial fossil-fuel-producing zones of the globe (Moodley 2021; Dayton and Foust 2020).

Biomass can potentially be turned into a range of biofuels, such as liquid fuels like biomethanol, biodiesel, and bioethanol, gaseous fuels such as methane and hydrogen. The oxygen content is the primary distinction between biomass-generated and petroleum-based fuels.

2 Generation of Biofuel

Biofuels are categorized into four types depending on the biomass sources/ feedstock types, as shown in the Fig. 3.

2.1 1st Generation

Sugar, starch, and vegetable oil are all used to produce 1st generation biofuels. Biofuels, including propanol, butanol, and ethanol, are generated by fermentation of sugars, starches, or cellulose by the action of microorganisms & enzymes. Although first-generation biofuel processes are advantageous, they cannot generate

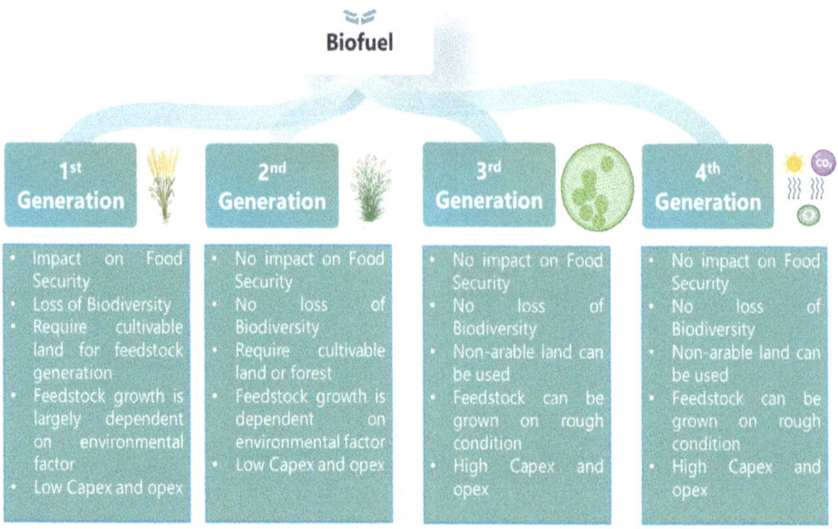

Fig. 3 Generation of biofuel

enough biofuel without jeopardising food supplies and ecosystems. Many first-generation biofuels are subsidized, are not economically equivalent to traditional fossil fuels like oil, and deliver only minimal reductions in greenhouse gas emissions (Nanda et al. 2018).

2.2 2nd Generation

Lignocellulosic crops are employed to produce second generation biofuels. Second generation biofuels can be extracted from non-food crops such as food crop waste, waste cooking oil, agricultural leftovers, and chips of wood (Bux 2013). The production method of this generation separates a plant's lignin and cellulose to ensure that the cellulose could be processed into alcohol (Groves et al. 2018). These kinds of biofuels are capable of being produced from a wide range of biomass, which can be characterised as any form of organic carbon. As a constituent of the carbon cycle, this may be swiftly replenished. Due to the fact that first-generation biofuel development has significant constraints, second-generation biofuel technologies have been developed (Jeswani and Azapagic 2012).

Second-generation biofuels can overcome plant diversity depletion obstacles and deliver a greater volume of ethanol in a more sustainable and cost-effective manner while giving additional environmental benefits. The goal of second-generation biofuel approaches is to enhance the amount of ethanol that can be produced sustainably from biomass.

Characteristics and examination of biofuels in particular to this Generation, biomass refers to the non-food components of contemporary crops that remain after the food crop has been removed, such as stems, leaves, and husks. Furthermore, crops that are not utilised for food, such as rice straw, switch grass, oats, wheat straw, etc. can be used for biofuel extraction. Biofuels of the first and the majority of the second Generation are commercially produced, accounting for around 4% and 96% of total production in 2019 (Abdullah et al. 2019).

2.3 *3rd Generation*

Since its initiation in the 1950s, third-generation biofuels have been regaining their path back into the industry. Algae has been making an impressive case since it has the ability to transform the energy sector radically. Algae have several advantages, including the fact that they simply require water, CO_2, nutrients, and sunlight to develop. Certain algae species can flourish in wastewater, lowering expenses even further by eliminating the necessity for a freshwater medium. Furthermore, renewable biofuels are being determined to have a far greater energy yield than conventional fuels, implying that the energy gained vs. the energy required is substantially higher despite having a significantly smaller carbon footprint (Dahman et al. 2019).

High oil costs, conflicting needs for foods and alternative biofuel sources, and the global food crisis have sparked interest in algal culture (growing algae) for producing biofuels on places unfavourable for agriculture. Algae were additionally found to have substantially higher lipid levels than typical biofuel feedstock and to develop much quicker, with incubation durations ranging from 1 to 10 days, giving it a nearly 200-fold benefit compared to the next highest-yielding oil crop (Abbaszaadeh et al. 2012). This means that algae have a considerably greater chance of becoming a key role in world energy use.

2.4 *4th Generation*

Fourth generation biofuels are still in the early stages of development. Bioengineered microorganisms such as microalgae, yeast, fungus, and cyanobacteria are utilized for producing 4th generation biofuels. Genetically engineered crops absorb more CO_2 from the atmosphere than they emit. These bacteria are utilized to make ethanol, butanol, hydrogen, methane, vegetable oil, biodiesel, isoprene, petrol, and jet fuel, among other things.

4th generation biofuels impart renewable energy and take in and retain CO_2. As second-generation biofuels, the same techniques are employed to convert biomass sources which absorb carbon dioxide all while developing into fuel. In contrast to second and third-generation biofuel production, carbon dioxide is captured at all phases of production using techniques such as oxy-fuel combustion.

Table 1 summarizes the four Generations of biofuel; each has been summarised on food security, cultivable land and water requirement, environmental factors, sustainability, harvesting nutrition, and economic aspect.

Table 1 Pros and cons of the 1st, 2nd, 3rd, 4th generation of biofuel on various factor (Chisti 2007; Sheppard et al. 2011; Qiu et al. 2012; Zhao et al. 2014; Barros et al. 2015; Behera et al. 2015; Salama et al. 2017; Leong et al. 2018; Markou et al. 2018)

Generations/Factor			
1st	2nd	3rd	4th
Impact on Food Security			
✓	x	x	x
Land footprint (Land for cultivation or woods requirement)			
✓	✓	x	x
Conversion to biofuels			
Easy	Advanced downstream processing methods required	Easy	Easy
Commercialization			
✓	✓	x	x
Nutrient requirements			
Pesticides and fertilization requirement	No need	Carbon, Nitrogen, solar energy requirement	Carbon, Nitrogen, solar energy requirement
Regulation			
Pretty clear	Pretty clear	Not available for marine cultivation	Not available for marine cultivation
Environmental condition (Temperature, Ph, humidity, salinity)			
Parameters within an allowable range	Parameters within an allowable range	It can be grown in extreme environmental conditions	It can be grown in harsh environmental conditions
Environment-friendliness (Pesticides, expenditure on fertilizer)			
✓	x	x	x
Sustainability (social, economic, environment)			
x	X (Ecology)	X (Economic)	X (Environment)
Harvesting			
Hand or machine	Hand or machine	costly and complex	costly and complex
Financial input			
Low	Low	High	High
Water footprint			
Potable water	Potable water	Non-potable water, waste, saline water	Non-potable water, waste, saline water

3 Current Status of Biofuel in the World

According to the International Energy Agency's most recent World Energy Investment report, approximately USD 2.8 trillion is expected to be invested worldwide in the energy sector in 2023, with more than USD 1.7 trillion going to clean technologies such as electric vehicles, renewables, low-emissions fuels, nuclear power, storage, grids, efficiency improvements, and heat pumps. The remaining, or a little over $1 trillion, will be used for coal, gas, and oil (Birol 2014). Figure 4 shows the investments in clean energy and fossil fuels worldwide from 2015 to 2023. Investments in fossil fuel show a decreasing trend and an increasing trend in clean fuel, which depicts the world's concern on new fuel development for the future.

According to International Energy Agency (IEA), the global biofuel demand will increase by 28% in 2021–26 (main case) (Renewables 2021 Analysis and forecast to 2026, 2020). Government policies are mostly held accountable for the rise, while other factors such as widespread demand for transportation fuel, prices, and the conceptualization of specific programmes. During the period of prediction, the combination of these dynamics drives Asian biofuel output to far surpass that of Europe (Renewables 2021 *Analysis and forecast to 2026*, 2020).

Figure 5 shows the leading nations in terms of the proportion of global production of biofuels in 2022. With more than 38% of the world's biofuel output, the United States was the leading producer in 2022. Brazil came in second place, accounting for about 21.4% of global output. Germany, which produced 3.2 percent of the world's biofuels that year, came in fifth place overall (Institute 2022).

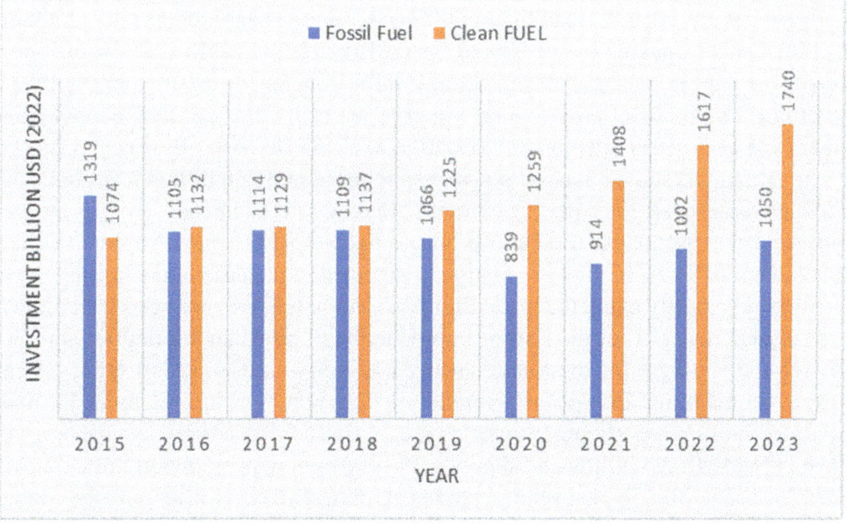

Fig. 4 Investments in clean energy and fossil fuels worldwide from 2015 to 2023 (Adapted from IEA (Birol 2014))

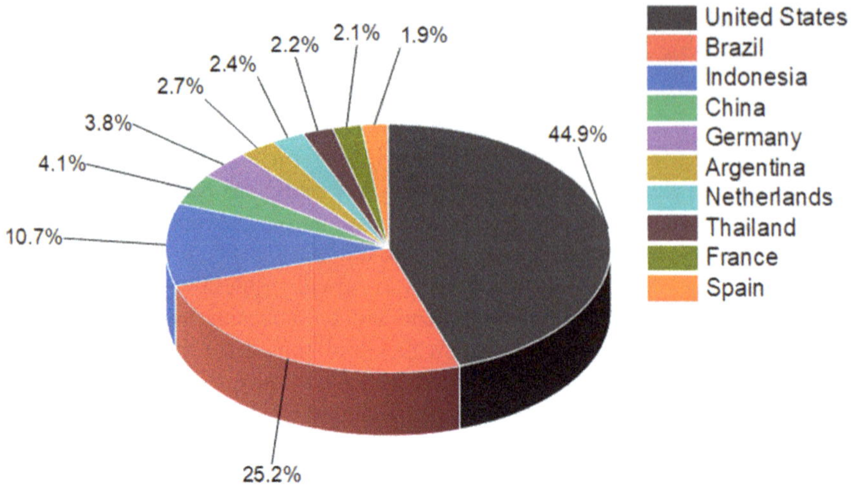

Fig. 5 Leading nations in terms of the proportion of global production of biofuels in 2022 (Institute 2022)

In 2021, the United States generated around 17.5 billion gallons of biofuels and used approximately 16.8 billion gallons (Biofuels explained 2022). In 2021, the United States was a net exporter of around 0.8 billion gallons of biofuels, with gasoline-ethanol representing for the majority of net biofuel exports (Biofuels explained 2022). The United States mostly makes use of maize as the primary raw material for bioethanol production and soybeans for biodiesel production.

Brazil has served as the leading producer and user of ethanol for decades although in 2006 it was surpassed by the United States. Brazil is the world's largest exporter. Brazil was the foremost country to commercially generate bioethanol through their government's Proalcool programme, which began in 1975 to use fuel alcohol from sugar cane as a petrol alternative in reaction to rising oil costs (Robert and Brown 2023). Brazilian total ethanol output was anticipated at 31.66 billion litres in 2022, a 6% increase over 2021 due to expected increases in sugarcane production and consistent increases in maize ethanol production (USDA 2022a). With more Brazilian government policies, biofuel production is expected to increase in the future.

Ethanol is mostly extracted in the European Union (EU). Sugar beets have shown to be a great material for the European production of bioethanol. For 2020, the EU has set a 10% target for renewable energy consumption in transport (JRC - Joint Research Center and European Union 2022). This aim rises to 14% by 2030, with advanced biofuels contributing twice as much. Biofuel has a direct contribution to EU GDP of EUR 2.3 billion and supports 250 000 employment opportunities across the value chain. The EU ranks third in terms of innovative firms, after only the United States and Japan (JRC - Joint Research Center and European Union 2022).

In 2018, Argentina generated 70.6 thousand barrels of biofuels per day. This comprised 2.7% of all global industrial capacity (Argentina: Biofuels Annual 2021).

The country mostly manufactures biodiesel and bioethanol. Argentina at present has 33 biodiesel facilities in operation. The largest facility has the capacity to generate up to 700 million litres per year (Argentina: Biofuels Annual 2021).

After the United States, Brazil, and the European Union (EU), China is the world's fourth-largest producer and consumer of ethanol. China has been using 10% ethanol-gasoline combination by 2020. It also intends to transition from renewable fuel generation to commercial-scale cellulosic ethanol production by 2025 (The Potential of Biofuels in China 2017).

From the beginning of 2000, the Australian government has backed gasoline-ethanol with a variety of tax breaks and production incentives. Biofuel projects and businesses might be an appealing investment possibility. The investment required in Australia for production facilities alone is projected to be between $25–$30 billion (ARENA 2019). Recent announcements from Australian government may stimulate increasing production of environmentally conscious aviation fuel in upcoming years.

Currently, Canada generates only a small amount of ethanol. A lot of projects are presently underway to greatly enhance output (Littlejohns et al. 2018). If ongoing announced biofuel projects are executed, production is expected to rise substantially in the coming years.

Thailand increased its aim for lowering greenhouse gas emissions at the Climate Change Conference of the parties. The Thai government is still working on finalizing the new biofuel usage objectives. Due to the protracted outbreak of COVID-19 and the delayed economic recovery following the pandemic created by Russia's invasion of Ukraine, biofuel usage in 2021 and 2022 remains significantly below current objectives (Prasertsri 2022).

Japan is amending its transportation biofuel rules, which are set to expire in March 2023. Japan vowed in May 2022 to take all feasible efforts to quadruple the demand for bioethanol by 2030 (USDA 2022b). Japan's principal aim for reaching that pledge is increasing locally renewable aviation fuel (SAF) production, especially alcohol-to-jet SAF (USDA 2022b).

In Asia, Indonesia has planned 40% ethanol blending for 2022, which would drive up biodiesel output, while India has planned to incorporate 20% ethanol by 2025, which will push up global ethanol consumption. The median ethanol mixing proportion with petroleum in India is anticipated to be 11.5% in calendar year 2023 (India: Biofuels Annual 2023).

In order to meet its E20 objective, India has been pushing second-generation bioethanol. It can contribute to the goal of transforming waste into electricity while also reducing agricultural waste burning. The second-generation facility will be critical in producing ethanol accessible for blending. According to the secretary of the Union Ministry of Petroleum and Natural Gas, ethanol output in India will triple by 2025 to around 10 billion litres per year. The Hindustan Petroleum Corporation Ltd. intends to build four 2nd generation ethanol facilities that will turn waste from agriculture into biofuel (HPCL Leads the Way in India's Ethanol Blending Program with Successful Pilot Study on E27 Fuel and Ethanol Blended Diesel Fuel 2023).

Almost all the globe is concerned about the depletion of fossil fuel and aims to develop alternative fuel, particularly utilizing things that are abundant in mother earth.

4 Sustainability Impact of Biofuel

Sustainability involves fulfilling current generations requirements without risking future generations requirements while maintaining a balance of economic growth, environmental stewardship, and social well-being (Sustainability Science 2014). Figure 6 shows the sustainable impact of biofuel on social, environmental, and economic aspects.

4.1 Economics of Biofuels

Economic sustainability is an organization's capacity to manage its resources and sustain long-term profitability. The usage of biofuels as a substitute of fossil fuels has a chance of substantially mitigate a number of the negative consequences of fossil fuel production and use, particularly conventional and greenhouse gas pollutant emissions, shortage of scarce resources, and dependability on unsteady foreign suppliers. The market for biofuels could additionally bolster agricultural revenue. However, due to the fact that many feedstocks for biofuel need water, land, and other

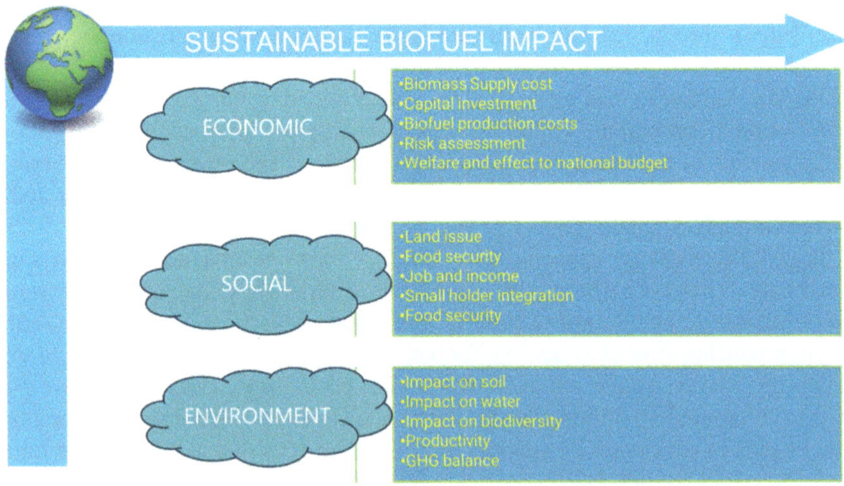

Fig. 6 Sustainable biofuel impact

resources, research indicates that the production of biofuel could have a variety of adverse implications.

Changes in land use patterns may result in greater emissions of greenhouse gases, increase load on water resources, contamination of the water and air and increased costs for food, a few potential negative consequences. Depending on the material used for production, the technique of production, and the time horizon of the research, biofuels can emit even more greenhouse gases (GHGs) than certain fossil fuels on an energy-equivalent basis. Furthermore, in order for biofuels to compete economically with fossil fuels, subsidies and other market interventions are usually needed, culminating in economic deadweight losses (Hasan et al. 2023).

The United States, Canada, Brazil, Indonesia, and India account for 80% of global growth in the use of biofuels, all of which have robust, growth-supporting governmental frameworks (Transport biofuels 2022).

Job possibilities in various industries, including manufacturing and agriculture, could develop due to the development and use of biofuels. Every 100 million gallons of biodiesel production support around 3,200 jobs, and under the Renewable Fuel Standard, the manufacturers are ready to grow output and hire more people (The Economic Impact of Biodiesel 2023). The production of biofuel which is present in practically every state in the country, supports a wide range of economic sectors, including industry, agriculture, transportation, and retail (The Economic Impact of Biodiesel 2023).

4.2 Social Impact of Biofuel

Different organizations and stakeholders may be impacted by biofuels in both favorable and detrimental ways. They can raise smallholder farmers standard of living and revenues, particularly if they work with processing, creating, and distributing biofuels. If women, young people, and other marginalized groups have equitable access to biofuel possibilities and assets, they can also improve the social inclusion and empowerment of these groups (Wu et al. 2018). However, societal injustices and disputes between various parties and interests can also be made worse by the use of biofuels.

Local communities and indigenous peoples may have their rights and customs violated if forced off their lands and territories. They could also worsen the way that employees, particularly migrant and undocumented ones, are exploited and mistreated in the biofuel industry. Additionally, they can make it worse how migrants and illegal workers are taken advantage of and abused in the biofuel sector.

4.3 Environment Impact of Biofuel

Biofuels have been positioned as a carbon–neutral substitute to fossil fuels as a way of reducing greenhouse gas (GHG) emissions and the impact of mobility on climate change. The focus of the majority of LCA research on biofuels is on GHG emissions and savings compared to fossil fuels. Studies on the life cycle assessment (LCA) of biofuels also take into account other environmental effect categories, such as acidification, photochemical smog, eutrophication, human toxicity, and ecotoxicity (Wu et al. 2018).

Without taking into account land use change (LUC), first-generation biofuels may often release less greenhouse gases (GHGs) than fossil fuels, but the reductions for the bulk of feedstocks are insufficient to reach the GHG savings required by the EU Renewable Energy Directive (RED) (Jeswani, Chilvers and Azapagic 2020). In general, second-generation biofuels can cut emissions better, assuming there is no land use change (Dey et al. 2022). Due to their high GHG emissions than fossil fuels, 3rd-generation biofuels do not yet provide an efficient alternative.

In general, the tools that may be used to assess the effects of biofuels include life cycle analysis (LCA), sustainability certification (SC), socio-economic analysis (SEA), and multi-stakeholder dialogue (MSD) (Mukrimaa et al. 2019). Although there is no one-size-fits-all answer to energy and climate concerns, biofuels can contribute if they are created and used wisely. Making enlightened judgements about whether to go along with or condemn efforts in one's area requires knowing biofuel production's economic, environmental, and social effects.

5 Feedstock and Production Technologies of Biofuel

Plants produced specifically for energy production as well as plant remnants, components, processing wastes, and materials from human and animal activities, can all be used as biomass feedstocks for energy generation (Alalwan, Alminshid and Aljaafari 2019). Figure 7 shows the general classification of biomass, which depicts that biomass can be broadly classified into three categories crops, natural vegetables, and organic waste and residue.

Globally, there is an increase in the extraction of biofuels employing various technological advances and biological methods from various bioresources. Waste management barriers could potentially be resolved and ecological burdens reduced by using agricultural crop leftovers and biomass waste to make biofuels (Ambaye et al. 2021). Figure 8 shows the broad classification of biomass conversion technologies, namely thermochemical, physiochemical, and biochemical conversion.

First-generation biofuels' Feed stock sources are extracted from eatable energy-producing crops like sugar beet, sugarcane, sorghum, corn, barley, wheat, sunflower, rapeseed, and canola (Inamuddin et al. 2021). Technologies used for the production

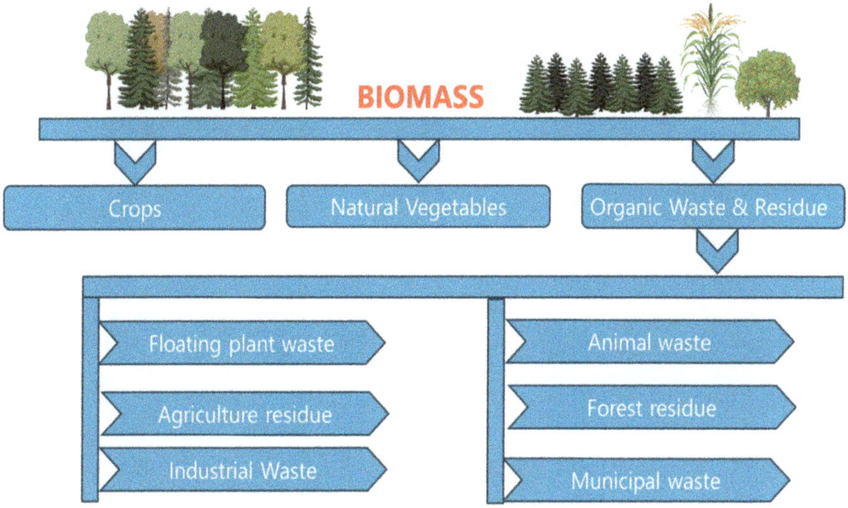

Fig. 7 Classification of biomass

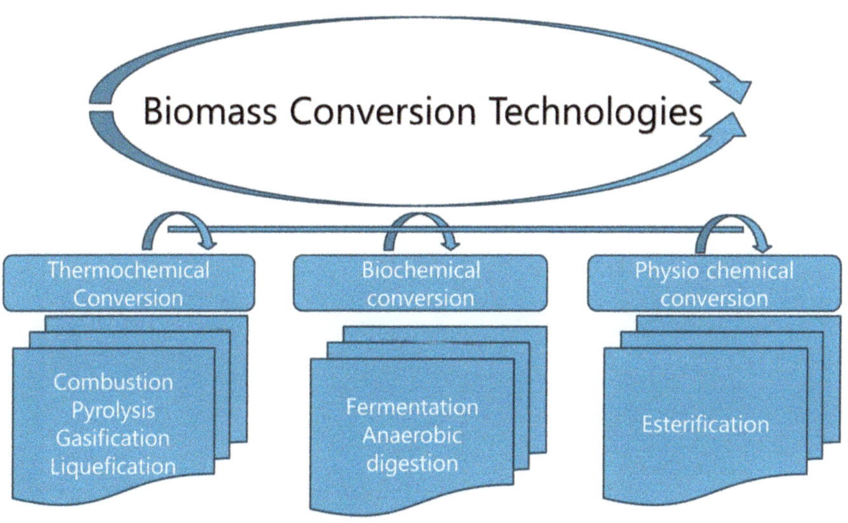

Fig. 8 Classification of biomass conversion technologies

or Generation of the 1st Generation of biofuel are fermentation and transesterification, as depicted in Fig. 9.

Non-food crops, corn, energy crops, wheat straw, wood, waste oils, agricultural and municipal wastes, and inedible oils are some of the feedstocks used to produce second-generation biofuel (Sachin Powar et al. 2022). On the other hand, second-generation processing systems also employ grain and sugar crops as feedstocks. As

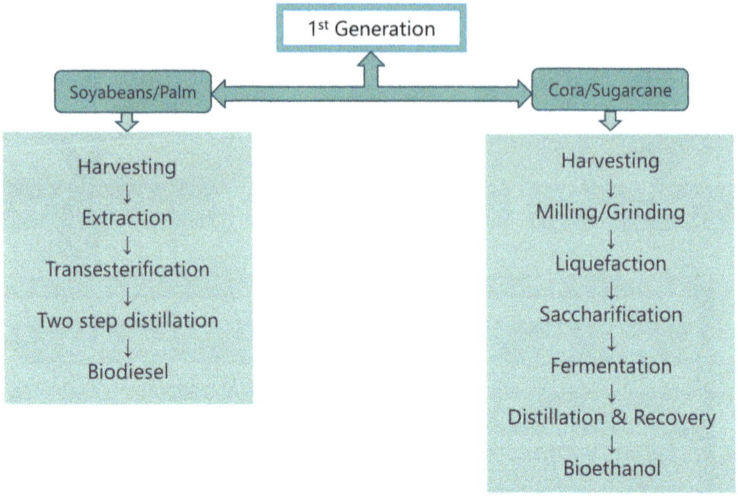

Fig. 9 Technologies involved in the production of 1st generation biofuel (adapted from (Dutta et al. 2014))

shown in Fig. 10, the conversion methods employed for the development of the 2nd generation of biofuel are biochemical, thermochemical processes.

The plant cell wall is made up of cellulose, hemicellulose, lignin, and trace amounts of inorganic material (Chauhan et al. 2023). Thus, it is known as lignocellulosic biomass; nevertheless, the mass balance and biochemical makeup of this

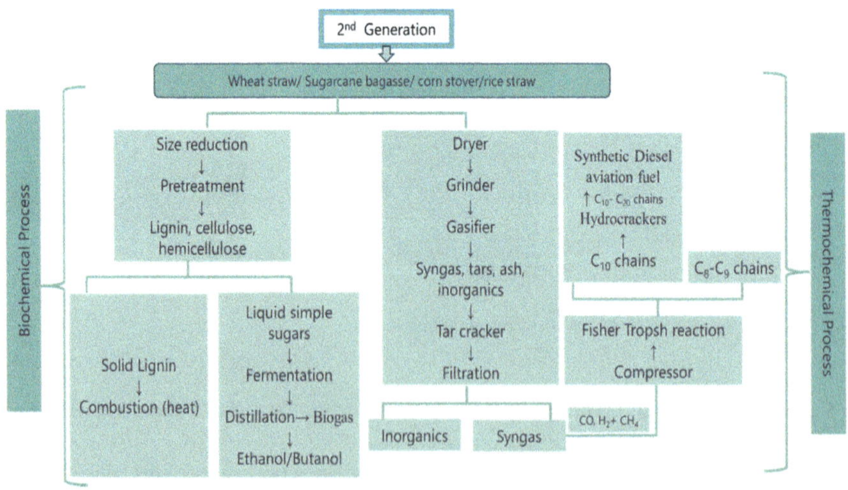

Fig. 10 Technologies involved in production of 2nd generation biofuel (adapted from (Dutta et al. 2014))

biomass vary depending on the kind of plant (Ambaye et al. 2021). Table 2 displays the amounts of cellulose, hemicellulose, and lignin in sorted wastes, chemical pulps, animal dung, and agricultural leftovers.

Algae serve as the feedstock for third-generation biofuel. Microalgae offer great biofuel potential as a third-generation feedstock because of their quick development, high biomass production, and high lipid and carbohydrate content (Chowdhury et al. 2019). The lucrative biofuels biogas, bioethanol, biodiesel, and biomethane are all produced by algae (Sachin Powar et al. 2022). As shown in Fig. 11, the conversion methods employed for the development of the 3rd generation of biofuel are biochemical, thermochemical, and physiochemical.

Table 2 Amounts of cellulose, hemicellulose, and lignin in various sources (Prasad et al. 2007)

Agricultural residue	Lignin	Cellulose	Hemicellulose
Waste papers (chemical pulps)	5–10	60–70	10–20
Wheat straw	15–20	33–40	20–25
Rice straw	55	40	18
Solid cattle manure	2.7–5.7	1.6–4.7	1.4–3.3
Corn Cobs	15	45	35
Nutshells	30–40	25–30	25–30
Cottonseed hairs	0	80–90	5–20
Leaves	0	15–20	80–85
Swine waste		6.0	28
Primary wastewater solids	24–29	8–15	–
Paper	0–15	85–99	0
Sorted refuse	20	60	20

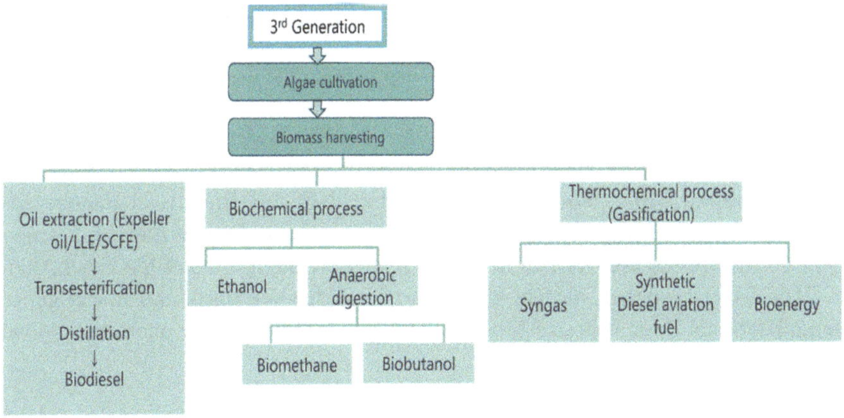

Fig. 11 Technologies involved in production of 3rd generation biofuel (adapted from Dutta et al. 2014)

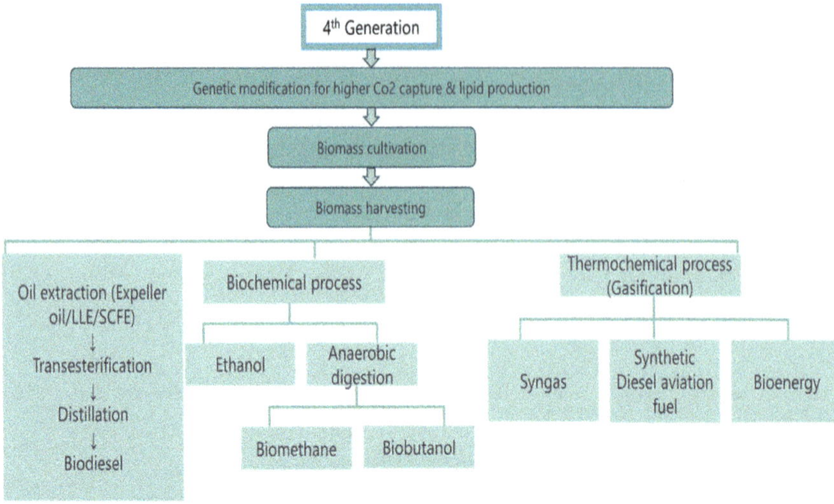

Fig. 12 Technologies involved in the production of 4th generation biofuel (adapted from Dutta et al. 2014)

Genetically engineered organisms for the improvement of biohydrogen production processes are an integral component of the fourth Generation of feedstock. After genetic alteration, fourth-generation feedstock includes biomass from the second and third generations (Goria et al. 2022). As shown in the Fig. 12, the conversion methods employed for the development of the fourth generation of biofuel are biochemical, thermochemical, and physiochemical. Closing this, the authors of this chapter have considerable experience in the field of bio-composites (Arpitha et al. 2022; Bharath et al. 2020; Chaturvedi et al. 2023; Dogra et al. 2021; Jain et al. 2019, 2023; Marichelvam et al. 2021; Prabhakaran et al. 2022; Sati et al. 2023; TG et al. 2023; Verma et al. 2017, 2018, 2019Verma et al. 2023).

6 Conclusion

Due to its renewability, biodegradability, and ecologically friendly characteristics, biofuel has attracted significant interest as a practical and flexible fuel replacement to the currently used non-renewable fuels. Depending on the kind of feedstock, there are four distinct types of biofuel generation: first, second, third, and fourth generation. There are three types of biomass conversion technologies: thermochemical, physiochemical, and biochemical.

7 Future Trends

Global demand for biofuels is steadily increasing. The environmental benefits of growing or expanding biofuel production and usage, especially the ability to reduce emissions such as greenhouse gases (GHG), are of particular importance. This section highlights potential future research areas that should be prioritised.

- It is challenging to use nanotechnology to the industrial production of biofuels since nano-catalyst-based biofuel manufacturing is still in its infancy. Future effort should be focused on implementing nanotechnology to industrial production of biofuel.
- Future studies should focus on developing sources for fourth-generation biofuels.
- Future research should focus on optimizing commercial biofuel production technologies to meet the need for a high biofuel production rate.
- Future studies should concentrate on using agricultural waste as a feedstock for the production of biofuels to solve the problem of global pollution potentially.
- Future studies should concentrate on biomass's biochemical conversion for biofuel production.

Acknowledgements Author "Akarsh Verma" is grateful to the monetary support provided by the University of Petroleum and Energy Studies (UPES)-SEED Grant program.

Declarations

Conflicts of Interest "There are no conflicts of interest to declare by the authors."

Ethical Approval The authors hereby state that the present work is in compliance with the ethical standards.

Competing Interests The authors declare no competing interests.

Availability of Data and Materials Not applicable.

References

Abbaszaadeh, A. *et al.* (2012) 'Current biodiesel production technologies: A comparative review', *Energy Conversion and Management*, 63, pp. 138–148. Available at: https://doi.org/10.1016/j.enconman.2012.02.027.

Abdullah, B. *et al.* (2019) 'Fourth generation biofuel: A review on risks and mitigation strategies', *Renewable and Sustainable Energy Reviews*, 107(November 2018), pp. 37–50. Available at: https://doi.org/10.1016/j.rser.2019.02.018.

Alalwan, H.A., Alminshid, A.H. and Aljaafari, H.A.S. (2019) 'Promising evolution of biofuel generations. Subject review', *Renewable Energy Focus*, 28(00), pp. 127–139. Available at: https://doi.org/10.1016/j.ref.2018.12.006.

Ambaye, T.G. *et al.* (2021) 'Emerging technologies for biofuel production: A critical review on recent progress, challenges and perspectives', *Journal of Environmental Management*, 290. Available at: https://doi.org/10.1016/j.jenvman.2021.112627.

ARENA, C. and (2019) *Biofuels and Transport : An Australian opportunity*. Available at: https://apo.org.au/node/268686.

Arpitha, G.R., Verma, A., MR, S., Gorbatyuk, S., Khan, A., Sobahi, T.R., Asiri, A.M. and Siengchin, S., 2022. Bio-composite film from corn starch based vetiver cellulose. *Journal of Natural Fibers*, *19*(16), pp.14634-14644.

Argentina: Biofuels Annual (2021). Available at: https://www.fas.usda.gov/data/argentina-biofuels-annual-6.

Barros, A.I. *et al.* (2015) 'Harvesting techniques applied to microalgae: A review', *Renewable and Sustainable Energy Reviews*, 41, pp. 1489–1500. Available at: https://doi.org/10.1016/j.rser.2014.09.037.

Behera, S. *et al.* (2015) 'Scope of Algae as Third Generation Biofuels', *Frontiers in Bioengineering and Biotechnology*, 2(February), pp. 1–13. Available at: https://doi.org/10.3389/fbioe.2014.00090.

Bharath, K.N., Madhu, P., Gowda, T.G., Verma, A., Sanjay, M.R. and Siengchin, S., 2020. A novel approach for development of printed circuit board from biofiber based composites. *Polymer Composites*, *41*(11), pp.4550-4558.

Biofuel (no date) *Wikipedia*. Available at: https://en.wikipedia.org/wiki/Biofuel.

Biofuels explained (2022). Available at: https://www.eia.gov/energyexplained/biofuels/.

Birol, F. (2014) 'World Energy Investment', pp. 1–183. Available at: https://www.iea.org/data-and-statistics/charts/global-energy-investment-in-clean-energy-and-in-fossil-fuels-2015-2023, IEA.

Brito Cruz, C.H., Souza, G.M. and Barbosa Cortez, L.A. (2014) 'Biofuels for Transport', in *Future Energy*. Elsevier, pp. 215–244. Available at: https://doi.org/10.1016/B978-0-08-099424-6.00011-9.

Bux, F. (ed.) (2013) *Biotechnological Applications of Microalgae*. CRC Press. Available at: https://doi.org/10.1201/b14920.

Chandel, A.K. and Sukumaran, R.K. (2017) *Sustainable biofuels development in India, Sustainable Biofuels Development in India*. Available at: https://doi.org/10.1007/978-3-319-50219-9.

Chaturvedi, S., Kataria, A., Chaudhary, V., Verma, A., Jain, N., Sanjay, M.R. and Siengchin, S., 2023. Bionanocomposites reinforced with cellulose fibers and agro-industrial wastes. In *Cellulose Fibre Reinforced Composites* (pp. 317–342). Woodhead Publishing. https://doi.org/10.1016/B978-0-323-90125-3.00017-3

Chauhan, K. *et al.* (2023) 'Lignin extraction from lignocellulosic biomass (sugarcane bagasse) and its potential application as a feedstock for fuel production', *Materials Today: Proceedings*, 78(xxxx), pp. 688–694. Available at: https://doi.org/10.1016/j.matpr.2022.12.190.

Chauhan, K. and Pratap Singh, V. (2023) 'Prospect of biomass to bioenergy in India: An overview', *Materials Today: Proceedings* [Preprint], (xxxx). Available at: https://doi.org/10.1016/j.matpr.2023.01.419.

Chen, J. (2022) *Biofuel: Definitions, Examples, and Limitations*. Available at: https://www.investopedia.com/terms/b/biofuel.asp.

Chisti, Y. (2007) 'Biodiesel from microalgae', *Biotechnology Advances*, 25(3), pp. 294–306. Available at: https://doi.org/10.1016/j.biotechadv.2007.02.001.

Chowdhury, H. *et al.* (2019) *Algae for biofuels: The third generation of feedstock, Second and Third Generation of Feedstocks: The Evolution of Biofuels*. Elsevier Inc. Available at: https://doi.org/10.1016/B978-0-12-815162-4.00012-4.

Dahman, Y. *et al.* (2019) *Biofuels: Their characteristics and analysis, Biomass, Biopolymer-Based Materials, and Bioenergy: Construction, Biomedical, and other Industrial Applications*. Elsevier Ltd. Available at: https://doi.org/10.1016/B978-0-08-102426-3.00014-X.

Dayton, D.C. and Foust, T.D. (2020) 'Lifecycle Assessment of Advanced Biofuels Processes', in *Analytical Methods for Biomass Characterization and Conversion*. Elsevier, pp. 117–127. Available at: https://doi.org/10.1016/B978-0-12-815605-6.00008-1.

Dey, S. *et al.* (2022) 'Second-Generation Biofuels Toward Environmental Sustainability', in *Encyclopedia of Green Materials*. Singapore: Springer Nature Singapore, pp. 1–8. Available at: https://doi.org/10.1007/978-981-16-4921-9_104-1.

Dogra, V., Kishore, C., Verma, A., Rana, A.K. and Gaur, A., 2021. Fabrication and experimental testing of hybrid composite material having biodegradable bagasse fiber in a modified epoxy resin: evaluation of mechanical and morphological behavior. *Applied Science and Engineering Progress*, *14*(4), pp.661-667.

Dutta, K., Daverey, A. and Lin, J.G. (2014) 'Evolution retrospective for alternative fuels: First to fourth generation', *Renewable Energy*, 69, pp. 114–122. Available at: https://doi.org/10.1016/j.renene.2014.02.044.

Goria, K. *et al.* (2022) 'Biohydrogen: potential applications, approaches, and hurdles to overcome', in *Handbook of Biofuels*. Elsevier, pp. 399–418. Available at: https://doi.org/10.1016/B978-0-12-822810-4.00020-8.

Groves, C., Sankar, M. and Thomas, P.J. (2018) 'Second-generation biofuels: exploring imaginaries via deliberative workshops with farmers', *Journal of Responsible Innovation*, 5(2), pp. 149–169. Available at: https://doi.org/10.1080/23299460.2017.1422926.

Hasan, M. *et al.* (2023) 'Sustainable biofuel economy: A mapping through bibliometric research', *Journal of Environmental Management*, 336(February). Available at: https://doi.org/10.1016/j.jenvman.2023.117644.

HPCL Leads the Way in India's Ethanol Blending Program with Successful Pilot Study on E27 Fuel and Ethanol Blended Diesel Fuel (2023) *HPCL*. Available at: https://www.hindustanpetroleum.com/NewsroomDetails/401.

Inamuddin *et al.* (eds) (2021) *Biodiesel Technology and Applications*. Wiley. Available at: https://doi.org/10.1002/9781119724957.

India: Biofuels Annual (2023) *USDA*. Available at: https://www.fas.usda.gov/data/india-biofuels-annual-8.

Institute, E. (2022) *Leading biofuel producing countries worldwide in 2022, by share of global production*. Available at: https://www.statista.com/statistics/274169/top-biofuel-producing-countries-worldwide/.

Jain, N., Verma, A. and Singh, V.K., 2019. Dynamic mechanical analysis and creep-recovery behaviour of polyvinyl alcohol based cross-linked biocomposite reinforced with basalt fiber. *Materials Research Express*, *6*(10), p.105373.

Jain, N., Sharma, P., Verma, A. and Gupta, J., 2023. Enhancement of thermo-mechanical, creep-recovery, and anti-microbial properties in PVA-based biodegradable films through cross-linking with oxalic acid: implications for packaging application. *Biomass Conversion and Biorefinery*, pp.1–10. DOI: https://doi.org/10.1007/s13399-023-04748-y

Jeswani, H.K. and Azapagic, A. (2012) 'Life cycle sustainability assessment of second generation biodiesel', in *Advances in Biodiesel Production*. Elsevier, pp. 13–31. Available at: https://doi.org/10.1533/9780857095862.1.13.

Jeswani, H.K., Chilvers, A. and Azapagic, A. (2020) 'Environmental sustainability of biofuels: A review: Environmental sustainability of biofuels', *Proceedings of the Royal Society A: Mathematical, Physical and Engineering Sciences*, 476(2243). Available at: https://doi.org/10.1098/rspa.2020.0351.

JRC - Joint Research Center and European Union (2022) *Advanced Biofuels in the European Union Issn 1831-9424*. Available at: https://doi.org/10.2760/938743.

Leong, W.H. *et al.* (2018) 'Third generation biofuels: A nutritional perspective in enhancing microbial lipid production', *Renewable and Sustainable Energy Reviews*, 91(April), pp. 950–961. Available at: https://doi.org/10.1016/j.rser.2018.04.066.

Littlejohns, J. *et al.* (2018) 'Current state and future prospects for liquid biofuels in Canada', *Biofuel Research Journal*, 5(1), pp. 759–779. Available at: https://doi.org/10.18331/BRJ2018.5.1.4.

Marichelvam, M.K., Manimaran, P., Verma, A., Sanjay, M.R., Siengchin, S., Kandakodeeswaran, K. and Geetha, M., 2021. A novel palm sheath and sugarcane bagasse fiber based hybrid composites for automotive applications: An experimental approach. *Polymer Composites*, *42*(1), pp.512-521.

Markou, G. *et al.* (2018) 'Using agro-industrial wastes for the cultivation of microalgae and duckweeds: Contamination risks and biomass safety concerns', *Biotechnology Advances*, 36(4), pp. 1238–1254. Available at: https://doi.org/10.1016/j.biotechadv.2018.04.003.

Moodley, P. (2021) *Sustainable biofuels: opportunities and challenges, Sustainable Biofuels: Opportunities and Challenges*. INC. Available at: https://doi.org/10.1016/B978-0-12-820297-5.00003-7.

Mukrimaa, S.S. *et al.* (2019) *Advances in Eco-Fuels for a Sustainable Environment, Jurnal Penelitian Pendidikan Guru Sekolah Dasar*. Elsevier. Available at: https://doi.org/10.1016/C2017-0-04211-8.

Mukrimaa, S.S. *et al.* (2023) *Handbook of Biofuels Production, Jurnal Penelitian Pendidikan Guru Sekolah Dasar*. Elsevier. Available at: https://doi.org/10.1016/C2020-0-03711-4.

Nanda, S. *et al.* (2018) 'A broad introduction to first-, second-, and third-generation biofuels', *Recent Advancements in Biofuels and Bioenergy Utilization*, pp. 1–25. Available at: https://doi.org/10.1007/978-981-13-1307-3_1.

Oil Consumption by Country (2023). Available at: https://www.worldometers.info/oil/oil-consumption-by-country/.

Prabhakaran, S., Sharma, S., Verma, A., Rangappa, S.M. and Siengchin, S., 2022. Mechanical, thermal, and acoustical studies on natural alternative material for partition walls: A novel experimental investigation. *Polymer Composites*, *43*(7), pp.4711-4720.

Prasad, S. and Ingle, A.P. (2023) 'The emerging role of nanotechnology in ethanol production', in *Nanotechnology for Biorefinery*. Elsevier, pp. 235–256. Available at: https://doi.org/10.1016/B978-0-323-95965-0.00001-9.

Prasad, S., Singh, A. and Joshi, H.C. (2007) 'Ethanol as an alternative fuel from agricultural, industrial and urban residues', *Resources, Conservation and Recycling*, 50(1), pp. 1–39. Available at: https://doi.org/10.1016/j.resconrec.2006.05.007.

Prasertsri, P. (2022) *Biofuels Annual (Thailand), Unites States Department of Agriculture Foreign Agriculture Service*. Available at: https://www.fas.usda.gov/data/malaysia-biofuels-annual-3.

Qiu, H. *et al.* (2012) 'Liquid biofuels in China: Current status, government policies, and future opportunities and challenges', *Renewable and Sustainable Energy Reviews*, 16(5), pp. 3095–3104. Available at: https://doi.org/10.1016/j.rser.2012.02.036.

Renewables 2021 Analysis and forecast to 2026 (2020) *International Energy Agency (IEA)*. Available at: https://doi.org/10.1002/peng.20026.

Robert, B. and Brown, E.B. (2023) *Cellulases in the Biofuel Industry*. Elsevier. Available at: https://doi.org/10.1016/C2021-0-02078-2.

Roberts, L.G. and Smagala, T. (2022) 'Biofuels', in *Reference Module in Biomedical Sciences*. Elsevier. Available at: https://doi.org/10.1016/B978-0-12-824315-2.00332-8.

Sachin Powar, R. *et al.* (2022) 'Algae: A potential feedstock for third generation biofuel', *Materials Today: Proceedings*, 63, pp. A27–A33. Available at: 0.1016/j.matpr.2022.07.161.

Salama, E.S. *et al.* (2017) 'Recent progress in microalgal biomass production coupled with wastewater treatment for biofuel generation', *Renewable and Sustainable Energy Reviews*, 79(July 2016), pp. 1189–1211. Available at: https://doi.org/10.1016/j.rser.2017.05.091.

Sati, P., Verma, A., Zindal, A., Chauhan, S. and Singh, V.K., 2023. PVA biopolymer-acidic functionalized graphene hybrid nano composite for vibration isolation application: An experimental approach with variable reflux and vacuum timings. *Chemical Physics Impact*, *6*, p.100212.

Sheppard, A.W. *et al.* (2011) 'Biosecurity and sustainability within the growing global bioeconomy', *Current Opinion in Environmental Sustainability*, 3(1–2), pp. 4–10. Available at: https://doi.org/10.1016/j.cosust.2010.12.011.

Sustainability Science (2014). Elsevier. Available at: https://doi.org/10.1016/C2012-0-06785-3.

TG, Y.G., Ballupete Nagaraju, S., Puttegowda, M., Verma, A., Rangappa, S.M. and Siengchin, S., 2023. Biopolymer-Based Composites: An Eco-Friendly Alternative from Agricultural Waste Biomass. *Journal of Composites Science*, *7*(6), p.242. DOI: https://doi.org/10.3390/jcs7060242

The Economic Impact of Biodiesel (2023) *Renewable Energy Group*. Available at: https://www.regi.com/resources/cleaner-fuel/the-economic-impact-of-biodiesel.

The Potential of Biofuels in China (2017) *IEA Bioenergy*. Available at: https://www.ieabioenergy.com/blog/publications/the-potential-of-biofuels-in-china/.

Transport biofuels (2022). Available at: https://www.iea.org/reports/renewables-2022/transport-biofuels.

USDA (2022a) *Biofuels Annual (Brazil), Unites States Department of Agriculture Foreign Agriculture Service*. Available at: https://www.fas.usda.gov/data/malaysia-biofuels-annual-3.

USDA (2022b) *Japn: Biofuels Annual*. Available at: https://www.fas.usda.gov/data/japan-biofuels-annual-6.

Velvizhi, G. *et al.* (2023) 'Emerging trends and advances in valorization of lignocellulosic biomass to biofuels', *Journal of Environmental Management*, 345(July), p. 118527. Available at: https://doi.org/10.1016/j.jenvman.2023.118527.

Verma, A., Gaur, A. and Singh, V.K., 2017. Mechanical properties and microstructure of starch and sisal fiber biocomposite modified with epoxy resin. *Materials Performance and Characterization*, *6*(1), pp.500-520.

Verma, A., Joshi, K., Gaur, A. and Singh, V.K., 2018. Starch-jute fiber hybrid biocomposite modified with an epoxy resin coating: fabrication and experimental characterization. *Journal of the Mechanical Behavior of Materials*, *27*(5-6), p.20182006. DOI: https://doi.org/10.1515/jmbm-2018-2006

Verma, A., Negi, P. and Singh, V.K., 2019. Experimental analysis on carbon residuum transformed epoxy resin: Chicken feather fiber hybrid composite. *Polymer Composites*, *40*(7), pp. 2690-2699.

Verma, A. and Ogata, S., 2023. Magnesium based alloys for reinforcing biopolymer composites and coatings: a critical overview on biomedical materials. *Advanced Industrial and Engineering Polymer Research*. DOI: https://doi.org/10.1016/j.aiepr.2023.01.002

Wu, Y. *et al.* (2018) 'Bioenergy production and environmental impacts', *Geoscience Letters*, 5(1), p. 14. Available at: https://doi.org/10.1186/s40562-018-0114-y.

Zhao, B. *et al.* (2014) 'Efficient anaerobic digestion of whole microalgae and lipid-extracted microalgae residues for methane energy production', *Bioresource Technology*, 161, pp. 423–430. Available at: https://doi.org/10.1016/j.biortech.2014.03.079.

Evaluating the Impact of Climate Change on Antimicrobial Resistance and Rise in Dysentery Using Next Generation Sequencing Based Approaches

Sarmishta Mukhopadhyay, Rupsha Karmakar, Santanu Chakrabarti, Mahashweta Mitra Ghosh, and Sayak Ganguli

1 Introduction

The world is experiencing an anthropogenic climate catastrophe, which threatens humanity's existence as well as the survival of millions of other species on the planet. Negative impact of climate change on human health have prompted World Health Organization (WHO) to declare it as one of the top ten global health hazards in 2019 (WHO, 2019). The incidence of infectious diseases and its sensitivity to climatic fluctuations, has gained a lot of international attention in recent years. Dysentery which is primarily categorized as a food or water borne infectious disease, is believed to be strongly influenced by climatic changes, which manipulates either the transmission process or the pathogen or vector survival itself. Studies have shown that overuse of antimicrobial drugs to curb the rapidly spreading infectious agents (*Shigella* sp., *Pseudomonas aeruginosa, Escherichia coli,* etc.) have selectively favoured the emergence of antimicrobial resistance (AMR) (Langford et al., 2021). The widespread incidence of communicable diseases and rapidly surfacing antibiotic resistance are interlinked in the sense that, both have been exacerbated by current trends in climate

These authors "Sarmishta Mukhopadhyay, Rupsha Karmakar" have equally contributed to the current work.

S. Mukhopadhyay · S. Ganguli (✉)
Post Graduate and Research Department of Biotechnology, St. Xavier's College (Autonomous), Kolkata, India
e-mail: sayakganguli@sxccal.edu

R. Karmakar · M. M. Ghosh
Post Graduate and Research Department of Microbiology, St. Xavier's College (Autonomous), Kolkata, India

S. Chakrabarti
Singur Government General Degree College, Singur, India

J. Gupta and A. Verma (eds.), *Microbiology-2.0 Update for a Sustainable Future*,
https://doi.org/10.1007/978-981-99-9617-9_17

changes. In developing countries like India, where diarrheal diseases remain as the most accountable cause of morbidity and mortality, proper analysis on the association of climatic factors with occurrence of diarrheal episodes and ascent of antimicrobial resistant strains, may help the government to formulate effective control strategies to keep the infections in check. Wastewater surveillance has gained significant attention in recent years for its potential to detect pathogens and residual antibiotics present in community wastewater representing the lifestyle patterns in a community. Wastewater monitoring is known to complement traditional epidemiological and environmental monitoring techniques, providing a broader understanding of communities' health and environmental status. Moreover, data from wastewater monitoring can be used in public health decision making, allocating resources, and implementing targeted interventions in response to specific health or environmental concerns. Among the variety of rapid in situ techniques of wastewater monitoring like Fluorescent in Situ Hybridization (FISH), Flow Cytometry, Immunofluorescence Assays, Biosensors, Microfluidic Devices, and many more, Next-Generation Sequencing (NGS) has emerged as a potent technique for its ability to provide comprehensive and high-throughput analysis of microbial communities and their functional genes present in wastewater. NGS not only allows extensive and in-depth microbial profiling with its high throughput capabilities, but also aids in real-time monitoring of microbial dynamics in wastewater treatment systems (Diamond et al., 2022). Climate change and antimicrobial resistance, now being acknowledged as interlinked health emergencies, this chapter attempts to identify the complex commonalities between them and puts forward the advantages, disadvantages and applications of next generation sequencing technologies as rapid and accurate detection methods for waterborne pathogens.

2 Spatio-Temporal Trends in Shigellosis

In context of diarrheal diseases, of particular importance is the gram-negative pathogen *Shigella* spp, which is the most prevalent etiologic agent in low-income countries, attributable to 165 million cases of diarrheal episodes, in children below 5 years of age (The et al., 2016). In the light of the massive morbidity and mortality of shigellosis in developing countries, it is of utmost relevance to focus on the temporal and seasonal trends in shigellosis incidence, primarily in tropical and subtropical regions. Although, we find a ubiquitous distribution of *Shigella* worldwide (Fig. 1), the different serotypes of the pathogen exhibit remarkable dynamicity in their spatial and temporal dispersal across the globe.

Several studies have established a socio-economic dependency of serogroup distribution, projecting *Shigella flexneri*, as the predominant pathogen in impoverished countries, and *Shigella sonnei* as the major serotype in transitional and high-income countries (Michael et al., 2020). Epidemiological studies, conducted across India, observed temporal alterations in the prevailing serogroups, nominating *Shigella flexneri* as the principal contributor from 1994–2002, followed by the dominance of

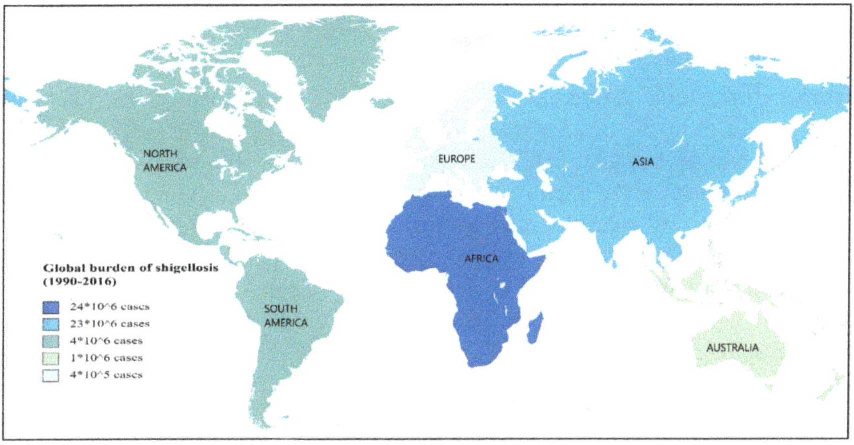

Fig. 1 Global burden of shigellosis in people of all ages, from the year 1990–2016 (Khalil et al., 2018)

Shigella dysenteriae for a single year in 2003 (Taneja and Mewara, 2016). This cyclic revolution in prevalence of specific serogroups in a defined geographical terrain, and the overall genome plasticity recorded worldwide, hint at the strong correlation between climatic variables and shigellosis incidence. The three key factors shaping up the evolutionary history of *Shigella*, viz.,

- the emergence of *Shigella* spp from ancestral strains of *Escherichia coli*
- the acquisition of virulence plasmid by lateral gene transfer events
- convergent evolution of similar genes by independent events

might be considered as consequences of environmental and adverse climatic events, moulding the pathogen to adopt to particular lifestyle features (The et al., 2016). *Shigella* has been isolated from varied sources starting from water bodies, to free living amoebas to wild animals and are mainly transmitted by ingestion of contaminated food and drink or by direct contact with an infected individual. Climatic parameters can impose bias on the extent of shigellosis occurrence at multiple steps, for example, a certain scale of temperature can serve as optimal for pathogen reproduction and survival, whereas variation in levels of precipitation can directly impact the rate of dissemination of the organism. The following segment will serve as a first-hand account of the orbit of individual climatic components, connected to increased shigellosis incidence, and will have growing implications in accurate predictions of future outbreaks and implementation of prevention protocols.

3 Meteorological Determinants of Shigellosis

Temperature: Temperature has always been on the priority list of environmental elements, affecting the incidence of infectious diseases, particularly dysentery, one of the very reasons being the temperature sensitivity of the pathogen itself, apart from favoring the spread of the microbe under certain conditions. Owing to the omnipresence of *Shigella* across the globe, the pathogen differentially regulates the expression of its outer membrane proteins to cope up in the wide confines of temperature. A study conducted by Harikrishnan et al. (2013), to compare the expression profile of different outer membrane proteins at 37 °C, 38.5 °C and 40 °C respectively, strikingly revealed a surge in the levels of majority of the outer membrane proteins at elevated temperatures of 38.5 °C and 40 °C respectively (Harikrishnan et al., 2013). This might be regarded as a stress response of the pathogen to ensure its continued survival and growth under extreme environmental conditions. However, the majority of the outer membrane proteins being potent virulence factors, influencing different facets of the pathogenesis cascade from host cell invasion to evading the innate immune response, their upregulation gives a boon to the pathogen by enhancing its infectivity and virulence. Similar findings were put forward by another group from China, which recorded a shoot-up in the levels of the invasion plasmid antigens (IpaA, IpaB, IpaC, IpaD) in *Shigella flexneri*, on a shift from 30 to 37 °C (Zhu et al., 2010), which can be attributed to their intrinsically disordered structures (Chakrabarti and Ganguli, 2013). The positive association between temperature and shigellosis incidence thus inferred, were further reinforced by a substantial body of epidemiological data, that established the interrelation between temperature and shigellosis prevalence using different statistical models. Recent research in Taiwan has enumerated an 8% increase in reported cases for each 1 °C rise in temperature, with a consistent growth in cases at temperatures above 21 °C (Chen et al., 2019a, b). Equivalent data were projected by another survey carried out in Yazd province in Iran, where incidence rates were found to peak in every summer, during the years of study (Aminharati et al., 2018). In one of the first studies estimating shigellosis occurrence in regions with a winter seasonality (Korea), a 13.6% climb in cases have been quantified with each 1 °C rise in temperature and a lag of 2 weeks for the events from pathogen growth to notification of infection in the official records (Song et al., 2018). Although these are conclusive evidences to establish the positive association between temperature and shigellosis incidence, the rate of increase in cases for unit rise in temperature, are found to vary for each season and across different territories.

Relative Humidity: While assessing the sustainable impact of climatic variables on the incidence of bacillary dysentery, a large collection of data was found to highlight the tie up between temperature and relative humidity in fostering the multiplication and spread of the pathogen. In Binyang County, China, a group of researchers tried to classify the yearly shigellosis cases on the basis of individual weather determinants, keeping the values of each parameter, constant for a particular season. The number of cases were found to peak dramatically, during a particular period of the year, characterized by a temperature and relative humidity of 26.7 °C and 86% respectively, for all the years under investigation (Li et al., 2019). Another study, reported 1.103 times increase in incidence rate of outbreaks with one percent rise in relative humidity, in Yazd province, Iran. Similar trends were outlined by another survey at Taiwan, where a gradual rise in dysentery episodes were noted at a relative humidity of 70–74%, which were eventually worked out as a 6% climb in cases for every 5% raise in relative humidity (Aminharati et al., 2018). That, vulnerability to dysentery, rises with increasing temperature and relative humidity, were further validated from reports of investigation executed in India, experiencing a spike in cases during hot and humid summers, while winters recorded infrequent cases (Moors et al., 2013).

Following the similar trend of Shigellosis, gradual warming of climate has been found to favour global resurgence and spread of several other vector-borne, water, air and food borne infectious illnesses due to their association with environmental factors. Rising incidence of communicable diseases necessitates added antibiotic treatments (Table 1). Extensive use of antibiotics followed by their rapid discharge into the environment can build up reservoir for antibiotic resistant bacteria. Thus, climate change also fosters the emergence of AMR, which is one of the most impending existential threats to mankind. This has led the WHO to put up a list of Global Priority Pathogens (GPP) in 2017, mentioning 12 bacterial species classified into three categories (critical, high, medium) along with the antibiotics against which they have become resistant (Fig. 2) (WHO, 2017).

4 Emerging Anti-Microbial Resistance with Climatic Fluctuations

In reaction to changes in their environment, microbial pathogens have a remarkable capability to adapt which often occurs in the form of acquiring antibiotic resistance. Inactivation of antimicrobial agents, reduction in intracellular drug accumulation and alteration in drug binding sites are a few of the noticeable mechanisms through which bacteria acquire antibiotic resistance. Environmental variables such as temperature, pH, and nutrient availability have a substantial influence on all of these resistance mechanisms (Rodríguez-Verdugo et al., 2020). As per the recent reports, about 58,000 lives are lost in India every year due to AMR. Climate change is anticipated to result in an additional 250,000 deaths and 2–4-billion-dollar loss every year between 2030 and 2050 (Sweileh, 2020). Hence, though antibiotic resistance in microorganisms is thought to evolve primarily as a result of selection pressure from antibiotic misuse, other factors intimately connected to climate change have potentials to worsen the situation.

Local temperature: In 2018, MacFadden et al. demonstrated that increases in local temperature had a favorable effect on antibiotic resistance in common infections. Their studies concluded that antibiotic resistance for several broadly used antibiotic classes increased by 4.2 percent, 2.2 percent and 2.7 percent in three common nosocomial pathogens- *Escherichia coli, Klebsiella pneumoniae,* and *Staphylococcus*

Table 1 Routinely used antibiotics against common diseases, indicating the countries where they are overused

Antibiotics	Diseases	Countries with most usage	References
Aminoglycosides	Severe urinary and abdominal infections, bacteremia, endocarditis	Mongolia, India, Russia, South Africa, United Kingdom, United States, China, Canada, Brazil, Korea, Serbia, Romania, Bulgaria, Ecuador	(Gonzalez and Spencer, 1998; WHO, 2021; CDDEP, 2018)
Broad spectrum penicillins	Endocarditis, urinary tract infections, prostatitis	Mongolia, Tunisia, Spain, Algeria, Turkey, Romania, Lebanon, France, Belgium, Greece	(Acar, 1997; WHO, 2021; CDDEP, 2018)
Carbapenems	Blood stream infections, intra-abdominal infections, severe infections caused by multidrug resistant gram-negative bacteria (ESBL, *P. aeruginosa*, carbapenem-susceptible *A. baumannii* and)	Japan, Puerto Rico, Spain, Taiwan, Australia, Germany, Italy,	(Patrier and Timsit, 2020; WHO, 2021)
Cephalosporins	Skin and soft tissue infections, meningitis, urinary tract infections, strep throat, pneumonia, ear infections, gonorrhea	Mongolia, Turkey, Greece, Vietnam, Korea, India, Bulgaria, Germany, Romania, Saudi Arab, Slovakia,	(WHO, 2021; CDDEP, 2018; Jordan et al., 2020)
Chloramphenicols	Conjunctivitis, meningitis, plague, cholera, and typhoid fever	Kazakhstan, Indonesia, Ukraine, Vietnam, Russia	(Bartlett, 1982; WHO, 2021; CDDEP, 2018)

(continued)

Table 1 (continued)

Antibiotics	Diseases	Countries with most usage	References
Fluoroquinolones	Several microbial infections like bacterial bronchitis, typhoid fever, pneumonia, septicemia and intraabdominal infections, bone joint infections, skin infections, gastroenteritis etc	Belgium, Brazil, Bulgaria, Egypt, Greece, Italy, Lebanon, Kazakhstan, Pakistan, Puerto Rico, Romania, Turkey, Venezuela, Vietnam, Spain	(Majalekar et al., 2020; WHO, 2021)
Macrolides	Diffuse panbronchiolitis (DPB), chronic obstructive lung disease, cystic fibrosis, and bronchiolitis obliterans syndrome (BOS), inflammatory airway diseases	Mongolia, Greece, Serbia, Slovakia, Turkey, Australia, Belgium, Bulgaria, Czech Republic. France, Ireland, Japan, Italy, Korea, New Zealand,	(Kwiatkowska and Maślińska, 2012; CDDEP, 2018)
Tetracyclines	Pneumonia, respiratory tract infections; infections of lymphatic systems, eye, intestinal and genital tract	New Zealand, Mongolia, Australia, Finland, South Africa, Taiwan, Unite Kingdom, United States	(Totsuka, 2007; CDDEP, 2018)
Glycopeptides	Infective endocarditis, Enterocolitis, Clostridium difficile-associated diarrhoea etc	United States, Puerto Rico, United Kingdom, Italy, Taiwan	(Blaskovich, 2018; CDDEP, 2018)

aureus, respectively, as temperatures rose by 10 °C across United States (MacFadden et al., 2018). Although reasons for the link between higher temperatures and greater antibiotic resistance need to be investigated further, a number of variables might help explain the association, for example, warmer temperatures may affect bacterial growth rates, resulting in an increase in the carriage and transmission of drug

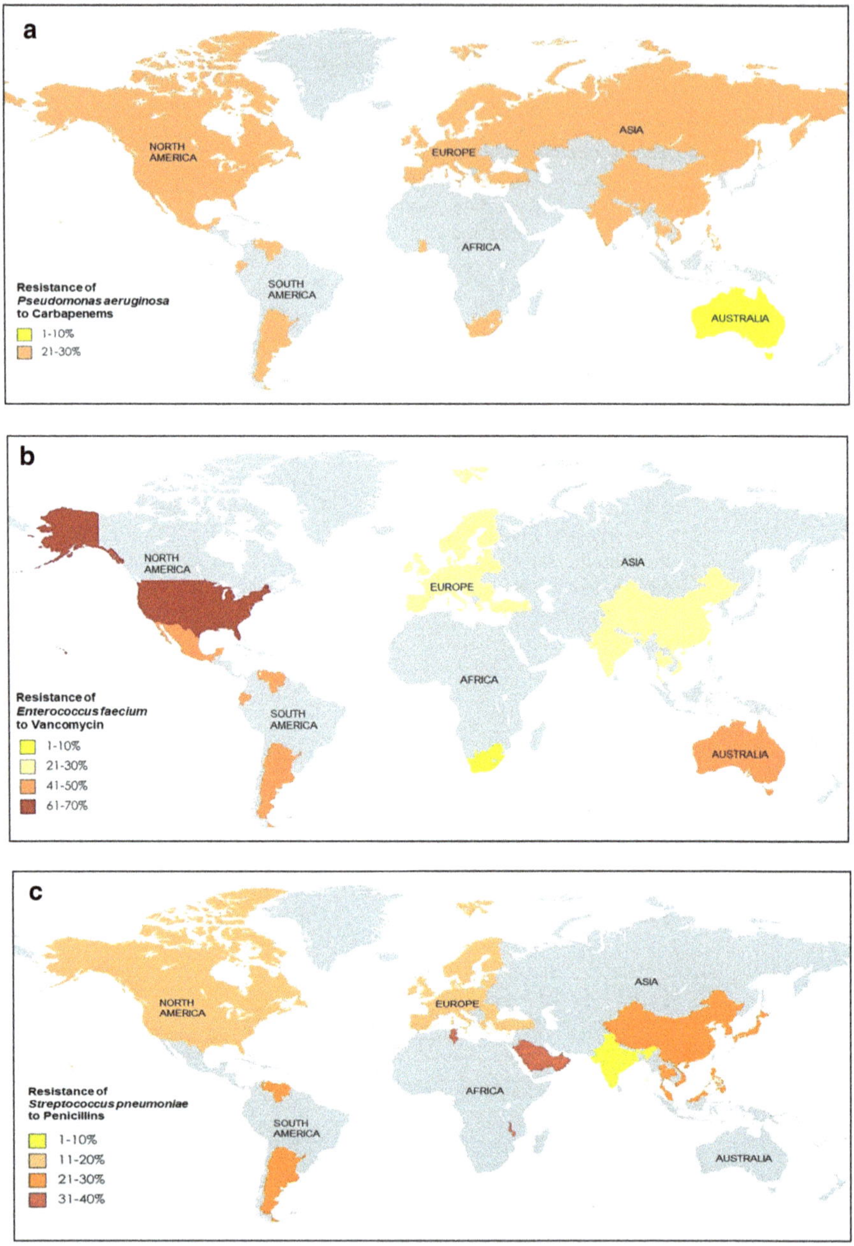

Fig. 2 Maps showing the trends in resistance of priority pathogens (one from each category, as listed by WHO) across the globe, in the year 2017. **a** *Pseudomonas aeruginosa* (ranked 'Critical') **b**. *Enterococcus faecium* (ranked 'High') **c**. *Streptococcus pneumoniae* (ranked 'Medium') (WHO, 2017; WHO, 2021; CDDEP, 2018)

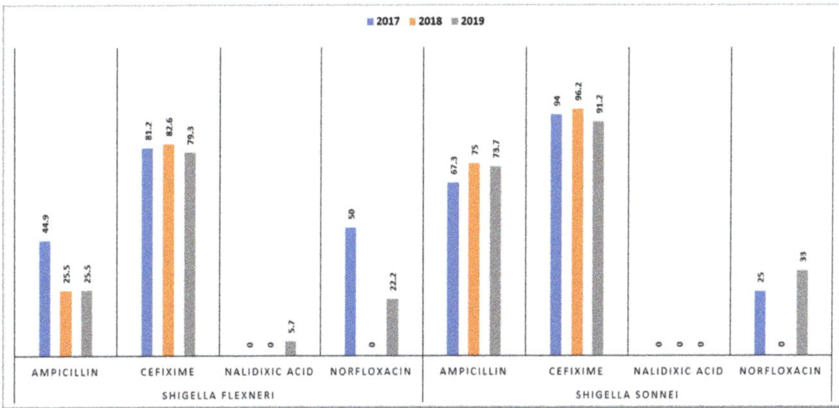

Fig. 3 Graph illustrating the gradual fall in the susceptibility percentage of antimicrobial agents against *Shigella flexneri* and *Shigella sonnei*, over a period of 3 consecutive years (AMR surveillance Network, 2019)

resistant strains. A study conducted in Göttingen, Germany, showed higher prevalence of Carbapenem-resistant *Pseudomonas aeruginosa,* multidrug-resistant *E coli*, and Methicillin-resistant *Staphylococcus aureus* during the warmer season (Kaba et al., 2020). According to the reports published by the Antimicrobial Resistance Surveillance and Research Network team of ICMR, 58 out of 75 *Shigella* isolates tested in the year 2019, were found to harbour antimicrobial resistance genes, viz., dhfrA, sulII, bla_{OXA}, bla_{TEM}, $bla_{CTX\text{-}M\text{-}1}$, AmpCs and qnrA/B/S (AMR surveillance Network, 2019). This molecular analysis was consistent with the diminishing trend in the susceptibility of various antimicrobials, when tested against *Shigella flexneri* and *Shigella sonnei*, the two major serogroups sequestered from India (Fig. 3) (AMR surveillance Network, 2019).

The alarming rate of emergence of resistant strains, point that the rising temperature might be directly proportional to the rate of transfer of horizontal genes. Unstable environmental factors can even fuel the emergence of several new mobile genetic elements responsible for antibiotic resistance. Horizontal transfer of New Delhi metallo-β-lactamase 1 (NDM-1) gene, OXA-48-like carbapenemases has been proved to be highly influenced by local ambient temperature (Mohsin et al., 2020).

Droughts and Floods: The frequency, severity, and consequences of severe weather events have been predicted to intensify as a result of climate change. Water sources serve as a great reservoir for antimicrobial resistant bacteria, owing to the unrestricted dumping of industrial, pharmaceutical, agricultural, municipal and hospital effluents into them. In a study conducted to monitor the presence of antimicrobial resistant strains in the Cauvery river of Karnataka, 100% of the isolated *Escherichia coli* strains were found resistant to third generation cephalosporins (Skariyachan et al., 2015). Striking reports were disclosed by another survey, over the two most sacred rivers of India, viz., Yamuna and Ganges, where 17.4% of the sampled Gram-negative bacteria were found to express extended spectrum beta lactamases (ESBLs) (Azam et al., 2016). Flooding plays a prime role in dissemination of antimicrobial resistant microbial members from waterways to different sectors in a community through several means. Flooding induced admixture of riverine system with anthropogenic community sewer systems, and freshwater aquifers results in contamination of drinking water resources in the community. As a consequence, individuals who come into contact with this contaminated water may unknowingly become carriers of antimicrobial resistant pathogens. Flood induced exodus of local inhabitants and subsequent refuge leading to overcrowding and inadequate sanitation measures can promote the transmission of various infectious diseases. Inappropriate use of antimicrobial agents to combat such disease outbreak can fuel the development of antimicrobial resistance further. Moreover, floods can create conducive breeding environments for disease-carrying vectors like mosquitoes and rodents. Control of these vectors often involves the use of insecticides and rodenticides, leading to the potential selection of resistant strains. In research, carried out with the purpose of assessing the potential impact of flood on bacillary dysentery in Baise (Guangxi Province of China) from 2004–2012, mild and extreme floods were evidenced to enhance the relative risk of shigellosis by 1.40 and 1.78 respectively (Liu et al., 2017). Agricultural runoffs containing nitrogen fertilizers are another route through which floodwater accelerates antibiotic resistance. Increased agricultural runoffs enhances bacterial blooms in aquatic systems, which in turn increase the possibilities for antibiotic resistance genes to be transferred (Chen et al., 2019a, b). In contrast to floods, drought-induced water shortage increases the sharing of same water resources, resulting in poor sanitation. A diarrheal outbreak following a period of drought in the South Pacific Island Nation of Tuvalu, was reported by an epidemiological inspection, which mainly emphasized on the scarcity of water due to negligent rainfall and poor sanitary hygiene as the important contributors behind the sudden rise in dysentery cases (Emont et al., 2017). Antibiotic-resistant enteric pathogens are more likely to be acquired as a result of poor sanitation (Burnham, 2021). Spread of diarrheal infections caused by multidrug resistant *E. coli* was reported to be aided by unclear water and inadequate sanitation in Africa. ARGs encoded resistance to macrolides, tetracyclines, vancomycin, β-lactams, aminoglycoside and fluoroquinolones showed high abundances in the stool samples of young children carrying enteric infections (Berendes et al., 2019).

Synthetic Pollution: Microplastic pollution has emerged as one of the world's most pressing environmental issues. According to the United Nations Environment Assembly (UNEP), the rate of discharge of plastics into the ocean is 4.8–12.7 million metric tons per year. During the current COVID-19 epidemic, global disposal of face masks, as well as other plastic laboratory and medical equipment, has expanded dramatically, adding to the huge plastic and microplastic waste in the environment. When plastic debris from coastal cities, ports, coastal landfills and dumping sites reaches the water, they get degraded into microplastics by various photolytic, mechanical and biological processes (Alfaro-Núñez et al., 2021). Microplastics has the ability to adsorb antibiotics like ampicillin, tetracycline, triclosan etc. (Wang et al., 2021). On the other hand, microplastics are preferred by bacteria for biofilm development due to their vast surface area. Hence, in a nutshell microplastics may act as a repository of antibiotic resistant bacteria (ARB) and a hotspot for horizontal transfer of antibiotic resistant genes (Logar-Henderson et al., 2019). According to Logar-Henderson et al., 2019, the prevalence of *Vibrio* species is increasing day by day as the ocean warms. Microplastic pollution has exacerbated the situation, with the emergence of antibiotic resistant *Vibrio* spp. leading to the outbreak of antibiotic-resistant cholera (Logar-Henderson et al., 2019).

5 Application of NGS Methodologies in Wastewater Microbial Profiling

With the rapid advancement in sequencing technologies enabling high throughput massively parallel sequencing of millions of DNA fragments, genomic analysis has become increasingly accessible for application in environmental research. Since their inception in 2005, next generation sequencing (NGS) has successfully overcome its limitations in terms of read length, accuracy and cost, promising for large scale adoption of NGS technologies in environmental research (Shokralla et al., 2012). The term NGS collectively refers to a variety of platforms, capable of recovering and reading thousands of DNA sequences in parallel from environmental samples, highlighting the improvement from the ancient first-generation sequencing methods. With NGS reinvigorating the interest in ecology and biodiversity research, the key applications of NGS in wastewater surveillance can be explored under four major categories, viz. whole genome sequencing (WGS), shotgun metagenomics sequencing, targeted metagenomics sequencing (amplicon sequencing), and meta-transcriptomic sequencing (Qin, 2019).

Whole genome sequencing: Whole Genome Sequencing (WGS) refers to the sequencing, assembly and analysis of the genome of a single microbial organism (Garner et al., 2021). Whole genome sequencing typically involves the isolation and extraction of the genomic DNA from pure culture plates, following which the DNA is randomly sheared into fragments and labelled with adapters to generate a sequencing library, which are then sequenced using the platform of choice (Hilt and Ferrieri, 2022). The sequencer generated reads are quality checked and trimmed by removing the adapters, before assembling them for taxonomic assignment. Alignment and assembly of raw reads can be accomplished by either de novo or reference guided assembly pipelines, offered by sophisticated analytical programs. De novo assembly generates contiguous contigs by aligning raw reads at areas of overlap, which are eventually binned into draft genomes, whereas reference guided assembly relies on using a reference genome to map the reads (Hilt and Ferrieri, 2022). While different assembly programs use different algorithms for contigs assembly, tools like SPAdes, MEGAHIT are among the finest performing programs in terms of accuracy of mapping and percentage of reads covered (Chan et al., 2019). In contrast to the currently existing methods for detection of antimicrobial resistance in wastewater limited by either phenotypic studies with bacterial culture plates or detection of specific marker genes using targeted sequencing, WGS allows for reconstruction of genomes of previously uncharacterized species and offers a comprehensive antimicrobial resistance profile of the organism, thus making it possible to track transmission of resistance genes based on events of horizontal gene transfer and presence of mobile genetic elements (Chan et al., 2019). WGS can be accomplished by both short read and long read sequencing platforms, with each approach having its own set of advantages and drawbacks.

Whole genome sequencing has found extensive application in unraveling pathogenic microbes and virulence genes from environmental isolates. In a related study, Jiang et al. (2019) employed whole genome sequencing to furnish comprehensive resistance profiles of 18 strains of *Escherichia coli*, producing extended spectrum beta lactamases (ESBL), isolated from waste water treatment plants receiving pharmaceutical effluents in China. All the strains were found to harbour at least one beta lactamase encoding gene, with blaCTX–M being the most regular member. The intricate sequencing details covered by whole genome sequencing also reported the presence of mobile genetic elements and transposons in near proximity of blaCTX–M gene, thus underscoring the transfer potential of these resistance genes and their role in emergence of resistant pathogens (Jiang et al., 2019). With an attempt to investigate the genetic relatedness between clinically and environmentally isolated *Klebsiella pneumoniae* strains, Runcharoen et al. (2017) performed WGS analysis of 77 *Klebsiella pneumoniae* strains (48 clinical and 29 environmental isolates) (Runcharoen et al., 2017). They found variegated distribution of strains from the two groups across the phylogenetic tree, hinting at their recent divergence from common ancestors, which might be attributed to recurrent infections originating from environmental reservoirs. These findings underline the budding prospects of WGS in tracking antimicrobial resistance and epidemic outbreaks from environmental samples.

Metagenomic Shotgun Sequencing: Metagenomic Shotgun Sequencing involves sequencing of the entire DNA content including microbial genomes present in an environmental sample. Total DNA extraction, DNA fragmentation, adapter ligation, amplification of the prepared DNA library is few of the initial steps that are performed in this method that follows the final step of high-throughput sequencing using next-generation sequencing (NGS) platforms which may involve both short and long read sequencing technology, depending on the nature of the output needed for downstream analysis. This method offers a holistic view of the microbial community in a sample, including both known and potentially novel microorganisms without the need for culture-based techniques, enabling early detection of potential health risks (Chen et al., 2022).

One of the pioneering instances of metagenomic surveillance using shotgun approach took place during 2013–2014 involving around 475 samples collected from 24 sites in subway systems and 12 different locations in the Gowanus canal across New York City, United States. Multiple downstream bioinformatic analysis of the substantial 10.4 billion paired-end sequence data generated through shotgun sequencing platforms resulted in the identification of city-wide prevalence of *Pseudomonas stutzeri* followed by strains from *Enterobacter* and *Stenotrophomonas*. Particularly noteworthy was the consistent dominance of bacteriophages among the most abundant viruses. This baseline metagenomic mapping of New York City underscored the efficacy of shotgun metagenomic profiling in foretelling potential disease threats and enhancing health management within the urban infrastructure. (Afshinnekoo et al., 2015). In line with the same approach, the team leading the global sewage surveillance initiative, performed metagenomic shotgun sequencing of untreated sewage samples obtained from 76 cities across 60 nations utilizing the Illumina HiSeq3000 platform. In the entirety of this analysis, a total of 1,625 distinct antibiotic resistance determinants such as blaNDM, mcr, optrA etc. distributed among 408 gene classes showing resistance to majority of the last-resort antibiotics were identified from the assembled sequences. By highlighting the variation in overall abundance of AMR genes across different study sites, this study evidenced the potential efficacy of this approach as a valuable tool for monitoring antibiotic resistance at a population level (Hendriksen et al., 2019).

However, it's worth noting that Whole genome shotgun sequencing stands as a more costly approach when compared with alternative microbial monitoring methods due to its reliance on expertise of advanced bioinformatics tools and databases for accurate analysis. Additionally, longer turnaround time for whole genome shotgun sequencing might limit its immediate application for real-time wastewater monitoring and decision-making in certain situations (Chen et al., 2022).

Amplicon Sequencing: Amplicon sequencing is a powerful genomics technique used to study specific regions of DNA in a sample through PCR amplification and analyzing targeted DNA sequences. It is commonly employed in various fields, including genetics, genomics, microbiology, and ecology providing valuable insights into genetic variations, microbial diversity, and many more. Followed by the amplification, amplicons are subjected to high-throughput sequencing using second and third generation sequencing platforms such as Illumina, Ion Torrent, PacBio etc. (Callahan et al., 2019). The targeted sequencing approach of amplicon sequencing is beneficial in terms of cost effectivity and time consumption, mainly attributed to the multiplexing of samples. Concentrating on short specific regions in amplicon sequencing method provide higher coverage depth, resulting in increased sensitivity for detecting microbial members in wastewater samples.

In the field of wastewater monitoring, few of the commonly used amplicon sequencing methods involves 16S rRNA gene sequencing, 18S rRNA gene sequencing, and Internal Transcribed Spacer (ITS) Sequencing for targeting bacterial, eukaryotic and fungal communities in wastewater respectively. Researchers combine these approaches with bioinformatic analysis to identify pathogenic microbes in an environmental niche by comparing the obtained sequences to the reference databases containing known pathogen sequences (Callahan et al., 2019). In a recent investigation conducted by Ríos-Castro et al. in 2023, amplicon sequencing method was employed on Illumina MiSeq platform using a universal primer targeted against the V3 and V4 region of the 16S rRNA gene for thorough profiling of bacterial communities present in the raw community wastewater collected from the influent channel of a Wastewater Treatment plant in Ría de Vigo, Spain. Around 6339 Operational Taxonomic Units (OTUs) were identified through taxonomic classification against the 16S SSU Silva reference database. The majority of these OTUs were found to be associated with the human gut microbiome including numerous potentially pathogenic organisms such as *Vibrio, Prevotella, Staphylococcus, and Acinetobacter* (Ríos-Castro et al., 2023). Singh et al. (2019) adopted the similar kind of high-throughput metagenomic sequencing strategy in one of their pilot studies to assess the abundance of microbial populations, as well as potential pathogenic bacterial members, within the effluent samples discharged from clinical facilities situated in both rural and urban setups (Singh et al., 2019). Presently, the widespread utilization of various long and short-read platforms for high-throughput sequencing of amplicons derived from diverse hypervariable regions of the 16S rRNA gene such as V3, V4, V5, V6 etc., has become a standard approach in NGS-based Microbial Source Tracking methods which is known for its significant potential of identifying the sources of fecal contamination in freshwater resources in a community adding an avenue to wastewater surveillance system. With the aid of the 16S metagenomic approach, Unno et al., 2018 was able to successfully identify the sources of faecal pollution and, consequently, pathogenic microbial pollution load in the recreational water by comparing the similarity and dissimilarities index of the microbial community present in the water resources to the effluents from community sewer networks and runoff from localized animal facilities contaminated with pathogenic bacteria (Unno et al., 2018). Role of amplicon sequencing technologies in Wastewater Based Epidemiology (WBE) as a rapid public and environmental health monitoring tool was eminent in a study by Harrington et al., 2022 where community level data generated using second generation Illumina sequencing platform was leveraged to uncover potential correlations between antibiotic misuse and the rise of AMR in an urban community during the Covid-19 surge in December 2020. Comprehensive analysis of microbial communities and AMR profiles in sewage samples obtained from both a wastewater treatment plant (WWTP) and a community shelter located in Las Vegas, Nevada using the high-throughput sequencing panel could identify the prevalence of various nosocomial pathogenic phyla in the community under study. Abundance of genes associated with fluoroquinolone, beta-lactam, and colistin resistance indicated the potential occurrence of multidrug-resistant bacterial infections shedding light on the microbial dynamics within the community during the Covid-19 surge (Harrington et al., 2022).

In spite of having multiple advantages, amplicon sequencing also comes with few of the significant drawbacks. The targeted sequencing strategy of this method can miss rare or novel pathogens that do not have closely related sequences in the reference databases (Stefanini and Cavalieri, 2018).

Meta-transcriptomic sequencing: Meta-transcriptome sequencing refers to the sequencing of the entire pool of RNA transcripts present in an environmental sample at a given point in time (Garner et al., 2021). While it is principally similar to shotgun metagenomic sequencing, meta-transcriptomics differs by targeting RNA instead of DNA. Like other NGS methods, analysis of RNA sequence data also involves the steps of quality control, de novo assembly and taxonomic profiling following which the RNA transcripts are functionally annotated. The RNA sequencing reads are quality checked to remove erroneous reads by using any of the many available quality control software like FastQC, Trimmomatic, etc. (Shakya et al., 2019). The clean reads are then subjected to de novo assembly using tools like IDBA-MT, Transcript Assembly Graph (TAG), IDBA-MTP, that are particularly devised to handle the complex yet unique hallmark of each RNA transcript. The de novo assemblers for meta-transcriptomes, assemble reads by presenting typically longer segments of genomic regions as reference scaffolds, thus allowing for more efficient matching of reads to homologs and appropriate taxonomic origin (Shakya et al., 2019). Functional roles of the transcripts can be explored by using either the reads or the contigs as input for different tools like MetaCLADE, UProC, FragGeneScan, etc. (Shakya et al., 2019). The conventional sequencing technologies discussed so far, all target DNA, either in its entirety or using targeted genetic markers, and have been effective in capturing detailed insights of microbial diversity and genes in environmental samples while simultaneously elucidating their functional contributions. However, none of the approaches could identify the functionally operative genes within the community. Meta-transcriptome sequencing not only makes it possible to infer the actively contributing members in the community by directly measuring the actively expressed genes but also identifies novel transcripts and sequence variants, which might have important implications in conferring resistance to antimicrobial drugs. However, this technique also comes with its share of shortcomings that limits large scale application of meta-transcriptome sequencing in waste water monitoring. The short half-life of RNA in environmental samples, unwanted background noise due to excess ribosomal RNAs, and inability to capture the entire spectrum of transcripts due to enormous diversity, are some of the challenges that needs to be addressed for plausible implementation of this method in environmental research (Garner et al., 2021).

The application of metatranscriptomic approaches for detecting antimicrobial resistance determinants has grown to be commonly used for research in varied areas, ranging from environmental samples to human gut. In a study conducted by Vikram et al., alterations in transcriptomic profiles of microbial communities from wastewater produced by the largest gas deposit in USA, were explored to identify the active metabolic pathways. RNA sequencing conducted on three wastewater samples collected from the natural gas wellhead in USA, demonstrated elevated expression of genes related to diverse stress protection mechanisms and biofilm formation, having significant implications in rendering biocides ineffective (Vikram et al., 2016). Another study conducted in Taiwan, adopted a combined metagenomic and metatranscriptomic approach to evaluate the diversity and expression of antibiotic resistance genes (ARGs) in activated sludge from three waste water treatment plants (WWTPs). Out of the 360 ARGs detected using metagenomic studies, metatranscriptomic analysis revealed siginificant transcript abundance for 65.8% genes, thus highlighting the active microbial fraction in WWTPs (Liu et al., 2019). A similar study designed to monitor the real time incidence and dissemination of SARS-CoV-2 through waste water surveillance in Maryland, USA, described cooccurrence of specific microbial taxas and resistance genes with SARS-CoV-2 RNA, suggesting a positive correlation between certain opportunistic bacterial genera and SARS-CoV-2 infection (Brumfield et al., 2022). These studies point towards the in-depth and high flexibility of metatranscriptomic approaches in waste water assessment and improving disease management.

With the discussion on the state of the art of different sequencing technologies available for waste water assessment, we have presented a comprehensive pipeline, summarizing the workflow for each different NGS based approaches, that may be optimized by the researcher as per the specific requirements of the study (Fig. 4). All the detection methods discussed, essentially begins with environmental water sample as the starting point and culminates in different depths of sequencing information that are eventually analyzed using different bioinformatics tools to assist in pathogen detection and identification of resistance genes.

We envision that such targeted and untargeted genomic analysis of wastewater samples offered by next generation sequencing platforms will soon emerge as a rich resource for tracking infection dynamics in population.

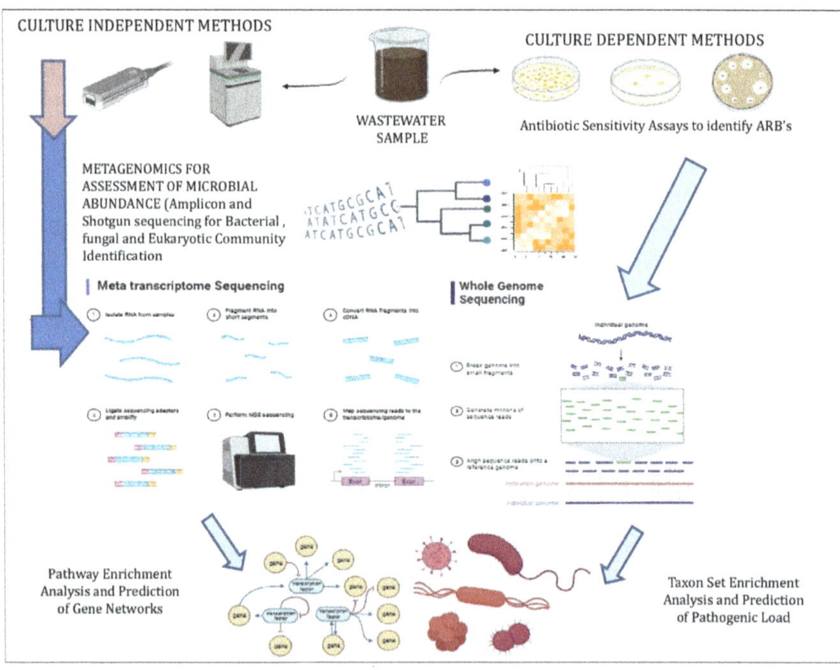

Fig. 4 Flowchart summarizing the steps involved in wastewater monitoring using different NGS based approaches (generated using biorender.com)

6 Conclusion

The current trend in climate change, largely contributed by global warming due to increasing emissions of green-house gases worldwide, has manifested itself in the form of both long-term and short-term alterations in weather conditions and more frequent drastic weather events (droughts, floods). With increasing evidences underscoring the intricate balance between climatic factors and incidence of bacillary dysentery, analysis of climatic risk windows and accurate forecasting of future outbreaks has become a budding area of research. The waxing and waning pattern of occurrence of shigellosis, as governed by the complex interplay between separate weather entities, can be effectively used as a warning signal for future epidemics and for implementing proper preventive measures. However, some limitations still remain in the quantification of such impacts, made complex by the spatial heterogeneity in the effects of weather variables, divergent impact on populations of different age groups, disparity in socio-economic status dominant across distinct provinces and finally the life style and behaviour of the community. As sequencing technologies continue to progress and become more cost-effective, NGS can be explored for detailed pathogenic profiling of wastewater samples, emerging as an effective means of screening large populations without overburdening the healthcare facilities. Researchers and wastewater monitoring practitioners should carefully weigh the benefits and drawbacks of using next generation sequencing technologies and consider it to obtain a comprehensive and reliable understanding of the trends in antimicrobial resistance and microbial community dynamics. Advanced research, incorporating all these aspects together with climatic factors, should be able to project future shigellosis outbreaks more confidently. As the pace of climate change accelerates, one of the key foundations of health care system i.e., availability of efficient antibiotics for infectious diseases, is facing serious paucity of alternatives. Collective endeavours of governmental and non-governmental authorities should be initiated along with awareness generation for countering climatic fluctuations and a wide range of environmental risks.

Acknowledgements The authors hereby express their sincere gratitude to the authorities of their respective institutions for their constant support and for providing the necessary infrastructure for completion of the work. Authors acknowledge the funding provided by the DBT Grant (BT/INF/22/SP41296/2020) from the Department of Biotechnology, Government of India and the Intramural Grant (IMSXC2023-24/003) awarded to Dr. Sayak Ganguli by St. Xavier's College (Autonomous) Kolkata.

Conflict of Interest The authors declare that they do not have any identifiable competing financial interests or personal relationships that could have potentially influenced the content of this chapter.

References

Acar, J., 1997. Broad-and narrow-spectrum antibiotics: an unhelpful categorization. *Clinical Microbiology and Infection*, *3*(4), pp.395-396.

Afshinnekoo, E., Meydan, C., Chowdhury, S., Jaroudi, D., Boyer, C., Bernstein, N., Maritz, J.M., Reeves, D., Gandara, J., Chhangawala, S. and Ahsanuddin, S., 2015. Geospatial resolution of human and bacterial diversity with city-scale metagenomics. *Cell systems*, *1*(1), pp.72-87.

Alfaro-Núñez, A., Astorga, D., Cáceres-Farías, L., Bastidas, L., Soto Villegas, C., Macay, K., and Christensen, J.H. 2021. Microplastic pollution in seawater and marine organisms across the Tropical Eastern Pacific and Galápagos. *Scientific reports*, *11*(1), pp.6424.

Aminharati, F., Soltan Dallal, M.M., Ehrampoush, M.H., Dehghani-Tafti, A., Yaseri, M., Memariani, M. and Rajabi, Z., 2018. The effect of environmental parameters on the incidence of Shigella outbreaks in Yazd province, Iran. *Water Science and Technology: Water Supply*, *18*(4), pp.1388-1395.

AMR surveillance Network, Indian Council of Medical Research. 2019.

Azam, M., Jan, A.T. and Haq, Q.M., 2016. Bla CTX-M-152, a novel variant of CTX-M-group-25, identified in a study performed on the prevalence of multidrug resistance among natural inhabitants of river Yamuna, India. *Frontiers in microbiology*, *7*, p.176.

Bartlett, J.G., 1982. Chloramphenicol. *Medical Clinics of North America*, *66*(1), pp.91-102.

Berendes, D., Knee, J., Sumner, T., Capone, D., Lai, A., Wood, A., Patel, S., Nalá, R., Cumming, O. and Brown, J., 2019. Gut carriage of antimicrobial resistance genes among young children in urban Maputo, Mozambique: Associations with enteric pathogen carriage and environmental risk factors. *PLoS One*, *14*(11), p.e0225464.

Blaskovich, M.A., Hansford, K.A., Butler, M.S., Jia, Z., Mark, A.E. and Cooper, M.A., 2018. Developments in glycopeptide antibiotics. *ACS Infectious Diseases*, *4*(5), pp.715-735.

Brumfield, K.D., Leddy, M., Usmani, M., Cotruvo, J.A., Tien, C.T., Dorsey, S., Graubics, K., Fanelli, B., Zhou, I., Registe, N. and Dadlani, M., 2022. Microbiome analysis for wastewater surveillance during COVID-19. *MBio*, *13*(4), pp.e00591-22.

Burnham, J.P., 2021. Climate change and antibiotic resistance: a deadly combination. *Therapeutic Advances in Infectious Disease*, *8*, p.2049936121991374.

Callahan, B.J., Wong, J., Heiner, C., Oh, S., Theriot, C.M., Gulati, A.S., McGill, S.K. and Dougherty, M.K., 2019. High-throughput amplicon sequencing of the full-length 16S rRNA gene with single-nucleotide resolution. *Nucleic acids research*, *47*(18), pp.e103-e103.

Center for Disease Dynamics, Economics & Policy, 2018. ResistanceMap: antibiotic resistance. *The Center for Disease Dynamics Economics & Policy.*

Chakrabarti, S. and Ganguli, S., 2013. Structural analyses of Shigella invasion proteins reveals non-conserved; intrinsically unstructured regions. *International Letters of Natural Sciences*, *5.*

Chan, A.W., Naphtali, J. and Schellhorn, H.E., 2019. High-throughput DNA sequencing technologies for water and wastewater analysis. *Science progress*, *102*(4), pp.351-376.

Chen, Y., Fan, L.C., Chai, Y.H. and Xu, J.F., 2022. Advantages and challenges of metagenomic sequencing for the diagnosis of pulmonary infectious diseases. *The Clinical Respiratory Journal*, *16*(10), pp.646-656.

Chen, C.C., Lin, C.Y. and Chen, K.T., 2019. Epidemiologic features of shigellosis and associated climatic factors in Taiwan. *Medicine*, *98*(34).

Chen, J., McIlroy, S.E., Archana, A., Baker, D.M. and Panagiotou, G., 2019. A pollution gradient contributes to the taxonomic, functional, and resistome diversity of microbial communities in marine sediments. *Microbiome*, *7*, pp.1–12.

Diamond, M.B., Keshaviah, A., Bento, A.I., Conroy-Ben, O., Driver, E.M., Ensor, K.B., Halden, R.U., Hopkins, L.P., Kuhn, K.G., Moe, C.L. and Rouchka, E.C., 2022. Wastewater surveillance of pathogens can inform public health responses. *Nature Medicine*, *28*(10), pp.1992-1995.

Emont, J.P., Ko, A.I., Homasi-Paelate, A., Ituaso-Conway, N. and Nilles, E.J., 2017. Epidemiological investigation of a diarrhea outbreak in the South Pacific island nation of Tuvalu during a severe

La Niña–associated drought emergency in 2011. *The American journal of tropical medicine and hygiene*, *96*(3), p.576.

Garner, E., Davis, B.C., Milligan, E., Blair, M.F., Keenum, I., Maile-Moskowitz, A., Pan, J., Gnegy, M., Liguori, K., Gupta, S. and Prussin II, A.J., 2021. Next generation sequencing approaches to evaluate water and wastewater quality. *Water Research*, *194*, p.116907.

GONZALEZ III, U.S. and Spencer, J.P., 1998. Aminoglycosides: a practical review. *American family physician*, *58*(8), pp.1811-1820.

Harikrishnan, H., Ismail, A. and Banga Singh, K.K., 2013. Temperature-regulated expression of outer membrane proteins in Shigella flexneri. *Gut pathogens*, *5*(1), pp.1-7.

Harrington, A., Vo, V., Papp, K., Tillett, R.L., Chang, C.L., Baker, H., Shen, S., Amei, A., Lockett, C., Gerrity, D. and Oh, E.C., 2022. Urban monitoring of antimicrobial resistance during a COVID-19 surge through wastewater surveillance. *Science of The Total Environment*, *853*, p.158577.

Hendriksen, R.S., Munk, P., Njage, P., Van Bunnik, B., McNally, L., Lukjancenko, O., Röder, T., Nieuwenhuijse, D., Pedersen, S.K., Kjeldgaard, J. and Kaas, R.S., 2019. Global monitoring of antimicrobial resistance based on metagenomics analyses of urban sewage. *Nature communications*, *10*(1), p.1124.

Hilt, E.E. and Ferrieri, P., 2022. Next generation and other sequencing technologies in diagnostic microbiology and infectious diseases. *Genes*, *13*(9), p.1566.

Jiang, X., Cui, X., Xu, H., Liu, W., Tao, F., Shao, T., Pan, X. and Zheng, B., 2019. Whole genome sequencing of extended-spectrum beta-lactamase (ESBL)-producing Escherichia coli isolated from a wastewater treatment plant in China. *Frontiers in Microbiology*, *10*, p.1797.

Jordan, E., Voide, C., Petignat, P.A. and Gobin, N., 2020. Cephalosporins in clinical practice. *Revue Medicale Suisse*, *16*(710), pp.1906-1911.

Kaba, H.E., Kuhlmann, E. and Scheithauer, S., 2020. Thinking outside the box: Association of antimicrobial resistance with climate warming in Europe–A 30 country observational study. *International journal of hygiene and environmental health*, *223*(1), pp.151-158.

Khalil, I.A., Troeger, C., Blacker, B.F., Rao, P.C., Brown, A., Atherly, D.E., Brewer, T.G., Engmann, C.M., Houpt, E.R., Kang, G. and Kotloff, K.L., 2018. Morbidity and mortality due to shigella and enterotoxigenic Escherichia coli diarrhoea: the Global Burden of Disease Study 1990–2016. *The Lancet infectious diseases*, *18*(11), pp.1229-1240.

Kwiatkowska, B. and Maślińska, M., 2012. Macrolide therapy in chronic inflammatory diseases. *Mediators of inflammation*, *2012*.

Langford, B.J., So, M., Raybardhan, S., Leung, V., Soucy, J.P.R., Westwood, D., Daneman, N. and MacFadden, D.R., 2021. Antibiotic prescribing in patients with COVID-19: rapid review and meta-analysis. *Clinical microbiology and infection*, *27*(4), pp.520-531.

Li, C., Wu, X., Ji, D., Liu, J., Yin, J. and Guo, Z., 2019. Climate change impacts the epidemic of dysentery: determining climate risk window, modeling and projection. *Environmental Research Letters*, *14*(10), pp.104019.

Liu, X., Liu, Z., Zhang, Y. and Jiang, B., 2017. The effects of floods on the incidence of bacillary dysentery in Baise (Guangxi Province, China) from 2004 to 2012. *International journal of environmental research and public health*, *14*(2), p.179.

Liu, Z., Klümper, U., Liu, Y., Yang, Y., Wei, Q., Lin, J.G., Gu, J.D. and Li, M., 2019. Metagenomic and metatranscriptomic analyses reveal activity and hosts of antibiotic resistance genes in activated sludge. *Environment international*, *129*, pp.208-220.

Logar-Henderson, C., Ling, R., Tuite, A.R. and Fisman, D.N., 2019. Effects of large-scale oceanic phenomena on non-cholera vibriosis incidence in the United States: implications for climate change. *Epidemiology & Infection*, *147*.

MacFadden, D.R., McGough, S.F., Fisman, D., Santillana, M. and Brownstein, J.S., 2018. Antibiotic resistance increases with local temperature. *Nature Climate Change*, *8*(6), pp.510-514.

Majalekar, P.P. and Shirote, P.J., 2020. Fluoroquinolones: blessings or curses. *Current Drug Targets*, *21*(13), pp.1354-1370.

Michael, L., Bennish, Sabeena, Ahmed. 2020. Shigellosis. In: Edward, T. Ryan, David R. Hill, Tom Solomon, Naomi E. Aronson, Timothy P. Endy (Eds.), *Hunter's Tropical Medicine and Emerging Infectious Diseases, 10th ed.*, pp. 492–499.

Mohsin, M., Shad, A.A., Ali, J., and Sajjad-ur-Rahman. 2020. Antimicrobial Resistance, Food Systems and Climate Change. In Panwar, H., Sharma, C., Lichtfouse, E (Eds.). *Sustainable Agriculture Reviews* (46). Springer, Cham. doi:https://doi.org/10.1007/978-3-030-53024-2_3

Moors, E., Singh, T., Siderius, C., Balakrishnan, S. and Mishra, A., 2013. Climate change and waterborne diarrhoea in northern India: Impacts and adaptation strategies. *Science of the Total Environment*, *468*, pp.S139-S151.

Patrier, J. and Timsit, J.F., 2020. Carbapenem use in critically ill patients. *Current Opinion in Infectious Diseases*, *33*(1), pp.86-91.

Qin, D., 2019. Next-generation sequencing and its clinical application. *Cancer biology & medicine*, *16*(1), p.4.

Quince, C., Walker, A.W., Simpson, J.T., Loman, N.J. and Segata, N., 2017. Shotgun metagenomics, from sampling to analysis. *Nature biotechnology*, *35*(9), pp.833-844.

Ríos-Castro, R., Cabo, A., Teira, E., Cameselle, C., Gouveia, S., Payo, P., Novoa, B. and Figueras, A., 2023. High-throughput sequencing as a tool for monitoring prokaryote communities in a wastewater treatment plant. *Science of The Total Environment*, *861*, p.160531.

Rodríguez-Verdugo, A., Lozano-Huntelman, N., Cruz-Loya, M., Savage, V. and Yeh, P., 2020. Compounding effects of climate warming and antibiotic resistance. *IScience*, *23*(4).

Runcharoen, C., Moradigaravand, D., Blane, B., Paksanont, S., Thammachote, J., Anun, S., Parkhill, J., Chantratita, N. and Peacock, S.J., 2017. Whole genome sequencing reveals high-resolution epidemiological links between clinical and environmental Klebsiella pneumoniae. *Genome medicine*, *9*(1), pp.1-10.

Shakya, M., Lo, C.C. and Chain, P.S., 2019. Advances and challenges in metatranscriptomic analysis. *Frontiers in genetics*, *10*, p.904.

Shokralla, S., Spall, J.L., Gibson, J.F. and Hajibabaei, M., 2012. Next-generation sequencing technologies for environmental DNA research. *Molecular ecology*, *21*(8), pp.1794-1805.

Singh, M., Ganguli, S. and Ghosh, M.M., 2019. Comparative metagenomic dataset of hospital effluent microbiome from rural and urban hospitals in West Bengal. *Data in brief*, *25*, p.104264.

Skariyachan, S., Mahajanakatti, A.B., Grandhi, N.J., Prasanna, A., Sen, B., Sharma, N., Vasist, K.S. and Narayanappa, R., 2015. Environmental monitoring of bacterial contamination and antibiotic resistance patterns of the fecal coliforms isolated from Cauvery River, a major drinking water source in Karnataka, India. *Environmental monitoring and assessment*, *187*, pp.1-13.

Song, Y.J., Cheong, H.K., Ki, M., Shin, J.Y., Hwang, S.S., Park, M., Ki, M. and Lim, J., 2018. The epidemiological influence of climatic factors on shigellosis incidence rates in Korea. *International journal of environmental research and public health*, *15*(10), pp.2209.

Stefanini, I. and Cavalieri, D., 2018. Metagenomic approaches to investigate the contribution of the vineyard environment to the quality of wine fermentation: potentials and difficulties. *Frontiers in Microbiology*, *9*, p.991.

Sweileh, W.M., 2020. Bibliometric analysis of peer-reviewed literature on climate change and human health with an emphasis on infectious diseases. *Globalization and health*, *16*(1), pp.1-17.

Taneja, N. and Mewara, A., 2016. Shigellosis: epidemiology in India. *The Indian journal of medical research*, *143*(5), p.565.

The, H.C., Thanh, D.P., Holt, K.E., Thomson, N.R. and Baker, S., 2016. The genomic signatures of Shigella evolution, adaptation and geographical spread. *Nature Reviews Microbiology*, *14*(4), pp.235-250.

Totsuka, K. 2007. Nihon rinsho. *Japanese journal of clinical medicine, 65 Suppl 3*, pp. 675–680.

Unno, T., Staley, C., Brown, C.M., Han, D., Sadowsky, M.J. and Hur, H.G., 2018. Fecal pollution: new trends and challenges in microbial source tracking using next-generation sequencing. *Environmental microbiology*, *20*(9), pp.3132-3140.

Vikram, A., Lipus, D. and Bibby, K., 2016. Metatranscriptome analysis of active microbial communities in produced water samples from the Marcellus Shale. *Microbial ecology*, 72, pp.571-581.

Wang, Z., Gao, J., Zhao, Y., Dai, H., Jia, J. and Zhang, D., 2021. Plastisphere enrich antibiotic resistance genes and potential pathogenic bacteria in sewage with pharmaceuticals. *Science of The Total Environment*, *768*, p.144663.

WHO Media Centre. 2017. News Release. WHO publishes list of bacteria for which new antibiotics are urgently needed. http://www.who.int/mediacentre/news/releases/2017/bacteria-antibiotics-needed/en/.

World Health Organization. 2019. Food Safety—Climate Change and the Role of WHO. https://www.who.int/foodsafety/publications/all/Climate_Change_Document.pdf?ua=1

World Health Organization, 2021. Global antimicrobial resistance and use surveillance system (GLASS) report: 2021.

Zhu, L., Zhao, G., Stein, R., Zheng, X., Hu, W., Shang, N., Bu, X., Liu, X., Wang, J., Feng, E. and Wang, B., 2010. The proteome of Shigella flexneri 2a 2457T grown at 30 and 37 C. *Molecular & Cellular Proteomics*, *9*(6), pp.1209-1220.

Microbial Desalination Cell (MDC): A Next-Generation Environmental Technology for Wastewater Treatment and Bioelectricity Generation

Karan Saini, Anwesha Bandyopadhyay, Sibiraj Murugesan, Dayakant Kashyap, Rahul Yadav, Ambuj Mishra, and Gaurav Saxena

1 Introduction

Water shortages for drinking and agriculture are a serious problem, particularly in many nations in the Southeastern region of Asia, Latin America, and the northern section of Africa (Ahuja 2019; Manzoor et al. 2017). According to estimates, only 2.5% of the 71% of water on Earth's surface is freshwater fit for human use, with the remainder being seawater or brackish water that require desalination for further use (Shannon et al. 2008). To lessen the shortage of freshwater, desalination plant building has been rising internationally and every day (Amma and Ashraf 2020). Desalination is necessary in those areas where freshwater is scarce, but seawater is plentiful and provides enormous amounts of freshwater for humans.

Desalination of saline water is done using commercially accessible thermal and membrane-based techniques; however, they are inefficient since they are expensive and energy-intensive, which results in high water production costs (Mathioulakis et al. 2007). Additionally, adopting renewable energy sources to power desalination systems, such as wind, solar, geothermal, and hydroelectricity, may provide alternatives to the fossil fuels now used (Mathioulakis et al. 2007). However, the cost

K. Saini · A. Bandyopadhyay · S. Murugesan · D. Kashyap · G. Saxena (✉)
EMBL-Environmental Microbiology and Biotechnology Laboratory, EERG-Ecotoxicology and Environmental Remediation Group, School of Biotechnology, Shoolini University, Solan, Himachal Pradesh 173 229, India
e-mail: gaurav10saxena@gmail.com; gauravsaxena@shooliniuniversity.com

R. Yadav
School of Biological and Environmental Sciences, Shoolini University, Kasauli Hills, Solan, Himachal Pradesh 173 229, India

A. Mishra
Plant Ecology Laboratory, School of Environmental Sciences, Jawaharlal Nehru University, New Delhi 110067, India

J. Gupta and A. Verma (eds.), *Microbiology-2.0 Update for a Sustainable Future*,
https://doi.org/10.1007/978-981-99-9617-9_18

of existing renewable energy-powered desalination systems is higher than that of traditional energy sources, but the long-term environmental benefits might justify the high cost (Karagiannis and Soldatos 2008).

In addition to maximizing environmental advantages, combining waste treatment with desalination enables us to collect energy from waste before it is released into the environment and utilize that energy to run desalination systems. It is more advantageous to produce energy from organic waste using microbial fuel cells (MFC) than anaerobic digestion to convert waste into biogas during waste treatment (Rozendal et al. 2008). MFC is a BES that generates electrical energy from the microbial digestion of organic wastes while removing salts to preserve ionic equilibrium and desalination in the anode chamber, although the reason was unclear (Logan et al. 2006; Rozendal et al. 2006; Mohanakrishna et al. 2010).

MFCs have been transformed into microbial desalination cells (MDCs) according to current research that remove salts (Cao et al. 2009a, b). Three main goals drove the creation of MDCs in 2009 includes: desalination, production of bioelectricity, and water/wastewater treatment (Cao et al. 2009a, b). A conventional MFC consists of three chambers: anode and cathode separated by ion-exchange membrane (IEM) and a desalination chamber. Recently, several MDC system designs have been introduced and developed (Chen et al. 2011; Gude et al. 2013a, b; Jingyu et al. 2017; Mehanna et al. 2010a, b; Zuo et al. 2016; Das 2017). The modularized design enabled a large-scale MDC system of 105 L based on a tubular configuration consisting individual MDC units have been expanded up to 10 L (Chen et al. 2017; Zhang and He 2015). MDCs coupled with other membrane technologies such as forward osmosis, microfiltration, and bipolar membrane have also been developed to improve treatment performance (Chen et al. 2017; Zuo et al. 2018; Yuan et al. 2015).

However, MDC systems are also reported to experience some problems with the need for different solutions for different chapters, biofouling, imbalance in pH, high internal resistance, reduced desalination rate due to diluted saline solution, reduced conductivity of saline water due to water flux into the middle chamber, and poor quality of treated effluent (<85%) with inefficient desalination (Jingya et al. 2017; Luo et al. 2012a, b, 2010, 2015; Saeed et al. 2015; Davis et al. 2013; Chen et al. 2012b; Jacobson et al. 2011b; Zuo et al. 2016). These limitations and others have been thoroughly examined and resolved in many researches.

Exoelectrogens, which are primarily bacteria, are microscopic organisms that are crucial parts of the MDC. They are called "exo" because they move electrons to an electron acceptor outside of their cells to oxidize organic materials and produce energy. They can be employed as pure or mixed cultures and can be isolated from anaerobic sediment, anaerobic sludge from residential or commercial ETPs, farm soil, and primary industrial or municipal effluent (Utami et al. 2015). Exoelectrogens can generate electricity from organic or chemical sources, including waste from wastewater treatment facilities, simple sugars like glucose, and other pollutants (Utami et al. 2015; Kalleary et al. 2014). Due to the self-replicating nature of exoelectrogens, which is one of the key factors determining the effectiveness of MDC systems, they must continue to grow and survive.

Exoelectrogens can be distinguished by a variety of characteristics, including salt content, temperature, pH, and media, and their careful selection can improve MDCs' ability to produce energy (Logan 2008). They are classified into several functional groups based on types of anaerobic respiration which include denitrifying Ochrobactrum and Pseudomonas, dissimilatory metal-reducing Geopsychrobacter, Shewanella, Geothrix, Geobacter, sulfate-reducing Desulfobulbus and Desulfuromonas, and fermentative Clostridium and *Escherichia coli* (Park et al. 2001; Rabaey et al. 2004; Holmes et al. 2004; Zuo et al. 2008; Bond et al. 2002; Bond and Lovley 2003, 2005; Kim et al. 2002; Zhang et al. 2006; Logan 2009).

The scaling-up effort, however, has been progressing slowly, leaving a knowledge gap for implementing MDC systems in the real world. This chapter discusses the fundamentals of MDC, its microbiology, functioning, various configurations, and environmental applications in wastewater treatment and bioelectricity generation to serve as a reference for future studies investigating effective ways to desalinate water and wastewater and produce bioelectricity. To keep readers informed, the developments, challenges, and opportunities in MDC technology are also described.

2 Overview of Microbial Desalination Cell (MDC)

The concept of MFC led to the creation of MDCs, which have since made significant advancements to accommodate unique multipurpose responses to environmental, water source, and energy challenges. Unlike a standard MFC, MDC consists of an extra desalination chamber between anode and cathode and IEMs at specific distance (Logan 2008) (Fig. 1). The only desalination capacity of MDC technology at this stage of research is the de-salting of NaCl solution, which has been used to simulate seawater (Imoro et al. 2021); however, the literature has seen the emergence of a variety of MDC designs that are categorized based on the orientations, types, membrane-based processing, and membrane shapes (Table 1).

Different MDC configurations have been developed over time, including osmotic MDCs (OsMDCs) that combine forward osmosis (FO) and CEM, microbial electrodialysis and chemical production cells (MEDCCs) that combine a bipolar membrane (BPM) and IEM, submersible MDCs (SMDCs) that align the IEM with a source of bulk waste or wastewater, microbially reverse electrodialysis cells (M (HFM-MDC)).

IEMs, namely cationic exchange membranes (CEMs) and anion exchange membranes (AEMs) have been used to create a three-chambered MDC (Jafary et al. 2018). Different MDC configurations have been developed over time, including osmotic MDCs (OsMDCs) that combine forward osmosis (FO) and CEM, microbial electrodialysis and chemical production cells (MEDCCs) that combine a bipolar membrane (BPM) and IEM, submersible MDCs (SMDCs) that align the IEM with a source of bulk waste or wastewater, membrane-based microbially reverse electrodialysis cell (MREC), capacitive deionization (CDI) based microbial capacitive desalination cells (MCDCs), and membrane-based hollow fiber membrane MDC

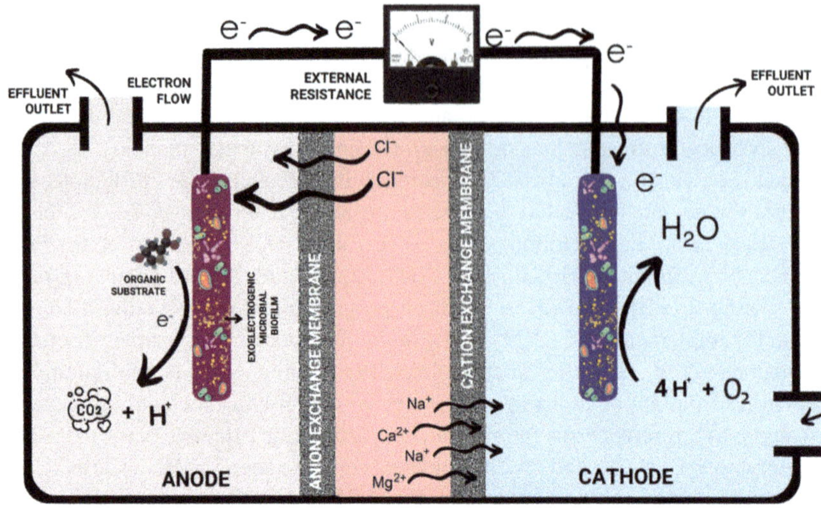

Fig. 1 A typical design of microbial desalination cell (MDC)

(HFM-MDC) (Zhang and He 2012b; Ye et al. 2017; Zhang and Angelidaki 2015b; Cusick et al. 2013; Won et al. 2014; Zuo et al. 2018).

MDC relies on the idea of the water-energy nexus, in which energy produced by microbes during effluent treatment is used to drive effluent/seawater into MDC, hence negating the requirement for external energy for its functioning. MDCs, like MFCs, can only generate power from thermodynamically viable reactions; however, any excess energy produced, if any, can be used to power other desalination systems, like as EDI, CDI, and RO.

3 Components of MDC

Electroactive bacteria (EAB) are a crucial component of MDCs that produce energy from effluents by transferring electrons across their membranes into extracellular habitats (Rabaey 2009). Generation of electrons and protons in anode is accompanied by the transportation of various ions (NH_4^+, HPO_4^-, PO_4^{-2}, K^+, Cl^-) into the desalination chamber under the influence of concentration gradient and electric potential while oxidation of substrates (Al-Mamun et al. 2018a, b; Salehmin et al. 2021). Thus, the transfer of these ions into the desalination chamber could occur in conjunction with the treatment of actual wastewater that contains complicated components (such as $CaCO_3$, $Mg(NO_3)_2$, $Ca(NO_3)_2$, $MgSO_4$, $MgCO_3$, and $CaSO_4$). However, when these ions enter the desalination chamber, IEM and treated water may be contaminated, necessitating the need for additional water treatment (Al-Mamun et al. 2018a,

Table 1 Microbial desalination cell in wastewater treatment and bioelectricity generation

MDC configuration	Substrate	Anode	Cathode	Anion exchange membrane	Cation exchange membrane	Exoelectrogen	Mode of operation	Analyte	Catholyte	COD removal	Desalination	Power output	References
Multi-Chambered MDC	Steel Plant Wastewater	Carbon felt	Carbon felt	–	–	*Pseudomonas putida* with activated sludge	Batch	Filtered raw wastewater	Tap water	70 ± 1.8%	–	10.2 mW/m^2	Shinde et al. (2018)
Conventional MDC	Glucose	Graphite rod	Graphite rod	AMI-7001, Membrane International, Inc	CMI-7000, Membrane International, Inc	*Debaryomyces hansenii*	Batch	Phosphate buffer	NaCl 30 g/L or seawater	–	55.03%	488 mW/m^3	Utami et al. (2015)
MCDC with CDI module inside DC	local municipal Wastewater	Carbon fiber brush as bioanode (Gordon Brush, CA)/23 mL	Carbon cloth as air–cathode (30% Teflon coated)	No AEM was used, another CEM was inserted at place of AEM	CMX-SB, Astom Corporation, Japan	Micro-organisms from Activated Sludge	Fed batch	Phosphate buffered Sodium Acetate (1.6 g/L)	Sodium phosphate buffer	–	70	–	Stoll et al. (2015)
MDC-MCDI	Local wastewater	Carbon felt	Carbon felt	AMI-7001S/ Membrane s International, USA	CMI-7000S/ Membranes International, USA	Anaerobic microorganism	Batch	Phosphate buffered sodium acetate (1.6 g/L)	Phosphate buffered sodium acetate (1.6 g/L)	–	74.4	–	Wen et al. (2014)
rMDC	Activated carbon granules	Polyvinyl chloride (PVC)	Polyvinyl chloride (PVC)	1.8 mol/kg, Shanghua, China	2.0 mol/kg, Shanghua, China	Electrochemically active bacteria (activated carbon granules)	Batch	Phosphate buffered Sodium Acetate (1.64 g/L)	Phosphate buffered Sodium bicarbonate + ammonium chloride	–	93.4	11,800 mW/m^3	Zuo et al. (2014)
Conventional MDC	0.1% yeast extract with Malachite green dye	Carbon cloth	Carbon cloth	Membranes International, USA	Membranes International, USA	*Bacillus subtilis* moh3	Fed-Batch	Azo dye	Phosphate (50 mM) buffered ferricyanide	Complete decolorization	62.24%	0.15–0.05 W/m^3	Kalleary et al. (2014)

(continued)

Table 1 (continued)

MDC configuration	Substrate	Anode	Cathode	Anion exchange membrane	Cation exchange membrane	Exoelectrogen	Mode of operation	Analyte	Catholyte	COD removal	Desalination	Power output	References
Biocathode MDC	Dewatered sludge	Graphite fiber brush/ 1400 mL	Graphite brush embedded in graphite granules/500 mL	Ultrex AMI-7001	CEM, Ultrex CMI-7000	Hydrogen-oxidizing bacteria from soil	Fed-Batch	Dewatered Sludge	Soil solution in deionized water	25.71 ± 0.15	–	3178 mW/m^3	Meng et al. (2014)
Stacked MDC	Xylose	Carbon graphite fiber brushes/28 mL	Carbon cloth as air–cathode (30% wet proofed, BASF, USA)/14 mL	DF120/ Tianwei Membrane	Ultrex CMI7000/ Membrane International	aerobic microorganisms	Fed-Batch	Xylose (1 g/ L) in 50 mM PBS	50 mM PBS buffer	60 ± 2–59 ± 2	76 ± 1–97 ± 1	860 ± 11 mW/m^2	Qu et al. (2013)
Stacked MDC	Synthetic seawater	Graphite fiber brush (Mill-Rose Lab Inc., USA)/160 mL	Carbon cloth as air–cathode (30% wet proofed)/53 mL	AMV/Asahi glass, Japan	Selemion CMV	Mixed culture community of microorganisms	Fed-Batch	Sodium acetate (1 g/ L) in 50 mM PBS	Synthetic seawater (35 g/L NaCl)	55.2 ± 1.7–62.8 ± 0.4	26 ± 0.5	685 mW/m^2	Davis et al. (2013)
IER-MDC	Effluent from an existing MFC	Graphite fiber brush (Mill-Rose Lab Inc., USA)/30 mL	Platinum nanoparticles used as air–cathode/ 18 mL	AMV/Asahi glass, Japan	Selemion CMV	Bacteria	Fed-batch	Phosphate buffered sodium acetate (1 g/ L)	Brackish water	78 ± 2	93–100	650 ± 40 mW/m^3	Shehab et al. (2013)
PMDC	Aerobic sludge from wastewater	Graphite paper/ 180 mL	Graphite paper/ 180 mL	AMI 7001/ Membranes International	CMI 7000/ Membranes International	Chlorella vulgar	Fed-batch	Synthetic wastewater with aerobic sludge	Mineral solution with microalgae as biocatalyst	65.6	40	84 mW/m^3	Kokabian and Gude (2013)

(continued)

Table 1 (continued)

MDC configuration	Substrate	Anode	Cathode	Anion exchange membrane	Cation exchange membrane	Exoelectrogen	Mode of operation	Analyte	Catholyte	COD removal	Desalination	Power output	References
Conventional MDC	Domestic Waste Water	Graphite brushes	Carbon cloth	AEM, AMI 7001, Membranes International, NJ	CEM, CMI 7000, Membranes International, NJ	Biofilm predominantly Proteobacteria	Fed-Batch	Raw wastewater	Phosphate (50 mM) buffered ferricyanide	55%	<66%	3.6 W/ m^3	Luo et al. (2012a, b)
Conventional MDC	Municipal Waste Water	Graphite brushes	Carbon cloth	AEM, AMI 7001, Membranes International, NJ	CEM, CMI 7000, Membranes International, NJ	Biofilm predominantly Actinobacteria	Batch	Raw-wastewater	Phosphate (50 mM) buffered ferricyanide	52%	52%	8.01 W/ m^3	Luo et al. (2012a, b)
Biocathode MDC	Wastewater	Carbon Felt/49 Ml	Carbon Felt/ 40 mL	AMI-7001S/ Membranes International	CMI-7000S/ Membranes International	Anaerobic microorganisms	Fed-batch	Sodium acetate (1.6 g/L)	Sodium acetate	56.2 ± 6.0	92	960 mW/ m^2	Wen et al. (2012)
rMDC	Domestic wastewater	Carbon graphite fiber brushes/28 mL	Carbon cloth as air–cathode (30% wet proofed, BASF, US)/ 14 mL	DF120/ Tianwei Membrane	Ultrex CMI7000/ Membrane International	Anaerobic microorganisms	Fed-batch	Xylose (1 g/ L) in 25 or 50 mM PBS	25 or 50 mM PBS	79 ± 4, 78 ± 3	34 ± 1, 37 ± 2	931 ± 29 mW/m^2	Qu et al. (2012)
IER-MDC	Local municipal wastewater	Carbon felt as bioanode/ 300 mL	Same as anode electrode functioning as biocathode/ 30 0 mL	AMI-7001/ Membrane International, Inc	CMI-7000/ Membrane International, Inc	Microbial consortium	Fed-batch	Phosphate buffered Sodium Acetate (1.64 g/L)	Phosphate buffered Sodium bicarbonate (1.9 g/L)	–	36 (desalination ratio)	–	Morel et al. (2012)
IER-MDC	Local wastewater	Carbon fiber felt/ 100 mL	Same as anode electrode/ 100 mL	AMI-7001/ Membranes International Inc., New Jersey	CMI-7000/ Membranes International Inc., New Jersey	Micro-organisms from synthetic wastewater	Fed-batch	Acetateladen synthetic wastewater with phosphate buffer (90 mL)	50 mM potassium ferricyanide	–	95–98	360 mW/ m^2	Zhang et al. (2012)

(continued)

Table 1 (continued)

MDC configuration	Substrate	Anode	Cathode	Anion exchange membrane	Cation exchange membrane	Exoelectrogen	Mode of operation	Analyte	Catholyte	COD removal	Desalination	Power output	References
cMDC	Anaerobic sludge	Activated Carbon Cloth (ACC) (Chemviron Carbon, UK)/23 mL	Same as anode electrode/27 mL	Assembly: AEM/Ni or Cu/ACC (AMX, Astom Corporation, Japan)/Ni or Cu mesh current collector/ activated carbon cloth	Assembly: CEM/Ni or Cu/ACC (CMX, Astom Corporation, Japan)/Ni or Cu mesh current collector/ activated carbon cloth	Micro-organisms from anaerobic sludge	Fed-batch	Phosphate buffered Sodium Acetate (1.6 g/L)	50 mM Ferricyanide solution	–	69.4	–	Forrestal et al. (2012a)
cMDC	Anaerobic sludge from the wastewater treatment plant	Graphite brush (Golden Brush, CA)/23 mL	Carbon cloth as air–cathode (30% wet proofed)/27 mL	Assembly: CEM/ACC/ Ni or Cu (CMXSB, Astom Corporation, Japan/ activated carbon cloth (McMaster Carr, IL)/Ni or Cu mesh current)	Assembly: Ni or Cu/ ACC/CEM (Ni or Cu mesh current/ activated carbon cloth/ CMXSB, Astom Corporation, Japan)	Micro-organisms from anaerobic sludge	Fed-batch	Phosphate buffered sodium acetate (1.6 g/L)	Phosphate buffered potassium chloride (10 g/L)	–	88	–	Forrestal et al. (2012b)
Stacked MDC	Synthetic seawater	Graphite fiber brush, (Mill-Rose Lab Inc., USA)/30 mL	Carbon cloth as air–cathode (30% wetproofed, ETek, Type B)/18 mL	AMV/Asahi glass, Japan	Selemion CMV	Micro-organisms from Synthetic Seawater	Fed-batch	Sodium acetate (1 g/ L)	Synthetic Seawater	–	98	1140 mW/m^2	Kim and Logan (2011a)

(continued)

Table 1 (continued)

MDC configuration	Substrate	Anode	Cathode	Anion exchange membrane	Cation exchange membrane	Exoelectrogen	Mode of operation	Analyte	Catholyte	COD removal	Desalination	Power output	References
Stacked MDC	Wastewater	Carbon felt as bioanode/ 21.2 mL	Carbon cloth as air–cathode (30% wetproofed, ETek, Type B, BASF)/ 7.1 mL	DF120/ Tianwei Membrane	Ultrex CMI7000/ Membra ne International	Biofilm	Fed-batch	Sodium acetate (1.64 g/L)	Potassium phosphate salts solutions	–	99.4	–	Chen et al. (2011)
UMDC	Synthetic wastewater	Graphite granules (Carbon Activated Corp., Compton, CA, USA)/ 500 mL	Carbon cloth (Zoltek Companies, Inc., St. Louis, MO, USA)/ 350 mL	AMI-7001/ Membrane International, Inc., Glen Rock, NJ, USA	CMI-7000/ Membrane International, Inc	Microorganisms from anaerobic sludge from local wastewater	Fed-batch	Synthetic wastewater (phosphate buffered sodium acetate)	Acidified water	–	>99	30,800 mW/m^3	Jacobson et al. (2011a)
UMDC	Synthetic wastewater	Carbon brushes (Gordon Brush Mfg Co., Inc., Commerce, CA)/1.9 L	Outer surface of CEM coated with layers of Pt and Nafoin solution, and covered by two layers of carbon cloth	AMI-7001/ Membrane International, Inc., Glen Rock, NJ	CMI-7000/ Membrane International, Inc	Microorganisms from anaerobic sludge from local wastewater	Fed-batch	Synthetic wastewater containing acetate (35 g/L)	Acidified water	–	94.3 ± 2.7 and 73.8 ± 2.1	28,900 mW/m^3	Jacobson et al. (2011b)
Three-chamber MDC	Wastewater	Carbon Cloth, (BASF, NJ)	30% Wetproofed Carbon Cloth (0.5 $mgcm^{-2}$ Pt, 4 layers of PTFE), (type B-1B, E-TEK	AMI-7001/ Membrane International + Experimental AEM/ Aminated Radel R-5500	CMI-7000/ Membrane International + Experimental CEM Sulfonated Radel R-5500	Bacteria from an existing MFC	Fed-batch	Sodium acetate (1 or 2 g/L)	50 mM PBS Buffer	–	43–67	480 mW/m^2	Mehanna et al. (2010b)

b). According to spectroscopic tests, AEM is more contaminated by organic contaminants than CEM while removing complex chemical pollutants from domestic effluent (Luo et al. 2012b). However, more study is required to make the MDC operation the only method for completely removing salt and organic components while creating deionized freshwater.

4 Microbiology of MDC

Exoelectrogens are frequently used in MDCs, either in pure strains or cultures or mixed cultures, and respire by directly transferring electrons from their cells to external environments, a substance or material, and depending on the environmental conditions, they react differently in MDCs (Logan 2009; Feng et al. 2014; Shinde et al. 2018; Kalleary et al. 2014; Zhang et al. 2010a, b). Two main methods allow the transmission of electrons outside the cell: transfer via nanowires linked to the Geobacter and Shewanella species, and transfer mediated by mediators like pyocyanin produced by *Pseudomonas aeruginosa* and some bacteria (Logan 2008). Although a large variety of bacteria has this capability, Shewanella and Geobacter spp. are two of the most researched species (Rabaey et al. 2005).

Most of the research has found that EAB can consume and produce energy from complex organic compounds, live in biofilms, and increase their resistance to system disruptions (changes in anolyte conditions), as well as metabolic byproducts or intermediate products of one group of bacteria may act as a source of carbon or energy for another (Guang et al. 2020). Exoelectrogens isolated from anodic biofilms include *Lactobacillus brevis, Pseudomonas xanthomarina, Enterococcus avium, Pseudomonas caeni, Lactobacillus pentosus, Acinetobacter venetianus, Pediococcus acidilactici, Acinetobacter rudis, Clostridium butyricum, Acinetobacter brisouii, Halomonas salicampi, Acinetobacter gandensis, Arcobacter skirrowii, and Campylobacter upsaliensis* (Beenish et al. 2016). However, it is crucial to remember that for an MDC to function at its best, biofilms must successfully form, and to make sure this happens, ideal anodic chamber conditions must be established and sustained, including surfaces that are good for attachment, the proper growth temperatures, and the right pH levels (Saeed et al. 2015; Kumar et al. 2016).

5 Working of MDC

MDCs use an external wire to link the electrodes that enable anode to cathode electrons transport unlike MFCs that based on electroactive sludge to oxidize organic substrate. Upon the introduction of organic wastewater, electrogenic bacteria typically form biofilms on anodes and biocatalyze the oxidation of organic molecules to produce electrons and protons. Desalting salt-ions and cations that travel onto the cathode through CEM to interact with oxygen and electrons under the potential gradient between the two chambers could result in the production of clean water.

Exoelectrogens, which oxidize organic matter in effluent to produce electricity, transport electrons to an anode to make energy, and build up protons that attract negatively charged Cl^- desalination into anodic chamber to balance the charge form the basis for desalination in MDCs (Abubakari et al. 2019). The transport of cations (Na^+) from the desalination chamber to the cathodic chamber, along with the transfer of electrons from the anode surface to the cathode, which reduces the species of oxidized electron acceptors on the cathode, is necessary for the desalination of saltwater, effluent treatment, and the production of bioelectricity in MDCs (Cao et al. 2009a, b).

6 Configurations of MDC

Ferricyanide was used as a catholyte in the initial attempts at MDCs because of its capacity to generate larger cathode potentials and achieve faster reduction kinetics; but, due to its high cost and hazardous properties, it is not appropriate for large-scale processes. Thus, to lower the salt of the water by 63% in a single cycle, a carbon cloth electrode with platinum as a catalyst may be utilized in an air cathode MDC, which uses oxygen as a terminal electron acceptor (TEA) due to its high reduction potential and affordability (Gude et al. 2013a, b; Mehanna et al. 2010a, b). Alternately, activated carbon and cobalt tetramethoxyphenylporphyrin can be utilized to further enhance MDC performance by increasing surface area while cobalt tetramethoxyphenylporphyrin functions as a catalyst. Additionally, as oxygen has relatively few hazardous effects in comparison to other compounds, using ambient oxygen as the electron acceptor is particularly advantageous in terms of environmental sustainability.

As was previously mentioned, ferricyanide catholyte or air cathodes with platinum catalysts are typically used as cathodes in conventional MDCs; however, other materials that have been investigated as cathodic catalysts include pyrolyzed iron (II) phthalocyanine and cobalt tetramethylphenylporphyrin (Clauwaert et al. 2007). Another problem with these cathodes is the use of chemicals as catholytes, which is not a sustainable strategy and requires continual replacement, and platinum is too expensive to be employed in a large-scale operation. According to Wen et al. (2012), the usage of biocathode MDC with an aerobic biocathode that contained carbon felt and bacterial catalysts was taken into consideration. They asserted that using 0.441 L of the anode solution with approximately 96.2% coulombic efficiency and 2.83 mg/h total desalination rate produced an increase in voltage produced of 136 mV in comparison to an air cathode MDC and a reduction of 92% in the salty content of seawater, showing that the effectiveness and overall performance of the desalination cell would be enhanced with a biocathode MDC (Wen et al. 2012).

A stack structure MDC system is one way to increase the desalination performance of an MDC by inserting multiple pairs of ion-exchange membranes (IEMs) between the anode and cathode chambers to increase the charge transfer efficiency and allow the flow of saline water through several MDCs, increasing the amount of salt removed (Gude et al. 2013a, b). In the desalination chamber, the stack's alternating AEMs and CEMs create alternating pairs of concentrating and desalting cells, increasing the separation of ion pairs from the seawater with each electron traversing the circuit (Saeed et al. 2015). The parallel electrode connection in series flow mode produced 15–20% more electricity and 80–85% more coulombic efficiency than the case of series connection, according to Choi and Ahn's (2013) testing of five different types of stacked connections for treating domestic wastewater: parallel in parallel flow, series in series flow, parallel in flow, and individual in series flow. Extremely thin stacks were utilized in another case study to maximize the extent and efficiency of desalination and to enable high power densities across the stacks by lowering the ohmic resistance, as Kim and Logan (2011a, b, c) explored the stacked MDC and its impacts on DR and external resistance. Additionally, according to the findings of Chen et al. (2011), the highest DR for the stacked MDC was 0.0252 g/h, which is around 1.4 times greater than that for the normal MDC unit using a 10 Ω external resistance.

Recently, a more sophisticated method known as "re-circulation microbial desalination cells" (rMDCs) was developed that works by gradually recirculating the catholyte and anolyte solutions through the cell to neutralize the pH, which increases the efficiency of both power generation and desalination (Kim et al. 2013; Luo et al. 2011). A recent study found that rMDC operating with 50 mM of phosphate buffer solution (PBS) produced 33% more power density than a cell running normally, and a cell operating with 25 mM of PBS produced a 53% increase in power density (Luo et al. 2011). However, increasing the buffer solution's concentration is not always necessary to boost power density, thus an ideal buffer concentration should be given. Recirculation additionally increased desalination efficacy by 13 and 48% for the 33 mM PBS solution and 25 mM PBS solution, respectively (Luo et al. 2011). Furthermore, the coulombic efficiency of the cell is negatively impacted by the recirculation of the anolyte and catholyte solutions. It is suggested to use hydrogen evolution at the cathode rather than oxygen reduction to address this problem and to circulate solutions in very thin tubes to avoid the creation of equal potentials in the cathode and anode chambers (Kim and Logan 2013a, b; Luo et al. 2011; Freguia et al. 2008).

The microbial electrolysis desalination and chemical-production cell (MEDCC), which can simultaneously produce sodium hydroxide, hydrochloric acid, and desalinate seawater, is created by inserting an acid-production chamber and a bipolar membrane (BPM) into the microbial electrolysis desalination cell (MEDC), which combines the electrodialysis and microbial electrolysis cell (MEC) (He et al. 2008; Chen et al. 2012a, b, c, 2013; Mehanna et al. 2010a, b). The issues of pH changes and Cl^- accumulation in a typical MDC are solved by MEDCC because OH is produced at the anode chamber (Chen et al. 2013). Usually, a conventional cell using a single desalination chamber desalinates at a rate that is approximately 1.4 times the rate of the stacked MDC and MEDCC (Chen et al. 2011, 2012a, b, c). In this case, the bipolar membrane's ability to dissociate water is facilitated by an electric field, which also permits the OH ions to flow into the anode chamber for pH regulation and the H + ions to flow into the acid-production chamber to make acids (Chen et al. 2012a, b, c. It has been shown that MEDCC desalination outperforms MEDC and the electrolysis desalination and chemical production cell (EDCC) and that its performance can be enhanced by using stacks of desalination membranes that increase the charge transferred per electron released (Qu et al. 2012; Chen et al. 2012a, b, c). The development of capacitive MDC (cMDC) took salt removal into account by adding capacitive deionization into the system (Yuan et al. 2012a, b, c; Zhang et al. 2012). A double-layer capacitor in cMDC deionizes the saltwater via "electrochemical salt adsorption" by a double-layer capacitor on the surface electrodes without polluting the anode and cathode chambers and the ions are then allowed to flow back into the liquid after the potential gradient is erased (Yuan et al. 2012a, b, c; Zhang et al. 2012).

Forward osmosis (FO) is a water desalination method in which a water flux is created between two solutions, one of which is referred to as the feed solution and the other as the draw solution; however, the feed solution has a higher water potential than the draw solution, and a water flux is created because of a concentration gradient (Zhang et al. 2011, 2012; Werner et al. 2013; Kim et al. 2013; Zhao et al. 2012). By swapping out the ion-exchange membranes (IEMs) for FO membranes, attempts to incorporate FO in microbial fuel cells (MFCs) and microbial desalination cells (MDCs) have been made and are known as osmotic MDCs or osmotic MFCs (Kim et al. 2013; Cath et al. 2006; Zhang et al. 2013). Zhang et al. (2011) found that the FO membrane orientation causes more water flux and dilution than the PRO mode, albeit the results are contradictory with past research on the FO membrane orientation since osmotic MDCs are not a typical osmotic unit (Cath et al. 2006). The difference in electrical potential in the middle chamber drives osmotic MDCs, as opposed to a regular osmotic unit, which is powered by the gradient in salt concentration (Cath et al. 2006). An additional MDC variant is the bipolar membrane MDC, which consists of layers that are both cation- and anion-selective that have been laminated (heat-pressed or adhered) into a single membrane (Buck 2014). High permselectivity (the capacity to permit ions of just one sign to pass through), low water splitting voltage drop and electrical resistance, and long-life length are membrane characteristics that affect the MDC performance (Buck 2014; Alvarez et al. 1997). A decoupled MDC system was also developed, where the anode (plate configuration with an AEM on either side) and cathode units (contains the CEM on both sides) are placed in a salt solution as opposed to being placed in separate chambers and electrode functions as both a supporting structure and a current collector and is made of stainless-steel mesh wrapped in carbon cloth (Ping et al. 2013). The decoupled MDC has far higher flexibility than a standard MDC unit and offers numerous scaling-up opportunities because it is relatively simple to manage, repair, and adjust the liquid volume ratios as necessary (Chen et al. 2012a, b, c).

Along with the decoupling MDC, a cylindrical up-flow MDC (UMDC) has been developed, consisting of an external annular desalination chamber sandwiched between an AEM and a CEM, as well as an internal cylindrical anode chamber (Jacobson et al. 2011a). Additionally, the potential of the UMDC for pre-treatment in traditional desalination processes was evaluated, and the outcomes showed that it can produce 16.5% of the energy needed for desalination by the downstream RO system or reduce 45.3% of the energy needed for desalination by the downstream electrodialysis system (Jacobson et al. 2011b; Zhang et al. 2012). The overall liquid capacity of the system increased to 105 L when 30 UMDCs were integrated, and the TDR was up to 9.2 kg NaCl/m^3 per day with an applied voltage of 1.1 V (Zhang and He 2015). A variety of MDC-based configurations have been built and tested to realize new functionalities or increase the effectiveness of the treatment. The development of a BES with altered the position of the AEM and CEM for the treatment of saline wastewater resulted in the elimination of 21–84% of salt and 72–94% of COD with an initial conductivity of 40 mS/cm (sodium acetate supply, 2–8 g/L) (Kim and Logan 2013a, b). Meanwhile, a spatial-decoupling anode and cathode were built to increase the flexibility of MDCs, and the MDC displayed significant variations in power output when connected to different external circuits (Ping and He 2013). Additionally, a traditional MDC using a forward osmosis (FO) membrane was replaced by an osmotic MDC (OsMDC) because of its permeability to collect freshwater from wastewater (Zhang and He 2012a, b). Moreover, the BES and FO membrane assemblies have been employed in numerous research to treat saline water, however, the desalination approach used in those studies uses a FO membrane to move water molecules rather than the internal electrical field of the BES to promote directed ion transport (Lu and He 2015; Zhang et al. 2011).

Different types and amounts of saline water affected the desalination performance of various MDC configurations, but the fundamental three-chamber topologies served as proof-of-concepts for a variety of functional applications. The TDRs and CTEs of three-chamber MDCs were situated around the relatively low level because of the small scale and straightforward structure; as a result, there is a significant potential to raise the desalination rate and treatment capacity as well as integrate desalination with other activities. The SMDC arrangement effectively used the current generated by a BES and elevated the TDRs and CTEs. The SMDC configuration produced the highest TDR (107.4 mg/h) and CTE (450%) of all the reported MDC research (Davis et al. 2013; Ge et al. 2014). A promising arrangement that might find practical application is the SMDC, which combines MDCs and electrodialysis modules.

7 Environmental Applications of MDC

7.1 MDC in Wastewater Treatment

The use of MDCs has grown in importance because of their low energy requirements and self-sustaining design; they can be a successful alternative in the treatment of huge amounts of wastewater, producing electricity, polishing of water etc. (Fig. 2). With acetate used as the bacteria's food source, Cao et al. (2009a, b) showed for the first time the capacity of MDCs to desalinate saltwater at various initial salt concentrations (5, 20, and 35 g/L) that produced a maximum of 2 W/m^2 (31 W/m^3) while at the same time removing about 90% of the salt in a single desalination cycle. Ge et al. (2014) successfully desalinated water at a rate of 90.8 ± 8.3 mg/h using the cell pair (anion exchange membrane-cation exchange membrane pair) MDC. Most of the substrates employed in MDC are readily biodegradable materials including butyrate, oxalate, glucose, and others (Zhang et al. 2010a, b; Catal et al. 2009; Rikame et al. 2012). Domestic, commercial, and industrial wastewater has also primarily been utilized as a substrate for exoelectrogens in the anode chamber of MDC (Luo et al. 2012a, b; Utami et al. 2015; Shinde et al. 2018). The configuration of the MDC system and the research's primary objectives will determine the mode of operation (batch, fed-batch semi-continuous, and continuous modes) at the anode chamber (Cao et al. 2009a, b; Chen et al. 2011; Jingyu et al. 2017; Luo et al. 2012a, b; Utami et al. 2015; Shinde et al. 2018; Kalleary et al. 2014).

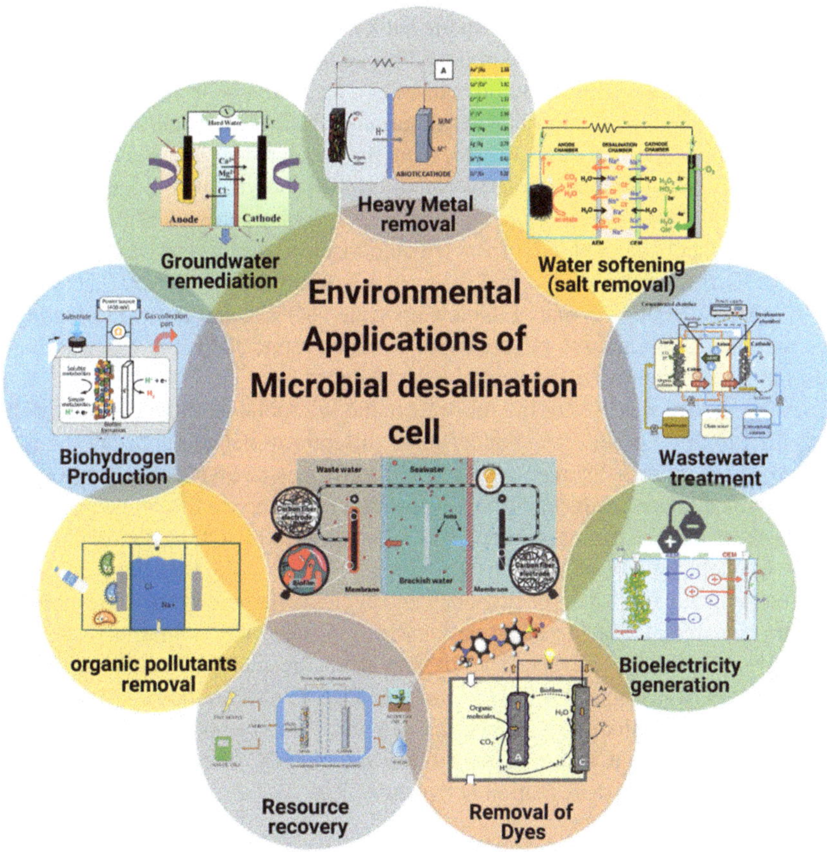

Fig. 2 Applications of microbial desalination cell (MDC) in environmental management

Exoelectrogenic bacteria are substrate-specific, therefore the type of substrate utilized in an MDC system will dictate the exoelectrogens that must be used or that are most prevalent, as well as how well the system performs overall. According to Luo et al. (2012a, b), a mixed culture with Proteobacteria as a dominating species at the anode chamber was able to remove 55% of COD from home wastewater Actinobacteria predominated at the anode chamber in another study by Luo et al. (2012a, b) using urban wastewater, which removed 52% of COD. *Pseudomonas putida* in activated sludge at the anode chamber was used by Shinde et al. (2018) to treat the wastewater from a steel plant in a traditional MDC system, and they saw a COD removal rate of roughly 70%. Pure cultures should have well-defined substrates when utilized in MDC systems to considerably improve their efficiency in removing COD. In a typical example, Kalleary et al. (2014) employed Bacillus subtilis moh3 in the anode chamber of a traditional MDC that included 0.1% yeast extract with Malachite green dye led to a complete decolorization. This was performed with 0.1% yeast extract and Sunset yellow dye as the anolyte, and both times *Bacillus subtilis* moh3 decolored it (Kalleary et al. 2014). Anode and cathode potential differences are used to drive NO_3^- and Na^+ from contaminated groundwater outside the reactor through IEMs and into the anode and cathode chambers, respectively, in Zhang and Angelidaki's (2013) submerged microbial desalination-de nitrification cell (SMDDC). The SMDDC removed 90.5% of the nitrates from the groundwater in the anode chamber in less than 12 h that were subsequently released with anolyte into the cathode, where they underwent autotrophic denitrification to create nitrogen gas. With the help of the UMDC, Jacobson et al. (2011b) were able to successfully remove 92.0 $\pm$ 0.4% of the COD from anolytes that had received a loading rate of 6.78 $\pm$ 0.36 g COD $L^{-1}d^{-1}$. This UMDC was one of the few up-scaled MDCs, and it was fascinating to note that its COD reduction effectiveness could be replicated using either artificial seawater or salt solution in the desalination chamber.

Sewage sludge was used as the substrate in the Microbial desalination cell (MDC) study by Ebrahimi et al. (2018) under three different initial salt concentrations (5, 20, and 35 gL^{-1}) and the maximum salt removal rates of 50.6%, 64%, and 69.6% were obtained under batch condition, respectively, with the maximum power density of 47.1 Wm^3 and the averaged chemical oxygen demand (COD) removal of 58.2 $\pm$ 0.89% when the initial COD was 6610 $\pm$ 83 mg L^{-1}. According to Liang et al. (2016), electrochemical impedance spectroscopy and cyclic voltammetry data, the increased electron transfer rate at the electrode/biofilm interface was primarily responsible for the rise in PMDC current. The effluent from the middle chamber had a salt concentration of less than 1.4 mg L^{-1}, and the PMDC's salt removal efficiency was consistently higher than 96%. Since the estimated number of electrons harvested and the removal of NaCl were in good agreement, PMDC effectively provides both energy generation and desalination at the same time.

7.2 MDC in Bioelectricity Production

Exoelectrogens discharge electrons into the environment, which go down the circuit to the cathode and change oxygen into water in the cathode chamber, producing a power output that is measurable and storable. The power output in an MDC system is observed to be influenced by the rate of substrate oxidation by the exoelectrogens; the higher the rate of substrate oxidation, the higher the power output. Furthermore, the MDC system's power output depends on the anode's capacity to receive electrons from exoelectrogens. A study found that employing the main species, Actinobacteria, in the mixed culture employed in the anode chamber resulted in a power production of 8.01 W/m^3, much higher than the 3.6 W/m^3 from household wastewater in a similar conventional MDC (Luo et al. 2012a, b). A high electrical output of 1.55 kW/m^3 has been attained in different BES, such as air–cathode MFC with a mixed culture in the anode, which is higher than the use of pure culture (Fan et al. 2007). According to a study, *Debaryomyces hansenii*, a pure culture, was able to desalinate water in a typical MDC with an electrical output as low as 488 mW/m^3 (Utami et al. 2015).

The open- and closed-circuit operations were originally contrasted as a proof-of-concept for the MDC for water softening; however, earlier experiments utilizing MDCs to treat saline water also discovered salt loss under the open-circuit potential (Mehanna et al. 2010a, b). We may deduce that hardness removal was mostly caused by the current generation and that natural diffusion/ion exchange/scale formation had a minimal impact by comparing electricity generation and hardness removal between the open- and closed-circuit operations. Rusyn et al. (2021) showed that multi-electrode plant–microbial fuel cells with 12 anodes and 11 cathodes with *Caltha palustris* plants at 10 Ω produced a current of 40.98 mA which is 10.1 times larger than mono-electrode plant–microbial fuel cells. The current generated by the two multi-electrode electro-biosystems connected in parallel was 2.1 times more than the current produced by the single multi-electrode electro-biosystem. Three multi-electrode plant-microbial fuel cells were connected serially, which resulted in a 2.9-fold increase in bioelectric potential but no increase in voltage. Ebrahimi et al. (2018) claim that using treated sludge as a catholyte increased COD removal and power density to 87.3% and 54.4 Wm^3, respectively, with a pH variation in the treated effluent that balanced out the effects.

8 Factors Affecting Efficiency of MDC

There are several factors which affect the working of MDCs (Fig. 3). The force of the electrical field acts as the ion transport driver in a BES, and the transport rate is dependent on the current density; however, better ion transport can be achieved by maximizing the current density, and the energy produced by the BES should be allocated as much as possible to drive the internal ion transport (Chen et al. 2011). Therefore, to achieve better operating outcomes, the external resistance should be as low as possible. This association indicates that the majority of the BES used to drive ion transport has been run with a low external resistance of 1–10 X (Zuo et al. 2014). The internal resistance must be lowered accordingly to further increase the current in the BES. Intermembrane distances of 3, 10, and 20 mm were examined in studies of the early MDC stage (Cao et al. 2009a, b; Chen et al. 2011). These reactors had total ohmic resistances of 25, 21, and 449 X, respectively, at the start of the desalination process. At the end of the cycle, the desalination-related decreases in the conductivity of the salt solution caused an additional increase in internal resistance that could reach more than 30 times. The anode and cathode chamber thicknesses and salt solution concentrations varied, despite the reactors' identical diameters, which made it difficult to directly compare the internal resistance.

In a bench-scale MDC with constant operating parameters for HRT and influent flow rate, Ping and He (2014) looked at the impact of intermembrane distance (0.3–2.5 cm) on desalination effectiveness. For initial salt concentrations of 10 and 30 g/

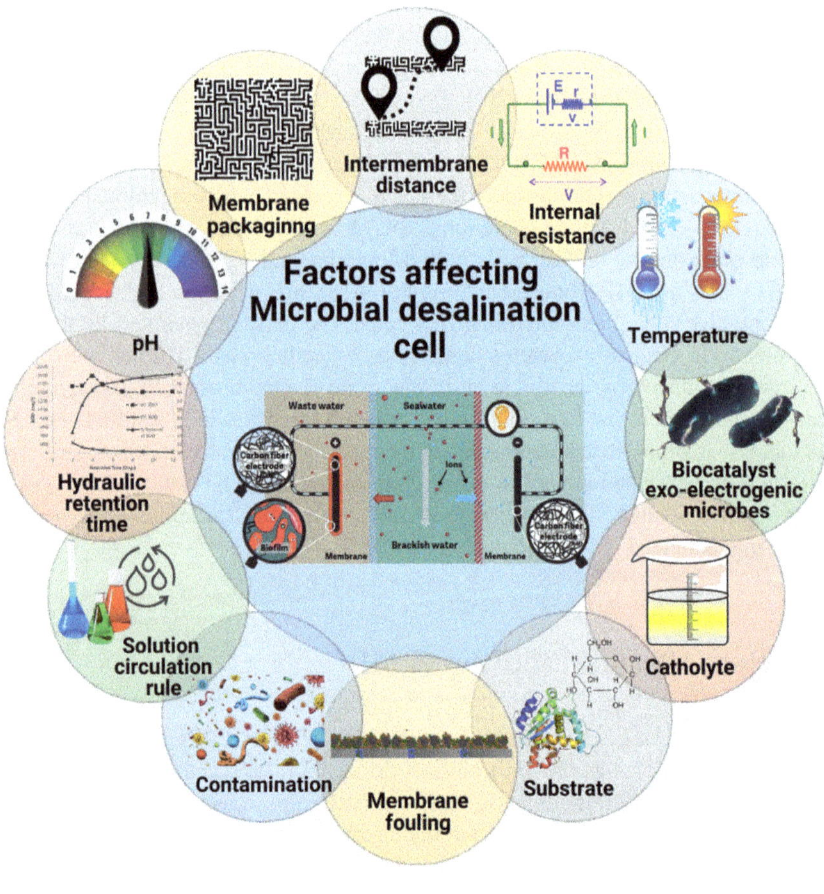

Fig. 3 Factors affecting the efficiency of microbial desalination cell (MDC)

L, respectively, desalination rates using an inter-membrane spacing of 0.3 cm were twelve and seven times higher than those using 2.5 cm. In the early stages of MDC investigations, several MDC procedures were carried out at starting resistances of 25, 21, and 449 Ω with intermembrane lengths of 3, 10, and 20 mm. The accumulated internal resistance can be very high when using the stacked desalination chamber layout. Ion exchange resin is a useful method for lowering the reactor's internal resistance because it helps ions move through the desalination chamber more easily. This increases the chamber's conductivity and lowers the resistance of the system. The MDC filled with mixed anion and cation exchange resins was known as R-MDC, and when used to treat 2–10 g/L NaCl solutions, the R-MDC displayed an internal resistance of 3.0–4.7 X, 26.9–64.5%, and a desalination rate of 1.5–8 times that of the non-filled control MDC (calculated according to the data presented in the reference) (Morel et al. 2012).

The findings were confirmed by an additional MDC with a mixture of ion exchange resins (Zhang et al. 2012). The subsequent studies have shown that anions and cations are competitively transported through R-MDC (Zuo et al. 2013). The trend for ion transport was $SO_4^{2+} > NO_3^- > Cl^-$ for anions and $Ca^{2+} > Mg^{2+} > NH_4^+$ for cations; however, after extended use, calcium and magnesium ions are deposited on the membrane surface and bonded to sulfonic acid groups on the resin to reduce its conductive properties, indicating that the device is unsuitable for the treatment of water with high hardness. They also created a 10-L stacked R-MDC with three desalination chambers, which they utilized to desalinate a 0.5 g/L NaCl solution that represented the secondary effluent from a municipal wastewater treatment facility (Zuo et al. 2014). As a result, the operating circumstances have a direct impact on the MDC's ability to produce electricity and desalinate water. The rate of desalination, the desalination endpoint, and the amount of freshwater produced can all be impacted by the solution circulation rate (SCR), hydraulic retention time (HRT), and concentration/dilution solution volume ratio. The SCR influences the thickness of the solution boundary layer, which in turn influences the desalination rate and mass transfer resistance. The depth of desalinated water continuously increases as desalination advances, while the existing levels of MDC continue to decline. A sensible selection for HRT is required to achieve adequate desalination because the impacts of water osmosis and salt ion diffusion against the direction of the electrical field will cause the desalination rate and its effect to gradually drop.

Another aspect that has to be improved is the concentration/dilution solution volume ratio. The optimum condition from the standpoint of freshwater production is to gather more freshwater while using fewer concentrated solutions. A mathematical model combining the Nernst-Monod equation and the desalination process has been developed to guide the optimization and scale-up of the configuration and operation of the MDC. Various conditions, including the operation and bioelectrochemistry of the reactor, significantly affect the operating results (Ping et al. 2014). The mass balances for the anodic substrate, microorganisms, and electron mediator as well as the mass and electrical balances of the salt ions were established by this model, which connected the MDC power generation and desalination processes. The mathematical model's predictions were quite similar to the experimental data acquired under various situations, but because it disregarded the decrease in current efficiency under low-current conditions, the estimates of salt concentration were off. They further updated and enhanced the model by taking into account the junction potential, current effectiveness, diffusion limitation, and osmosis flux (Ping et al. 2015a, b). Strong evidence was provided to support the future optimization of operating conditions by the updated model, which yielded results that were exactly aligned with the experimental findings.

The MDC's performance was also governed by its operational mode. Due to the rising resistance, the desalination in batch mode reduced over time (Ping et al. 2013). Contrary to the continuous mode with an air cathode, where the pH changes reduced conductivity and overall MDC performance (Chen et al. 2011). However, utilizing mixed inoculum could improve energy production while also desalinating wastewater (Gude et al. 2013a, b). The constant distribution of substrate in up-flow and stacked MDC improved the performance of MDC in terms of current generation and desalination (Jacobson et al. 2011a).

9 Challenges in MDC Technology

The use of simulated wastewater (e.g., acetate) and seawater (e.g., NaCl) and field settings where actual wastewater with an unknown composition can come from various sources, such as dewatered sludge, sewage sludge, domestic wastewater, industrial wastewater, leachate, and engine oil, cannot accurately represent the performance of MDC. The system performance is frequently influenced by a variety of characteristics that typically have a diverse experimental setup and where accurate information is typically lacking, making it difficult to compare the claimed performance of MDCs across the globe. The complexity of wastewater, high quantities, and the presence of resistant chemicals make the development of MDCs for field applications particularly difficult (Salehmin et al. 2021).

Although wastewater with different compositions also contributes to the varied MDC performance among the reported studies concerning COD removal, hydraulic retention time, and power generation, pH fluctuations in MDC anodes and cathodes are a serious limiting factor impeding the optimization and scale-up of this technology (Salehmin et al. 2021). Normally, the anolyte's pH decreases to an acidic state while the catholyte's pH rises to an alkaline state; however, in MDC anode chambers, pH drops result from the slower movement of protons relative to electrons, whereas pH increases in cathodic solutions are brought on by the reduction of oxygen (Qu et al. 2012; Luo et al. 2012b). Potential losses in BES have been linked to an increase in pH in the cathode chamber (Zhang et al.2010a, b). Numerous research studies have linked pH fluctuations to MDCs' poor desalination and electricity-generating capabilities. Among these is a study by Luo et al. (2011) that observed a decline in the production of electricity and hydrogen gas and mostly attributed it to the anolyte's pH dropping from 7.0 to between 5 and 6. The pH of electrolytes should be kept at neutral (pH 7.0) or close to neutral pH settings for the greatest MDC performances (Imoro et al. 2021).

Several studies have used various methods to regulate or moderate the effects of pH on MDC performances. Osmotic water movement in stacked MDCs should be kept to a minimum to prevent water losses into the concentrate cells, which would raise the salinity of the dilute and lower water recoveries (Kim et al. 2011). It was demonstrated in testing with a 5-cell-paired stack that water transport constrained the amount of desalination that the MDCs could potentially achieve, hence preventing this water loss is crucial (Kim et al. 2011). Hydraulic residence times (HRTs) in the stack should be lowered by raising salt water flow rates to lessen osmotic water transport (Kim et al. 2013). Current densities would also need to be raised to sustain adequate desalination in the diluted effluent (Kim et al. 2013). The establishment of a biofilm on the surface of the AEM, which is near the anode chamber and exposed to organic chemicals and saltwater creatures, is inevitable and could modify the chemistry of the functional groups of the membrane or jeopardize its polymeric structure. It is challenging to estimate the long-term efficacy of IEMs under these circumstances because there is minimal experience with their use in conventional ED for conditions characteristic of wastewater and biological treatment applications.

We can assess the likelihood of contaminants in wastewater moving to desalinated water even though concerns regarding MDCs' access to safe water are not yet known. The AEM will serve as the main barrier when IEMs are utilized, hence the desalinated water's safety will be impacted by the AEM's characteristics (Strathmann 2004). IEMs must have incredibly small pore sizes, in the region of 1 nm, to be exclusively for co-ions (for example, cations for AEMs), and because of this small pore size, even counter-ions (for example, anions for AEMs) cannot travel through an IEM if their molecular weight is more than 350 Da (Strathmann 2004). Given that most viruses have a size between 20 and 300 nm (Madigan and Martinko 2006), AEMs should act as a reliable defense against viruses or larger pathogenic bacteria during MDC operation, so long as the membrane's physical integrity is preserved; however, the AEM will be less effective at preventing the passage of organic substrates or inorganic ions through the AEM. Osmotic water transfer to the saltwater chamber can speed up the rate of transport of dissolved organic molecules with negative or neutral charges, such as glucose (180 Da) and acetate (60 Da), via the AEM. The movement of organic material through the AEM may encourage microbial growth in the stream of desalinated water, perhaps promoting the growth of any pathogenic microbes that may be present. Investigations of organic matter crossover through AEMs in MDC studies have not been conducted, however, acetate crossover is known to occur in MFCs (Kim et al. 2007).

10 Advances in MDC Technology

Several improvements have been implemented to increase the effectiveness of MDC. Due to the current generation and water flux, Zhang and He (2012a, b) substituted AEM with FO membrane and introduced a novel osmotic MDC that had a greater salt reduction and effective water recovery from wastewater than the available standard MDC and FO. The created combination was economical since the FO membrane is thought to be less expensive than AEM, however, salt removal proved ineffective due to the FO membrane's restriction on chloride ion movement (Zhang and He 2012a, b). The FO in FO-MDC used the anodic organic solution as a feed solution, recovering water from it, and feeding the desalinated draw solution back to the MDC after it had been diluted.

Cathodic alkalinization of boron in MDC was unsuccessful, hence MDCs are combined with Donnan dialysis to remove boron (Ping et al. 2015b). The primary causes were the delayed operation of the Donnan dialysis system caused by the driving force of the concentration gradient and the high pH of the effluent that contained 2 mg/L of boron, prompting further research (Ping et al. 2015b). The constraint of the MDC-Donnan dialysis system was removed by employing a single membrane MFC with the cathode and anode compartments separated by AEM and the pH of the solution enhanced by an oxygen reduction process in the cathode to transform boric acid to borate ion and achieve maximal removal. Additionally, to improve the performance of MDC by utilizing sunlight, Qian et al. (2014) used hematite nanowire photoanodes, Liu et al. (2015) used nanowire bacterium hybrids, and Zang et al. (2014) used nanowire photocathodes. Kokabian and Gude were the first to describe photosynthetic MDC, which improved the rate of desalination, COD reduction, and overall performance when compared to single-chambered MFC (Kokabian and Gude 2013). Given that hematite (–Fe_2O_3) has a lower bandgap than graphite, Liang et al. (2016) created a photosynthetic MDC with hematite (α-Fe_2O_3) as the bio-photocatalyst anode (2.2 eV).

Additionally, based on electricity production in MDCs, Dong et al. (2017) proposed the ED system as a post-desalination treatment for water recovery following sedimentation. Zuo et al. (2018) combined a biocathode MDC with a hollow fiber microfiltration membrane to treat actual domestic wastewater and get rid of organic contaminants. They postulated that the effluent's quality would be improved by using an anaerobic/aerobic process, filtration, and then migration to the central membrane stack. Hollow fiber microfiltration membranes with biocathodes were stable and less expensive than Pt cathodes in MDCs for the purification of wastewater, recovering 98.7% of the total phosphorus and a portion of the total nitrogen (Zuo et al. 2018).

The removal efficiency of chemical oxygen demand, ammonia nitrogen, total nitrogen, and total phosphorous was tested using a dual anode MDC that was also built in parallel and had a hollow fiber microfiltration membrane (Liu et al. 2019b). Further investigation led to the construction of two dual anode-MDCs linked with hollow fiber microfiltration membrane modules, and the overall best performance was optimized at 10 L/d treatment capacity with 6 h HRT (Liu et al. 2019a). Wen et al. (2014) created a system that used membrane capacitive deionization with MDC, a less energy-intensive method, to desalinate concentrated seawater into lower concentrations (1 g/L). The 264.8 mol/g electro-sorption capacity of MDC, used as a power source for membrane capacitive deionization, was deemed to be the best of all. The desalination rate was seen to be 3.7 mg/h in one cycle, hence membrane capacitive deionization in conjunction with MDC might be regarded as an efficient technique for total salt removal (Wen et al. 2014).

Huang et al. (2018) created a bio-electro-Fenton linked MDC system in which polypyrrole-coated polyacrylonitrile-based carbon fiber served as the cathode for selective oxygen reduction that encouraged the H_2O_2 electro-production that aids in the decomposition of methylene blue. The performance of the system was found to be stable with an 87.2–96% desalination ratio thanks to the up-flow–stacked MDC that Wang et al. (2020) built by combining the up-flow and stacked designs. Proteobacteria and Bacteroides, which are functional bacteria present in the anode chamber and oversee the removal of nitrogen and organic matter, as well as Geobacter, which can generate electricity, are present in the cathode of the MDC-built wetland, which was used as part of a composite system to treat salt-containing water (Mengqi et al. 2021).

11 Future Research Needs

According to the current situation of large-scale MDC systems, there aren't as many upscaling studies as there are for MFC and MEC systems, and there isn't enough information offered, which delays the systems' applicability in meeting the growing needs of humans. The current manuscript outlines numerous recommendations for upcoming up-scaling investigations, led by well-accepted and ongoing work of other bioelectrochemical systems such as MFC and MEC. The development of the flat plate reactor should proceed in tandem with the design of the tubular reactor in terms of technological challenges and techno-economic issues, to ensure proper comprehension of the challenges and potential solutions for each design in the literature. The diameter, length, and quantity of the brushes, as well as the orientation and spacing of the electrodes, should all be optimized if the anode is made of a brush to reduce internal resistance and hence increase power generation.

Reduced internal resistance, which nonconductive granular materials do not encourage, should be achieved by using conductive granular materials. It is important to identify the elements that influence the mass and charge transfer from electrolytes to electrodes and vice versa, such as the structural and supporting components of a reactor. Optimizing feed spacers will improve the production of conductive membranes, enabling the use of PEF technology with low power consumption to clean membranes on an industrial scale. Future up-scaling experiments should utilize real wastewater and seawater to hasten the MDC technology's viability. It is initially necessary to evaluate different types of wastewater and seawater to see if the influents are suitable for operating MDC systems.

MDC activities should be continued for a longer period (>1 year) to evaluate system performance and durability as well as other crucial factors such as operation and maintenance costs, which are a component of the capital cost of a large-scale operation. It should be developed in conjunction with an RO device of a similar size to assess whether it is practical to employ a large-scale MDC reactor with current wastewater treatment plants. Using current waste-to-energy and water reclamation technologies, the capital cost of operating MDC systems must be calculated to achieve the goals of high-efficiency wastewater treatment, a high desalination rate, and high energy output.

12 Conclusions and Recommendations

MDCs based on the microbial metabolism to generate bioelectricity while removing salts from water, are a promising technology for sustainable and economical desalination; nevertheless, there is still potential for advancement in terms of energy efficiency and scalability. According to recent studies, MDCs can produce electricity and achieve large salt content reductions at the same time; however, the effectiveness of MDCs is affected by several variables, including the electrode material, microbial community, and operational conditions. Therefore, it is essential to optimize these settings to increase the MDCs' effectiveness and stability. The long-term stability and scalability of MDCs need to be improved, and this can be done by producing more effective electrode materials, enhancing microbial populations, and investigating novel operational procedures. To compare the viability and sustainability of MDCs to traditional desalination methods, it is crucial to conduct economic and environmental studies.

Declaration of Interest: There is no conflict of interest.

References

Al-Mamun A, Ahmad W, Baawain MS, Khadem M, Dhar BR (2018) A Review of Microbial Desalination Cell Technology: Configurations, Optimization and Applications, Journal of Cleaner Production

Abubakari Z.I, Mensah M, Buamah R, Abaidoo R.C (2019) Assessment of the electricity generation, desalination and wastewater treatment capacity of a plant microbial desalination cell (PMDC). Int J. Energy Water Resour. 3:213– 218

Ahuja S (2019) Global water challenges and solutions. Advances in Water Purification Techniques. Elsevier, pp. 17–39.

Al-Mamun A, Ahmad W, Baawain MS, Khadem M, Dhar BR (2018b) A review of microbial desalination cell technology: Configurations, optimization, and applications. Journal of Cleaner Production 183:458–480.

Alvarez F, Alvarez R, Coca J, Sandeaux J, Sandeaux R, & Gavach C. (1997). Salicylic acid production by electrodialysis with bipolar membranes. Journal of membrane science, 123(1):61-69.

Amma L, Ashraf F (2020) Brine management in reverse osmosis desalination: a UAE perspective. In: 2020 Advances in Science and Engineering Technology International Conferences (ASET).

Beenish S, Ann D.C, Zhongtang Y, Anne C.C, Tansol P (2016) Simultaneous power generation and desalination of microbial desalination cells using Nannochloropsis salina (marine algae) versus potassium ferricyanide as catholytes. Environ. Eng. Sci., 1–12

Bond D.R, Holmes D.E, Tender L.M, Lovley D.R (2002) Electrode-reducing microorganisms that harvest energy from marine sediments. Science 295:483–485.

Bond D.R, Lovley D.R (2003) Electricity production by Geobacter sulfurreducens attached to electrodes. Appl Env. Microbiol. 1548–1555.

Bond D.R, Lovley D.R (2005) Evidence for involvement of an electron shuttle in electricity generation by Geothrix fermentans. Appl. Environ. Microbiol. 71:2186–2189.

Buck R. P. (2014). Ion-selective membranes and electrodes, Access Science.

Cao X, Huang X, Liang P, Xiao K, Zhou Y, Zhang X, Logan B.E (2009a) A New Method for Water Desalination Using Microbial Desalination Cells. Environmental Science & Technology 43:7148–7152.

Cao X, Huang X, Liang P, Xiao K, Zhuo Y, Zhang X, Logan B.E (2009b). A new method for water desalination using microbial desalination cells. Environ. Sci. Technol. 43:7148–7152.

Catal T, Bermek H, Liu H (2009) Removal of selenite from wastewater using microbial fuel cells. Biotechnol. Lett. 31:1211–1216.

Cath T.Y, Childress A.E, & Elimelech M. (2006). Forward osmosis: Principles, applications, and recent developments. Journal of membrane science, 281(1-2):70-87.

Chen S, Liu G, Zhang R, Qin B, & Luo Y. (2012a). Development of the microbial electrolysis desalination and chemical-production cell for desalination as well as acid and alkali productions. Environmental science & technology, 46(4):2467–2472.

Chen S, Liu G, Zhang R, Qin B, Luo Y, & Hou Y. (2012b). Improved performance of the microbial electrolysis desalination and chemical-production cell using the stack structure. Bioresource technology, 116:507–511.

Chen S, Luo H, Liu G, Zhang R, Wang H, Qin B, & Hou Y. (2013). Integrated utilization of seawater using a five-chamber bioelectrochemical system. Journal of membrane science, 444:16-21.

Chen X, Liang P, Wei Z, Zhang X, & Huang X. (2012c). Sustainable water desalination and electricity generation in a separator coupled stacked microbial desalination cell with buffer free electrolyte circulation. Bioresource Technology, 119:88–93.

Chen X, Xia X, Liang P, Cao X, Sun H, & Huang X. (2011). Stacked microbial desalination cells to enhance water desalination efficiency. Environmental science & technology, 45(6):2465-2470.

Chen X, Zhou H, Zuo K.C, Zhou Y, Wang Q.Y, Sun D.Y, Gao Y.F, Liang P, Zhang X.Y, Ren Z.J, Huang X (2017) Self-sustaining advanced wastewater purification and simultaneous in situ nutrient recovery in a novel bioelectrochemical system. Chem. Eng. J. 330:692–697.

Choi J, & Ahn Y. (2013). Continuous electricity generation in stacked air cathode microbial fuel cell treating domestic wastewater. Journal of environmental management, 130:146-152.

Clauwaert P, Van der Ha D, Boon N, Verbeken K, Verhaege M, Rabaey K. and Verstraete W. (2007). Open air biocathode enables effective electricity generation with microbial fuel cells. Environmental science & technology, 41(21):7564-7569.

Cusick RD, Hatzell M, Zhang F, Logan BE (2013) Minimal RED cell pairs markedly improve electrode kinetics and power production in microbial reverse electrodialysis cells. Environmental science & technology 47:14518-14524.

Das D (2017) Microbial Fuel Cell: A Bioelectrochemical System That Converts Waste to Watts; Springer International Publishing: Berlin/Heidelberg, Germany, pp. 1–506.

Davis R.J, Kim Y, Logan B.E (2013) Increasing desalination by mitigating anolyte pH imbalance using catholyte effluent addition in a multi-anode bench scale microbial desalination cell. ACS Sust. Chem. Eng. 1(9):1200–1206.

Dong Y, Liu J, Sui M, Qu Y, Ambuchi JJ, Wang H (2017) A combined microbial desalination cell and electrodialysis system for copper-containing wastewater treatment and highsalinity- water desalination. Journal of Hazardous materials 321:307-315.

Ebrahimi A, Kebria D.Y. and Najafpour G.D (2018) Co-treatment of septage and municipal wastewater in a quadripartite microbial desalination cell. Chemical Engineering Journal, 354:1092-1099.

Fan Y, Miguez-Macho G, Weaver C.P, Walko R, Robock A (2007) Climate and Dynamics-D10125-Incorporating water table dynamics in climate modeling: 1. Water table observations and equilibrium water table simulations.

Feng C, Li J, Qin D, Chen L, Zhao F, Chen S, Hu H, Yu C.P (2014) Characterization of exoelectrogenic bacteria enterobacter strains isolated from a microbial fuel cell exposed to copper shock load. PLoS ONE 9:113379.

Forrestal C, Xu P, Jenkins P.E, Ren Z, (2012a) Microbial desalination cell with capacitive adsorption for ion migration control. Bioresource technology 120:332–336.

Forrestal C, Xu P, Ren Z (2012b). Sustainable desalination using a microbial capacitive desalination cell. Energy & Environmental Science 5:7161–7167.

Freguia S, Rabaey K, Yuan Z, & Keller J. (2008). Sequential anode–cathode configuration improves cathodic oxygen reduction and effluent quality of microbial fuel cells. Water research, 42(6-7):1387-1396.

Ge Z, Dosoretz C.G, He Z (2014) Effects of number of cell pairs on the performance of microbial desalination cells. Desalination 341:101-106.

Guang L, Ato-Koomson D, Jingyu H, Ewusi-Mensah D, Miwornunyuie N (2020) Performance of exoelectrogenic bacteria used in microbial desalination cell technology. Int. J. Environ. Res. Public Health 17:1121.

Gude V.G, B. Kokabian, V. Gadhamshetty (2013a) Beneficial bioelectrochemical systems for energy, water, and biomass production, Microb. Biochem. Technol. S6:1–14.

Gude V.G, Kokabian B, & Gadhamshetty V. (2013b). Beneficial bioelectrochemical systems for energy, water, and biomass production. Journal of Microbial & Biochemical Technology, 6:2.

He Z, Huang Y, Manohar A. K, & Mansfeld F. (2008). Effect of electrolyte pH on the rate of the anodic and cathodic reactions in an air-cathode microbial fuel cell. Bioelectrochemistry, 74(1):78-82.

Holmes D.E, Bond D.R, Lovley D.R (2004) Electron Transfer by Desulfobulbus propionicus to Fe (III) and Graphite Electrodes Electron Transfer by Desulfobulbus propionicus to Fe (III) and Graphite Electrod. Appl. Environ. Microbiol. 70:1234–1237.

Huang G, Wang H, Zhao H, Wu P, Yan Q (2018) Application of polypyrrole modified cathode in bio-electro-Fenton coupled with microbial desalination cell (MDC) for enhanced degradation of methylene blue. J. Power Sources 400:350–359.

Imoro A. Z, Mensah M & Buamah R (2021). Developments in the microbial desalination cell technology: A review. Water-Energy Nexus, 4:76-87.

Jacobson K. S, Drew D. M & He Z (2011a). Efficient salt removal in a continuously operated upflow microbial desalination cell with an air cathode. Bioresource technology, 102(1):376–380.

Jacobson K.S, Drew D.M, He Z (2011b). Use of a liter-scale microbial desalination cell as a platform to study bioelectrochemical desalination with salt solution or artificial seawater. Environ. Sci. Technol. 45:4652–4657.

Jafary T, Daud WRW, Aljlil SA, Ismail AF, Al-Mamun A, Baawain MS (2018) Simultaneous organics, sulphate and salt removal in a microbial desalination cell with an insight into microbial communities. Desalination 445:204-212.

Jingyu H, Ewusi-Mensah D, Norgbey E (2017) Microbial desalination cells technology: A review of the factors affecting the process, performance and e_ciency. Desalin. Water Treat. 87:40–159.

Kalleary S, Mohammed Abbas F, Ganesan A, Meenatchisundaram S, Srinivasan B, Packirisamy A.S.B, Kesavan R.K, Muthusamy S (2014) Biodegradation and bioelectricity generation by Microbial Desalination Cell. Int. Biodeterior. Biodegrad. 92:20–25.

Karagiannis I.C, Soldatos P.G (2008) Water desalination cost literature: review and assessment. Desalination 223(1–3):448–456.

Kim H.J, Park H.S, Hyun M.S, Chang I.S, Kim M, Kim B.H (2002) A mediator-less microbial fuel cell using a metal reducing bacterium, Shewanella putrefaciens. Enzyme Microb. Technol. 30:145–152.

Kim J. R, Cheng S, Oh S. E, & Logan B. E. (2007). Power generation using different cation, anion, and ultrafiltration membranes in microbial fuel cells. Environmental science & technology, 41(3):1004-1009.

Kim Y, & Logan B. E. (2011a). Series assembly of microbial desalination cells containing stacked electrodialysis cells for partial or complete seawater desalination. Environmental science & technology, 45(13):5840–5845.

Kim Y, Logan BE (2011b) Hydrogen production from inexhaustible supplies of fresh and salt water using microbial reverse-electrodialysis electrolysis cells. Proceedings of the National Academy of Sciences 108:16176–16181.

Kim Y, Logan BE (2011c) Microbial reverse electrodialysis cells for synergistically enhanced power production. Environmental science & technology 45:5834–5839.

Kim S.K, Kim Y.C, Lee S, Kim J.C, Yun M.Y, Kim I.S (2011) Insecticidal activity of rhamnolipid isolated from Pseudomonas sp. EP-3 against green peach aphid (Myzus persicae) J. Agric. Food Chem. 59:934–938.

Kim Y & Logan B.E (2013a). Microbial desalination cells for energy production and desalination. Desalination, 308:122–130.

Kim Y, Logan B.E (2013b) Simultaneous removal of organic matter and salt ions from saline wastewater in bioelectrochemical systems. Desalination 308:115–121.

Kim E.J, Fathoni A, Jeong G.T, Jeong H.D, Nam T.J, Kong I.S, Kim J.K (2013) Microbacterium oxydans, a novel alginatev and laminarin-degrading bacterium for the reutilization of brown-seaweed waste. J. Environ. Manage. 130(10):153–159.

Kokabian B, Gude V.G (2013). Photosynthetic microbial desalination cells (PMDCs) for clean energy, water and biomass production. Environ. Sci.: Processes & Impacts 15:2178–2185.

Kumar R, Singh L, Zularisam A.W (2016). Exoelectrogens: recent advances in molecular drivers involved in extracellular electron transfer and strategies used to improve it for microbial fuel cell applications. Renew. Sustain. Energy Rev. 56:1322–1336.

Liang Y, Feng H, Shen D, Li N, Long Y, Zhou Y, Gu Y, Ying X, Dai Q (2016). A high-performance photo-microbial desalination cell. Electrochim. Acta 202:197–202

Liu C, Gallagher J.J, Sakimoto K.K, Nichols E.M, Chang C.J, Chang M.C, Yang P (2015) Nanowire–bacteria hybrids for unassisted solar carbon dioxide fixation to value-added chemicals. Nano Lett. 15:3634–3639.

Liu F, Luo S, Wang H, Zuo K, Wang L, Zhang X (2019a) Improving wastewater treatment capacity by optimizing hydraulic retention time of dual-anode assembled microbial desalination cell system. Separation and Purification Technology 226:39–47.

Liu F, Wang L, Zuo K, Luo S, Zhang X, Liang P (2019b) A novel operational strategy to enhance wastewater treatment with dual-anode assembled microbial desalination cell. Bioelectrochemistry 126:99–104.

Logan B.E (2008) Microbial fuel Cells. John Wiley and Sons Inc, New Jersey.

Logan B.E (2009) Exoelectrogenic bacteria that power microbial fuel cells. Nat. Rev. Microbiol. 7:375–381.

Logan B.E, Hamelers B, Rozendal R.A, Schroder U, Keller J, Freguia S, Aelterman P, Verstraete W, Rabaey K (2006) Microbial fuel cells: methodology and technology. Environmental Science and Technology 40(17):5181–5192.

Lu Y, He Z (2015) Mitigation of salinity buildup and recovery of wasted salts in a hybrid osmotic membrane bioreactor – electrodialysis system. Environ. Sci.Technol. 49 (17).

Luo H, Jenkins P. E, & Ren Z. (2011). Concurrent desalination and hydrogen generation using microbial electrolysis and desalination cells. Environmental science & technology, 45(1):340-344.

Luo H, Xu P, Jenkins P.E, Ren Z (2012a) Ionic composition and transport mechanisms in microbial desalination cells. Journal of membrane science 409:16–23.

Luo H, Xu P, Roane T.M, Jenkins P.E, Ren Z (2012b) Microbial desalination cells for improved performance in wastewater treatment, electricity production, and desalination. Bioresour. Technol.105:60–66.

M.T. Madigan, J.M. Martinko (2006) Brook Biology of Microorganisms, 11th ed. Pearson Education, Upper Saddle River, 2006.

Manzoor K, Khan S, Jamal Y, Shahzad M (2017) Heat extraction and brine management from salinity gradient solar pond and membrane distillation. Chem. Eng. Res. Des. 118: 226–237.

Mathioulakis E, Belessiotis V, Delyannis E (2007). Desalination by using alternative energy: review and state-of-art. Desalination 203:346–365.

Mehanna M, Kiely P. D, Call D. F, & Logan B. E. (2010a). Microbial electrodialysis cell for simultaneous water desalination and hydrogen gas production. Environmental science & technology, 44(24):9578–9583.

Mehanna M, Saito T, Yan J, Hickner M, Cao X, Huang X, & Logan B. E. (2010b). Using microbial desalination cells to reduce water salinity prior to reverse osmosis. Energy & Environmental Science, 3(8):1114–1120.

Meng F, Jiang J, Zhao Q, Wang K, Zhang G, Fan Q, Wei L, Ding J, Zheng Z (2014) Bioelectrochemical desalination and electricity generation in microbial desalination cell with dewatered sludge as fuel. Bioresource Technology 157:120-126.

Mengqi C, Yanan L, Xinshan S, Yuhui W, Zhang Y (2021) A new constructed wetland combined with microbial desalination cell and its application.

Mohanakrishna G, Venkata Mohan S, Sarma P.N (2010) Bio-electrochemical treatment of distillery wastewater in microbial fuel cell facilitating decolorization and desalination along with power generation. Journal of Hazardous Materials 177(1–3):487–494.

Morel A, Zuo K, Xia X, Wei J, Luo X, Liang P, Huang X (2012). Microbial desalination cells packed with ion-exchange resin to enhance water desalination rate. Bioresource Technology 118:43-48.

Park H.S, Kim B.H, Kim H.S, Kim H.J, Kim G.T, Kim M, Chang I.S, Park Y.K, Chang H.I (2001) A novel electrochemically active and Fe(III)-reducing bacterium phylogenetically related to Clostridium butyricum isolated from a microbial fuel cell. Anaerobe 7:297–306.

Ping Q, He Z (2013) Improving the flexibility of microbial desalination cells through spatially decoupling anode and cathode. Bioresour. Technol. 144:304–310.

Ping Q, & He Z. (2014). Effects of inter-membrane distance and hydraulic retention time on the desalination performance of microbial desalination cells. Desalination and Water Treatment, 52(7-9):1324-1331.

Ping Q, Abu-Reesh I.M, He Z (2015a) Boron removal from saline water by a microbial desalination cell integrated with Donnan dialysis. Desalination 376:55–61.

Ping Q, Cohen B, Dosoretz C, He Z (2013) Long-term investigation of fouling of cation and anion exchange membranes in microbial desalination cells. Desalination 325:48-55.

Ping Q, Huang Z, Dosoretz C, He Z (2015b) Integrated experimental investigation and mathematical modeling of brackish water desalination and wastewater treatment in microbial desalination cells. Water Research 77:13–23.

Ping Q, Zhang C, Chen X, Zhang B, Huang Z, He Z (2014) Mathematical Model of Dynamic Behavior of Microbial Desalination Cells for Simultaneous Wastewater Treatment and Water Desalination. Environmental Science & Technology 48:13010-13019.

Qian F, Wang H, Ling Y, Wang G, Thelen M.P, Li Y (2014). Photoenhanced electrochemical interaction between Shewanella and a hematite nanowire photoanode. Nano Lett. 14:3688–3693.

Qu Y, Feng Y, Liu J, He W, Shi X, Yang Q, Lv J, Logan B.E (2013) Salt removal using multiple microbial desalination cells under continuous flow conditions. Desalination 317:17-22.

Qu Y, Feng Y, Wang X, Liu J, Lv J, He W, & Logan B. E. (2012). Simultaneous water desalination and electricity generation in a microbial desalination cell with electrolyte recirculation for pH control. Bioresource technology, 106:89-94.

Rabaey K (2009) Bioelectrochemical systems: a new approach towards environmental and industrial biotechnology. Bioelectrochemical systems: From extracellular electron transfer to biotechnological application. IWA Publishing, 1-16.

Rabaey K, Boon N, Hofte M, Verstraete W (2005) Microbial phenazine production. enhances electron transfer in biofuel cells. Environ. Sci. Technol. 39(9):3401–3408.

Rabaey K, Nico B, Steven D.S, Marc V, Marc V, Verstraete W (2004) Biofuel Cells Select for Microbial Consortia That Self-Mediate Electron Transfer Korneel. Appl. Environ. Microbiol. 70:5373–5382.

Rikame S.S, Mungray A.A, Mungray A.K (2012) Electricity generation from acidogenic food waste leachate using dual chamber mediator less microbial fuel cell. Int. Biodeterior. Biodegradation 75:131–137.

Rozendal R.A, Hamelers H.V, Rabaey K, Keller J, Buisman C.J (2008) Towards practical implementation of bioelectrochemical wastewater treatment. Trends in Biotechnology 26 (8):450–459.

Rozendal R.A, Hamelers H.V.M, Buisman C.J.N (2006) Effects of membrane cation transport on pH and microbial fuel cell performance. Environmental Science and Technology 40(17):5206–5211.

Rusyn I (2021) Role of microbial community and plant species in performance of plant microbial fuel cells. Renew. Sustain. Energy Rev. 152:111697.

Saeed H.M, Husseini G.A, Yousef S, Saif J, Al-Asheh S, Abu Fara A, Azzam S, Khawaga, R, Aidan A (2015) Microbial desalination cell technology: A review and a case study. Desalination 359:1–13. https://doi.org/10.1016/j.desal.2014.12.024

Salehmin M.N.I, Lim S.S, Satar I. and Daud W.R.W (2021) Pushing microbial desalination cells towards field application: Prevailing challenges, potential mitigation strategies, and future prospects. Science of the Total Environment, 759, p.143485.

Shannon M.A, Bohn P.W, Elimelech M, Georgiadis J.G, Marinas B.J, Mayes A.M (2008) Science and technology for water purification in the coming decades. Nature 452 (7185): 301–310.

Shehab N.A, Logan B.E, Amy G.L, Saikaly P.E (2013) Microbial Electrodeionization Cell Stack for Sustainable Desalination, Wastewater Treatment and Energy Recovery. Proceedings of the Water Environment Federation 222–227.

Shinde O.A, Bansal A, Banerjee A, Sarkar S (2018) Bioremediation of steel plant wastewater and enhanced electricity generation in Microbial desalination cell. Water Sci. Technol. 77: 2101–2112.

Stoll Z.A, Forrestal C, Ren Z.J, Xu P (2015). Shale gas produced water treatment using innovative microbial capacitive desalination cell. Journal of hazardous materials 283:847-855.

Strathmann H. (2004). Ion-exchange membrane separation processes. Elsevier.

Utami T.S, Arbianti R, Manaf B.N (2015) Sea water desalination using Debaryomyces hansenii with microbial desalination cell technology. Int. J. Technol. 6:1094–1100.

Wang Y, Xu A, Cui T, Zhang J, Yu H, Han W, Shen J, Li J, Sun X, Wang L (2020) Construction and application of a 1-liter upflow-stacked microbial desalination cell. Chemosphere 248:126028

Wen Q, Zhang H, Chen Z, Li Y, Nan J, & Feng Y. (2012). Using bacterial catalyst in the cathode of microbial desalination cell to improve wastewater treatment and desalination. Bioresource Technology, 125:108-113.

Wen Q, Zhang H, Yang H, Chen Z, Nan J, Feng Y (2014) Improving desalination by coupling membrane capacitive deionization with microbial desalination cell. Desalination 354:23–29.

Werner C.M, Logan B.E, Saikaly P.E, & Amy G. L. (2013). Wastewater treatment, energy recovery and desalination using a forward osmosis membrane in an air-cathode microbial osmotic fuel cell. Journal of membrane science, 428:116-122.

Won DH, Choi CH, Chung J, Woo SI (2014) Photoelectrochemical production of formic acid and methanol from carbon dioxide on metal-decorated CuO/Cu2O-layered thin films under visible light irradiation. Applied Catalysis B: Environmental 158:217- 223.

Ye B, Luo H, Lu Y, Liu G, Zhang R, Li X (2017) Improved performance of the microbial electrolysis desalination and chemical-production cell with enlarged anode and high applied voltages. Bioresource Technology 244:913-919

Yuan H.Y, Abu-Reesh I.M, He Z (2015) Enhancing desalination and wastewater treatment by coupling microbial desalination cells with forward osmosis. Chem. Eng.J. 270:437–443.

Yuan L, Yang X, Liang P, Wang L, Huang Z. H, Wei J, & Huang X. (2012a). Capacitive deionization coupled with microbial fuel cells to desalinate low-concentration salt water. Bioresource technology, 110:735–738.

Yuan L, Yang X, Liang P, Wang L, Huang Z-H, Wei J (2012b) Capacitive deionization coupled with microbial fuel cells to desalinate low-concentration salt water. Bioresource Technology 110:735–738.

Yuan Z.G, Pratt S, Batstone D.J (2012c) Phosphorus recovery from wastewater through microbial processes. Curr. Opin. Biotech. 23:878–883.

Zang G.-L, Sheng G.-P, Shi C, Wang Y.-K, Li W.-W, Yu H.-Q (2014) A bio-photoelectrochemical cell with a MoS 3-modified silicon nanowire photocathode for hydrogen and electricity production. Energy Environ. Sci. 7:3033–3039.

Zhang B, & He Z. (2012a). Energy production, use and saving in a bioelectrochemical desalination system. RSC advances, 2(28):10673–10679.

Zhang B, & He Z. (2012b). Integrated salinity reduction and water recovery in an osmotic microbial desalination cell. Rsc Advances, 2(8):3265–3269.

Zhang C, Liu G, Zhang R, Luo H (2010a) Electricity production from and biodegradation of quinoline in the microbial fuel cell. J. Environ. Sci. Heal. Part A Toxic/Hazardous Subst. Environ. Eng. 45:250–256.

Zhang F, Brastad K.S, & He Z. (2011). Integrating forward osmosis into microbial fuel cells for wastewater treatment, water extraction and bioelectricity generation. Environmental science & technology, 45(15):6690-6696..

Zhang F, Chen M, Zhang Y, Zeng R.J (2012) Microbial desalination cells with ion exchange resin packed to enhance desalination at low salt concentration. Journal of Membrane Science 417–418, 28-33.

Zhang F, He Z (2015) Scaling up microbial desalination cell system with a post-aerobic process for simultaneous wastewater treatment and seawater desalination. Desalination 360:28–34.

Zhang T, Cui C, Chen S, Ai X, Yang H, Shen P, Peng Z (2006) A novel mediatorless microbial fuel cell based on direct biocatalysis of Escherichia coli. Chem. Commun. 2006: 2257–2259.

Zhang X, Cheng S, Huang X, Logan BE (2010b) Improved performance of single-chamber microbial fuel cells through control of membrane deformation. Biosensors and Bioelectronics 25:1825–1828.

Zhang Y, Angelidaki I (2013) A new method for in situ nitrate removal from groundwater using submerged microbial desalination–denitrification cell (SMDDC). Water Res. 47 (5): 1827–1836.

Zhang W, Wei C, Yan B et al (2013) Identification and removal of polycyclic aromatic hydrocarbons in wastewater treatment processes from coke production plants. Environ. Sci. Pollut. Res. 20:6418–6432.

Zhang Y.F and Angelidaki I (2015a) Submersible microbial desalination cell for simultaneous ammonia recovery and electricity production from anaerobic reactors containing high levels of ammonia. Bioresource Technology 177:233–239.

Zhang Y.F. and Angelidaki I (2015b) Counteracting ammonia inhibition during anaerobic digestion by recovery using submersible microbial desalination cell. Biotechnology and Bioengineering 112(7):1478–1482.

Zhang, Y.F. and Angelidaki, I. (2015c). Bioelectrochemical recovery of waste-derived volatile fatty acids and production of hydrogen and alkali. Water Research 81:188–195.

Zhao S, Zou L, Tang C. Y, & Mulcahy D. (2012). Recent developments in forward osmosis: Opportunities and challenges. Journal of membrane science, 396:1-21.

Zuo K, Liu F, Ren S, Zhang X, Liang P, Huang X (2016) A novel multi-stage microbial desalination cell for simultaneous desalination and enhanced organics and nitrogen removal from domestic wastewater. Environ. Sci. Water Res. Technol. 2:832–837.

Zuo K, Yuan L, Wei J, Liang P, Huang X (2013). Competitive migration behaviors of multiple ions and their impacts on ion-exchange resin packed microbial desalination cell. Bioresource Technology 146:637-642

Zuo K.C, Cai J.X, Liang, S, Wu S.J, Zhang C.Y, Liang P, Huang X (2014) A ten liter stacked microbial desalination cell packed with mixed ion-exchange resins for secondary effluent desalination. Environ. Sci. Technol. 48:9917–9924.

Zuo K.C, Chen M.S, Liu F.B, Xiao K, Zuo J.L, Cao X.X, Zhang X.Y, Liang P, Huang X (2018) Coupling microfiltration membrane with biocathode microbial desalination cell enhances advanced purification and long-term stability for treatment of domestic wastewater. J. Membr. Sci. 547:34–42.

Zuo Y, Xing D, Regan J.M, Logan B.E (2008) Isolation of the exoelectrogenic bacterium Ochrobactrum anthropic YZ-1 by using a U-tube microbial fuel cell. Appl. Environ. Microbiol. 74:3130–3137.

Material Models for Finite Element Analysis of Soft Tissues

Nishank Verma and Mythravaruni Pullela

1 Introduction

Any solid body part consisting from the bones to the cell can be considered a living tissue. Based on the soft tissue's high flexibility and soft mechanical properties, they can be distinguished from the other kinds of tissues available in the body. They connect, surround, and support the internal organs and bones. These include fibrous tissues, tendons, ligaments, muscles, fibrous tissues, blood vessels, lymph, etc. Soft tissues are non-homogeneous, are multiphasic, and exhibit nonlinear, time-dependent behaviour. They are generally considered as the complex structures of the body in the aspect of mechanical analysis. Looking toward the Finite Element Analysis (FEA) provides an upper edge for the investigation of the mechanical properties of the soft tissues.

2 Properties of Different Soft Tissues

2.1 Cartilage

Articular cartilage comprises 68–85% water, 10–20% of type-II collagen fibrils, and the remaining proteoglycans, glycoproteins, and cartilaginous cells. About 30% of cartilage water is primarily found in the interstitial fluid. During compression, cartilage experiences fluid movement from the tissue into the surrounding solution,

N. Verma (✉)
School of Nanoscience and Technology, Indian Institute of Technology Kharagpur, Kharagpur 721302, India
e-mail: nishank.verma@kgpian.iitkgp.ac.in

M. Pullela
Rubber Technology Centre, Indian Institute of Technology Kharagpur, Kharagpur 721302, India

J. Gupta and A. Verma (eds.), *Microbiology-2.0 Update for a Sustainable Future*,
https://doi.org/10.1007/978-981-99-9617-9_19

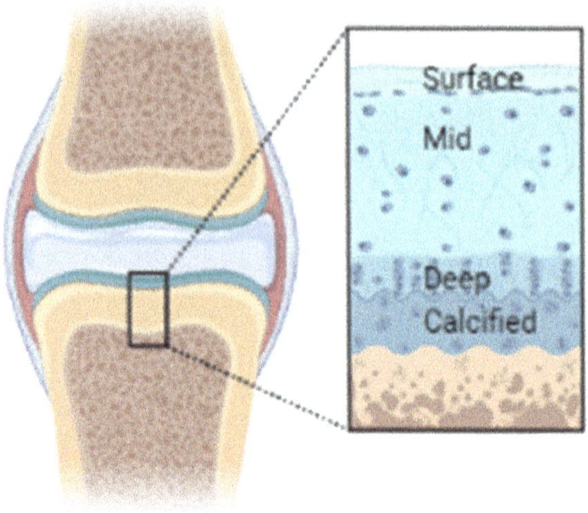

Fig. 1 Cartilage structure

and when the load is removed, fluid flows back to the tissue, analogous to the osmosis process. The friction between the fluid and the tissue is responsible for the viscoelastic behaviour and the energy dissipation mechanism. The structure of cartilage can be divided into superficial, middle, deep, and calcified, as shown in Fig. 1 (Hong et al. 2015).

Cartilage exhibits anisotropic behavior due to the presence of four distinct divisions with varying structural characteristics throughout its depth. Joint cartilages come into contact with one another. Electrostatic charges, fluid, and proteoglycan molecules combine to create lubrication and stress absorption. The friction coefficient between two cartilage surfaces can be as low as 0.002 (Malcom 1976), which offers the well working of joint. Studies have shown that the mechanical properties of cartilage depend on anatomical regions. For example, femoral cartilage is stiffer than tibial cartilage (Li et al. 2021). Another study found that human costal cartilage has anisotropic elastic behaviour and significant age-related changes (Weber et al. 2021).

2.2 *Ligaments and Tendons*

Tendons are flexible, fibrous structures that attach muscle to bone. Their main function is to transfer the force produced by the muscles to the bone, making movement around a joint easier. The majority of tendon tissue is composed of collagen, which is arranged hierarchically parallel to the tendon's long axis and gives it a high tensile strength (Thorpe and Screen 2016). Type I collagen fibres dominate the tendons and ligaments' structure, encased in a web of loose connective tissue (Zschäbitz 2005). The primary function of the tendon and ligament can be understood from the Fig. 2a.

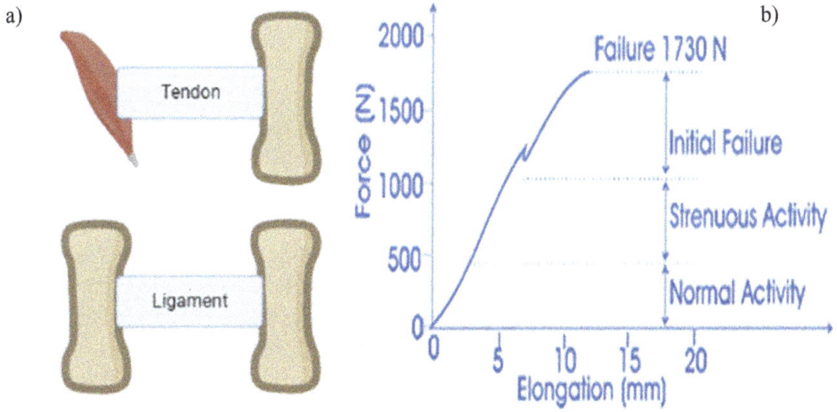

Fig. 2 **a** The primary function of tendons and ligaments; **b** Force–elongation of ligament during the tensile test

Despite sharing the same basic building blocks, tendons and ligaments have different compositions and arrangements of the matrix macromolecules that give each structure the unique mechanical characteristics it needs to function properly (Rumian et al. 2007). Soft tissue classification and function are significantly influenced by its structure rather than its chemical makeup. Variables like temperature, strain rate, hydration, maturation, immobilization, aging, exercise, and healing might affect the characteristics of these tissues.

Tendons exhibit viscoelastic behaviour and nonlinear "J"-shaped stress–strain curves. Tendons are least stiff (lowest slope) at low stresses and stiffer under high loads (stresses between 0 and 20 MPa). Tendons maintain a reasonably constant modulus (1.2–1.5 GPa) with higher loads and strains (from 0.02 to 0.09) (Biewener 2008). Tendons are elastic tissues that attach muscles to the skeleton and transmit force proportionally to their degree of stretch. Since it is challenging to evaluate the mechanical characteristics of human tendons non-invasively, generic values are frequently considered in musculoskeletal models to represent all participants (Morrison et al. 2015). Force elongation during the tensile test has been shown in Fig. 2b (Pal 2014).

Ligaments show time-dependent and history-dependent viscoelastic properties, meaning their mechanical properties are affected by load conditions and frequent intense activities. They are also sensitive to temperature changes, with peak stresses increasing as temperature decreases. The "crimp pattern" of ligaments allows for easy joint movement within a specific range, but resistance to elongation increases significantly once the joint is moved beyond its normal range of motion (Kastelic et al. 1978).

2.3 *Skeletal Muscles*

Muscles are unique soft tissues that can contract and act as flexible shortening actuators. Skeletal, smooth, and cardiac muscles are the three categories of muscles. While they share comparable anatomical and mechanical characteristics, their orientation, strength, size, and activation characteristics vary (Fung 2013). The strongest muscles are those found in the skeleton, which are connected to the bones. Smooth muscles are associated with organs and soft tissues, while cardiac muscles serve as a pump. Muscle fibres are arranged in parallel, between tendons of origin and insertion, with few muscles having a layout of fibres at an acute angle to the tendons called pennation (Zajac 1989). Muscle is made of fascicles containing bundles of fibres, which consist of parallel bundles of myofibrils. The structure of skeletal muscle is shown in Fig. 3. The myofibril is the primary muscle fibre, consisting of sarcomeres with actin and myosin myofilaments. The muscle fibre is about 100 μm in diameter, the myofibril about 1 μm, and the myofilaments about 100 Å (Winter 2009). The three types of muscle connective tissue are the endomysium, which surrounds each individual muscle fibre, the perimysium, which surrounds bundles of muscle fibres, and the epimysium, which surrounds the entire muscle. The sarcomere is the basic contractile unit of muscle, with thick filaments mainly composed of myosin molecules arranged so that a bulbous region projects out from the filament. These regions momentarily combine with thin actin filaments, creating a shearing force that shortens the sarcomere.

The contractile element shortens, and the series elastic element somewhat lengthens during isometric contractions. Tension increases non-linearly in the parallel part as the muscle lengthens. The loss of tension that occurs as the cross-bridges in the contractile element break and reform in a shorter condition and passive damping brought on by fluid viscosity in the contractile element and connective tissues cause a muscle to lose tension as it shortens under stress (Winter 2009).

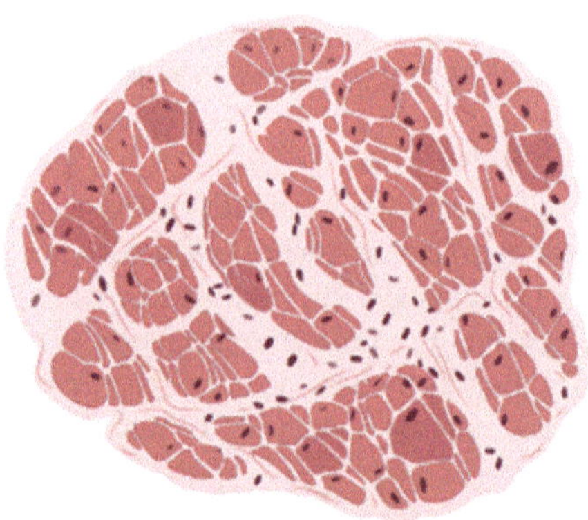

Fig. 3 Typical structure of skeletal muscle

3 Soft Tissue Constitutive Models

Soft tissues are known to have a complex and multiphasic structure, as was covered above in Sect. 2 of the chapter. Depending on the type of soft tissue and the body component it belongs to, the soft tissue's physical characteristics change. The earliest anatomical model was made in India in 800 B.C., where a flap of skin was utilized to recreate the nose. Before the development of imaging technology, a century ago, the only way to see inside a body was through an operation or a dissection. High-end graphics that enable the simulation of medical operations and body parts have been introduced as a result of the rapid advancements in hardware and software over the past ten years. An in-depth knowledge of the fundamental physical laws is required to compute natural tissue forces, interactions, and deformations using the continuum mechanics approach. Soft biological tissues have complicated properties, including significant deformations, highly nonlinear stress–strain relationships, heterogeneity, mechanical anisotropy and viscoelasticity, which make it difficult to simulate their mechanical behaviour (Fung 2013). Due to this complexity, simplistic material models are insufficient for accurate simulations, and various strains applied to the tissue must be considered.

In terms of biomechanics, simulating soft tissue entails creating a constitutive equation that connects the material's load to the deformation it undergoes. Therefore, the tissue's stress may be predicted using this relationship, which can also mimic the deformation of the tissue under external stress. It is crucial to pick the right factors to consider during the modelling process. It is crucial to guarantee that the models are accurate outside the scope of the tests for a 3D application that enables the modelling of any mechanical event on the tissues. This section will describe the constitutive equations, governing model equations, and their parameters relating tissue strain to tissue stress in light of the different material models as per the application area.

3.1 Strain Energy Density Function (SEDF)

SEDF represents the stress–strain properties of hyperelastic materials. Finite element analysis has proven to be beneficial in solving the mechanical problems related to soft tissues by discretization and application of SEDF to each element. Different ways of applying SEDF in soft tissues, such as analytical forms or using data-driven and machine learning (ML) approaches, are available. Analytical forms are generally based on experimental data/observations. At the same time, there is a recent data-driven approach called the Gaussian process (GP), which is a probabilistic model fitting in experimental data for providing estimated uncertainty, which can be applied to the planar, hyperelastic, incompressible soft tissues (Aggarwal et al. 2023). Consider a deformable body that undergoes deformation characterised by a deformation gradient $\boldsymbol{F} = \frac{\partial \tilde{x}}{\partial \tilde{X}}$, where $\tilde{X}$ and $\tilde{x}$, are the position vectors in the original configuration, and deformed configuration respectively (Weinberg and Kaazempur-Mofrad 2006). Hyperelasticity of material is evidenced by the existence of such an SEDF (W), which can be used to express stress components:

$$\boldsymbol{P} = \frac{\partial W}{\partial \boldsymbol{F}} \boldsymbol{F}^{\mathrm{T}} \text{ using the first Piola-Kirchoff stress tensor, or} \tag{1}$$

$$\boldsymbol{\sigma} = \frac{1}{J} \boldsymbol{P} \boldsymbol{F}^{-\mathrm{T}} \text{ using the Cauchy stress tensor,} \tag{2}$$

where **P** is the first Piola-Kirchoff stress tensor, W is the strain energy density function, $\boldsymbol{\sigma}$ is the Cauchy stress tensor, and J is the determinant of **F**, depending on the complexity of W, the features like nonlinearity and anisotropy can be included in the model.

3.2 *Linear Elastic Model*

The stress–strain relationship of soft tissues is assumed to be linear, isotropic, and independent of the deformation history in a simplified model known as the linear elastic model of soft tissues. This demonstrates that the mechanical response is independent of the loading or unloading path, the stress is proportionate to the strain, and the material properties are the same in all directions (Korhonen and Saarakkala 2011). Two material parameters, such as Young's modulus and Poisson's ratio, which can be derived from experimental data, can provide a linear elastic model of soft tissues. The model can be used where the soft tissues show small deformations without showing anisotropic and viscoelastic behaviour (Gladilin et al. 2001). Instead of the Green strain tensor, the linear elastic model employs the Green-St Venant strain tensor, which is defined as:

$$\boldsymbol{E}_{l(linear)} = \frac{1}{2}\left(\nabla \tilde{U} + \nabla \tilde{U}^{T}\right) \text{ Green-St Venant strain tensor} \tag{3}$$

where $\tilde{U}$ is the displacement vector given by $\tilde{x} - \tilde{X}$. Hooke defined the elastic energy of a linearly elastic deformable object which is obtained from a displacement vector (Knops and Ciarlet 1994)

$$W = \frac{\lambda}{2}(tr\boldsymbol{E})^2 + \mu tr\boldsymbol{E}^2 \tag{4}$$

with λ and μ beingthe Lame's coefficients characteristic of the material's stiffness. These parameters are related to Young modulus (E) and Poisson's ratio (ν) by Eqs. (5) and (6).

$$\lambda = \frac{E\nu}{(1+\nu)(1-2\nu)} \tag{5}$$

$$\mu = \frac{E}{2(1+\nu)} \tag{6}$$

The second Piola-Kirchoff stress tensor (S), sometimes also called as pull-back of two point and spatial stress tensors, is defined as

$$\boldsymbol{S} = \lambda(tr\boldsymbol{E})\boldsymbol{I} + 2\mu\boldsymbol{E} \text{ and,} \tag{7}$$

the constitutive equation

$$S_{ij} = D_{ijkl}E_{kl} \text{ relating } \mathbf{S} \text{ with } \boldsymbol{E}. \tag{8}$$

where D is the 4th order tensor with 81 elastic constants, which gets reduced to 21 constants for anisotropic material due to the symmetry of strain tensor, stress tensor, and D (Bonet and Wood 1997; Picinbono et al. 2001). The physical behaviour of soft tissues is not generally linearly elastic as the linear elastic model is for small deformations and displacements, which limits its application.

Certain assumptions are there for the linear elastic model, such as.

- The response is directly proportional to the load applied for which certain conditions must be satisfied. The stress–strain curve has a linear range, where the straight line starts from the origin. The material's behaviour becomes nonlinear when the stress increases beyond a certain point. This assumption states that the stress should remain below this point, and the maximum displacement is considerably smaller than the characteristic dimension of the model.
- The loads do not bring on permanent deformation. In other words, it is presumed that the model is perfectly elastic. The removal of the loads causes a perfectly elastic model to resume its original shape.

As discussed above, soft tissues often do not behave in a linear fashion, and the deformation condition for the linear elastic behaviour is not satisfied. Due to the limitation of the linear elastic model, researchers have moved forward and created nonlinear elastic models that can accurately predict significantly large strains and deformations.

3.3 Nonlinear Elastic Model

The basis for the nonlinear elastic model is that the relation between the soft tissues' stress and strain is not linear. There are many ways to represent this phenomenon of the soft tissues, but the most common way is by expressing it in the form of strain energy density function. Nonlinear elastic models are based on the microstructure, which has been adopted in recent studies of soft tissues. Due to features and most importantly, the consideration of more realistic conditions in the nonlinear elastic models, it is supposed to provide more accurate results in comparison to the linear elastic model. These models also have applications in imaging and simulation, such as image registration (Gladilin and Eils 2008), deformation study of the soft tissues (Picinbono et al. 2001), and simulation of surgicalprocedures (Misra et al. 2010). The vast and critical area applications are only possible due to including realistic factors and conditions during modelling.

Nonlinear stress–strain behaviour is generally treated as a hyperelastic material with a low elastic modulus and a high bulk modulus. Different models can be employed to study soft tissues, such as the Ogden model, Mooney-Rivlin model, Neo-Hookean, and so on.

3.3.1 Mooney-Rivlin Model

Mooney-Rivlin is a hyperelastic model that describes complex materials' nonlinear behaviour, including soft tissues. The model was proposed by Mooney in 1940 (Mooney 1940) and later expressed in terms of invariants by Ronald Rivlin in 1948 (Rivlin 1948). This model's (SEDF) is a linear combination of the two invariants of the left Cauchy-Green deformation tensor. The strain measures for the hyperelastic model are defined in terms of the Cauchy-Green deformation tensor derived from the deformation gradient (Hackett and Hackett 2018). The right Cauchy-Green deformation tensor or Green deformation tensor '**C**' or related to the deformation gradient '**F**' by

$$\mathbf{C} = \mathbf{F}^{\mathrm{T}}\mathbf{F} \text{ (Hackett and Hackett 2018)} \tag{9}$$

and the left Cauchy-Green tensor or Finger tensor 'B' is related to the deformation gradient by

$$\mathbf{B} = \mathbf{F}\mathbf{F}^{\mathrm{T}} \text{ (Hackett and Hackett 2018)} \tag{10}$$

In 1948, Mooney-Rivlin defined the SEDF of the material as

$$W = C_{10}(I_1 - 3) + C_{01}(I_2 - 3) \text{ (Mooney 1940)} \tag{11}$$

Later in 1948, Rivlin contributed to the Mooney postulate by expressing it in terms of the invariants of the left Cauchy-Green tensor. Where C_{01} and C_{10} are the experimentally determined material constants, and I_1 and I_2 are the first and second invariants of the left Cauchy-Green deformation tensor. In order to find out the material constants for soft tissues using the Mooney-Rivlin model, it is necessary to conduct mechanical tests on the specific soft tissue whose behaviour is being modelled. These tests will provide data forthe tissue's stress–strain behaviour, which can then be used to determine the material constants C_{10} and C_{01} through a process of curve-fitting or optimization using the Mooney-Rivlin model.

The Mooney-Rivlin model can also be represented in different versions of two-parameter, three-parameter, five-parameter, and nine-parameter models (Sussman and Bathe 1987). Two-parameter model is described in Eq. 11 where, C_{10} and C_{01} are the stiffness of material at small strain and nonlinearity of material at large strain. Mooney-Rivlin Model can be further represented with different numbers of parameters as:

$$W = C_{10}(I_1 - 3) + C_{01}(I_2 - 3) + C_{11}(I_1 - 3)^2 \text{ three-parameter model} \quad (12)$$

$$W = C_{10}(I_1 - 3) + C_{01}(I_2 - 3) + C_{11}(I_1 - 3)^2 + C_{20}(I_1 - 3)^2 + C_{02}(I_2 - 3)^2 \text{ five-parameter model} \quad (13)$$

$$\begin{aligned} W = {} & C_{10}(I_1 - 3) + C_{01}(I_2 - 3) + C_{11}(I_1 - 3)^2 + C_{20}(I_1 - 3)^2 + C_{02}(I_2 - 3)^2 \\ & + C_{21}\left[(I_1 - 3)^2 * (I_2 - 3)\right] + C_{12}\left[(I_1 - 3) * (I_2 - 3)^2\right] \\ & + C_{30}(I_1 - 3)^3 + C_{03}(I_2 - 3)^3 \quad \text{nine-parameter model} \end{aligned} \quad (14)$$

The constants in the model represent the material's stiffness, and nonlinearity. Additional material constants are added to higher-order Mooney-Rivlin models to represent the contribution of higher-order terms to the strain energy density function. These extra constants make it possible to represent the nonlinear stress–strain behaviour of complicated materials more precisely.

3.3.2 Neo-Hookean Model

In 1948, Ronald Rivlin proposed the hyperelastic neo-Hookean model. This model can be used to predict the nonlinear behaviour of the materials undergoing large deformation. This model is based on the statistical thermodynamics of the crosslinked polymer chains, making it promising for rubber-like materials and biological tissues. In contrast to linear elasticity, the neo-Hookean model does not have a linear stress–strain connection, which presumes a linear relationship between stress and strain. Instead, it applies a strain energy function to explain how the material's energy is stored when it deforms.

The neo-Hookean's SEDF can be defined for the compressible (Belytschko et al. 2014) and incompressible materials (Rivlin 1948) as:

$$W = \frac{\mu}{2}\left(\frac{1}{2}(tr(\boldsymbol{F}\boldsymbol{F}^T) - 3) - \ln(J)\right) + \frac{\kappa}{2}(\ln(J))^2 \text{ compressible material} \quad (15)$$

$$W = \frac{\mu}{2}(I_1 - 3) \text{ incompressible material} \quad (16)$$

where μ is the material constant/initial shear modulus, I_1 is the first invariant (trace) of the Cauchy-Green tensor, **F** is the deformation gradient, $J = det(\boldsymbol{F})$, and κ is the bulk modulus of the material. Anisotropy in soft tissues can be accounted for by altering the neo-Hookean model by using a soft-composite-based hyperelastic model (Chanda et al. 2020), a hyperelastic model for anisotropic tissue characterization (Chanda and Singh 2023), and a modified anisotropic model (Nolan et al. 2014).

3.3.3 Ogden Model

Ogden in 1972 introduced the model, which is today referred to as the Ogden model, which is (a) accurate for isotropic, incompressible rubber-like material under large deformations; (b) more susceptible to mathematical analysis than models based on strain invariants (Ogden 1972). The Ogden model is a general hyperelastic model for studying the nonlinear behaviour of a wide range of soft tissues. In the past decades, it has emerged as a promising model in the biomechanics of soft tissues, as soft tissues violate the assumption of perfect elasticity (Lohr et al. 2022; Weickenmeier and Jabareen 2014). The Ogden model helps in deducing the SEDF, which can be used to calculate the stress and deformation of the soft tissues under different loading conditions; using the principal stretches as independent variables provides a more accurate representation of the behaviour of soft tissues (Lohr et al. 2022).

The strain energy density function for the Ogden model (Ogden 1972) is represented in a mathematical form that is based on the Helmholtz free energy per unit reference volume:

$$W(\lambda_1, \lambda_2, \lambda_3) = \sum_{i=1}^{n} \frac{\mu_i}{\alpha_i} \left[\lambda_1^{\alpha_i} + \lambda_2^{a_i} + \lambda_3^{\alpha_i}\right] \tag{17}$$

where $\lambda_1, \lambda_2, \lambda_3$ are the principal stretches and μ_i and α_i are the Odgen material parameters. 'n' represents the order of the Ogden model. μ_i and α_i are dependent on the material to be modeled, the number of material constants, and the order of the model. The material constants in the Ogden model are determined by adjusting the model's parameters to match experimental data obtained from the specific material being studied. As part of the fitting procedure, the values of μ_i and α_i are changed to reduce the error between the model predictions and the experimental data. With the higher order of the model, more accuracy can be obtained, and for determining these material constants more experimental data under different loading conditions will be required.

The original Ogden model is unsuited for tissues with an organized microstructure, particularly those with collagen. Such types of tissues have anisotropic behaviour. There are assumptions on which the model works, such as homogeneity of the material, the material is assumed to be isotropic, incompressible, and strain rate independent (Lohr et al. 2022). The Ogden model can be used to model the collagenous tissues, but for those with axial in vivo loading and thus can be analyzed in 1-D, like

Table 1 Comparing the three nonlinear hyperelastic models

Model	Strain energy density function	Advantages
Ogden model	Expressed by the principal stretch ratios	Versatile and provides better agreement with test data
Mooney-Rivlin model	Developed from invariants of the left Cauchy-Green deformation tensor	Allows a simple definition of temperature-dependent shear modulus
Neo-Hookean model	Developed from invariants of the left Cauchy-Green deformation tensor	Simple, as only two parameters are required

tendons (Cheng and Gan 2008a), ligaments (Cheng and Gan 2008b), and tendineae (Smith et al. 2021).

Table 1 compares the strain energy density function and the possible applications of the three nonlinear hyperelastic models for their application. The Ogden model is a better choice than the Mooney-Rivlin model for studying soft tissues because the stretch ratio component of the Ogden model is limited to natural numbers, which is more realistic for soft tissues. However, since the number of parameters required in the Ogden model is more, it makes the computational process slower and more complex.

3.4 Viscoelastic Models

If a material can regain its shape once external forces are removed, it is considered to be (hyper)elastic. In addition, regardless of the way in which material is loaded, the material's deformation or stress pattern remains constant for elastic material (Bonet and Wood 1997). On the other hand, Viscoelastic materials behave in a time-dependent manner, indicating that the history of the deformation impacts how they react. As a result, when the applied forces are released, a viscoelastic object does not instantly assume its original shape. The simplest of the available models are based on the assumption of the linear elastic model, even though there is substantial evidence suggesting the viscoelastic behaviour of the soft tissues (Leeman and Jones 2009). Several viscoelastic models are available to represent soft tissues, such as the Maxwell, Kelvin-Voigt, Kelvin, and quasi-linear viscoelastic models which are discussed below.

3.4.1 Maxwell Model

In 1867, Maxwell established the model constructed using the spring and dashpot in series (Maxwell 1867). The spring and dashpot help represent the tissue's elastic and viscous components respectively, as shown in Fig. 4. To describe the model

Fig. 4 Maxwell model

mathematically, it is required to formulate the constitutive equation relating the stress and strain of the material.

The total stress and strain of the Maxwell component can be determined using the below equations

$$\sigma = \sigma_e = \sigma_v \tag{18}$$

$$\varepsilon = \varepsilon_e + \varepsilon_v \Rightarrow \dot{\varepsilon} = \dot{\varepsilon}_e + \dot{\varepsilon}_v \tag{19}$$

where σ_e of the spring and σ_v for the dashpot.

And it is possible to write in the first derivative form based on the component's coefficient as

$$\dot{\varepsilon} = \frac{\dot{\sigma}}{E} + \frac{\sigma}{\eta} \tag{20}$$

To identify the Maxwell model's constitutive elements, such as creep compliance and relaxation modulus, Eq. 20 is integrated for $\sigma = \sigma_0$ and by implementing the creep compliance definition to the above equation, it gives the creep compliance (J) for the Maxwell model as

$$J(t) = \frac{\varepsilon(t)}{\sigma_0} = \frac{1}{E} + \frac{t}{\eta} \tag{21}$$

When the material is subjected to constant strain, the Maxwell model leads to Eq. 22, which gives the stress relaxation response.

$$\sigma(t) = \sigma_0 e^{(-t/\lambda)}; \lambda = \frac{\eta}{E} \tag{22}$$

where λ is the relaxation time.

This stress–strain relationship can be used in the FEA framework to mimic the behaviour of soft tissues over time while considering both instantaneous elastic and time-dependent viscous responses.

Notably, the Maxwell model assumes linear viscoelasticity and is the simplest model that captures the stress relaxation response in viscoelastic fluids. However, in fact, soft tissues frequently display nonlinear and time-dependent behaviour characteristic of viscoelastic solids. Due to these distinct viscoelastic characteristics of soft tissues, such as their elastic modulus and viscosity coefficient, the parameters in the Maxwell model for soft tissues are different from those for other materials. By including more material properties and using more sophisticated viscoelastic models, such as the Prony series or the linear viscoelastic solid model, these complex behaviours can be described.

3.4.2 Kelvin-Voigt Model

The Kelvin-Voigt model is a parallel spring and dashpot viscoelastic model appropriate for soft tissues, as shown in Fig. 5. In the nineteenth century, Lord Kelvin and Woldemar Voigt proposed it (Voigt 1888). Soft tissues' rubbery characteristics and energy dissipation can be described, but not stress relaxation and finite deformation responses.

The strains in each of the two model components are the same because they are connected in parallel:

$$\varepsilon = \varepsilon_e = \varepsilon_v \tag{23}$$

and the total stress in the system is the sum of the stress developed in each component.

$$\sigma = \sigma_e + \sigma_v \tag{24}$$

The governing equation of the Kelvin-Voigt model can be represented in terms of stress, strain, and their rate of change with respect to time:

$$\sigma(t) = E\varepsilon(t) + \eta \frac{d\varepsilon(t)}{dt} \tag{25}$$

where E is the modulus, and the η is the viscosity.

The creep function $D(t)$ and relaxation function $G(t)$, explain how the strain varies over time when a constant stress σ_0 is applied, and how the stress changes over time when a constant ε_0 is applied, respectively, that define the Kelvin-Voigt model. For the Kelvin-Voigt model, the creep function and relaxation function are provided by:

$$D(t) = \frac{\sigma_0}{E} + \frac{\sigma_0}{\eta} t \tag{26}$$

$$G(t) = Ee^{-\frac{E}{\eta}t} \tag{27}$$

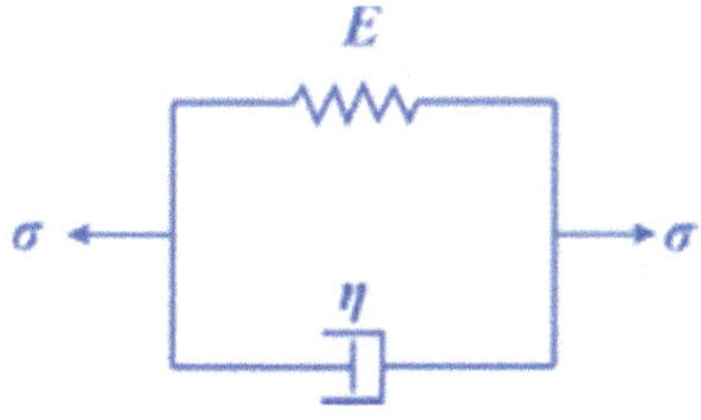

Fig. 5 Kelvin-Voigt model

and the relaxation time can be represented as $\lambda = \frac{\eta}{E}$.

The model can be used to study the physical interaction between the cell and the local surrounding, and how it affects the spatial organization of cells in tissues (Villa et al. 2021), to analyse the hysteresis and energy dissipation of soft tissues subjected to cyclic loading (Zhang et al. 2021), etc. the model exhibits a reversible strain creep, but without stress relaxation.

Apart from the above mentioned models, some of the advanced models are also available, which can be used to study multi-modes of deformation, stress relaxation, creep with nonlinearity, and anisotropicity. The comparison between the different models available apart from the Maxwell and Kelvin-Voigt models is shown in the Table 2.

3.4.3 Nonlinear Viscoelastic Model

Let κ_R and κ_t denotes the reference and current configuration of the viscoelastic solid. The motion is a mapping that assigns at every time t a position x in κ_t to every point X in κ_R (Fig. 6) (Mythravaruni and Ravindran 2022).

$$\tilde{x} = \chi_{\kappa_R}\left(\tilde{X}_{\kappa_R}, t\right), \tag{28}$$

$$\boldsymbol{F} = \frac{\partial \chi_{\kappa_R}}{\partial \tilde{X}_{\kappa_R}}, \tag{29}$$

is the deformation gradient.

The left and right Cauchy-green stress tensors are defined using the deformation gradient, such as

$$\boldsymbol{B}_{\kappa_R} = \boldsymbol{F}_{\kappa_R}\boldsymbol{F}^T_{\kappa_R}; \boldsymbol{C}_{\kappa_R} = \boldsymbol{F}^T_{\kappa_R}\boldsymbol{F}_{\kappa_R} \tag{30}$$

and similarly, the tensors and can be defined for the natural configuration associated with the current configuration such as

$$\boldsymbol{B}_{\kappa_{p(t)}} = \boldsymbol{F}_{\kappa_{p(t)}}\boldsymbol{F}^T_{\kappa_{p(t)}}; \boldsymbol{C}_{\kappa_{p(t)}} = \boldsymbol{F}^T_{\kappa_{p(t)}}\boldsymbol{F}_{\kappa_{p(t)}} \tag{31}$$

and $\mathbf{F}_{\kappa_R}$ is decomposed as

$$\boldsymbol{F}_{\kappa_R} = \boldsymbol{F}_{\kappa_{p(t)}}\boldsymbol{G} \tag{32}$$

where **G** is the mapping from reference to the current natural configuration. Here the I_1, I_2, I_3 are the first three strain invariants of $\boldsymbol{B}_{\kappa R}$ and $\boldsymbol{B}_{\kappa_{p(t)}}$.

For the incompressible solid, the forms of stored energy potential Ψ and the rate of dissipation ξ_{diss} are chosen to be,

Table 2 Linear Viscoelastic Models and their applications, constitutive equations, schematic representations, and advantages

Model	Application	Constitutive equation	Advantages	Schematic Diagram	Reference
Maxwell model	Liquids, polymers, and metals at high temperatures	$\sigma + \frac{\eta}{E}\frac{d\dot{\sigma}}{dt} = E\dot{\varepsilon}$	– Viscous flow and elastic resistance – Generalizable to complex models	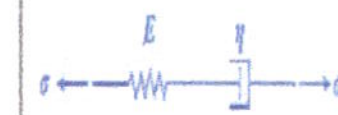	Maxwell (1867)
Kelvin-Voigt model	Solids, polymers, biological tissues	$\sigma = E\varepsilon + \eta\frac{d\varepsilon}{dt}$	– Simple and easy to use – It can describe hysteresis and damping behaviour	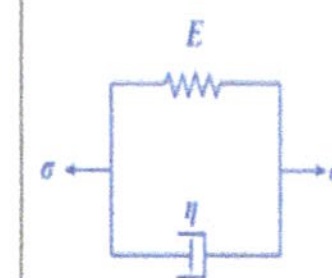	Voigt (1888)
Standard linear solid model (Kelvin representation)	Solids, polymers, biological tissues, hydrogels	$\sigma + \frac{\eta}{E_1+E_2}\frac{d\sigma}{dt} = \frac{E_1E_2}{E_1+E_2}\varepsilon + \frac{E_1\eta}{E_1+E_2}\frac{d\varepsilon}{dt}$	– More realistic than Maxwell or Kelvin-Voigt models – It can describe stress relaxation, creep, and hysteresis phenomena – Can be generalized to nonlinear cases	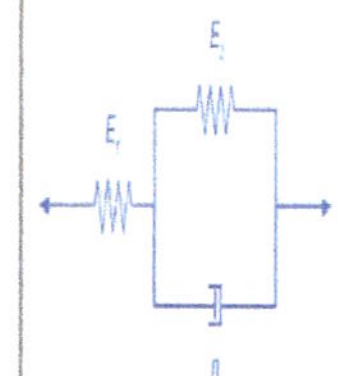	Lin (2022)
Burgers model	Fluids, polymers, biological tissues	$\sigma + \left(\frac{\eta_1}{E_1} + \frac{\eta_2}{E_1} + \frac{\eta_2}{E_2}\right)\dot{\sigma} + \frac{\eta_1\eta_2}{E_1E_2}\ddot{\sigma} = \eta_2\dot{\varepsilon} + \frac{\eta_1\eta_2}{E_1}\ddot{\varepsilon}$	– It can describe stress relaxation, creep, and hysteresis phenomena – Can be generalized to nonlinear cases	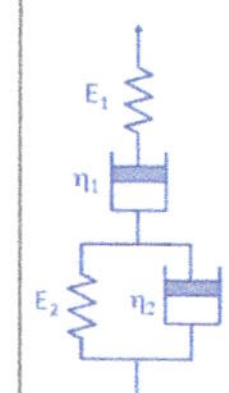	Huang et al. (2011)

(continued)

Table 2 (continued)

Model	Application	Constitutive equation	Advantages	Schematic Diagram	Reference
Standard linear solid model (Maxwell representation)	Solids, polymers, biological tissues	$\sigma = \frac{\eta}{E_2}\frac{d\sigma}{dt} = E_1\varepsilon + \frac{\eta(E_1+E_2)}{E_2}\frac{d\varepsilon}{dt}$	– More realistic than Maxwell or Kelvin-Voigt models – Can describe stress relaxation, creep and hysteresis phenomena – Can be generalized to nonlinear cases	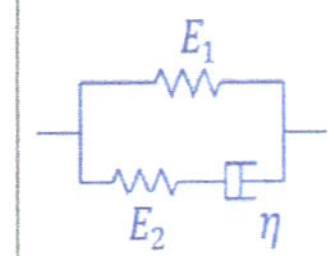	Zener and Siegel (1949)
Jeffreys model	Fluids, polymers, geophysics	$\left(1+\frac{\mu_s}{\eta_s}\right)\dot{\tau}_s + \frac{G_s}{\eta_s}\tau_s = G_s\dot{\gamma} + \mu_s\ddot{\gamma}$	– More accurate and flexible than integer-order models – Accounts for compressibility and coupled response of hydraulic systems – Lower dimension of parameter space	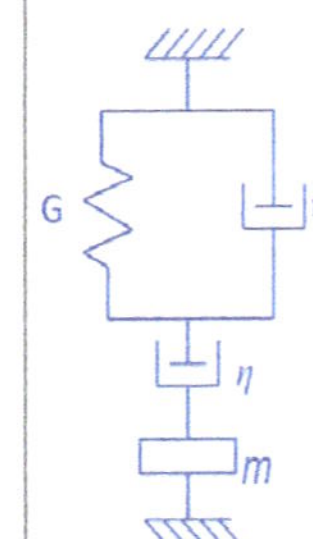	Jaishankar et al. (2011)

* where σ: stress, E: elastic modulus, ϵ: strain η: viscosity, t: time, G: shear modulus, γ: shear strain, m: inertial parameter, α: nonlinearity parameter

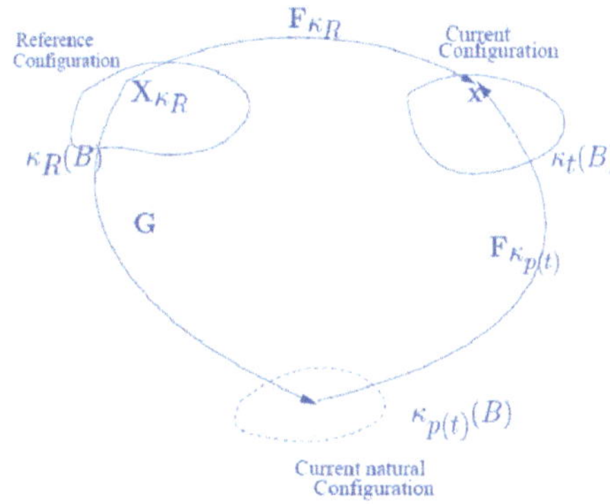

Fig. 6 Configurations of the viscoelastic body during deformation

$$\Psi = C_1\left(I_{1_{\kappa_R}} - 3\right) + C_2\left(I_{1_{\kappa_{p(t)}}} - 3\right), \tag{33}$$

$$\xi_{diss} = \eta(\boldsymbol{D}_{\kappa_{p(t)}} \cdot \boldsymbol{B}_{\kappa_{p(t)}} \boldsymbol{D}_{\kappa_{p(t)}}), \tag{34}$$

The reduced dissipation equation for an isothermal process is given by,

$$\boldsymbol{T} \cdot \boldsymbol{D} - \rho_{\kappa_t} \Psi = \xi_{diss} \geq 0 \tag{35}$$

The equation for stress is obtained by substituting these in the reduced dissipation equation.

$$\boldsymbol{T} = -p\boldsymbol{I} + 2\rho_{\kappa t}\left(C_1 \boldsymbol{B}_{\kappa_R} + C_2 \boldsymbol{B}_{\kappa_{p(t)}}\right), \tag{36}$$

where p is the Lagrange multiplier that accounts for incompressibility.

By maximising the rate of dissipation, ξ_{diss}, such that the second law of thermodynamics and incompressibility conditions are satisfied, the equation for the evolution of the natural configuration is obtained as shown,

$$-\frac{1}{2}\overset{\nabla}{\mathbf{B}}_{\kappa_{p(t)}} = \frac{2\rho_{\kappa_t} C_2}{\eta}\left(\boldsymbol{B}_{\kappa_{p(t)}} - \frac{3}{tr\boldsymbol{B}^{-1}_{\kappa_{p(t)}}} I\right). \tag{37}$$

$I_{1_{\kappa_R}}, I_{2_{\kappa_R}}, I_{3_{\kappa_R}}$ are the first three strain variants of $\boldsymbol{B}_{\kappa_R}$ and $I_{1_{\kappa_{p(t)}}}, I_{2_{\kappa_{p(t)}}}, I_{3_{\kappa_{p(t)}}}$ are the first three strain variants of $\boldsymbol{B}_{\kappa_{p(t)}}$.

where $I_{1_{\kappa_R}} = tr(\boldsymbol{B}_{\kappa_R})$; $I_{3_{\kappa_R}} = \det\left(\boldsymbol{B}_{\kappa_R}\right)$ and $J = \sqrt{I_{3_{\kappa_R}}}$, and $I_{1_{\kappa_{p(t)}}} = tr\left(\boldsymbol{B}_{\kappa_{p(t)}}\right)$; $I_{3_{\kappa_{p(t)}}} = \det\left(\boldsymbol{B}_{\kappa_{p(t)}}\right)$ and $J = \sqrt{I_{3_{\kappa_{p(t)}}}}$.

While for the compressible solid, the forms of stored energy potential Ψ and the rate of dissipation ξ_{diss} are chosen to be,

$$\Psi = \frac{C_1}{\rho_{\kappa_R}}\left[tr\left(\overline{\boldsymbol{B}}_{\kappa_R}\right) - 3\right] + \frac{C_2}{\rho_{\kappa_{p(t)}}}\left[tr\left(\overline{\boldsymbol{B}}_{\kappa_{p(t)}}\right) - 3\right] + \frac{K_1}{\rho_{\kappa_R}}(J-1)^2 + \frac{K_2}{\rho_{\kappa_{p(t)}}}\left(J_{\kappa_{p(t)}} - 1\right)^2, \tag{38}$$

$$\xi_{diss} = \eta\left(\boldsymbol{D}_{\kappa_{p(t)}} \cdot \boldsymbol{B}_{\kappa_{p(t)}} \boldsymbol{D}_{\kappa_{p(t)}}\right), \tag{39}$$

substituting the derivatives of Ψ in the reduced dissipation equation Eq. (35), we get the constitutive relation for stress as,

$$\begin{aligned} \boldsymbol{T} = & \frac{2}{J} C_1 \left[\overline{\boldsymbol{B}}_{\kappa_R} - \frac{1}{3} tr\left(\overline{\boldsymbol{B}}_{\kappa_R}\right)\right] + 2K_1(J-1) \\ & + \frac{2}{J_{\kappa_p}} C_2 \left[\overline{\boldsymbol{B}}_{\kappa_p} - \frac{1}{3} tr\left(\overline{\boldsymbol{B}}_{\kappa_p}\right)\right] + 2K_2\left(J_{\kappa_{p(t)}} - 1\right). \end{aligned} \tag{40}$$

By maximising the rate of dissipation, ξ_{diss}, such that the second law of thermodynamics is satisfied, the equation for the evolution of the natural configuration is obtained as shown,

$$\overset{\nabla}{\mathbf{B}}_{\kappa\, p(t)} = \frac{-2}{\eta}\left[\phi_1 \boldsymbol{B}_{\kappa_{p(t)}} + \phi_3 \boldsymbol{I}\right]; \tag{41}$$

where, $\phi_1 = 2\rho_{\kappa_{p(t)}}\left(\frac{\partial\psi}{\partial I_{1_\Psi}}\right)$ and

$$\phi_3 = 2\rho_{\kappa t}\frac{\partial \Psi}{\partial I_{3_{\kappa p}}} - \frac{1}{J_{\kappa_p}}\left[C_3\left(tr\left(\overline{\boldsymbol{B}}_{\kappa_p}\right) - 3\right) + K_2\left(J_{\kappa_p} - 1\right)^2\right]; \tag{42}$$

and $\overset{\nabla}{\mathbf{B}}_{\kappa\, p(t)}$ is the upper convected derivative of $\mathbf{B}_{k_{p(t)}}$ and

$$\begin{aligned} \overline{\boldsymbol{B}}_{k_R} &= J^{-2/3}\boldsymbol{B}_{k_R}; \\ \overline{\boldsymbol{B}}_{k_{p(t)}} &= J_{k_{p(t)}}^{-2/3}\boldsymbol{B}_{k_{p(t)}} \end{aligned}$$

3.4.4 Quasi-Linear Viscoelastic Model

The quasi linear viscoelastic model introduced by Fung (2013) is a standard tool for the representation of the historical and time dependent nonlinear viscoelasticity of soft tissues. As stated by different researchers, soft tissues shows linear stress strain behaviour at low strains and shows nonlinear behaviour at large strains (Tan et al. 2013). Different consititutive models as described above can be applied but Fung's quasi linear viscoelastic model is the most common one in use due to its some of the major advantages such as, it enables one to model the nonlinear elastic and nonlinear viscous behaviour of soft tissues with a single set of parameters and also the model is nonlinear viscoelastic in nature which help describing the mechanical nature of soft tissues accurately. This theory's central tenet is that time-dependent and strain-dependent features can be distinguished by using an elastic function, which produces elastic coefficients, and a reduced relaxation function, which produces viscous coefficients. The time-dependent theory has evolved over time, therefore there are numerous implementation techniques to explore. The indirect method, which uses simply the relaxation data to solve for the elastic and viscous coefficients concurrently, is a popular approach. The time-dependent theory implies that the elastic and time-dependent features are separate and describes the resulting stress using a linear combination of these non-linear factors, as shown here.

$$\sigma(t, \varepsilon) = \int_0^t G(t-\tau)\frac{\partial \sigma^e(\varepsilon)}{\partial \varepsilon}\frac{\partial \varepsilon}{\partial \tau}\partial \tau \tag{43}$$

where $G(t)$ (Lockett 1972) is the reduced relaxation function depending on t and $\sigma^e(\varepsilon)$ is the elastic function dependent on ε (Fung 2013). Relaxation spectrum $S(\tau)$ (Neubert 1963) defines the reduced relaxation function, where C_1, τ_1, and τ_2 are viscous coefficients that indicate the amplitude of the viscous effects and the time values corresponding to the relaxation spectrum's frequency limits, respectively as

$$G(t) = \frac{1 + \int_0^\infty S(\tau)e^{-\frac{t}{\tau}}d\tau}{1 + \int_0^\infty S(z)d\tau} \tag{44}$$

$$S(\tau) = \begin{cases} \frac{C_1}{\tau} & \text{for } \tau_1 \leq \tau \leq \tau_2 \\ 0 & \text{for } \tau < \tau_1, \tau > \tau_2 \end{cases} \tag{45}$$

The model has its shortcomings as that the stress response resulting from any level of instantaneous strain must be proportionate to single reduced relaxation function. Due to this, it is difficult to obtain the requisite accuracy. The number of degrees of freedom can be increased to overcome this limitation, but this in turn compounds for the second limitation of the model.

4 Validation and Limitation

For studying the biomechanics of soft tissues, such as skin, muscle, cartilage, blood vessels, etc., FEA is an effective approach. However, FEA also has some restrictions and difficulties that must be dealt with and resolved. Some methods available for validating FEA results, and some sources of error and uncertainty in the FEA model are presented.

4.1 Methods for Validation of FEA

Validation is comparing FEA results to experimental data or analytical findings. Any errors or defects in the input parameters or the FEA model can be found and corrected with validation. The validation process may increase the credibility and assurance of the FEA results. The methodologies listed below can be used to validate FEA in soft tissues analysis:

- Plausibility check: The validity of the FEA results is checked by examining their logical and physical plausibility.
- Comparison with hand calculations: In idealized circumstances, the FEA findings can be compared with hand calculations or streamlined analytical answers.
- Mesh Quality Check: FEA results are also be validated by examining the finite element mesh's quality and running mesh refinement studies.
- Comparison with experimental data: For actual scenarios, the FEA results are compared with experimental data from in-vivo measurements or laboratory experiments.

4.2 Sources of Error and Uncertainty

Any numerical analysis, including FEA, will inevitably contain error and uncertainty (Zafar and Verma 2023). Different factors, such as assumptions, approximations, discretizations, and simplifications, can lead to error and uncertainty. In the examination of soft tissues using FEA, some potential sources of error and uncertainty include:

- Geometric complexity and boundary conditions: Soft tissue geometry and boundary conditions are frequently complicated and erratic, necessitating advanced techniques for constructing and discretizing the geometry and simulating the consequences of application of stress.
- Nonlinearity of material behaviour: Soft tissues frequently exhibit nonlinear, anisotropic, viscoelastic, poroelastic, etc., material behaviours that calls for sophisticated constitutive models and parameters that may change based on the conditions and properties of the tissue.
- Combination of solid and fluid phases: During deformation, soft tissue's solid and fluid phases frequently interact with one another. This necessitates modelling the fluid flow through the tissue borders or membranes and coupling the analysis of the solid and fluid phases.

5 Future Scope

The creation of extremely realistic and precise models that can simulate the behaviour of soft tissues under a variety of loading circumstances, including compression, tension, shear, bending, and torsion, is vital to the future of FEA of soft tissues. The potential of FEA includes investigating the effects of various boundary conditions, such as interactions with surgical instruments like needles or deformable membranes that control fluid flow. Additionally, FEA can be extremely useful in maximising the design and production of tissue engineering scaffolds, fostering stem cell proliferation and differentiation into valuable tissues.

Applications of FEA in soft tissues cover a wide range of topics:

- Understanding the biomechanics of healthy and diseased tissues, including those affected by osteoporosis, osteoarthritis, brain edema, and trauma.
- Improving the detection and treatment of soft tissue diseases and injuries, including tumors, fractures, wounds, and ulcers.
- Creating and testing new biomaterials and implants for tissue replacement and enhancement, such as skin substitutes, cartilage implants, and bone transplants.
- Developing organ-on-a-chip devices that duplicate the structures and operations of human organs to aid in the testing of drugs and the modelling of illness.
- Using innovative methods to personalise medicines for the tailored treatment based on the unique properties of each patient's tissues.

6 Conclusion

Finite Element Analysis (FEA) has significant promise for the future of soft tissue modelling. FEA can better replicate soft tissue behaviour under various loading circumstances and boundary interactions by constructing more realistic and precise models. By advancing FEA, we can better understand biomechanics, develop new biomaterials, enhance diagnostics, and enable personalised medicine. FEA is still a potent tool for unlocking new biomedical research insights, ultimately improving patient care and medical achievements, despite difficulties in modelling complexity and material behaviour.

Declarations

Conflicts of Interest

"There are no conflicts of interest to declare by the authors."

Ethical Approval

The authors hereby state that the present work is in compliance with the ethical standards.

Competing Interests

The authors declare no competing interests.
Not applicable.

References

Aggarwal, A., Jensen, B.S., Pant, S. and Lee, C.H., 2023. Strain energy density as a Gaussian process and its utilization in stochastic finite element analysis: Application to planar soft tissues. *Computer Methods in Applied Mechanics and Engineering*, *404*, p.115812.

Biewener, A.A., 2008. Tendons and ligaments: structure, mechanical behaviour and biological function. *Collagen: structure and mechanics*, pp. 269–284.

Bonet, J. and Wood, R.D., 1997. *Nonlinear continuum mechanics for finite element analysis*. Cambridge university press.

Belytschko, T., Liu, W.K., Moran, B. and Elkhodary, K., 2014. *Nonlinear finite elements for continua and structures*. John Wiley & Sons.

Chanda, A., Chatterjee, S. and Gupta, V., 2020. Soft composite based hyperelastic model for anisotropic tissue characterization. *Journal of Composite Materials*, *54*(28), pp. 4525–4534.

Chanda, A. and Singh, G., 2023. Hyperelastic Models for Anisotropic Tissue Characterization. *Mechanical Properties of Human Tissues*, pp. 73–83.

Cheng, T. and Gan, R.Z., 2008a. Experimental measurement and modelling analysis on mechanical properties of tensor tympani tendon. *Medical engineering & physics*, *30*(3), pp. 358–366.

Cheng, T. and Gan, R.Z., 2008b. Mechanical properties of anterior malleolar ligament from experimental measurement and material modelling analysis. *Biomechanics and modelling in Mechanobiology*, *7*, pp. 387–394.

Fung, Y.C., 2013. *Biomechanics: mechanical properties of living tissues*. Springer Science & Business Media.

Gladilin, E. and Eils, R., 2008, March. Nonlinear elastic model for image registration and soft tissue simulation based on piecewise St. Venant-Kirchhoff material approximation. In *Medical Imaging 2008: Image Processing* (Vol. 6914, pp. 898–906). SPIE.

Gladilin, E., Zachow, S., Deuflhard, P. and Hege, H.C., 2001, May. Validation of linear elastic model for soft tissue simulation in craniofacial surgery. In *Medical imaging 2001: Visualization, display, and image-guided procedures* (Vol. 4319, pp. 27–35). SPIE.

Hackett, R.M. and Hackett, R.M., 2018. Strain Measures. *Hyperelasticity Primer*, pp. 5–8.

Hong, J., Evans, T.M. and Mente, P.L., 2015. Study on the damage mechanism of articular cartilage based on the fluid–solid coupled particle model. *Advances in Mechanical Engineering*, *7*(4), p.1687814015581264.

Huang, C.C., Wei, M.K. and Lee, S., 2011. Transient and steady-state nanoindentation creep of polymeric materials. *International Journal of Plasticity*, *27*(7), pp. 1093–1102.

Jaishankar, A., Sharma, V. and McKinley, G.H., 2011. Interfacial viscoelasticity, yielding and creep ringing of globular protein–surfactant mixtures. *Soft Matter*, *7*(17), pp. 7623–7634.

Kastelic, J., Galeski, A. and Baer, E., 1978. The multicomposite structure of tendon. *Connective tissue research*, *6*(1), pp. 11–23.

Knops, R.J. and Ciarlet, P., 1994. Mathematical elasticity, Vol. 1. Three-dimensional elasticity. *Bulletin of the American Mathematical Society*, *31*(2), pp. 246–251.

Korhonen, R.K. and Saarakkala, S., 2011. Biomechanics and modelling of skeletal soft tissues. *Theoretical biomechanics*, *6*.

Leeman, S. and Jones, J., 2009. Visco-elastic models for soft tissues. In Acoustical Imaging (pp. 369–376). Springer Netherlands.

Li, H., Li, J., Yu, S., Wu, C. and Zhang, W., 2021. The mechanical properties of tibiofemoral and patellofemoral articular cartilage in compression depend on anatomical regions. *Scientific reports*, *11*(1), pp. 1–11.

Lin, C.Y., 2022. Rethinking and researching the physical meaning of the standard linear solid model in viscoelasticity. *Mechanics of Advanced Materials and Structures*, pp. 1–16.

Lockett, F.J., 1972. Nonlinear viscoelastic solids.

Lohr, M.J., Sugerman, G.P., Kakaletsis, S., Lejeune, E. and Rausch, M.K., 2022. An introduction to the Ogden model in biomechanics: benefits, implementation tools and limitations. *Philosophical Transactions of the Royal Society A*, *380*(2234), p.20210365.

Malcom, L.L., 1976. An Experimental Investigation of the Frictional and Deformational Responses of Articular Cartilage Interfaces to Static and Dynamic Loading. University of California, San Diego.

Maxwell, J.C., 1867. IV. On the dynamical theory of gases. *Philosophical transactions of the Royal Society of London*, (157), pp. 49–88.

Misra, S., Ramesh, K.T. and Okamura, A.M., 2010. Modelling of nonlinear elastic tissues for surgical simulation. *Computer methods in biomechanics and biomedical engineering*, *13*(6), pp. 811–818.

Mooney, M., 1940. A theory of large elastic deformation. *Journal of applied physics*, *11*(9), pp. 582–592.

Morrison, S.M., Dick, T.J. and Wakeling, J.M., 2015. Structural and mechanical properties of the human Achilles tendon: Sex and strength effects. *Journal of Biomechanics*, *48*(12), pp. 3530–3533.

Mythravaruni, P. and Ravindran, P., 2022. Quantifying the Effect of Voids on the Response of a Viscoelastic Solid. In *Advances in Structural Integrity: Structural Integrity Over Multiple Length Scales* (pp. 217–226). Singapore: Springer Singapore.

Neubert, H.K.P., 1963. A simple model representing internal damping in solid materials. *Aeronautical Quarterly*, *14*(2), pp. 187–210.

Nolan, D.R., Gower, A.L., Destrade, M., Ogden, R.W. and McGarry, J.P., 2014. A robust anisotropic hyperelastic formulation for the modelling of soft tissue. *Journal of the mechanical behaviour of biomedical materials*, *39*, pp. 48–60.

Ogden, R.W., 1972. Large deformation isotropic elasticity–on the correlation of theory and experiment for incompressible rubber-like solids. *Proceedings of the Royal Society of London. A. Mathematical and Physical Sciences*, *326*(1567), pp. 565–584.

Pal, S., Design of Artificial Human Joints & Organs. Springer, 2014 [5].

Picinbono, G., Delingette, H. and Ayache, N., 2001, May. Nonlinear and anisotropic elastic soft tissue models for medical simulation. In *Proceedings 2001 ICRA. IEEE International Conference on Robotics and Automation (Cat. No. 01CH37164)* (Vol. 2, pp. 1370–1375). IEEE.

Rivlin, R.S., 1948. Large elastic deformations of isotropic materials IV. Further developments of the general theory. *Philosophical transactions of the royal society of London. Series A, Mathematical and physical sciences*, *241*(835), pp. 379–397.

Rumian, A.P., Wallace, A.L. and Birch, H.L., 2007. Tendons and ligaments are anatomically distinct but overlap in molecular and morphological features—a comparative study in an ovine model. *Journal of orthopaedic research*, *25*(4), pp. 458–464.

Smith, K.J., Mathur, M., Meador, W.D., Phillips-Garcia, B., Sugerman, G.P., Menta, A.K., Jazwiec, T., Malinowski, M., Timek, T.A. and Rausch, M.K., 2021. Tricuspid chordae tendineae mechanics: insertion site, leaflet, and size-specific analysis and constitutive modelling. *Experimental Mechanics*, *61*, pp. 19–29.

Sussman, T. and Bathe, K.J., 1987. A finite element formulation for nonlinear incompressible elastic and inelastic analysis. *Computers & Structures*, *26*(1–2), pp. 357–409.

Tan, K., Cheng, S., Jugé, L. and Bilston, L.E., 2013. Characterising soft tissues under large amplitude oscillatory shear and combined loading. *Journal of biomechanics*, *46*(6), pp. 1060–1066.

Thorpe, C.T. and Screen, H.R., 2016. Tendon structure and composition. *Metabolic influences on risk for tendon disorders*, pp. 3–10.

Villa, C., Chaplain, M.A., Gerisch, A. and Lorenzi, T., 2021. Mechanical models of pattern and form in biological tissues: The role of stress–strain constitutive equations. *Bulletin of Mathematical Biology*, *83*(7), p.80.

Voigt, W., 1888. Theorie des Lichtes für bewegteMedien. Annalen der Physik, 271(11), pp. 524–551.

Weber, M., Rothschild, M.A. and Niehoff, A., 2021. Anisotropic and age-dependent elastic material behaviour of the human costal cartilage. *Scientific Reports*, *11*(1), pp. 1–12.

Weickenmeier, J. and Jabareen, M., 2014. Elastic–viscoplastic modelling of soft biological tissues using a mixed finite element formulation based on the relative deformation gradient. *International journal for numerical methods in biomedical engineering*, *30*(11), pp. 1238–1262.

Weinberg, E.J. and Kaazempur-Mofrad, M.R., 2006. A large-strain finite element formulation for biological tissues with application to mitral valve leaflet tissue mechanics. *Journal of Biomechanics*, *39*(8), pp. 1557–1561.

Winter, D.A., 2009. *Biomechanics and motor control of human movement*. John Wiley & Sons.

Zafar, S. and Verma, A., 2023. Continuum Mechanics-Based Simulations in Coatings. In *Coating Materials: Computational Aspects, Applications and Challenges* (pp. 185–216). Singapore: Springer Nature Singapore. DOI: https://doi.org/10.1007/978-981-99-3549-9_9

Zajac, F.E., 1989. Muscle and tendon: properties, models, scaling, and application to biomechanics and motor control. Critical reviews in biomedical engineering, 17(4), pp.359-411.

Zener, C.M. and Siegel, S., 1949. Elasticity and Anelasticity of Metals. *The Journal of Physical Chemistry*, *53*(9), pp.1468-1468.

Zhang, W., Capilnasiu, A. and Nordsletten, D., 2021. Comparative analysis of nonlinear viscoelastic models across common biomechanical experiments. *Journal of Elasticity*, *145*(1-2), pp.117-152.

Zschäbitz, A., 2005. Anatomie und verhalten von sehnen und bändern. *Der Orthopäde*, *6*(34), pp.516-525.

Mechanical Stimulation Alters the Collagen Protein Structure in Skin Tissue

Krashn Kumar Dwivedi, Ravinder Jhorar, Sachin Kumar, and Navin Kumar

1 Introduction

Mechanical stimulation is one of the external factor which control the soft tissue remodelling (Sheng et al. 2020). Mechanical stimulations at tissue level are transmitted to molecular level through different micro and nano level deformation mechanics (Dwivedi et al. 2020a) and disturb the molecular structure equilibrium of the tissue. This disturbance in molecular structure causes activation of biochemical cascade, responsible for tissue remodelling. In most of soft tissues, applied mechanical signals transfers through collagen fibers. Also, in connective soft tissues, alteration in collagen molecules (protein structure) is found responsible to activate tissue remodelling cascade (Bonnans et al. 2014). In most of connective soft tissues, collagen fibers are main responsible to support the mechanical load (Fung 2013) and shows complex deformation response at different length scale (Yang et al. 2015). It is believed that transferring of mechanical signals from tissue to molecular level involves rotation, bending, stretching, shearing of collagen fibers with intera and inter fibers cross linkers breakage (Yang et al. 2015). These deformation mechanism of collagen fibers are specific to nature and rate of mechanical stimulation (Dwivedi et al. 2020a) and this complex deformation mechanism is the results of hierarchical structure of collagen fibers with in tissue (Sherman et al. 2015).

K. K. Dwivedi (✉)
Department of Biomedical Engineering, Indian Institute of Technology Ropar, Rupnagar, India
e-mail: kkdwivedi159@gmail.com

R. Jhorar
Delhi Skill and Entrepreneurship University, New Delhi, India

S. Kumar · N. Kumar
Department of Mechanical Engineering, Indian Institute of Technology Ropar, Rupnagar, India

J. Gupta and A. Verma (eds.), *Microbiology-2.0 Update for a Sustainable Future*,
https://doi.org/10.1007/978-981-99-9617-9_20

1.1 Structure of Collagen Fibers

The hierarchical schematic of collagen fiber from tropocollagen to collagen fiber bundle is provided in the Fig. 1. The assembly of the collagen fiber starts from the polypeptide strands of the protein, which are left handed alpha helix (Sherman et al. 2015). The assembly of three sets of these left hand alpha helix makes the procollagen, a macromolecular unit of collagen fibers (Hollingsworth and Karplus 2010). With the cleaving of C and N terminal, the procollagen becomes matured in tropocollagen, the basic load carrying unit of collagen fibers (Buehler 2006). It can be believed that mechanical stimulations (stretching dominated deformation) associated alteration in tropocollagen may activate the biochemical pathway of remodelling. The typical length and diameter of tropocollagen are 300 and 1.5 nm. Further, staggered arrangement of tropocollagen forms the collagen fibrils, a nano scale unit of collagen fibers. Within collagen fibrils, the tropocollagen are inter connected through enzymatic cross link. The total mechanical stimulation transferred to collagen fibrils is shared by stretching and shearing (Dwivedi et al. 2020a). The collagen fibrils are characterized by the d-periodicity (Fig. 1). The typical value of d-periodicity is 64–67 nm, and any change in this value due to mechanical shows the alteration in collagen fibrils structure. Interfibrillar cross linking of glycosaminoglycan assembled the collagen fibrils in collagen fibers, and along with proteoglycans chains collagen fibers take the form of bundles. The collagen fibers are main responsible to support the mechanical load and provide the strength to soft connective tissues. Assembly of collagen fibers bundles with other non-protein components such as proteoglycans and water forms the extracellular matrix of soft tissues.

1.2 Mechanism of Mechanical Stimulation Transferring from Tissue to Molecular Level

Figure 2 illustrates the step wise procedure of mechanical stimulation transfer from tissue to molecular level. The mechanical stimulation such as strain or load, which are applied at tissue level will initially shared by matrix (non-protein component) and collagen fibers (Fung 2013). Due compliance nature of matrix components, the amount of load carried by matrix is small, therefore most of the applied load is supported by collagen fibers. Load induced strain in collagen fibers is accumulated in fibers straighten and their stretching. Straighten and stretching of collagen fibers are two successive processes, first straighten starts and then straight collagen fibers start to stretch. After this step of collagen fibers deformation, stretching component of strain/load transferred to next hierarchy (collagen fibrils). The strain or load at collagen fibers level transferred to the collagen fibrils through interfibrillar shearing. The interfibrillar shearing or sliding causes stretching of interfibrillar cross linking. The stretching of cross linkers induced the tensile load or strain on the collagen fibrils. The amount of deformation/strain in collagen fibril can be measured by

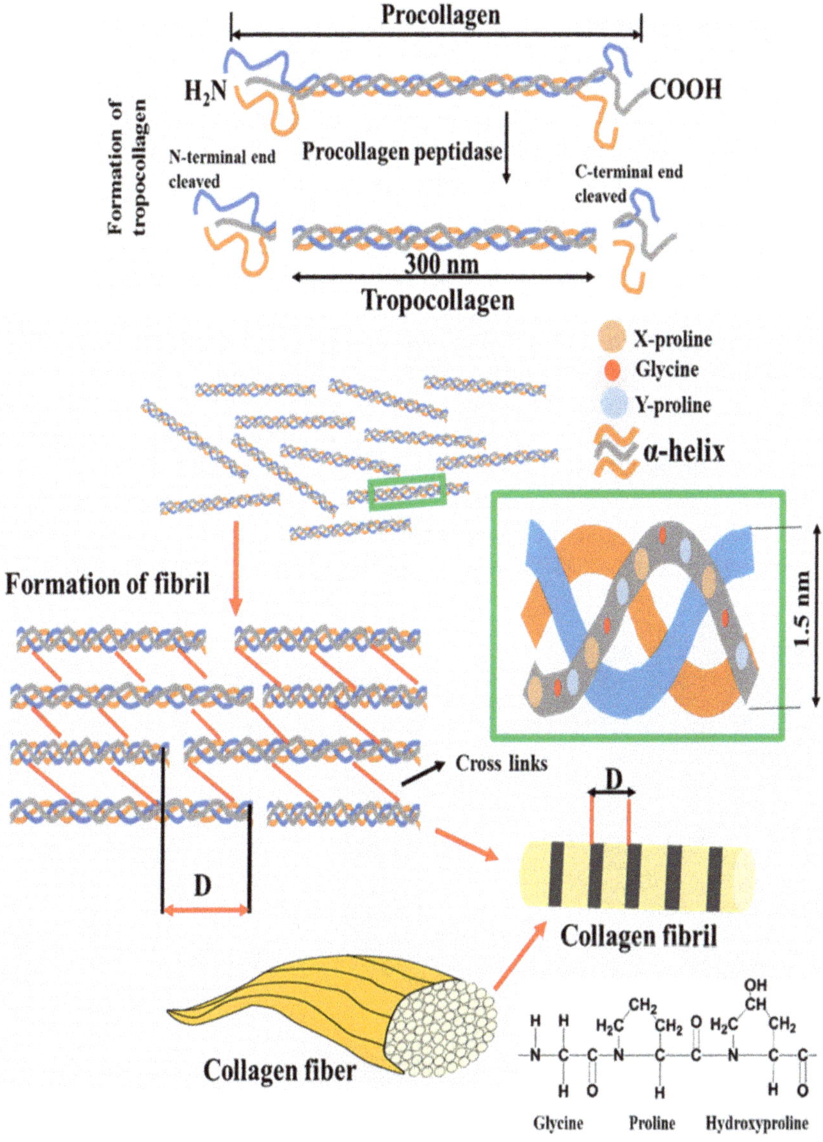

Fig. 1 Hierarchical schematic of collagen fiber from tropocollagen to collagen fiber. This figure was adopted with permission of Jhorar and Lamba (2022)

change in their d-periodicity (Dwivedi et al. 2020a). The strain or load on collagen fibril is further transferred to the stretching of tropocollagen through intrafibrillar shearing and stretching of intrafibrillar cross linkers. The stretching of tropocollagen may cause alteration in this molecular structure, which leads to activation of biochemical pathways, responsible for tissue growth and remodelling. The normal and abnormal tissue remodelling depends on the amount of applied mechanical stimulation transferred to stretching of tropocollagen.

Further, due to complex structure of soft tissues and collagen fibers, it can be believed that amount of load/strain transferred to the molecular scale may depend on the nature and rate of mechanical stimulation, which may also affect the tissue remodelling. However, it is well explored that the mechanical properties of soft tissue at different length scale depend on the rate of mechanical stimulations (Bose et al. 2022). For example, the stiffness, failure and modulus at tissue level, nano level and molecular level are specific to the rate of applied loading (Gautieri et al. 2012). However, the effect of rate of mechanical stimulation on the amount of applied strain/load transferred to the molecular level is not well known. Therefore, the aim of this study to investigate the effect of rate of mechanical stimulation on the alteration of tropocollagen structure, which may also affect the tissue remodelling. For,

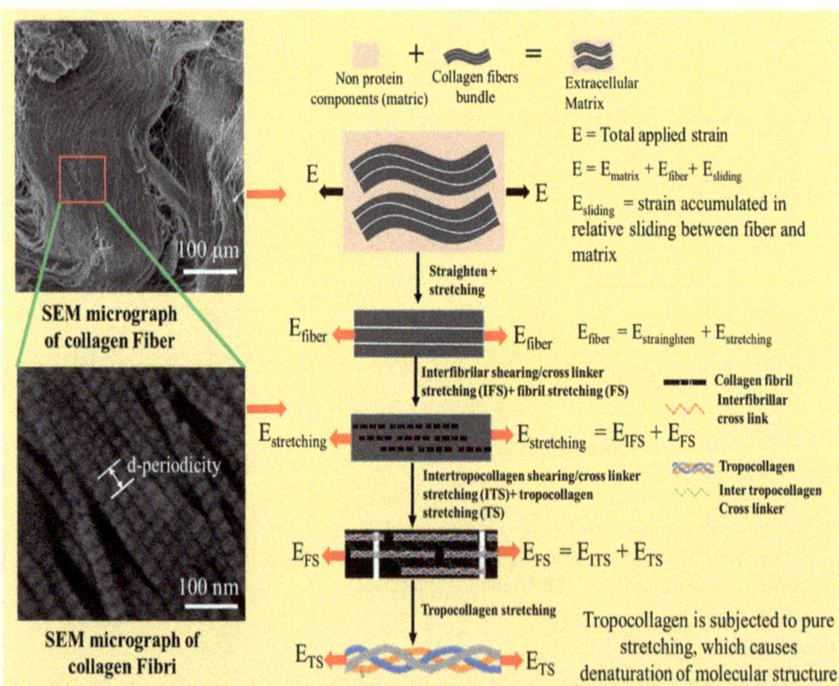

Fig. 2 Schematic presentation of mechanical stimulation transfer from tissue to molecular level. Representative SEM micrograph shows the native structure of collagen fiber and collagen fibrils. Schematics are based on the SEM micrographs

this purpose we explored the capacity ATR-FTIR to assess the mechanical induced alteration in protein structure.

1.3 ATR-FTIR

The ATR-FTIR is widely used to identify the alteration in collagen molecules caused by different pathological conditions (e.g. age, diseases etc.) and mechanical load. Some important peaks obtained using ATR-FTIR are summarized in Table 1. This data was borrowed from the work of Rabotyagova et al. (2008). In ATR-FTIR spectrum, the peak at 1590–1710 cm^{-1} present the secondary structure of protein and generally known as amide I band. This band carries are structural information of collagen molecules. The decovolution of amide-I in sub peaks provide the particular information about the native and altered structure of collagen protein. In general, made-I carries six sub peaks at wavelength location of 1610, 1630, 1645, 1660, 1678, and 1692 cm^{-1}. Uniyal et al. (2022) used these peaks to investigate the effect of temperature on the molecular structure of collagen. Whereas, Dwivedi et al. (2023) used this band to understand the effect of type-2 diabetes on the molecular structure of collagen. The representative FTIR-Spectrum for skin tissue (Dwivedi et al. 2023) is provided in Fig. 3.

1.4 Study Overview

In this study, the pig skin was cyclically stretched at three different frequencies: 0.1 Hz (physiological range), 2.0 Hz (moderate) and 5.0 Hz (over physiological range) for 1000 cycles and for three hours. The skin specimens for each experiments were different. After mechanical test the nano and molecular structure of untested and mechanical tested specimens were assessed using AFM and ATR-FTIR, respectively. The AFM was used measure the strain of collagen fibrils whereas ATR-FTIR data were used to measure alteration in collagen molecules. Different frequencies are used to understand the rate dependent alteration in collagen fibril and molecular structure. The skin tissue is selected purposely, as mechanical stimulation (tissue expansion) is applied to grow extra skin during plastic surgery. Also, oscillatory mechanical stimulations are used to enhance the wound care (Kuehlmann et al. 2020).

Table 1 Position of different ATR-FTIR spectrum peak associated with different organic components. Rabotyagova et al. 2008 Adopted from literature ()

Assignment	Position (cm^{-1})
Tyrosine side chain	1515
Amide II in a random coil	1530
Amide II in triple helix	1549
Carboxyl groups, N–H bending vibration and C–N stretching vibration	1565
Left-handed 3–10 helix in denaturated state, partially beta-sheet region	1630
Random coil conformation: imide residues (amide I in random coil)	1646–1650
Random chain contribution of the non-proline and non-hydroxyproline	1660
Collagen in triple helix, with contribution from α-helix and beta-turns (protein amide stretching vibration of C–O)	1675
Intermolecular association. Helices of aggregated collagen-like peptides β-turns	1690
CH_3– stretching of methyl (assimetric)	2876
CH_2– stretching of methylene	2946
N–H stretch (Amide-1)	3080
Asymmetric –CH_2–N=	3217–3180
N–H stretch	3330–3325
NH stretch coupled with H-bond from H_2O	3450
Intermolecular hydrogen bounds	3500–3470

2 Materials and Methods

2.1 *Sample Preparation*

Porcine skin was collected from the nearest slaughter house immediately after slaughtering. Animal was white Yorkshire breed and 9 months old with 120 kg weight. Skin was removed from the loin portion, along the spine direction and stored at 4 °C in phosphate buffer saline (PBS) soaked gauze. Before cutting of specimens, fat, hairs and epidermis layer were removed very carefully using a razor and surgical scalpel. Dog-bone shape specimens were cut parallel to the animal spine from dermis layer, using a punching machine (SCD manufacturer, Kolkata, India) with aluminum die (ASTM D412). During specimen preparation PBS was sprayed continuously to avoid the dehydration. All specimens from dermis were 30 mm long (gage length), 4 mm width (gage portion) and 2.75 ± 0.212 mm thick. The thickness of each specimen was measured at different sections over the gage portion using laser thickness measurement tool (ILD1320-10, Micro-Epsilon, Germany). All specimens were further stored in gauze soaked with PBS and refrigerated at 4 °C. Experimental

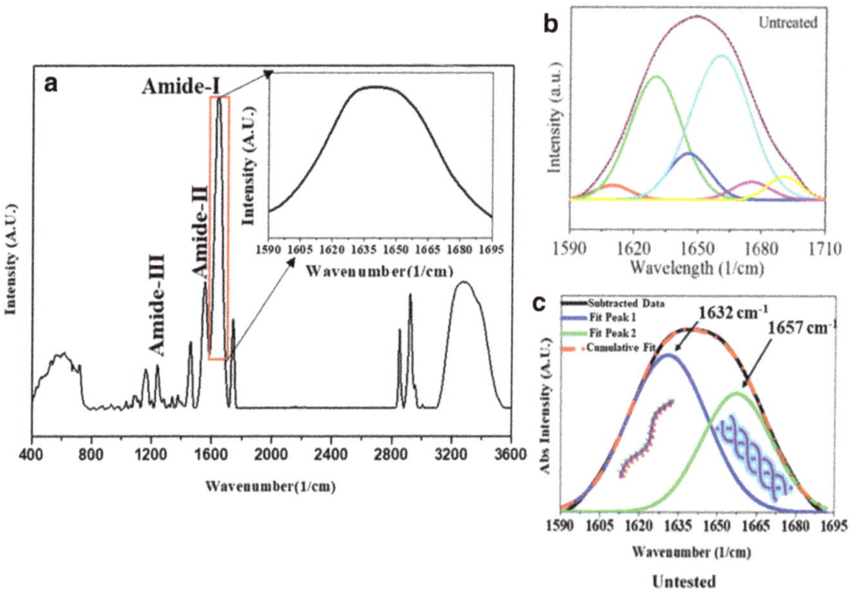

Fig. 3 **a** Representative of average FTIR-ATR spectrum of skin. **b** Deconvolution of amide I in six sub peaks. **c** Sub peaks associated with native and denatured structure of collagen tropocollagen

protocol is adopted from literature (Lakhani et al. 2020, 2021; Dwivedi et al. 2020b, 2022a, b).

2.2 *Mechanical Test*

Total number of specimens were divided in to six groups and each group had 6 specimens. First, second and third groups were tested for 1000 cycles at frequencies of 0.1 Hz, 2.0 Hz and 5.0 Hz respectively. During the cyclic tensile testing, specimens of each groups were stretch up to a peak displacement of 7.5 mm (i.e. 25% strain) and un-stretched to zero displacement. Stretching and un-stretching were performed at the same frequency.

2.3 *ATR-FTIR*

The FTIR spectra of dermis specimen were recorded using Bruker Tensor 27 FTIR spectrometer in Attenuated Total Reflection (ATR) mode. Spectra were scanned in transmission mode at 2 cm^{-1} and further transformed in absorption mode for sake of user convenience. Spectra were recorded before and just after the test at four different positions, along the gage length of the specimen and 32 scans were collected at each position. All spectra were collected in the range of 400–3600 cm^{-1} (wavenumber). After collecting the spectrum, amide band was isolated and deconvoluted in Gaussian sub peaks using second derivative method in Origin software (Origin 2018).

2.4 AFM

After ATR-ATR, the untested and mechanical tested skin specimen were embedded in optimum cutting temperature (OCT) liquid and frozen at −80 °C. The OCT frozen specimens were transferred to the cryostat section machine, where 50 μm thin sections of skin specimens were cut. These sections were floated in PBS and then scan under atomic force microscopy using triangular tip. Scanning of tissue was performed inside the water in tapping mode which ensures no dehydration of the tissue. Initially, an area of 5.0 μm^2 was scanned to choose the region of interest (ROI), and then finally ROI of 1.30 μm^2 area was imaged, which was utilized for all the analysis. The d-periodicity of each fibril at every location was measured through NanoScope analysis software and its average value was used for comparison. Further, a change in the value of d-periodicity in tested specimens as compared to the virgin specimens was used to calculate the strain accumulation in collagen fibrils. This procedure was adopted from the literature (Dwivedi et al. 2020a).

2.5 Statistical Analysis

All the statistical analysis were performed using SPSS 21.0 software. The normality of data distribution was evaluated by Kolmogorov–Smirnov test. Student's t test and one way ANOVA with post hoc Tukey HSD were utilized to make all comparisons. A confidence level of $p < 0.05$ implies a statistical significance among all three groups. All the results are reported as mean ± SD with p value.

3 Results

3.1 Mechanical Experiments

Figure 4a shows the first and 1000th cycle corresponding to all three frequencies, and *J*-shape stress–strain curves of dermis were obtained for all the frequencies i.e. 0.1, 2.0 and 5.0 Hz, which shows a nonlinear mechanical response of skin. These results show clear difference in the tissue level mechanical response of skin. This mechanical response can be the results of rate specific different response of complex structure of skin at different length scale. These results shows that the permanent or inelastic deformation in skin was more corresponding to lower value of frequency than moderate and high value of frequency. In general, the inelastic strain accumulation in the material is the result of microscopic damage or alteration in structure. The reduction in skin stiffness after 1000 cycles was found significantly larger at 0.1 Hz frequency than 2.0 Hz and 5.0 Hz frequency. The reduction in stiffness was about 40 ± 8%, 19 ± 4% and 15 ± 6% corresponding to 0.1, 2.0 and 5.0 Hz frequency, respectively

These results further direct about the rate specific alteration of collagen structure as collagen are main responsible for the stiffness of the soft tissue (Fung 2013).

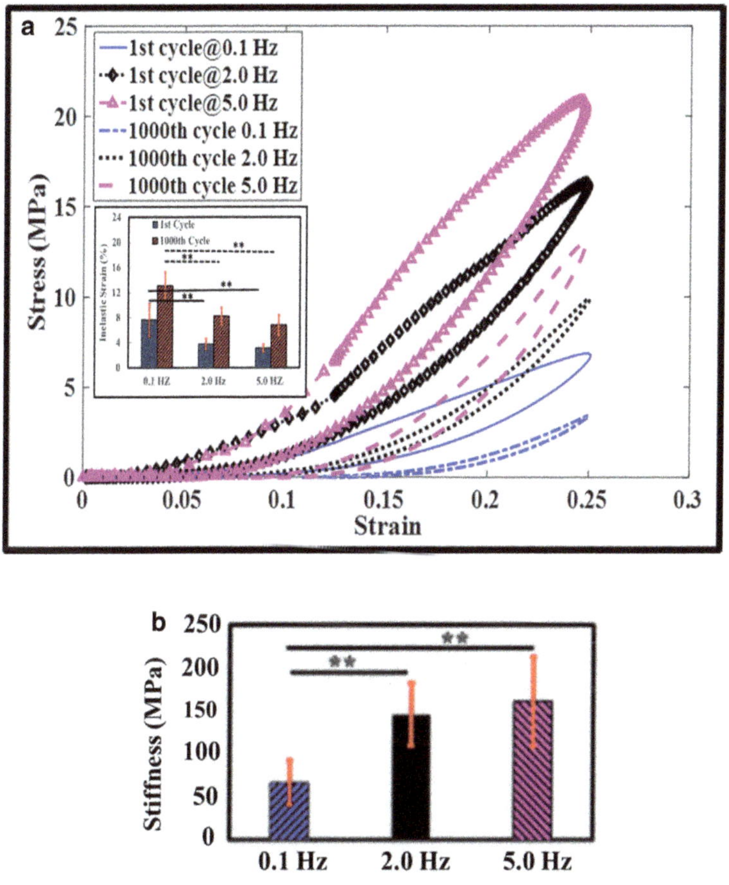

Fig. 4 **a** Stress strain response and inelastic strain after 1000 cycles of skin corresponding to different value of frequencies. **b** Comparison of stiffness among the frequencies. These values are corresponding to first cycle

3.2 AFM Results

Figure 5a illustrate the structure of collagen fibrils in untested and tested skin specimens corresponding to different cyclic frequencies. The fibrils inelastic strain was measured as change in d-periodicity after 1000 cycles between untested and tested specimens corresponding to all three frequencies. The comparison of fibril plastic strain is demonstrated in Fig. 5b. A significant difference was observed in plastic strain among all three frequencies ($p = 1.05 \times 10^{-3}$ between 0.1 and 2.0 Hz, $p = 10^{-3}$ between 0.1 and 5.0 Hz, and $p = 1.62 \times 10^{-2}$ between 2.0 and 5.0 Hz). The large value of fibril plastic strain corresponding to 0.1 Hz frequency indicate large share of applied strain transferred to fibril stretching. These results indicate that transfer of mechanical stimulation is rate dependent. The large value of plastic strain causes large alteration of nano and molecular structure

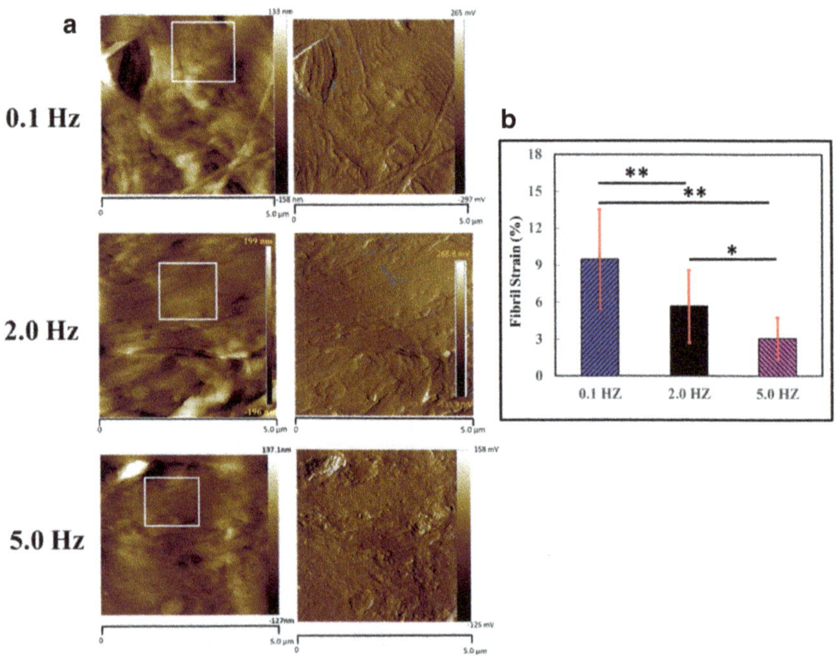

Fig. 5 **a** Shows the AFM height and peak force error images of virgin specimen and images after cyclic stretched at 0.1, 2.0 and 5.0 Hz. **b** Comparison of fibril plastic strain among three frequency

3.3 ATR-FTIR

The representative FTIR spectrum of porcine dermis is shown in supplementary Fig. 6a for untested and tested (at 0.1, 2 and 5 Hz frequency) specimens. The spectrum is dominated by a broad band (3150–3600 cm^{-1}) of (OH) vibration of water and (NH) vibration of proteins, amide I (at about 1645 cm^{-1}), amid II (at about 1550 cm^{-1}) and amid III (at about 1245 cm^{-1}). Here we utilized the amide I region since changes in relative intensity of this band is conformation dependents. As shown in Fig. 6b–d, deconvoluted band of amide I contains two main peaks: a peak at about 1631 cm^{-1}, which can be ascribed to collagen α-helix structure, and a peak about 1657 cm^{-1} can be assign to collagen triple helix structure (Dwivedi et al. 2020a). In our study, Intensity ratio (1631/1657 cm^{-1}) was thus explored to evaluate the relative changes between virgin and tested dermis specimens. Figure 6e shows the comparison of this intensity ratio between untested and tested specimen. An increase in intensity ratio was observed after the cyclic stretching: 2.03 ± 0.24 versus 1.384 ± 0.234 ($p = 0.000179$) for 0.1 Hz, 1.503 ± 0.194 versus 1.277 ± 0.158 ($p = 0.02044$) for 2 Hz and 1.38 ± 0.259 versus 1.283 ± 0.159 ($p = 0.3809$) for 5 Hz stretching frequency. One way ANOVA with post hoc Tukey HDS was conducted to compare the change in intensity ratio among the three groups. A significant difference was found between 0.1 and 2 Hz ($p = 0.001$) and between 0.1 and 5 Hz ($p = 0.001$). Whereas difference in intensity ratio between 2 and 5 Hz was not found significant ($p = 0.566$). To allow the standardization between scans all peaks were normalized to CH-stretch peak at 2922 cm^{-1}, as this peak was remain unchanged after test. The Pearson correlation test show a negative correlation with alteration in protein structure and frequency ($r = 0.74$, $p = 0.041$).

4 Discussion

This study also reveals that for the same number of cycles, the value of plastic strain in the collagen fibrils decreases significantly from lower to a higher frequency. These results suggest that the slow stretching causes irreversible rupture of glycosaminoglycans (GAGs) bridge in collagen fibrils and/or progressive unwinding of tropocollagen due to inter/intra fibrils sliding. This phenomenon may increase the gap portion of collagen fibrils which results in an increase in d-periodicity which may cause large irreversible strain accumulation in the collagen fibrils at low frequency. Whereas at high frequency, this mechanism of deformation does not happen due to lack of time. Hence, the permanent strain in fibrils is only due to progressive irreversible straightening of fibrils under repeated loading. Furthermore, the macroscopic inelastic strain is larger than the fibril's permanent strain, which indicates that the remaining part of macroscopic inelastic strain comes from the permanent straighten/realignment of collagen fibers and damage of interfiber matrix due to sliding between surrounding matrix and collagen fibers.

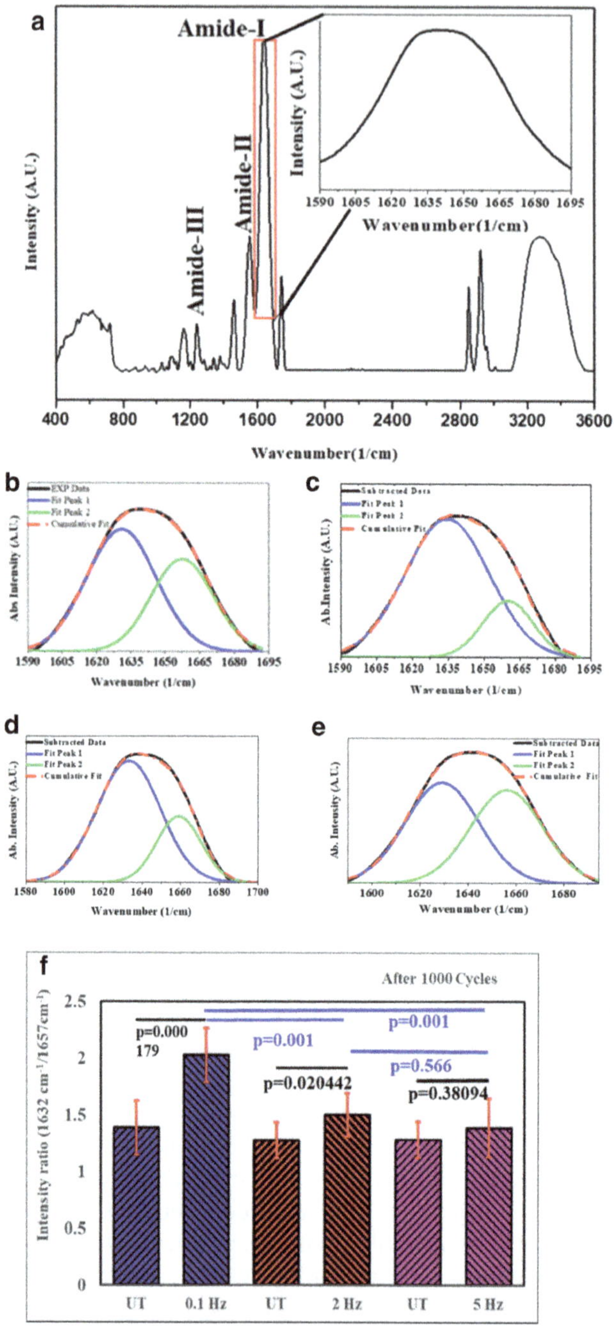

Fig. 6 **a** Representative ATR-FTIR spectrum of skin. Deconvoluted amide I spectrum **b** untested, **c** 0.1 Hz, **d** 2.0 Hz and **e** 5.0 Hz. **f** Comparison of alteration in molecular structure among three frequencies

ATR-FTIR is the best established technique to monitor the molecular structure of skin. In this study we focused Amide I region as it represents the secondary structure of protein, in particular, of collagen, and changes in the relative intensity is conformation dependent. This band of protein is related to the vibration of amide carbonyls along the polypeptide backbone. The deconvoluted amid I band with two sub peaks: peak at about 1657 cm^{-1} (collagen triple helix) and peak at about 1632 cm^{-1} (α-helix). The intensity of triple helix band decreases in denatured state of collagen whereas intensity of α-helix increases after the denaturing of collagen. From the best of our knowledge, this finding was first time utilized to monitor the alteration in collagen native structure of porcine dermis under cyclic loading at different frequencies. In this study, intensity ratio (1632/1657 cm^{-1}) was utilized to observe the relative changes in collagen conformation. The comparison between intensity ratios (1632/1657 cm^{-1}) of denature and native state of collagen. From the same it was observed that intensity ratio (1632/1657 cm^{-1}) increase after 1000 cycle of stretching. Therefore our results demonstrate that collagen are denatured under cyclic tensile test. Further for the same number of cycles, denaturing of collagen was found significantly more (i.e. intensity ratio (1632/1657 cm^{-1}) was significantly more) when specimen was stretched at low frequency whereas collagen were not found significantly denatured at high stretching frequency. Based on the obtained results we propose the following mechanism: at slow cyclic stretching, dermis remains stretch for long time, which cause transfer of the applied stress to molecular level and the stressed collagen molecule start to unfold and loosening, which causes denature of the collagen triple helix. The denatured state of triple helix may activate the biochemical pathways for remodeling of the tissue. This study indicates that the mechanical stimulation associated tissue remodeling may affect by the rate of mechanical stimulation as well as how long the mechanical stimulation is applied.

5 Conclusion

The obtained Results from the experiment show that the collagen denaturation is more at low stretching and negligible at high frequency which indicate rate dependent stretching of protein molecules and denature. The rate dependent alteration of collagen molecule may affect the remodelling of the tissue. Overall the result of this study will help to understand the effect cyclic mechanical stimulation at different frequency on the remodelling of collagen in dermis and other connective tissue.

Conflicts of Interest "There are no conflicts of interest to declare by the authors."

Ethical Approval The authors hereby state that the present work is in compliance with the ethical standards.

Competing Interests The authors declare no competing interests.

Availability of Data and Materials Not applicable.

References

Bonnans, C., Chou, J. and Werb, Z., 2014. Remodelling the extracellular matrix in development and disease. *Nature reviews Molecular cell biology*, *15*(12), pp.786-801.

Bose, S., Li, S., Mele, E. and Silberschmidt, V.V., 2022. Exploring the mechanical properties and performance of Type-I collagen at various length scales: a progress report. *Materials*, *15*(8), p.2753.

Buehler, M.J., 2006. Atomistic and continuum modeling of mechanical properties of collagen: elasticity, fracture, and self-assembly. *Journal of Materials Research*, *21*(8), pp.1947-1961.

Dwivedi, K.K, Lakhani, P., Sihota, P., Tikoo, K., Kumar, S. and Kumar, N., 2023. The multiscale characterization and constitutive modeling of healthy and type 2 diabetes mellitus Sprague Dawley rat skin. *Acta Biomaterialia*, *158*, pp.324-346.

Dwivedi, K.K., Lakhani, P., Kumar, S. and Kumar, N., 2020a. Frequency dependent inelastic response of collagen architecture of pig dermis under cyclic tensile loading: An experimental study. *Journal of the Mechanical Behavior of Biomedical Materials*, *112*, p.104030

Dwivedi, K.K., Lakhani, P., Kumar, S. and Kumar, N., 2020b. The effect of strain rate on the stress relaxation of the pig dermis: a hyper-viscoelastic approach. *Journal of Biomechanical Engineering*, *142*(9), p.091006.

Dwivedi, K.K., Lakhani, P., Kumar, S. and Kumar, N., 2022a. A hyperelastic model to capture the mechanical behaviour and histological aspects of the soft tissues. *Journal of the Mechanical Behavior of Biomedical Materials*, *126*, p.105013.

Dwivedi, K.K., Lakhani, P., Kumar, S. and Kumar, N., 2022b. Effect of collagen fibre orientation on the Poisson's ratio and stress relaxation of skin: an ex vivo and in vivo study. *Royal Society Open Science*, *9*(3), p.211301.

Fung, Y.C., 2013. *Biomechanics: mechanical properties of living tissues*. Springer Science & Business Media.

Gautieri, A., Vesentini, S., Redaelli, A. and Buehler, M.J., 2012. Viscoelastic properties of model segments of collagen molecules. *Matrix Biology*, *31*(2), pp.141-149.

Hollingsworth, S.A. and Karplus, P.A., 2010. A fresh look at the Ramachandran plot and the occurrence of standard structures in proteins.

Jhorar, R. and Lamba, C.S., 2022. A Review on the Deformation Mechanism of Soft Tissue Collagen Molecules: An Atomistic Scale Experimental and Simulation Approaches. *Forcefields for Atomistic-Scale Simulations: Materials and Applications*, pp.269–297.

Kuehlmann, B., Bonham, C.A., Zucal, I., Prantl, L. and Gurtner, G.C., 2020. Mechanotransduction in wound healing and fibrosis. *Journal of clinical medicine*, *9*(5), p.1423.

Lakhani, P., Dwivedi, K.K. and Kumar, N., 2020. Directional dependent variation in mechanical properties of planar anisotropic porcine skin tissue. *Journal of the mechanical behavior of biomedical materials*, *104*, p.103693.

Lakhani, P., Dwivedi, K.K., Parashar, A. and Kumar, N., 2021. Non-invasive in vivo quantification of directional dependent variation in mechanical properties for human skin. *Frontiers in bioengineering and biotechnology*, *9*, p.749492.

Rabotyagova, O.S., Cebe, P. and Kaplan, D.L., 2008. Collagen structural hierarchy and susceptibility to degradation by ultraviolet radiation. *Materials Science and Engineering: C*, *28*(8), pp.1420-1429.

Sheng, R., Jiang, Y., Backman, L.J., Zhang, W. and Chen, J., 2020. The application of mechanical stimulations in tendon tissue engineering. *Stem cells international*, *2020*.

Sherman, V.R., Yang, W. and Meyers, M.A., 2015. The materials science of collagen. *Journal of the mechanical behavior of biomedical materials*, *52*, pp.22-50.
Uniyal, P., Sihota, P. and Kumar, N., 2022. Effect of organic matrix alteration on strain rate dependent mechanical behaviour of cortical bone. *Journal of the Mechanical Behavior of Biomedical Materials*, *125*, p.104910.
Yang, W., Sherman, V.R., Gludovatz, B., Schaible, E., Stewart, P., Ritchie, R.O. and Meyers, M.A., 2015. On the tear resistance of skin. *Nature communications*, *6*(1), p.6649.

Atomistic Simulations for Mechanical Behaviour of Natural Biopolymers for Material Design

Jashveer Singh and Rajesh Kumar

1 Introduction

Natural biopolymers, encompassing a diverse range of molecules such as proteins, nucleic acids, and polysaccharides, play a crucial role in the structure and function of biological systems (Padalkar et al. 2010). These biopolymers encompass cellulose present in plants and bacteria, chitin found in animals, silk within the silks of spiders and silkworm cocoons, along with collagen forming the structural matrix in ligaments, the dermis of echinoderms, as well as tendons and bones in mammals (Postek et al. 2010; Raabe et al. 2005; Keten et al. 2010; Nair et al. 2013). These biopolymers exhibit intricate molecular architectures that contribute to their exceptional mechanical properties, making them intriguing candidates for materials science and engineering applications (Baranwal et al. 2022). For instance, spider silk fibers exhibit unparalleled toughness, surpassing even the strength of steel and Kevlar fibers by several magnitudes (Liu et al. 2016; Nova et al. 2010). Similarly, bones and exoskeletons in crustaceans exhibit exceptional resistance to defects (Wegst et al. 2015; Cranford and Buehler 2012), while woods demonstrate intelligent adaptive behaviour through their capacity for smart buckling (Fratzl and Weinkamer 2007). Understanding the mechanical behaviour of these biopolymers at the atomic scale is essential for designing innovative materials with tailored properties for various applications, including biomaterials, drug delivery systems, and tissue engineering scaffolds (Koons et al. 2020; Marrese et al. 2017; Matricardi et al. 2013; Gronau et al. 2012; Biswas, et al. 2021; Wu et al. 2014). These polymers often exhibit desirable attributes such as biodegradability, biocompatibility, and low toxicity, making them environmentally friendly and suitable for various biomedical applications. The mechanical properties of natural biopolymers are intricately linked to their structural

J. Singh · R. Kumar (✉)
Department of Mechanical Engineering, National Institute of Technology, Hamirpur, Himachal Pradesh 177005, India
e-mail: rajeshpawar4u@gmail.com

J. Gupta and A. Verma (eds.), *Microbiology-2.0 Update for a Sustainable Future*,
https://doi.org/10.1007/978-981-99-9617-9_21

arrangements and interactions. Flexibility, elasticity, and strength are fundamental attributes that influence their performance in material design applications (George et al. 2020). The capacity of biopolymers to undergo conformational alterations plays a crucial role in tasks such as engaging with ligands and catalyzing enzyme reactions (Grant et al. 2010).

Traditional experimental techniques, while invaluable, often face challenges in characterizing the mechanical properties of biopolymers due to their complex and dynamic nature (Kumar et al. 2011; Yakushevich 1993; Shahini et al. 2021). Atomistic simulations, particularly molecular dynamics simulations, have emerged as powerful tools to bridge this gap by providing detailed insights into the mechanical response of biopolymers at the atomic level (Heller et al. 1990; Zhang et al. 2020; Bejagam et al. 2022). By employing computational techniques to model the interactions between individual atoms and simulate their dynamic behavior over time, researchers can gain a deeper understanding of the underlying mechanisms governing them.

2 Atomistic Modelling Techniques

Numerous methods have been utilized by researchers to anticipate the mechanical characteristics of nanofillers. Despite the associated higher expenses and time requirements of experimental techniques, these methods are still utilized for the characterization of nanofillers. Transmission electron microscopy (TEM) and atomic force microscopy (AFM) (Golberg et al. 2007; Arenal et al. 2011) based techniques dominate experimental approaches. Since traditional methods are time-consuming and expensive, an alternative is developing in the form of computer methods based on the finite element method (FEM), density functional theory (DFT), and molecular dynamics (MD). Researchers often use finite element method (FEM) based atomic truss frame models for the modeling of carbon–carbon bonding in carbon-nanotube (CNT) and graphene (Parashar and Mertiny 2012; Li and Chou 2003). Researchers have also developed methods to predict hexagonal boron-nitride (hBN) nanofillers using molecular mechanics (MM) (Boldrin et al. 2011). It is believed that these numerical methods are adequate for simulating global characteristics. For evaluating the atomic level characteristics of nanofillers, researchers often use higher fidelity simulations like density functional theory (DFT) (Kudin et al. 2001; Andrew et al. 2012) in addition to the above numerical approaches. To evaluate the elastic characteristics of nanofillers, DFT-based techniques use a quantum mechanical framework (Peng et al. 2012). Using density functional theory is computationally expensive, hence it can only be done on a small number of atoms at a time.

In contrast to the more sophisticated numerical simulations and experimental approaches with higher accuracy, there is significant potential in the employment of molecular dynamics (MD) based simulations. Notably, advancements in interatomic potentials have played a vital role in the prediction of nanofiller behaviour (Tersoff 1986, 1988a, 1988b). Among the various material properties, Young's modulus stands out as a well-established and dependable parameter. The stress–strain behaviour and elastic modulus of the considered nanofiller can be anticipated using the strain energy (U), a common factor in MD-based simulations. The connection between strain energy (U) and strain (ε) is elucidated by the subsequent equations to derive Young's modulus (Y, measured in N/m2) and 2D Young's modulus (YS, measured in N/m).

$$Y = (\partial^2 U/\partial \varepsilon^2)/V \tag{1}$$

$$Y = (\partial^2 U/\partial \varepsilon^2)/S \tag{2}$$

Here, V is the volume and S is the surface area of the atomic system.

2.1 Interatomic Potentials

MD-based simulations often use interatomic potentials that have been experimentally obtained or established with the aid of experiments or higher-fidelity numerical simulations. Mathematical formulations representing the potential energy (E) of a system of atoms are called interatomic potentials. The most popular ways to categorize potentials are as one, two, or many bodies. A few examples of popular interatomic pair potentials include Tersoff (Tersoff 1989), LJ (Jones 1924), the embedded atom method (EAM) (Daw and Baskes 1984), REBO (Jeong et al. 2007), AIREBO (Stuart et al. 2000), CHARMM (MacKerell et al. 2000), and COMPASS (Sun et al. 1998). Potentials are further categorized as either bonded (Tersoff, REBO, and EAM) or non-bonded (van der Waals, *LJ*), electrostatic potentials. The mechanical and thermal behaviour of nanofillers like graphene, carbon nanotubes (CNTs), and hexagonal boron nitride (hBN) are often modeled using Tersoff-type interatomic potentials (Tersoff 1986, 1988a, 1988b). In Tersoff-type potentials, the potential energy (E) is expressed mathematically as:

$$E = \sum_i E_i = \left(\sum_{i \neq j} V_{ij}\right) \Big/ 2, \tag{3}$$

$$V_{ij} = f_c(r_{ij})[a_{ij} f_R(r_{ij}) + b_{ij} f_A(r_{ij})]. \tag{4}$$

Here, we have the system's total energy (E), conveniently broken down into its parts—the site energy (E_i) and the bond energy (V_{ij}). Both indices i and j extend across all particles in the system, while measuring the separation (r_{ij}) between them.

When the wave functions of two atoms overlap, a repulsive potential arises, represented by the function f_R, and an attracting potential arises, represented by the function f_A. These can be expressed as:

$$f_R(r) = A\ \exp(-\lambda_1 r), \tag{5}$$

$$f_A(r) = -B\ \exp(-\lambda_2 r). \tag{6}$$

These atomic interactions are only allowed to take place within a certain cutoff distance, defined by a cutoff function f_c, to keep the computational cost down.

$$f_c(r) = \begin{cases} 1, & r < R - D \\ (1/2) - \sin\left(\frac{\pi(r-R)}{2D}\right)\Big/2, & R - D < r < R + D \\ 0, & r > R + D \end{cases} \tag{7}$$

The cut-off radius (R) is typically selected such that just the first neighbour shell is considered. The function b_{ij} represents a measure of the bond order. In addition, the terms which act to limit the range of interaction to the first neighbour shell, are included in b_{ij}. The function a_{ij} consists solely of such range-limiting terms.

The mathematical equation used for potential energy in CHARMM27 potential for nucleic acids is expressed as:

$$\begin{aligned} V(r) = & \sum_{bonds} K_b(b - b_0)^2 + \sum_{angles} K_\theta(\theta - \theta_0)^2 \\ & + \sum_{dihedrals} K_\chi(1 + \cos(n\chi - \delta)) + \sum_{impropers} K_\phi(\phi - \phi_0)^2 \\ & + \sum_{Urey-Bradley} K_{UB}(S - S_0)^2 \\ & + \sum_{nonbondedatompairs} \left(\varepsilon_{ij}\left[\left(\frac{R_{min,ij}}{r_{ij}}\right)^{12} - \left(\frac{R_{min,ij}}{r_{ij}}\right)^{6}\right] + \frac{q_i q_j}{\varepsilon_D r_{ij}}\right) \end{aligned} \tag{8}$$

In Eq. (8), K_b, K_θ, K_χ, K_ϕ, and K_{UB} are the bond, valance angles, dihedral angle, improper dihedral angle force constant, and Urey-Bradley respectively; b, θ, χ, ϕ, and S are the bond length, bond angle, dihedral torsion angle, improper dihedral angle, and Urey-Bradley 1,3 distance respectively. The subscript value shows the values for individual terms at equilibrium. In addition, for the dihedral terms, δ is the phase and n are the multiplicity which tells about the min and max value.

For the atomic interaction in organic compound materials, the mathematical equation for the total energy of COMPASS force field potential is expressed as shown here.

$$\begin{aligned} E_{total} = & E_b + E_\theta + E_\phi + E_\chi + E_{bb\prime} + E_{b\theta} \\ & + E_{b\phi} + E_{\theta\theta\prime} + E_{\theta\theta\prime\phi} + E_{elec} + E_{LJ} \end{aligned} \tag{9}$$

These potential terms are divided into two categories: valance terms with diagonal and off-diagonal cross-coupling and nonbonded interaction terms. E_b, E_θ, E_ϕ, and E_χ are the valence terms for bond, angle, torsion, and out-of-plane angle coordinates. For predicting the vibrational frequency and structural variations associated with conformational variations, cross-coupling terms are useful. $E_{bb'}$, $E_{b\theta}$, $E_{b\phi}$, $E_{\theta\theta'}$, and $E_{\theta\theta'\phi}$ are the cross-coupling terms in between internal coordinates. On the other hand, for nonbonded interactions E_{elec} and E_{LJ} are used.

The use of MD-based simulations to forecast the characteristics of these novel materials is becoming more feasible. To further enhance these computations, researchers are also making improvements to these potentials. Brenner (1990) expanded upon the original Tersoff potential stated in the article (Tersoff 1986). Optimized Tersoff potential parameters have been suggested for increasing the fidelity of these atomistic simulations by many researchers, including Brenner (1990), Matsunaga et al. (2001, 2000), Albe et al. (1998, 1997), Nord et al. (2003), and Sevik et al. (2011).

3 Atomistic Studies of Mechanical Prosperties of Biopolymers

The mechanical characteristics of biopolymers play a significant role in a variety of fields, including biomedicine, biosensors, and biotechnology. Bejagam et al. (2022) conducted a study using molecular dynamics (MD) simulations to explore the mechanical behaviour of Polyhydroxy alkanoates (PHAs). Their findings indicated that an increased number of carbon atoms in both side chains and the backbone leads to the softening of PHAs, subsequently decreasing their mechanical properties. Rath et al. (2016) employed MD simulations to characterize the mechanical attributes of chitosan, a biodegradable biopolymer, which was reinforced with hydroxyapatite nanoparticles. Enhanced mechanical properties were observed in the resulting nanocomposite, attributed to hydrogen bonding and electrostatic interactions between chitosan and hydroxyapatite. However, this nanocomposite exhibited poor wear properties. Kulasinski et al. (2015) delved into the physical properties of two hydrophilic biopolymers, amorphous cellulose, and hemicellulose, using MD simulations. The researchers investigated the influence of water adsorption on the mechanical attributes of nanocrystalline biopolymers. These biopolymers' mechanical properties, when reinforced with nanocrystalline structures, are contingent on the interactions at their interface. Lai et al. (2018) utilized MD simulations to study the interfacial interactions of protein-mineral nanocomposites. Specifically, they investigated the reinforcement of osteopontin (OPN) with hydroxyapatite (HA), including defects on the HA surface. Results revealed that HA layers with more nano-sized grooves exhibited reduced pulling force and energy dissipation. In a different study, Patel et al. (2020) investigated the mechanical properties of Bombyx mori silk fibroin biopolymer using the CHARMM27 force field. They achieved improved elastic modulus (7.4 GPa) and tensile strength (340 MPa) compared to existing literature. To further enhance these biopolymers' mechanical attributes, researchers incorporated nanomaterials. Pandele et al. (2014) focused on chitosan-based graphene composites using MD simulations with COMPASS force fields, demonstrating an increase in the elastic modulus from 8.55 GPa (chitosan) to 11.55 GPa (chitosan/graphene). For the interfacial interactions of graphene-collagen biopolymer nanocomposites, Ebrahimi et al. (2013) conducted MD simulations, incorporating bond-stretch, bond-bending, and AIREBO potentials. Results indicated higher pull-out forces with rippled graphene compared to flat graphene within the nanocomposite. Rahman et al. (2013) enhanced the mechanical properties of cellulose biopolymer by reinforcing it with graphene nanosheets and utilizing molecular dynamics simulations with polymer-consistent force fields. Improved Young's modulus was observed in comparison to pure cellulose biopolymer. Bayoumy et al. (2019) introduced functionalized graphene quantum dots to chitosan biopolymer and performed DFT simulations for biophysical characterization. Their findings indicated that armchair hexagonal graphene quantum dots reinforced to chitosan formed the most stable structure. This nanocomposite exhibited potential for removing heavy metals from wastewater. Similarly, chitosan-based nanocomposites reinforced with graphene were explored by Menazea et al. (2020), utilizing DFT simulations. Masoumi et al. (2020) investigated the interaction between chitosan biopolymer and carbon nanotubes (CNTs) using DFT simulations. They revealed that chitosan tends to chemisorb onto CNT surfaces, suggesting potential applications in drug delivery and bone tissue engineering. In the realm of biosensors, Omar et al. (2022) reinforced functionalized graphene oxide with chitosan biopolymers. Computational models based on DFT calculations highlighted the stability of this structure for dopamine sensing. Kossovich et al. (2016) evaluated chitosan-based nanocomposites reinforced with various nanofillers using MD simulations, showcasing a tenfold improvement in Young's modulus upon reinforcement with graphene. Shende et al. (2021) emphasized the potential of biopolymers like chitin, cyclodextrins, and cellulose reinforced with graphene nanomaterials, predicting enhanced mechanical strength, biocompatibility, biodegradability, and lower toxicity. Introducing hexagonal boron nitride (hBN) reinforcement further enhanced properties, as demonstrated by Jayakumar et al. (2023) who used cellulose reinforced with hBN nanosheets for energy harvesting, indicating the promising potential for sensing technology. Anota et al. (2013) examined the impact of point defects on the adsorption properties of chitosan reinforced with hBN nanosheets through DFT calculations, revealing improved chemical adsorption capacity on hBN surfaces with divacancies.

4 Conclusions and Future Perspectives

Biopolymers are naturally existing polymers extracted from living organisms. They play a vital role as building blocks in humans and other living media. Their use in biomedical applications, ranging from drug delivery to bone tissue engineering, is getting more prevalent. At the one end, they are being used at the nano level for drug delivery while at the same time, they are useful as scaffolds in bone tissue engineering. The present chapter discusses their mechanical behavior estimated using atomistic simulations. Atomistic simulations are an excellent tool for capturing nanoscale occurring in living organisms which are hard to capture using experimental techniques. More and more use of atomistic techniques in exploring the mechanical of biopolymers and their nanocomposites will open the doors for their use in biomedical applications such as biocompatible heart stunts, scaffolds for tissue engineering, etc.

Declarations

Conflicts of Interest "There are no conflicts of interest to declare by the authors."

Ethical Approval The authors hereby state that the present work is in compliance with the ethical standards.

Competing Interests The authors declare no competing interests.

Availability of Data and Materials Not applicable.

References

K. Albe and W. Möller, "Modelling of boron nitride: Atomic scale simulations on thin film growth," *Comput. Mater. Sci.*, vol. 10, no. 1, pp. 111–115, 1998, doi: https://doi.org/https://doi.org/10.1016/S0927-0256(97)00172-9.

K. Albe, W. Möller, and K.-H. Heinig, "Computer simulation and boron nitride," *Radiat. Eff. Defects Solids*, vol. 141, no. 1–4, pp. 85–97, 1997, doi: https://doi.org/10.1080/10420159708211560.

R. C. Andrew, R. E. Mapasha, A. M. Ukpong, and N. Chetty, "Mechanical properties of graphene and boronitrene," *Phys. Rev. B*, vol. 85, no. 12, p. 125428, 2012, doi: https://doi.org/10.1103/PhysRevB.85.125428.

E. C. Anota, L. D. Rodríguez, and G. H. Cocoletzi, "Influence of point defects on the adsorption of chitosan on graphene-like BN nanosheets," *Graphene*, vol. 1, no. 2, pp. 124–130, 2013.

R. Arenal, M.-S. Wang, Z. Xu, A. Loiseau, and D. Golberg, "Young modulus, mechanical and electrical properties of isolated individual and bundled single-walled boron nitride nanotubes," *Nanotechnology*, vol. 22, no. 26, p. 265704, 2011, doi: https://doi.org/10.1088/0957-4484/22/26/265704.

J. Baranwal, B. Barse, A. Fais, G. L. Delogu, and A. Kumar, "Biopolymer: A Sustainable Material for Food and Medical Applications," *Polymers*, vol. 14, no. 5. 2022, doi: https://doi.org/10.3390/polym14050983.

A. M. Bayoumy *et al.*, "Functionalization of graphene quantum dots (GQDs) with chitosan biopolymer for biophysical applications," *Opt. Quantum Electron.*, vol. 52, no. 1, p. 16, 2019, doi: https://doi.org/10.1007/s11082-019-2134-z.

K. K. Bejagam, N. S. Gupta, K.-S. Lee, C. N. Iverson, B. L. Marrone, and G. Pilania, "Predicting the mechanical response of polyhydroxyalkanoate biopolymers using molecular dynamics simulations," *Polymers (Basel).*, vol. 14, no. 2, p. 345, 2022.

M. C. Biswas *et al.*, "Recent advancement of biopolymers and their potential biomedical applications," *J. Polym. Environ.*, pp. 1–24, 2021.

L. Boldrin, F. Scarpa, R. Chowdhury, and S. Adhikari, “Effective mechanical properties of hexagonal boron nitride nanosheets,” *Nanotechnology*, vol. 22, no. 50, p. 505702, 2011, doi: https://doi.org/10.1088/0957-4484/22/50/505702.

D. W. Brenner, “Empirical potential for hydrocarbons for use in simulating the chemical vapor deposition of diamond films,” *Phys. Rev. B*, vol. 42, no. 15, pp. 9458–9471, 1990, doi: https://doi.org/10.1103/PhysRevB.42.9458.

S. W. Cranford and M. J. Buehler, *Biomateriomics*, vol. 165. Springer Science & Business Media, 2012.

M. S. Daw and M. I. Baskes, “Embedded-atom method: Derivation and application to impurities, surfaces, and other defects in metals,” *Phys. Rev. B*, vol. 29, no. 12, pp. 6443–6453, 1984, doi: https://doi.org/10.1103/PhysRevB.29.6443.

S. Ebrahimi, A. Montazeri, and H. Rafii-Tabar, “Molecular dynamics study of the interfacial mechanical properties of the graphene–collagen biological nanocomposite,” *Comput. Mater. Sci.*, vol. 69, pp. 29–39, 2013, doi: https://doi.org/https://doi.org/10.1016/j.commatsci.2012.11.030.

P. Fratzl and R. Weinkamer, “Nature’s hierarchical materials,” *Prog. Mater. Sci.*, vol. 52, no. 8, pp. 1263–1334, 2007.

A. George, M. R. Sanjay, R. Srisuk, J. Parameswaranpillai, and S. Siengchin, “A comprehensive review on chemical properties and applications of biopolymers and their composites,” *Int. J. Biol. Macromol.*, vol. 154, pp. 329–338, 2020.

D. Golberg *et al.*, “Direct Force Measurements and Kinking under Elastic Deformation of Individual Multiwalled Boron Nitride Nanotubes,” *Nano Lett.*, vol. 7, no. 7, pp. 2146–2151, 2007, doi: https://doi.org/10.1021/nl070863r.

B. J. Grant, A. A. Gorfe, and J. A. McCammon, “Large conformational changes in proteins: signaling and other functions,” *Curr. Opin. Struct. Biol.*, vol. 20, no. 2, pp. 142–147, 2010.

G. Gronau *et al.*, “A review of combined experimental and computational procedures for assessing biopolymer structure–process–property relationships,” *Biomaterials*, vol. 33, no. 33, pp. 8240–8255, 2012.

H. Heller, H. Grubmüller, and K. Schulten, “Molecular dynamics simulation on a parallel computer,” *Mol. Simul.*, vol. 5, no. 3–4, pp. 133–165, 1990.

A. Jayakumar *et al.*, “Energy Harvesting Using High-Strength and Flexible 3D-Printed Cellulose/Hexagonal Boron Nitride Nanosheet Composites,” *ACS Appl. Nano Mater.*, vol. 6, no. 15, pp. 14278–14288, 2023, doi: https://doi.org/10.1021/acsanm.3c02233.

B.-W. Jeong, J.-K. Lim, and S. B. Sinnott, “Tensile mechanical behavior of hollow and filled carbon nanotubes under tension or combined tension-torsion,” *Appl. Phys. Lett.*, vol. 90, no. 2, 2007.

J. E. Jones, “On the determination of molecular fields.—II. From the equation of state of a gas,” *Proc. R. Soc. London. Ser. A, Contain. Pap. a Math. Phys. Character*, vol. 106, no. 738, pp. 463–477, 1924.

S. Keten, Z. Xu, B. Ihle, and M. J. Buehler, “Nanoconfinement controls stiffness, strength and mechanical toughness of β-sheet crystals in silk,” *Nat. Mater.*, vol. 9, no. 4, pp. 359–367, 2010.

G. L. Koons, M. Diba, and A. G. Mikos, “Materials design for bone-tissue engineering,” *Nat. Rev. Mater.*, vol. 5, no. 8, pp. 584–603, 2020.

E. L. Kossovich and R. A. Safonov, “Predictive analysis of chitosan-based nanocomposite biopolymers elastic properties at nano- and microscale,” *J. Mol. Model.*, vol. 22, no. 4, p. 75, 2016, doi: https://doi.org/10.1007/s00894-016-2942-z.

K. N. Kudin, G. E. Scuseria, and B. I. Yakobson, “${\mathrm{C}}_{2}\mathrm{F},$ BN, and C nanoshell elasticity from ab initio computations,” *Phys. Rev. B*, vol. 64, no. 23, p. 235406, 2001, doi: https://doi.org/10.1103/PhysRevB.64.235406.

K. Kulasinski, R. Guyer, S. Keten, D. Derome, and J. Carmeliet, “Impact of Moisture Adsorption on Structure and Physical Properties of Amorphous Biopolymers,” *Macromolecules*, vol. 48, no. 8, pp. 2793–2800, 2015, doi: https://doi.org/10.1021/acs.macromol.5b00248.

P. Kumar, K. P. Sandeep, S. Alavi, and V. D. Truong, "A review of experimental and modeling techniques to determine properties of biopolymer-based nanocomposites," *J. Food Sci.*, vol. 76, no. 1, pp. E2–E14, 2011.
Z. B. Lai, R. Bai, Z. Lei, and C. Yan, "Interfacial mechanical behaviour of protein–mineral nanocomposites: A molecular dynamics investigation," *J. Biomech.*, vol. 73, pp. 161–167, 2018, doi: https://doi.org/https://doi.org/10.1016/j.jbiomech.2018.03.044.
C. Li and T.-W. Chou, "A structural mechanics approach for the analysis of carbon nanotubes," *Int. J. Solids Struct.*, vol. 40, no. 10, pp. 2487–2499, 2003, doi: https://doi.org/https://doi.org/10.1016/S0020-7683(03)00056-8.
R. Liu, Q. Deng, Z. Yang, D. Yang, M. Han, and X. Y. Liu, "'Nano-Fishnet' Structure Making Silk Fibers Tougher," *Adv. Funct. Mater.*, vol. 26, no. 30, pp. 5534–5541, 2016.
A. D. MacKerell Jr, N. Banavali, and N. Foloppe, "Development and current status of the CHARMM force field for nucleic acids," *Biopolym. Orig. Res. Biomol.*, vol. 56, no. 4, pp. 257–265, 2000.
M. Marrese, V. Guarino, and L. Ambrosio, "Atomic force microscopy: a powerful tool to address scaffold design in tissue engineering," *J. Funct. Biomater.*, vol. 8, no. 1, p. 7, 2017.
M. Masoumi, M. Jahanshahi, M. G. Ahangari, and G. N. Darzi, "Density functional theory study on the interaction of chitosan monomer with TiO2, SiO2 and carbon nanotubes," *Mater. Chem. Phys.*, vol. 255, p. 123576, 2020, doi: https://doi.org/https://doi.org/10.1016/j.matchemphys.2020.123576.
P. Matricardi, C. Di Meo, T. Coviello, W. E. Hennink, and F. Alhaique, "Interpenetrating polymer networks polysaccharide hydrogels for drug delivery and tissue engineering," *Adv. Drug Deliv. Rev.*, vol. 65, no. 9, pp. 1172–1187, 2013.
K. Matsunaga, C. Fisher, and H. Matsubara, "Tersoff Potential Parameters for Simulating Cubic Boron Carbonitrides," *Jpn. J. Appl. Phys.*, vol. 39, no. Part 2, No. 1A/B, pp. L48–L51, 2000, doi: https://doi.org/10.1143/jjap.39.l48.
K. Matsunaga and Y. Iwamoto, "Molecular Dynamics Study of Atomic Structure and Diffusion Behavior in Amorphous Silicon Nitride Containing Boron," *J. Am. Ceram. Soc.*, vol. 84, no. 10, pp. 2213–2219, 2001, doi: https://doi.org/https://doi.org/10.1111/j.1151-2916.2001.tb00990.x.
A. A. Menazea *et al.*, "Chitosan/graphene oxide composite as an effective removal of Ni, Cu, As, Cd and Pb from wastewater," *Comput. Theor. Chem.*, vol. 1189, p. 112980, 2020, doi: https://doi.org/https://doi.org/10.1016/j.comptc.2020.112980.
A. K. Nair, A. Gautieri, S.-W. Chang, and M. J. Buehler, "Molecular mechanics of mineralized collagen fibrils in bone," *Nat. Commun.*, vol. 4, no. 1, p. 1724, 2013.
J. Nord, K. Albe, P. Erhart, and K. Nordlund, "Modelling of compound semiconductors: analytical bond-order potential for gallium, nitrogen and gallium nitride," *J. Phys. Condens. Matter*, vol. 15, no. 32, pp. 5649–5662, 2003, doi: https://doi.org/10.1088/0953-8984/15/32/324.
A. Nova, S. Keten, N. Pugno, A. Redaelli, and M. Buehler, "Molecular and nanostructural mechanisms of deformation, strength and toughness of spider silk fibrils," *Nat. Preced.*, p. 1, 2010.
A. Omar, A. M. Bayoumy, and A. A. Aly, "Functionalized Graphene Oxide with Chitosan for Dopamine Biosensing," *Journal of Functional Biomaterials*, vol. 13, no. 2. 2022, doi: https://doi.org/10.3390/jfb13020048.
S. Padalkar *et al.*, "Natural Biopolymers: Novel Templates for the Synthesis of Nanostructures," *Langmuir*, vol. 26, no. 11, pp. 8497–8502, 2010, doi: https://doi.org/10.1021/la904439p.
A. M. Pandele, M. IONIŢĂ, and H. Iovu, "Molecular modeling of mechanical properties of the chitosan based graphene composites," *UPB Sci Bull Ser B Chem Mater Sci*, vol. 76, pp. 107–112, 2014.
A. Parashar and P. Mertiny, "Study of mode i fracture of graphene sheets using atomistic based finite element modeling and virtual crack closure technique," *Int. J. Fract.*, vol. 176, no. 1, pp. 119–126, 2012, doi: https://doi.org/10.1007/s10704-012-9718-y.
M. Patel, D. K. Dubey, and S. P. Singh, "Phenomenological models of Bombyx mori silk fibroin and their mechanical behavior using molecular dynamics simulations," *Mater. Sci. Eng. C*, vol. 108, p. 110414, 2020, doi: https://doi.org/https://doi.org/10.1016/j.msec.2019.110414.

Q. Peng, W. Ji, and S. De, "Mechanical properties of the hexagonal boron nitride[1] Q. Peng, W. Ji, and S. De, 'Mechanical properties of the hexagonal boron nitride monolayer: Ab initio study,' Comput. Mater. Sci., vol. 56, pp. 11–17, 2012, doi: https://doi.org/10.1016/j.commatsci.2," *Comput. Mater. Sci.*, vol. 56, pp. 11–17, 2012, doi: https://doi.org/10.1016/j.commatsci.2011.12.029.

M. T. Postek *et al.*, "Development of the metrology and imaging of cellulose nanocrystals," *Meas. Sci. Technol.*, vol. 22, no. 2, p. 24005, 2010.

D. Raabe, C. Sachs, and P. Romano, "The crustacean exoskeleton as an example of a structurally and mechanically graded biological nanocomposite material," *Acta Mater.*, vol. 53, no. 15, pp. 4281–4292, 2005.

R. Rahman, J. T. Foster, and A. Haque, "Molecular Dynamics Simulation and Characterization of Graphene–Cellulose Nanocomposites," *J. Phys. Chem. A*, vol. 117, no. 25, pp. 5344–5353, 2013, doi: https://doi.org/10.1021/jp402814t.

A. Rath, S. Mathesan, and P. Ghosh, "Nanomechanical characterization and molecular mechanism study of nanoparticle reinforced and cross-linked chitosan biopolymer," *J. Mech. Behav. Biomed. Mater.*, vol. 55, pp. 42–52, 2016, doi: https://doi.org/https://doi.org/10.1016/j.jmbbm.2015.10.005.

C. Sevik, A. Kinaci, J. B. Haskins, and T. Çağın, "Characterization of thermal transport in low-dimensional boron nitride nanostructures," *Phys. Rev. B*, vol. 84, no. 8, p. 85409, 2011, doi: https://doi.org/10.1103/PhysRevB.84.085409.

M. H. Shahini, B. Ramezanzadeh, and H. E. Mohammadloo, "Recent advances in biopolymers/carbohydrate polymers as effective corrosion inhibitive macro-molecules: A review study from experimental and theoretical views," *J. Mol. Liq.*, vol. 325, p. 115110, 2021.

P. Shende and N. Pathan, "Potential of carbohydrate-conjugated graphene assemblies in biomedical applications," *Carbohydr. Polym.*, vol. 255, p. 117385, 2021, doi: https://doi.org/https://doi.org/10.1016/j.carbpol.2020.117385.

S. J. Stuart, A. B. Tutein, and J. A. Harrison, "A reactive potential for hydrocarbons with intermolecular interactions," *J. Chem. Phys.*, vol. 112, no. 14, pp. 6472–6486, 2000, doi: https://doi.org/10.1063/1.481208.

H. Sun, P. Ren, and J. R. Fried, "The COMPASS force field: parameterization and validation for phosphazenes," *Comput. Theor. Polym. Sci.*, vol. 8, no. 1, pp. 229–246, 1998, doi: https://doi.org/https://doi.org/10.1016/S1089-3156(98)00042-7.

J. Tersoff, "New empirical model for the structural properties of silicon," *Phys. Rev. Lett.*, vol. 56, no. 6, pp. 632–635, 1986, doi: https://doi.org/10.1103/PhysRevLett.56.632.

J. Tersoff, "New empirical approach for the structure and energy of covalent systems," *Phys. Rev. B*, vol. 37, no. 12, pp. 6991–7000, 1988, doi: https://doi.org/10.1103/PhysRevB.37.6991.

J. Tersoff, "Empirical Interatomic Potential for Carbon, with Applications to Amorphous Carbon," *Phys. Rev. Lett.*, vol. 61, no. 25, pp. 2879–2882, 1988, doi: https://doi.org/10.1103/PhysRevLett.61.2879.

J. Tersoff, "Modeling solid-state chemistry: Interatomic potentials for multicomponent systems," *Phys. Rev. B*, vol. 39, no. 8, pp. 5566–5568, 1989, doi: https://doi.org/10.1103/PhysRevB.39.5566.

U. G. K. Wegst, H. Bai, E. Saiz, A. P. Tomsia, and R. O. Ritchie, "Bioinspired structural materials," *Nat. Mater.*, vol. 14, no. 1, pp. 23–36, 2015.

S. Wu, X. Liu, K. W. K. Yeung, C. Liu, and X. Yang, "Biomimetic porous scaffolds for bone tissue engineering," *Mater. Sci. Eng. R Reports*, vol. 80, pp. 1–36, 2014.

L. V Yakushevich, "Nonlinear dynamics of biopolymers: theoretical models, experimental data," *Q. Rev. Biophys.*, vol. 26, no. 2, pp. 201–223, 1993.

C. Zhang, B. Coasne, R. Guyer, D. Derome, and J. Carmeliet, "Moisture-induced crossover in the thermodynamic and mechanical response of hydrophilic biopolymer," *Cellulose*, vol. 27, pp. 89–99, 2020.

A Brief Bibliometric Survey on Analysis of Wing Spar

Sonali Sahni, Abhishek Malhan, Harshil Jain, Singiredddy Siddhartha Reddy, and Rajesh Govindan

1 Introduction

1.1 Aircraft Wing

The word "aircraft" refers to an efficient mechanical invention with the ability to fly. The aircraft's two primary parts are its wing and fuselage (Tariq and Mazhar 2021; Elangovan et al. 2019; Bruce 2017; Venkatarthnaiah et al. 2013; Vartabedian 2018; Ajith et al. 2019; Girennavar et al. 2017; Arif 2020; Maheswaran et al. 2015; Madhu and Kadole 2015; Pavan and Manjunath 2018; Zakuan et al. 2019; Fang et al. 2019; Kleemann et al. 2020; Son et al. 2021; Wanhill et al. 1990). The wing of aircraft is made up of numerous airfoils. various types of flight loads are subjected to it i.e., shear stress, bending load, twisting load, etc. (Kumar et al. 2015; Kleemann et al. 2020; Son et al. 2021; Wanhill et al. 1990; Pak 2019; Toffol and Sergio 2023; Marangi and Salim 2020; Gandhi et al. 2021; Khan et al. 2023; Sun et al. 2017; Gupta et al. 2020; Jrad et al. 2017; Gokulnath et al. 2018; Zhang et al. 2018). It is discovered that the aerodynamic loads on the wing are what caused the stresses to be obtained (Kumar et al. 2013; Gokulnath et al. 2018; Zhang et al. 2018; Gupta et al. 2020; Xiao et al. 2022; Weckner 2021; Yu et al. 2019; Joels et al. 2021; Kontron and Dvořák 2020).

1.2 Wing Spar

The spar is a spanwise structural part that takes the lifting loads during flight as well as the self-weight and bending loads of the wing when the aircraft is on the

---

S. Sahni · A. Malhan · H. Jain · S. S. Reddy · R. Govindan (✉)
Aerospace Engineering, Chandigarh University, Mohali, India
e-mail: rajesh.e11762@cumail.in

J. Gupta and A. Verma (eds.), *Microbiology-2.0 Update for a Sustainable Future*,
https://doi.org/10.1007/978-981-99-9617-9_22

ground (Abbas et al. 2021; Joels et al. 2021; Kontron and Dvořák 2020; Stamatelos and George 2020; Khames et al. 2019; Alvarez-Montoya et al. 2020; Rodzewicz and Jerzy 2018; Eldwaib et al. 2018; Kilimtzidis et al. 2023; Augustin 2018; Irez et al. 2019; Black et al. 2018; Park et al. 2021). The principal bending loads that are applied to an aircraft's wings are normally carried by two spars in the wing. Sometimes the attachment for the landing gear uses an additional spar. The rib structures receive the lift force of aerodynamics operating on the bottom and top surfaces of the wing, which is then transmitted to the spar web as concentrated shear loads. The primary loading on the spars is caused by these shear loads, which bend the spars (Elangovans et al. 2019; Augustin 2018; Irez et al. 2019; Black et al. 2018; Park et al. 2021; Kumar and Selvaraj 2020; Jang et al. 2019; Das et al. 2018; Eldwaib et al. 2018; Cardoso 2020).

1.3 Static Loading

The strain and stresses experienced by the structure of the wing spar are referred to as Static Loading on a Wing Spar. These loads may originate from internal forces like lift and drag produced by the wing as well as external forces like the weight of the aircraft, fuel, cargo, and any other loads operating on the wing (Mukhopadhyay et al. 2019; Ahmed and Duncan 2021; Britto Maria et al. 2019; Yang et al. 2021; Zhang et al. 2018; Xiaozhe et al. 2018; Qiu et al. 2019; Clough et al. 2019; Stepanova et al. 2018).

1.4 Bibliometric Analysis

It is a computer-based analysis of different types of documents. In this study two type of analysis has been done i.e., Statistical Analysis and Numerical Analysis. Both analyses are based on Scopus data.

2 Materials and Approaches

2.1 Data Base Collection

Worldwide, there is numerous databases, including Scopus, etc. These documents include a huge collection of documents. For the analysis, the most well-known and significant database of these is Scopus. The keywords Aircraft, Wing, Spars, and Analysis are used. It returned a total of 345 publications. The various keywords are used to search databases across the world. Country, language, and other restrictions are not present. Every publication contains data on the author, the country, the citations, the documents, the sources, etc. This data is utilized in the analysis.

2.2 Outcomes Based on Initial Search

The language trends of the publications based on the Scopus database are presented in Table 1 and Fig. 1. The publications are found by using various keywords associated with the study. It is examined by considering the language. Chinese is the language with the second-highest number of publications and the English language has the highest publication (210). German has the lowest publication count.

Table 1 Publications of Language trends

Publishing language	No. of publication
ENGLISH	210
CHINESE	5
POLISH	2
KOREAN	2
GERMAN	1

Source http://www.scopus.com (assessed on 13th May 2023)

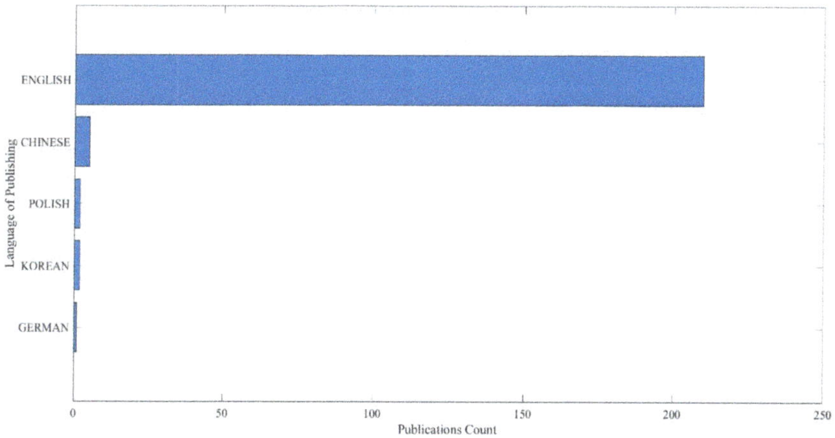

Fig. 1 Language Trends Based on Publications. *Source* http://www.scopus.com (assessed on 13th May 2023)

2.3 Top 10 Keywords on the Basis of Publication Outcome

The publications of the top 10 keywords based on the Scopus database are presented in Fig. 2. In addition to the basic keywords, there are numerous other keywords. The top 10 keywords are mentioned below. The highest publications of the keyword are Wings while Wing Spar is the keyword having the lowest publications (Table 2).

Table 2 Top 10 keywords based on highest publications

Keywords	No. of publications
Wings	103
Aircraft	76
Finite element method	38
Aircraft manufacture	30
Structural analysis	27
Stress analysis	18
Finite element analysis	12
Deformation	11
Wing spar	6
Aluminum alloys	5

Source http://www.scopus.com (assessed on 13th May 2023)

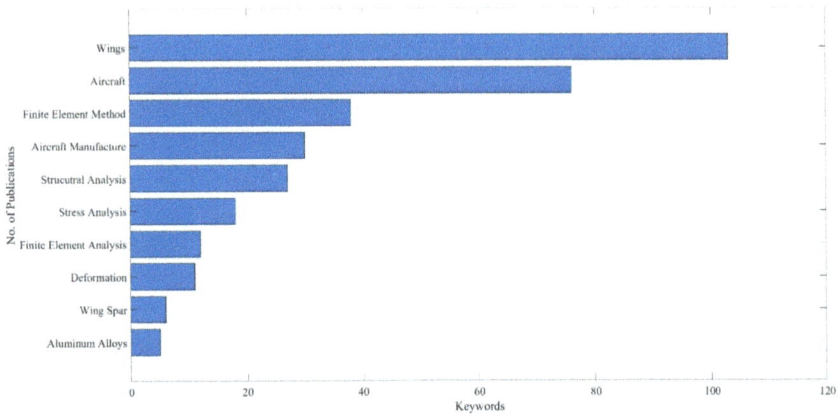

Fig. 2 Top 10 keywords based on highest number of publications. Source: http://www.scopus.com (assessed on 13th May 2023)

3 Performance Analysis

It is an analysis based on Scopus documents. The software named VOS Viewer is used for the analysis of the Database. It is quite an effective way to analyse.

Different types of analysis are performed which are mentioned below:

- Statistical Analysis of Databases.
- Network Analysis of Databases.

3.1 *Database Statistical Analysis*

1. Documents based on Source.
2. Documents based on year.
3. Documents based on the subject area.
4. Documents based on Type.
5. Documents based on Country.
6. Documents based on the author.
7. Documents based on affiliation.
8. Documents based on funding sponsor.

3.1.1 Documents Based on Source

The analysis of the document based on the source is presented in Fig. 3. The database shows distinct sources such as journals, Conference proceedings, Book series, and Book and Trade journals. The maximum number of documents published in the journal in the year 2013.

3.1.2 Documents Based on Year

This analysis of documents based on year is represented in Fig. 4. The documents collected through Scopus in the year 2009–2023 include distinct sources such as Book chapters, journals and Conferences from the graph. The limitation has been done for up to 10 years. It is found that 235 documents are left and it is recognized that the highest count of publications is in the years 2018 and 2019. The lowest number of publications is in the year 2023.

3.1.3 Documents Based on Subject Area

Document analysis depending on the subject area is represented in Fig. 5. The limitation of the subject area is Engineering and Mathematics because this subject area is related to our data and the rest of them are from different backgrounds such as

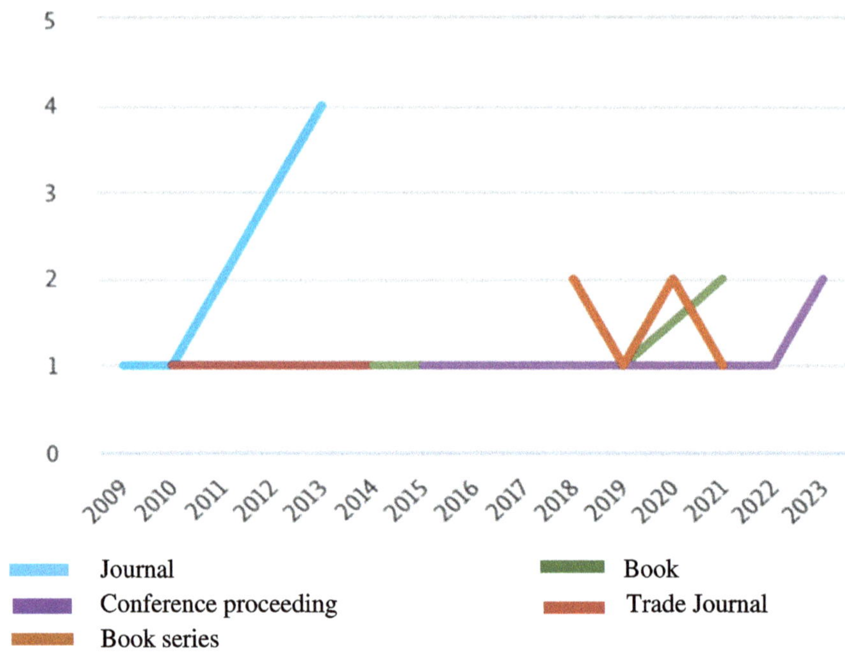

Fig. 3 Document analysis depending on the source. *Source* http://www.scopus.com (assessed on 13th May 2023)

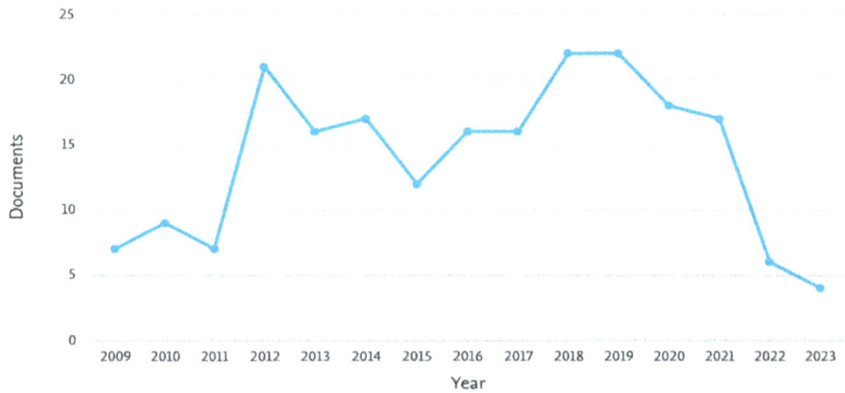

Fig. 4 Document analysis depending on the Year. *Source* http://www.scopus.com (assessed on 13th May 2023)

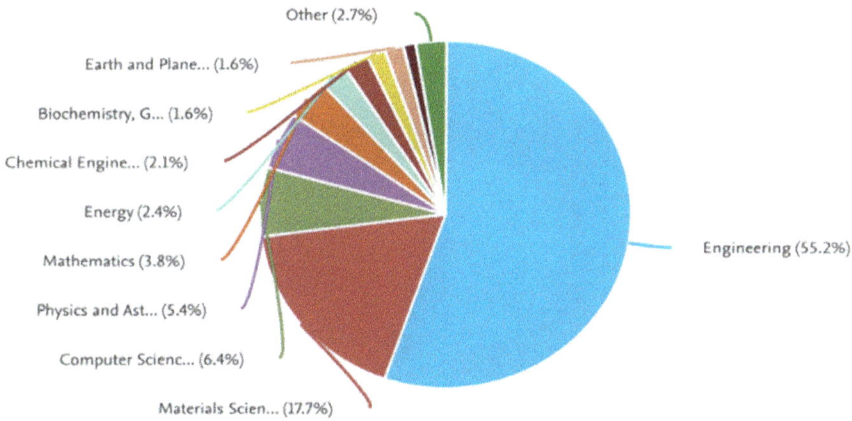

Fig. 5 Document analysis depending on the subject area. *Source* http://www.scopus.com (assessed on 13th May 2023)

health sciences and many more. After limiting the subject area, 217 documents are left. Engineering has 55.2% of documents, Material science has 17.7% of documents, Computer science has 6.4% of documents, Physics and astronomy has 5.4% of documents, Mathematics has 3.8% of documents, Energy has 2.4% of documents, Chemical engineering has 2.1% documents, Biochemistry has 1.6% documents, Earth and planetary has 1.6% documents and other has 2.7% documents.

3.1.4 Documents Based on Type

The analysis of the document based on type is represented in Fig. 6. This analysis is based on types of documents such as Articles conference papers, Book chapters, etc. It can be shown through the analysis. that most of the publications are conference papers. Conference paper has 47.6% of documents, article has 46.7% of documents, Book chapter has 2.4% of documents, conference review has 1.9% of documents and review has 1.4% of documents.

3.1.5 Documents Based on Country/Territory

This analysis of documents based on the country is represented in Fig. 7. The total count of documents published by various countries. It recognized that Serbia has the lowest count of publication of documents and United States has the highest count of publication of documents.

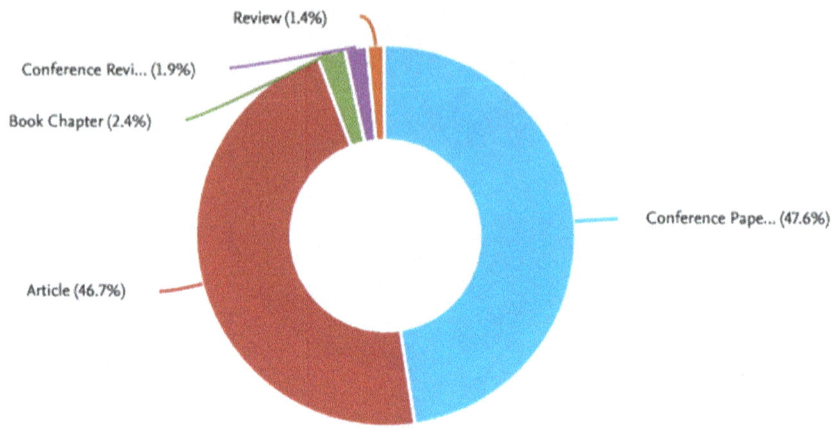

Fig. 6 Document analysis depending on type. *Source* http://www.scopus.com (assessed on 13th May 2023)

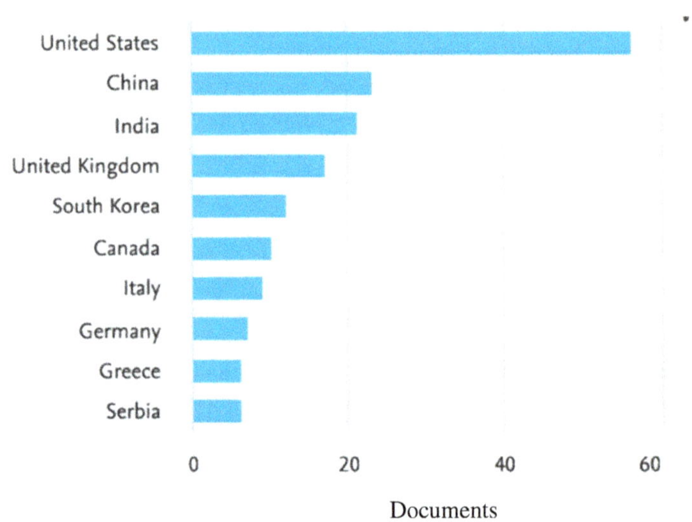

Fig. 7 Document analysis depending on Country. *Source* http://www.scopus.com (assessed on 13th May 2023)

3.1.6 Documents Based on Authors

The analysis of documents based on the author is represented in Fig. 8. In this analysis, the top 10 authors have been considered. It is observed that the maximum number of publications of Kapania R.K. and Kastratovic have the lowest number of publications.

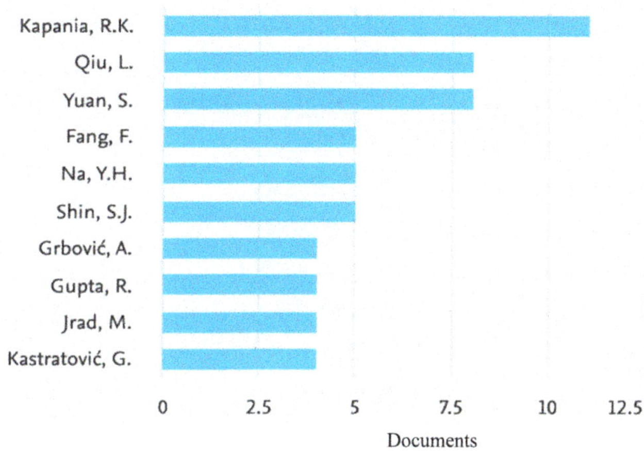

Fig. 8 Document analysis depending on Authors. *Source* http://www.scopus.com (assessed on 13th May 2023)

3.1.7 Documents Based on Affiliation

The analysis of documents based on affiliation is presented in Fig. 9. The top 10 affiliations have been considered. It is observed that Virginia Polytechnic has the maximum number of publications and University Libre de has the lowest number of publications.

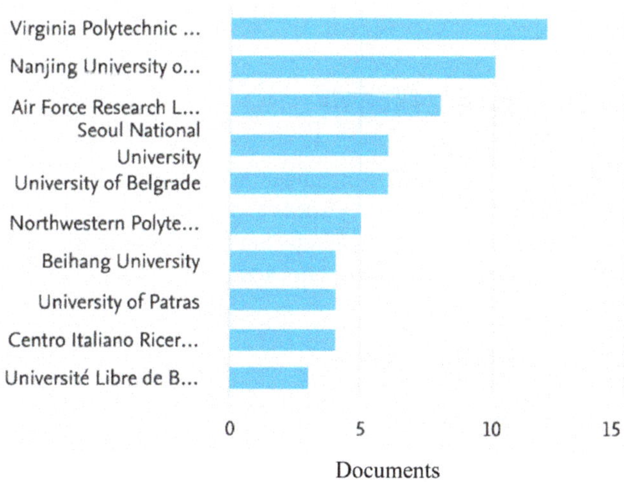

Fig. 9 Document analysis depending on Affiliation. *Source* http://www.scopus.com (assessed on 13th May 2023)

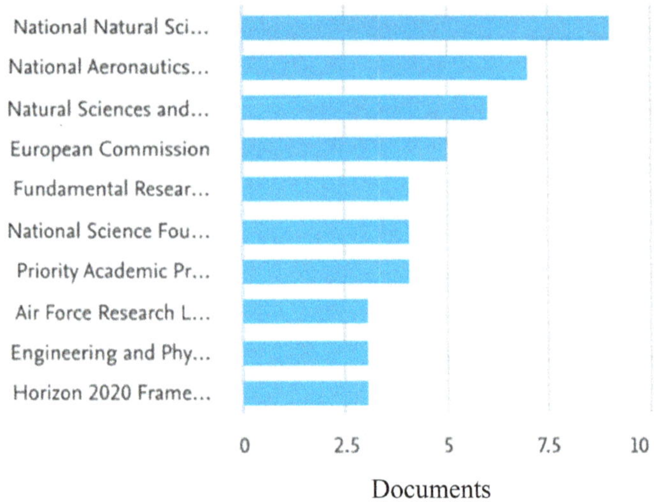

Fig. 10 Document analysis depending on Funding Sponsor. *Source* http://www.scopus.com (assessed on 13th May 2023)

3.1.8 Documents Based on Funding Sponsor

The analysis of documents based on funding sponsors is represented in Fig. 10. In this analysis, documents based on funding sponsors have been taken. It is observed that National Aeronautics has the highest funding and Horizon 2020 Frame has the lowest funding.

3.2 Databases Network Analysis

1. Co-authorship based on Authors, organizations, country.
2. Co-occurrence based on Author keywords and All keywords.
3. Citation Analysis based on organizations, Sources, country and authors.
4. Bibliographic coupling based on Authors and Documents.

3.2.1 Analysis of Co-Authorship

- Co-authorship based on Authors

Co-authorship analysis depending on authors is presented in Fig. 11. In this specification, three distinct parameters have been considered Such as organizations, the authors, and countries for this analysis. A tremendous number of authors have been neglected in this analysis. The highest count of authors per document is 25. The

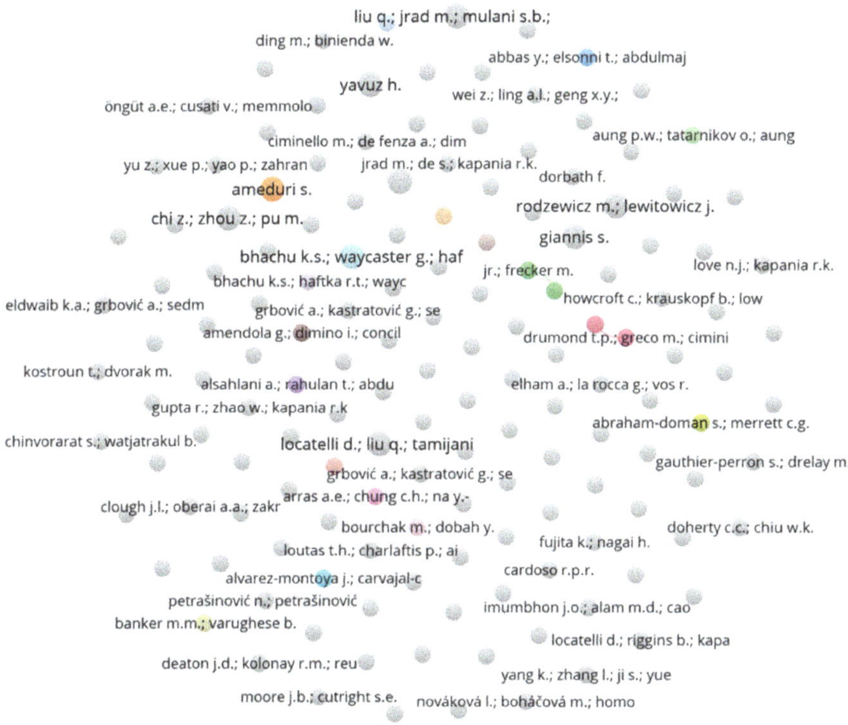

Fig. 11 Co-authorship analysis depending on authors. Source: http://www.scopus.com (Assessed on 13th May 2023)

count of documents of an author is 1. The count of citations of an author is 1 and It is observed that from the total of 199 authors, 151 authors met the criteria. Calculations will be made to determine the overall strength of the co-authorship links for each of the 151 authors. The highest count of links is 2 and the count of clusters is 149.

- Co-authorship based on country

The analysis of co-authorship based on the country is presented in Fig. 12. In this relation out of 19 countries,11 countries met the criteria Calculations will be made to determine the overall strength of the co-authorship links. The total strength of links is 21 and the highest count of links is 13.

- Co-authorship based on organizations

Co-authorship analysis depending on organizations is presented in Fig. 13. It is observed that from the total of 339 organizations, 205 organizations met the criteria. Calculations will be made to determine the overall strength of the co-authorship links for each of the 205 organizations. The highest count of links is 15 and the count of clusters is 1 (Fig. 13).

Fig. 12 Co-authorship analysis depending on country. Source: http://www.scopus.com (Assessed on 13th May 2023)

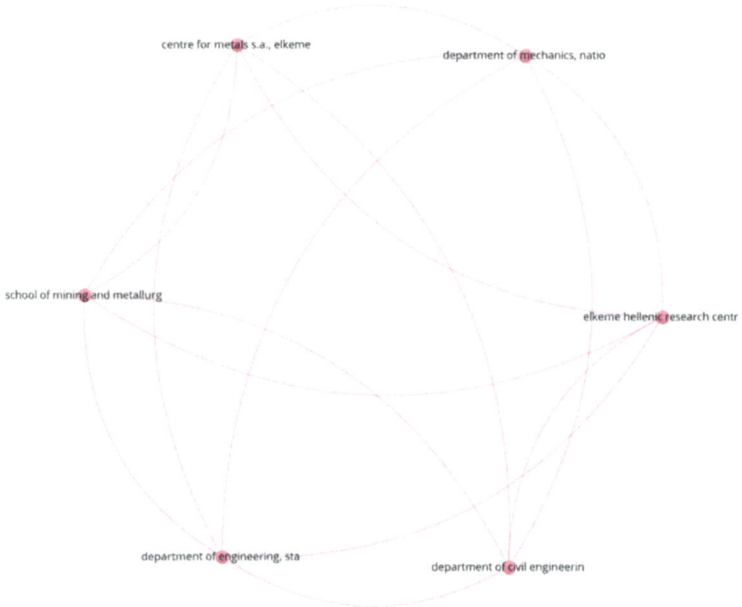

Fig. 13 Co-authorship analysis depending on organizations. *Source* http://www.scopus.com (assessed on 13th May 2023)

3.2.2 Co-Occurrence Analysis

- Co-Occurrence based on All keywords

Co-occurrence analysis depending on all keywords is presented in Fig. 14. It is discovered in this parameter 2004 keywords out of which 90 met with a threshold. The lowest count of occurrences of a keyword is 5. Calculations will be made to determine the overall strength of the co-occurrence links. The total strength of links is 3235 and the highest count of links is 1322.

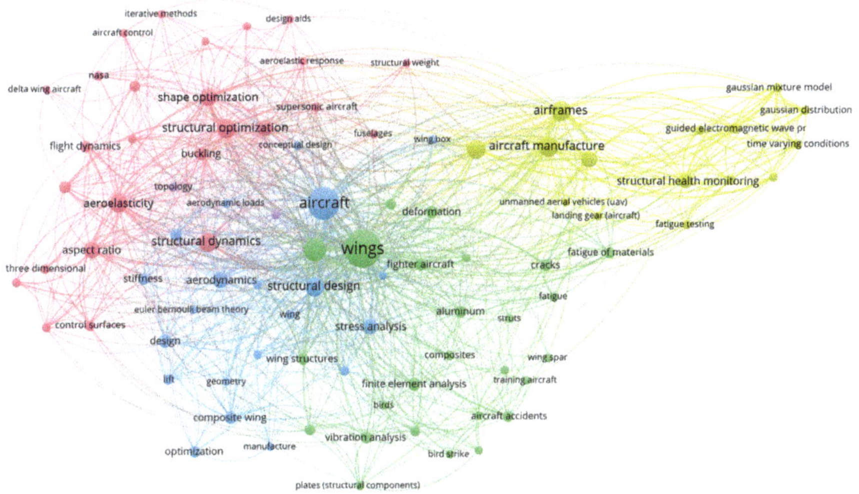

Fig. 14 Co-Occurrence based on Author keywords. *Source* http://www.scopus.com (assessed on 13th May 2023)

- Co-Occurrence based on Author keywords

Co-occurrence analysis depending on author keywords is presented in Fig. 15. It is observed that out of 445 Author keywords 12 met with threshold least count of occurrences of a keyword is 5. Calculations will be made to determine the overall strength of the co-occurrence links. The total strength of links is 41 and the highest count of links is 23.

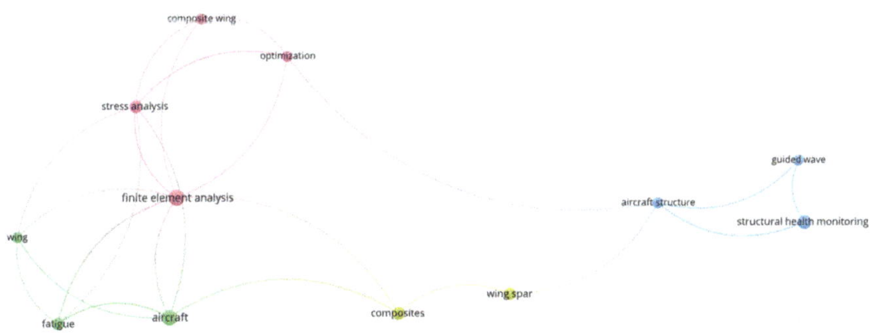

Fig. 15 Co-Occurrence analysis depending on Author keywords. *Source* http://www.scopus.com (assessed on 13th May 2023)

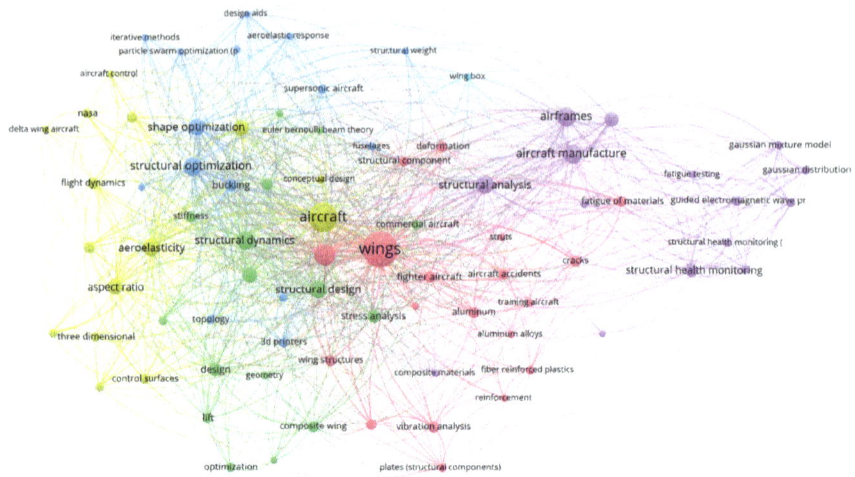

Fig. 16 Co-Occurrence analysis depending on index keywords. *Source* http://www.scopus.com (assessed on 13th May 2023)

- Co-occurrence based on Index keywords

The analysis of co-occurrence based on index keywords is presented in Fig. 16. It is discovered that in this parameter 1759 keywords out of which 81 met a threshold. The lowest count of occurrences of a keyword is 5. Calculations will be made to determine the overall strength of the co-occurrence links. The total strength of links is 2944 and the maximum count of links is 1154.

3.2.3 Citation Analysis

- Citation Analysis Based on Sources

Citation Analysis depending on sources is presented in Fig. 17. It is observed that of a total of 136 Sources, 18 met the threshold. The lowest count of occurrences of a keyword is 3. Calculations will be made to determine the overall strength of the Citation analysis. The highest number of links is 0 and the number of clusters is 18.

- Citation Analysis based on Authors

The analysis of citation analysis based on authors is presented in Fig. 18. It is discovered this parameter contains 199 out of which 151 met with thresholds. The lowest count of documents of an author is 1. Calculations will be made to determine the overall strength of the Citation analysis. The lowest count of citations of an author is 1. The number of links is 0 and the Clusters are 151.

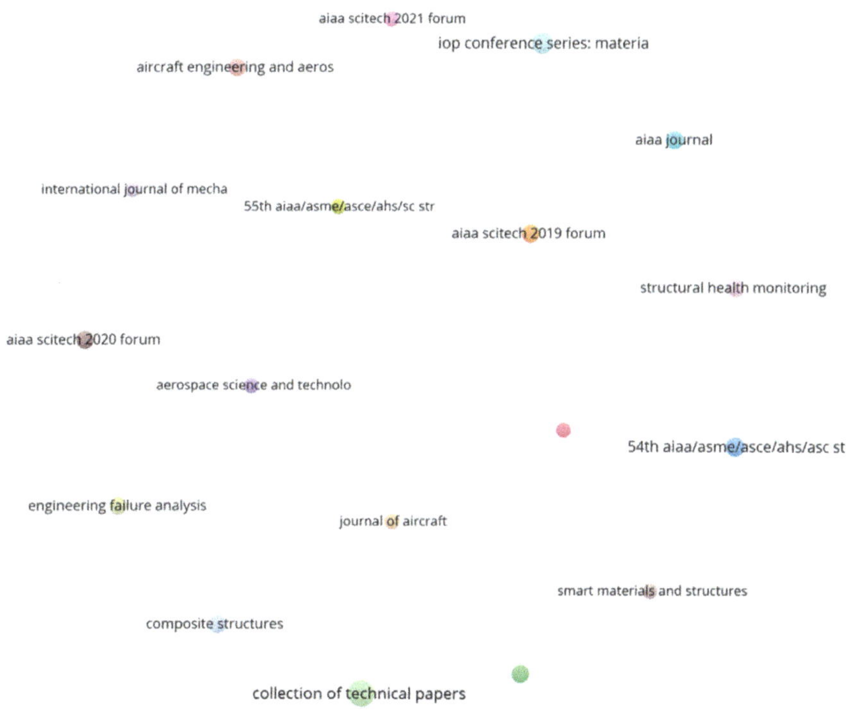

Fig. 17 Citation analysis depending on Sources. *Source* http://www.scopus.com (assessed on 13th May 2023)

- Citation Analysis Based on Organizations

Citation analysis depending on organizations is presented in Fig. 19. It is found that out of 339 organizations, 269 met with thresholds. The minimum number of documents for an organization is 1. The lowest count of citations in an organization is 1. Calculations will be made to determine the overall strength of the Citation analysis. Number of clusters is 269.

- Citation Analysis based on Country

The analysis of Citation analysis in Terms of Country is presented in Fig. 20. It is discovered this parameter contains 44 countries out of which 23 met with thresholds. The lowest count of documents in a country is 2. The lowest count of citations of a country is 1. The number of links is 0 and the Clusters are 23.

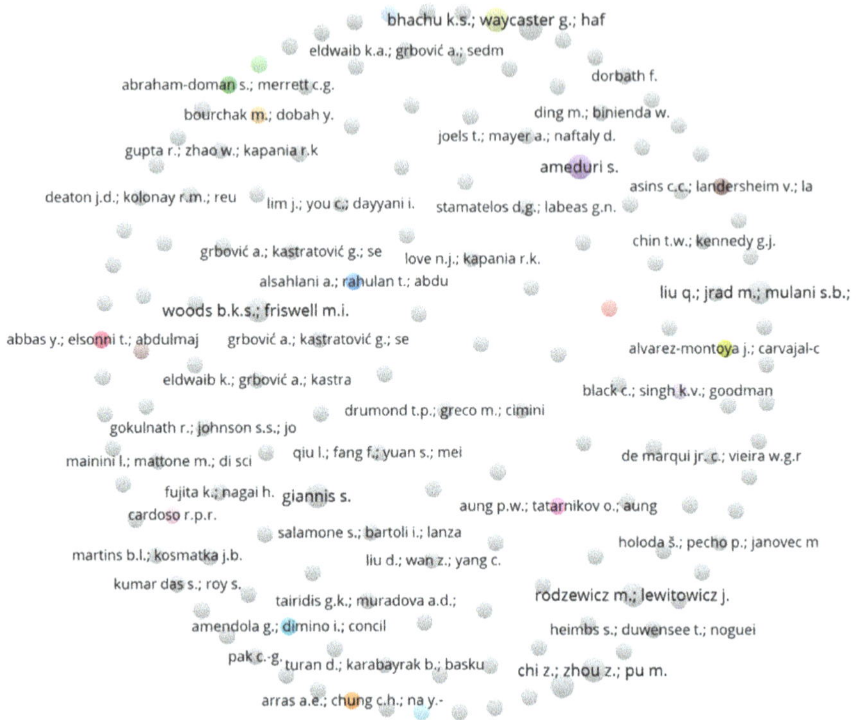

Fig. 18 Citation analysis depending on the Author. *Source* http://www.scopus.com (assessed on 13th May 2023)

3.2.4 Bibliographic Coupling

- Bibliographic Analysis based on documents

Bibliographic analysis depending on documents is presented in Fig. 21. It is observed that out of 210 documents, 113 met the threshold. The lowest number of citations of a document is 2. The highest number of links is 31. Total link Strength is 88.

- Bibliographic Analysis based on Authors

The bibliographic analysis based on authors is represented in Fig. 22. It is discovered that this parameter contains 199 authors out of which 151 met with thresholds. The lowest count of documents of an author is 1. The lowest count of documents in a country is 1. The Total link strength is 96 and the number of links is 35.

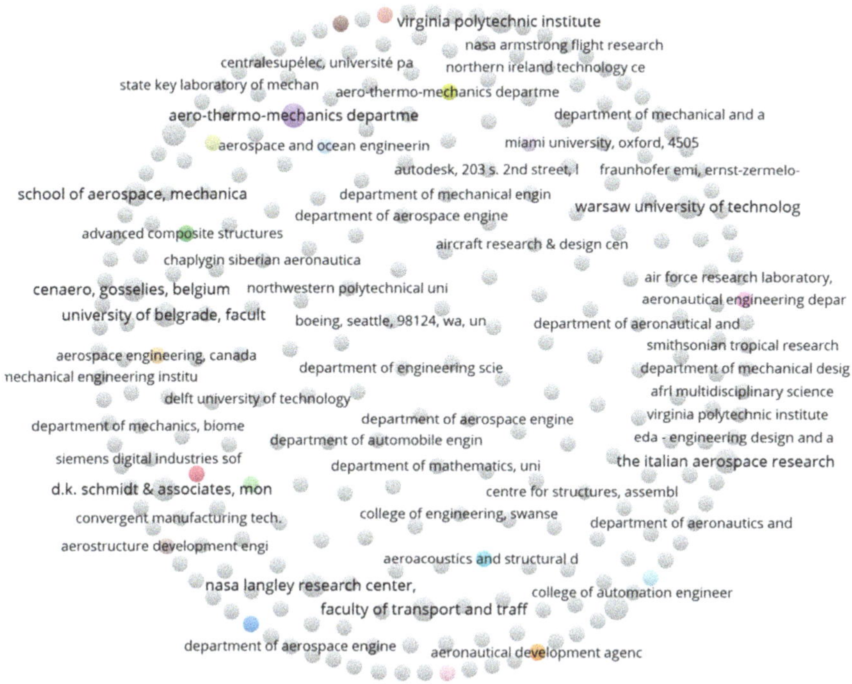

Fig. 19 Citation analysis depending on the organizations. *Source* http://www.scopus.com (assessed on 13th May 2023)

4 Conclusions

- In this chapter, a bibliometric analysis was conducted using various databases. Initially, 345 documents were collected by searching for four keywords: Aircraft, Wings, Spars, and Analysis. After applying limitations based on Years, Subject Area, and Language, the number of relevant documents was narrowed down to 210. The bibliometric analysis employed two methods: Statistical Analysis and Network Analysis, along with the utilization of the Scopus database. The analysis revealed that "Wings" was the most frequently used keyword, and English emerged as the predominant language in the selected documents.
- Through statistical analysis, diverse document types were identified, including sources, years of publication, and subject types. Emphasis on the subject area indicated a focus on mathematics and engineering, underscoring the technical nature of the research related to aircraft wings. Additionally, the analysis of document types indicated that a significant portion of the papers were published at conferences. Furthermore, a country-wise examination of publications highlighted the United States has highest count of publications in this field.

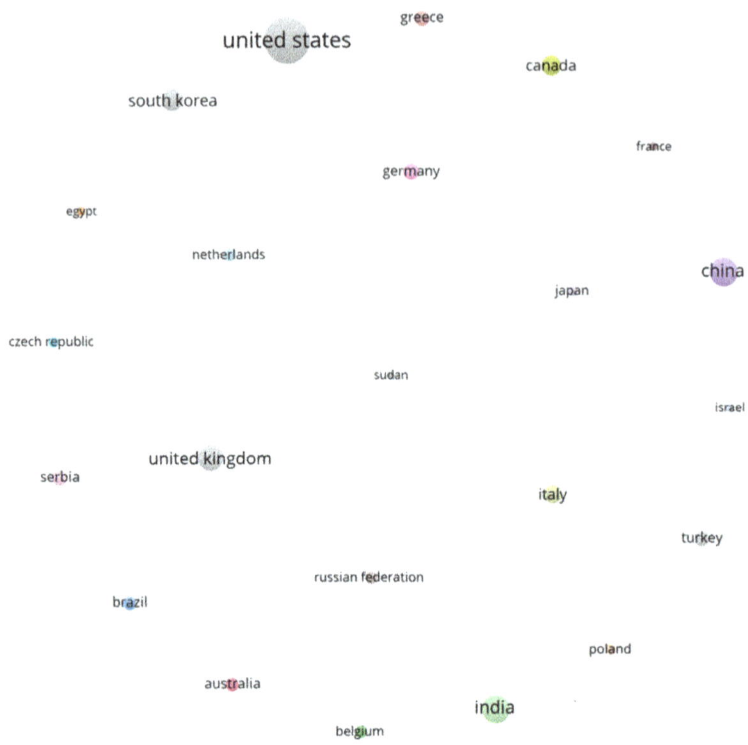

Fig. 20 Analysis of citation based on Terms of Country. *Source* http://www.scopus.com (assessed on 13th May 2023)

Fig. 21 Bibliographic analysis depending on documents. *Source* http://www.scopus.com (assessed on 13th May 2023)

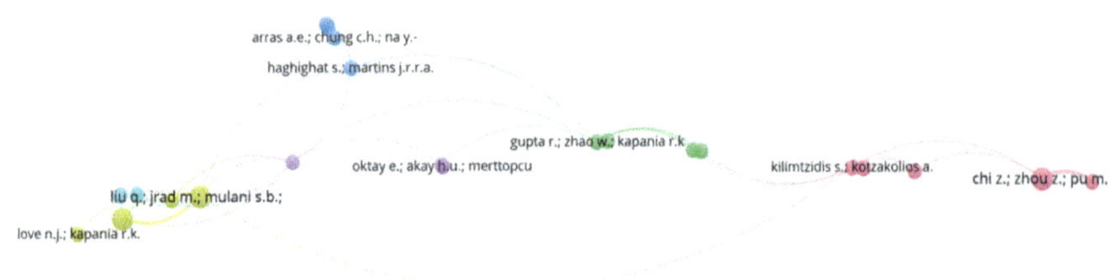

Fig. 22 Bibliographic Analysis based on Authors. *Source* http://www.scopus.com (assessed on 13th May 2023)

- Through network analysis, four types of analysis were identified: co-authorship, bibliographic coupling co-occurrence, citation analysis. The analysis reveals that author Ameduri S. has published the majority of the documents. Furthermore, it is evident that the keyword "Wings" has been extensively used, suggesting its significance in future publications. It is expected that major publications in this keyword will continue to emerge in the forthcoming years.

Declarations

Conflicts of Interest "There are no conflicts of interest to declare by the authors."

Ethical Approval The authors hereby state that the present work is in compliance with the ethical standards.

Competing Interests The authors declare no competing interests.

Availability of Data and Materials Not applicable.

References

Abbas, Yassir, Tariq Elsonni, Abdul Aziz Abdul Majid, Al nazir Khalafallah, and Mohammed Al nazir. "Structural analysis of a transport aircraft wing." *INCAS Bulletin* 13, no. 1 (2021): 3–9.

Ahmed, Mohammed A., and Duncan Koon. "Automated Section Freebody Loads Extraction Tool." In *AIAA Scitech 2021 Forum*, p. 1047. 2021.

Ajith, V. S., Ravikumar Paramasivam, and K. Vidhya. "Study of optimal design of spar beam for the wing of an aircraft." *International Journal of Engineering Development and Research* 5, no. 3 (2017): 179–193.

Alvarez-Montoya, Joham, Alejandro Carvajal-Castrillón, and Julián Sierra-Pérez. "In-flight and wireless damage detection in a UAV composite wing using fiber optic sensors and strain field pattern recognition." *Mechanical Systems and Signal Processing* 136 (2020): 106526.

Arif, Rahman. "Analysis Of Wing Spar Cross-Sectional Profile Variation On The Natural Frequency Of An Unmanned Aerial Vehicle." PhD diss., University of Andalas, 2020.

Augustin, P. "Fatigue life analysis of aging aircraft based on previous service usage." In *New Trends in Civil Aviation*, pp. 113–118. CRC Press, 2018.

Black, Charlene, Kumar V. Singh, Steven Goodman, Aaron Altman, and Raymond M. Kolonay. "Design, fabrication and testing of 3D printed wings for rapid evaluation of aeroelastic performance." In *2018 AIAA/ASCE/AHS/ASC Structures, Structural Dynamics, and Materials Conference*, p. 1997. 2018.

Bruce, Ralphin Rose. "Initial and progressive failure analysis of a composite wing spar structure/ Bruce Ralphin Rose. J and Priyadharshini. S." *Journal of Mechanical Engineering (JMechE)* 14, no. 2 (2017): 167-183.

Britto Maria, Rodrigo, Marcus de Freitas Leal, Edgard Sousa Junior, Vinicius Leite Lemos, Patrick Magalhães Cardoso, Leonardo Cosme de Oliveira, Saullo Giovani Pereira Castro, Marcelo de Lima Marcolin, and Darshan Joshi. "Applications of SPDM in aircraft structural analysis at Embraer." *Advanced Modeling and Simulation in Engineering Sciences* 6, no. 1 (2019): 1–16.

Cardoso, Rui PR. "A new beam element which blends the Euler-Bernoulli beam theory with idealised transverse shear flows for aircraft structural analysis." *Thin-Walled Structures* 157 (2020): 107118.

Clough, Justin L., Assad A. Oberai, and Andrew J. Zakrajsek. "Automated Wing Internal Structure Placement Guided by Finite Element Analysis." In *AIAA Aviation 2019 Forum*, p. 3369. 2019.
Das, Salu Kumar, and Sandipan Roy. "Finite element analysis of aircraft wing using carbon fiber reinforced polymer and glass fiber reinforced polymer." In *IOP conference series: Materials science and engineering*, vol. 402, no. 1, p. 012077. IOP Publishing, 2018.
Elangovans, suresh kumar c, divyabarathi p. "Design and analysis of aircraft wing spar with different materials using ansys" International Journal of Mechanical and Production Engineering Research and Development (IJMPERD) ISSN(P): 2249–6890; ISSN(E): 2249–8001 Vol. 9, Issue 3, (2019), 613–618.
Eldwaib, Khalid, Aleksandar Grbović, Gordana Kastratović, and Mustafa Aldarwish. "Design of wing spar cross section for optimum Fatigue life." *Procedia structural integrity* 13 (2018): 444-449
Eldwaib, Khalid Ahmed, Aleksandar Grbović, Aleksandar Sedmak, Gordana Kastratović, Danilo Petrasinović, and Simon Sedmak. "Fatigue life estimation of damaged integral wing spar using XFEM." *Tehnički vjesnik* 25, no. 6 (2018): 1837–1842.
Fang, Fang, Lei Qiu, Shenfang Yuan, and Yuanqiang Ren. "Dynamic probability modeling-based aircraft structural health monitoring framework under time-varying conditions: Validation in an in-flight test simulated on ground." *Aerospace Science and Technology* 95 (2019): 105467.
Gandhi, Viraj, John Joe, John F. Dannenhoffer, and Hamid Dalir. "Rapid design generation and multifidelity analysis of aircraft structures." *Aerospace Science and Technology* 112 (2021): 106612.
Girennavar, Mutturaj, H. Soumya, H. Subodh, Tanvi J. Heraje, and Deepak Raj PY. "Design, Analysis and Testing of Wing Spar for Optimum Weight." *International Journal of Research and Scientific Innovation (IJRSI)* 4, no. VII (2017): 104–112.
Gokulnath, R., S. S. Johnson, and J. J. Joshua. "Analysis of composite wing spar with cutouts for tensile load." *Int J Mech Eng Tech* 9 (2018): 173-179.
Gupta, Rikin, Wei Zhao, and Rakesh K. Kapania. "Controllability gramian as control design objective in aircraft structural design optimization." *AIAA Journal* 58, no. 7 (2020): 3199-3220.
Gupta, Rikin, Wei Zhao, Rakesh K. Kapania, and David K. Schmidt. "Incorporating Flight Dynamics and Control Criteria into MDAO of Composite Aircraft." In *AIAA Scitech 2020 Forum*, p. 1966. 2020a.
Irez, A. B., Emin Bayraktar, and Ibrahim Miskioglu. "Devulcanized Rubber Based Composite Design Reinforced with Nano Silica, Graphene Nano Platelets (GnPs) and Epoxy for "Aircraft Wing Spar" to Withstand Bending Moment." In *Mechanics of Composite, Hybrid and Multifunctional Materials, Volume 5: Proceedings of the 2018 Annual Conference on Experimental and Applied Mechanics*, pp. 9–22. Springer International Publishing, 2019.
Jang, Jun Hwan, and Sang Ho Ahn. "FE Modeling methodology for load analysis and preliminary sizing of aircraft wing structure." *International Journal of Aviation, Aeronautics, and Aerospace* 6, no. 2 (2019): 1.
Joels, Tsoof, Ariel Mayer, Lucy Edery-Azulay, and Daniella E. Raveh. "Design, Analysis, and Testing of the Active Aeroelastic Aircraft Testbed (A3TB) Platform." In *AIAA Scitech 2021 Forum*, p. 1390. 2021.
Jrad, Mohamed, Shuvodeep De, and Rakesh K. Kapania. "Global-local aeroelastic optimization of internal structure of transport aircraft wing." In *18th AIAA/ISSMO Multidisciplinary Analysis and Optimization Conference*, p. 4321. 2017.
Khames, Mohamed, Ahmed Embaby, and Abdelrahman Agha. "Comparison between the Use of Aluminum and Composites in the Design of a Wing-Spar of an Airplane." In *Materials Science Forum*, vol. 953, pp. 95–100. Trans Tech Publications Ltd, 2019.
Khan, Haris Ali, Armaghan Shahid, Sara Khushbash, Asad Hameed, and S. Zameer Abbas. "Investigation of failure and development of mitigation techniques of a cracked aircraft wing spar cap." *Engineering Failure Analysis* 147 (2023): 107149.
Kilimtzidis, Spyridon, Athanasios Kotzakolios, and Vassilis Kostopoulos. "Efficient structural optimisation of composite materials aircraft wings." *Composite Structures* 303 (2023): 116268.

Kleemann, Nils, Stanislav Karpuk, and Ali Elham. "Conceptual Design and Optimization of a Solar-Electric Blended Wing Body Aircraft for General Aviation." In *AIAA Scitech 2020 Forum*, p. 0008. 2020.

Kontron, T., and M. Dvořák. "Non-Destructive Evaluation of Small Aircraft Wing Adhesive Joints using Pulse Infrared Thermography Method." Proceedings of the 58th International Scientific Conference Experimental Stress Analysis, Sobotín, Czech Republic. 2020.

Kumar, Sanjay, and P. Selvaraj. "Finite Element Simulation and Statistical Analysis of Laminated Composite Shims for Gaps in Aircraft Structures Assembly." *International Journal of Vehicle Structures & Systems* 12, no. 1 (2020): 101–104.

Kumar, A. Ramesh, S. R. Balakrishnan, and S. Balaji. "Design of an aircraft wing structure for static analysis and fatigue life prediction." *Int. J. Eng. Res. Technol* 2, no. 5 (2013): 1154–1158.

Kumar, TS Vinoth, A. Waseem Basha, M. Pavithra, and V. Srilekha. "Static & dynamic analysis of a typical aircraft wing structure using MSC Nastran." *Int. J. Res. Aeronaut. Mech. Eng* 3 (2015): 1–12.

Maheswaran, N., S. Venkatesan, M. Sampath Kumar, G. Velmurugan, N. Sathishkumar, and M. Priya. "Study of weight optimization on spar beam for the wing of an aircraft." *International Journal of Software & Hardware Research in Engineering* 3, no. 3 (2015): 72–79

Marangi, Kevin, and Salim Mohamed Salim. "Numerical Investigation of an Unmanned Aircraft Vehicle (UAV) Using Fluid-Structure Interaction." In *Transactions on Engineering Technologies: World Congress on Engineering 2019*, pp. 1–14. Singapore: Springer Singapore, 2020.

Mukhopadhyay, Vivekanand, Erik D. Olson, and Thomas Ozoroski. "Structural analysis of a test flight vehicle with multi-functional energy storage." In *AIAA Scitech 2019 Forum*, p. 0551. 2019.

Pak, Chan-gi. "Shape Sensing for Wings with Spars and Ribs using Strain Data." In *AIAA Scitech 2019 Forum*, p. 1756. 2019.

Park, Dong-Cheol, Tianyu Yu, Soo-Jeong Park, Do-Hoon Shin, and Yun-Hae Kim. "Improvement of impregnation quality on out-of-autoclave processed CFRP aircraft wing spar through resin flow simulation." *Functional Composites and Structures* 3, no. 2 (2021): 025001.

Pavan G, Manjunath M V "Static Analysis on Wing Spar Joint For An Aircraft Using Finite Element Method," *IJEDR (2018)* | Volume 6, Issue 3 | ISSN: 2321–9939

Qiu, Lei, Fang Fang, Shenfang Yuan, Christian Boller, and Yuanqiang Ren. "An enhanced dynamic Gaussian mixture model–based damage monitoring method of aircraft structures under environmental and operational conditions." *Structural Health Monitoring* 18, no. 2 (2019): 524-545.

Rodzewicz, Mirosław, and Jerzy Lewitowicz. "Lightness factors and exertion factors related to its own mass of fiber reinforced plastics (FRP) applied in composite aeronautical structures (part 2)." *Journal of KONBiN* 47, no. 1 (2018): 23-43.

Son, Lovely, Arif Rahman, and Mulyadi Bur. "The Effect of Wing Spar Cross-Sectional Profile Variation on the Unmanned Aerial Vehicle (UAV) Natural Frequency." In *IOP Conference Series: Materials Science and Engineering*, vol. 1062, no. 1, p. 012032. IOP Publishing, 2021.

Stamatelos, Dimitrios, and George Labeas. "Towards the design of a multispar composite wing." Computation 8, no. 2 (2020): 24.

Madhu H & Kadole, Pavan. "Static and Dynamic Structural Response of Aircraft Wing." International Journal of Engineering Research & Technology (IJERT), vol. 3, no. 9, 2015

Stepanova, L. N., E. S. Petrova, and V. V. Chernova. "Strength tests of a CFRP spar using methods of acoustic emission and tensometry." *Russian Journal of Nondestructive Testing* 54 (2018): 243-248.

Sun, Ruisheng, Dawei Yan, Raj K. Nangia, and Jonathan E. Cooper. "Optimization design for aeroelastic morphing wings." In *2017 9th International Conference on Modelling, Identification and Control (ICMIC)*, pp. 25–30. IEEE, 2017.

Tariq, Usama, and Farrukh Mazhar. "Static Structural Analysis of Fighter Aircraft's Wing Spars." In *2021 International Bhurban Conference on Applied Sciences and Technologies (IBCAST)*, pp. 221–243. IEEE, 2021.

Toffol, Francesco, and Sergio Ricci. "A Meta-Model for composite wing box sizing in aircraft conceptual design." *Composite Structures* 306 (2023): 116557.

Vartabedian, T. "Design of an aircraft main wing spar design project 2 final report" 2018.

Venkatarthnaiah, & Venkategowda, C. & Girish, K.E. "Stress analysis of the spar and evaluation of shear buckling in the web region." *National Conference on challenges in research and technology.* (2013). 3.41–3.41. https://doi.org/10.1049/cp.2013.2560.

Wanhill, Russell, Simon Barter, Loris Molent, Russell Wanhill, Simon Barter, and Loris Molent. "Aermacchi MB-326H Wing Spar (1990): Exponential FCG Analysis." *Fatigue Crack Growth Failure and Lifing Analyses for Metallic Aircraft Structures and Components* (2019): 43–48.

Weckner, Olaf, Norair Ghazarian, Timothy Adams, and Vladimir O. Balabanov. "Large bay skin buckling of composite aircraft structure: method development and validation." In *AIAA Scitech 2021 Forum*, p. 1049. 2021.

Xiao, Liping, Haifeng Zhao, Zhao Xu, Xiaoyu Li, Chengcheng Shen, Ke Wang, and Lu Zhang. "A New Architecture of Morphing Wing Based on Hyperelastic Materials and Metastructures with Tunable Stiffness." *Frontiers in Mechanical Engineering* 7 (2022): 112.

Xiaozhe, W. A. N. G., W. A. N. Zhiqiang, L. I. U. Zhu, and Y. A. N. G. Chao. "Integrated optimization on aerodynamics-structure coupling and flight stability of a large airplane in preliminary design." *Chinese Journal of Aeronautics* 31, no. 6 (2018): 1258-1272.

Yang, Ye, Yan Jin, Mark Price, Gasser Abdelal, Colm Higgins, and Paul Maropoulos. "Investigation of point cloud registration uncertainty for gap measurement of aircraft wing assembly." In *2021 IEEE 8th International Workshop on Metrology for AeroSpace (MetroAeroSpace)*, pp. 164–169. IEEE, 2021.

Yu, Zeliang, Pu Xue, Peilun Yao, and M. S. Zahran. "Analytical determination of the critical impact location for wing leading edge under birdstrike." *Latin American Journal of Solids and Structures* 16 (2019).

Zakuan, M. A. M. B. M., Abdul Aabid, and Sher Khan. "Modelling and structural analysis of three-dimensional wing." *International Journal of Engineering and Advanced Technology* 9, no. 1 (2019): 6820–6828.

Zhang, Chi, Zhou Zhou, and Pu Meng. "Withdrawal: Nonlinear Static Aeroelastic Trim and Stability Analysis of Highly Flexible All-Wing Aircraft." In *2018 Multidisciplinary Analysis and Optimization Conference*, pp. 3574-c1. 2018.

Zhang, Chi, Zhou Zhou, Xiaoping Zhu, and Pu Meng. "Nonlinear static aeroelastic and trim analysis of highly flexible joined-wing aircraft." *AIAA Journal* 56, no. 12 (2018): 4988-4999.

GPSR Compliance
The European Union's (EU) General Product Safety Regulation (GPSR) is a set of rules that requires consumer products to be safe and our obligations to ensure this.

If you have any concerns about our products, you can contact us on

ProductSafety@springernature.com

In case Publisher is established outside the EU, the EU authorized representative is:

Springer Nature Customer Service Center GmbH
Europaplatz 3
69115 Heidelberg, Germany

www.ingramcontent.com/pod-product-compliance
Ingram Content Group UK Ltd.
Pitfield, Milton Keynes, MK11 3LW, UK
UKHW021410091026
15624UKWH00009B/201

9789819996162